T0135222

Lecture Notes in Electrical Engineering

Volume 806

The book series *Lecture Notes in Electrical Engineering* (LNEE) publishes the latest developments in Electrical Engineering - quickly, informally and in high quality. While original research reported in proceedings and monographs has traditionally formed the core of LNEE, we also encourage authors to submit books devoted to supporting student education and professional training in the various fields and applications areas of electrical engineering. The series cover classical and emerging topics concerning:

- Communication Engineering, Information Theory and Networks
- Electronics Engineering and Microelectronics
- Signal, Image and Speech Processing
- Wireless and Mobile Communication
- Circuits and Systems
- Energy Systems, Power Electronics and Electrical Machines
- Electro-optical Engineering
- Instrumentation Engineering
- Avionics Engineering
- Control Systems
- Internet-of-Things and Cybersecurity
- Biomedical Devices, MEMS and NEMS

For general information about this book series, comments or suggestions, please contact leontina.dicecco@springer.com.

To submit a proposal or request further information, please contact the Publishing Editor in your country:

China

Jasmine Dou, Editor (jasmine.dou@springer.com)

India, Japan, Rest of Asia

Swati Meherishi, Editorial Director (Swati.Meherishi@springer.com)

Southeast Asia, Australia, New Zealand

Ramesh Nath Premnath, Editor (ramesh.premnath@springernature.com)

USA, Canada:

Michael Luby, Senior Editor (michael.luby@springer.com)

All other Countries:

Leontina Di Cecco, Senior Editor (leontina.dicecco@springer.com)

** This series is indexed by EI Compendex and Scopus databases. **

More information about this series at https://link.springer.com/bookseries/7818

Rajeev R. Raje · Farookh Hussain ·
R. Jagadeesh Kannan
Editors

Artificial Intelligence and Technologies

Select Proceedings of ICRTAC-AIT 2020

 Springer

Editors
Rajeev R. Raje
School of Science, Computer
and Information Science
Indiana University–Purdue University
Indianapolis, IN, USA

Farookh Hussain
University of Technology Sydney
Sydney, NSW, Australia

R. Jagadeesh Kannan
School of Computer Science and
Engineering
Vellore Institute of Technology
Chennai, Tamil Nadu, India

ISSN 1876-1100 ISSN 1876-1119 (electronic)
Lecture Notes in Electrical Engineering
ISBN 978-981-16-6450-2 ISBN 978-981-16-6448-9 (eBook)
https://doi.org/10.1007/978-981-16-6448-9

This Springer imprint is published by the registered company Springer Nature Singapore Pte Ltd.
The registered company address is: 152 Beach Road, #21-01/04 Gateway East, Singapore 189721,
Singapore

Contents

About the Editors

Rajeev R. Raje is a Professor in the Department of Computer and Information Science (CIS) and is an Associate Dean in the School of Science at Indiana University-Purdue University Indianapolis (IUPUI). Dr. Raje's research interests are in distributed and service-oriented software systems, programming languages, and software engineering. Dr. Raje has authored over 140 publications. Dr. Raje is a Senior Member of the ACM and IEEE, and he also co-directs the software engineering and distributed systems group in the CIS department.

Farookh Hussain is a Professor in the School of Computer Science, University of Technology, Sydney, Australia, and is currently serving as Head of Software Engineering. Dr. Hussain is an Associate Member of the Advanced Analytics Institute and a Core Member of the Australian Artificial Intelligence Institute. He has written various books and chapters and has published widely in journals too. Dr. Hussain has over 500 papers in various international journals and conferences. He has completed over 50 funded projects for the government and various private organizations.

R. Jagadeesh Kannan is Professor and Dean of the School of Computer Science and Engineering at Vellore Institute of Technology, India. He completed his Ph.D. in Handwritten Character Recognition using Hybrid Techniques from Anna University, Chennai, India. He got his M.E. degree in Computer Science and Engineering from National Engineering College, Tamil Nadu, and B.E. in Instrumentation and Control Engineering from Madurai Kamaraj University, Tamil Nadu, India. Prof. Kannan has over 18 years of teaching and industrial experience in reputed organizations. Prof. Kannan has got several publications in conference proceedings and journals of national and international repute. His research interests are neural networks, fuzzy logic, neuro-fuzzy systems, soft computing tools, pattern recognition, natural language processing, image processing, networking, printed, handwritten and cursive character recognition, and artificial intelligence.

Design of Automatic Credit Card Approval System Using Machine Learning

S. Hemkiran⑩, G. Sudha Sadasivam⑩, A. Prasanna Rahavendra, and A. K. Anjhanna

1 Introduction

Machine learning (ML) helps to solve problems that are very difficult or even impossible to solve using conventional programming techniques. ML is typically utilized when identifying solutions require a lot of hand-tuning or long lists of rules. ML is also utilized to solve complex problems with large amounts of data, where no good solution using a traditional approach can be established [1]. Additionally, ML can adapt to new data, thus making it competent for fluctuating environments.

ML refers to the approach and art of programming in which computers can learn from data. ML is commonly categorized into supervised and unsupervised learning [2]. In supervised learning, classification is a major type of data segregation technique. It specifies the class to which data elements belong to and is best used when the output has finite and discrete values. It can be utilized to predict a class for an input variable as well.

Several ML algorithms such as k-nearest neighbours, logistic regression, support vector machines, decision trees, random forests and certain neural networks are employed for binomial or multiclass classification tasks. Such ML algorithms are utilized to solve real-world problems such as medical diagnosis, object recognition, product recommendation and financial analysis [3]. In particular, logistic regression is widely used for predicting the occurrence of binary responses based on one or more variables [4]. Numerous studies have established that artificial neural network (ANN) models are an efficient prediction tool used in multifarious applications such as medical diagnosis, financial forecasting and intelligent searching [5]. Specifically, ANN can be utilized for discovering nonlinear relationships and for

S. Hemkiran (✉) · A. Prasanna Rahavendra · A. K. Anjhanna
PSG Institute of Technology and Applied Research, Coimbatore 641062, India
e-mail: hemkiran@psgitech.ac.in

G. Sudha Sadasivam
PSG College of Technology, Coimbatore 641004, India

© The Author(s), under exclusive license to Springer Nature Singapore Pte Ltd. 2022
R. R. Raje et al. (eds.), *Artificial Intelligence and Technologies*,
Lecture Notes in Electrical Engineering 806,
https://doi.org/10.1007/978-981-16-6448-9_1

discerning the complex interactions between the multitudes of input parameters. ANN comprises of neurons which are interconnected by certain fractional numbers termed as weights, which are tuned in order to accurately predict the results [6].

Credit cards are issued by financial institutions. The banking sector receives numerous applications for credit card requests every day. Dealing with each request manually is time consuming and also prone to human errors [7]. Therefore, it would be advantageous for financial institutions if the historical data can be used to build competent models to shortlist candidates for approval. In this study, binomial classification systems are employed to automatically predict if a person can be issued a credit card or not as there exist only two classes for prediction—approved or not approved.

2 Methodology

In the existing credit card approval system, the process between when the customer applies online for a card to, when the verification process is completed, consumes multiple resources such as manpower and time.

As mentioned earlier, in this research, ML predictor models are built and compared using two techniques, namely logistic regression and ANN, to classify newer approvals by analysing the historical credit card approval data. Before feeding the data into both the ML algorithms, the data is pre-processed using certain well-known methods such as imputing missing values, encoding the labels and scaling the column values to fit the ML models. Encoding refers to converting the labels into numbers as, strings cannot be processed by ML algorithms. Scaling involves transforming all features to a same scale with zero mean and a variance of 1 to enable faster training of both the ML algorithms. Finally, the pre-processed dataset is split into a training (70% of the total dataset) and a testing set (30% of the total dataset). The training set is used to train the two models and its performance is evaluated using the testing set.

In order to compare the accuracy of prediction, three ML models are proposed in this study. The first model, henceforth referred as M1, utilizes LR without grid search. The second model, henceforth referred as M2, utilizes LR with grid search. Grid searching is a unique technique which is employed to enhance the model's performance by tuning its hyper-parameters. The third model, henceforth referred as M3, utilizes ANN. Finally, the accuracy of predictions of M1, M2 and M3 is studied and compared.

2.1 Description of Dataset

The UCI ML Repository Credit card dataset maintained by the University of California Irvine was used in this study, and a total of 690 observations were considered. These observations represent 690 historical applications requesting for credit cards. A total of 16 features exist for every application from an individual. Among the 16 features, 15 features cover diverse information about individuals such as their age, gender, income, marital status, number of years in employment, etc., and 1 attribute constitutes the final result, that is, whether the request was approved or rejected.

There exist in Fig. 1a, certain features such as 'Gender', 'DriversLicense', 'MaritalStatus', 'BankCustomer' where the impact level on determining the approval rate is bare minimum. Thus, various features of an applicant will be considered for issuing the credit card. Consequently, the ML models require training in a manner that it has to consider all necessary features while executing the entire process.

In the proposed ML-based implementations, the model is pre-trained with the intent to obtain accurate inferences consuming less time and processing power. Figure 1b depicts the flow chart of the proposed architecture for models M1, M2 and M3. It consists of three stages.

Stage 1—Collecting data from credit card approval dataset and pre-processing the data. The proposed method begins by collecting historic credit card approval data which will be used for the training of the three ML models M1, M2 and M3. The sample dataset is first extracted and the data is pre-processed in order to remove the missing values and outliers. Subsequently, all features of the data are scaled to a mean of zero and variance of one in order to permit the three ML algorithms to allot equal importance to every feature in the dataset.

Stage 2—Training the ML algorithms with the pre-processed data and tuning its parameters. The next stage involves training of the ML models to fit the dataset. In this stage, for M1 and M2, the widely used ML algorithm, namely logistic regression is selected and is trained with the pre-processed data. Training a model involves a few processes. First, the ML algorithm attempts to find a function that best describes the data. This function will separate the two classes in this study—

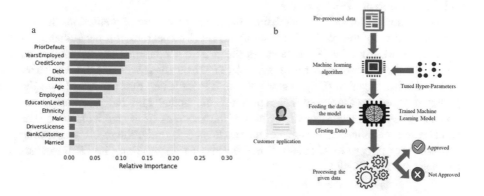

Fig. 1 **a** Relative importance of features determining approval rate, **b** proposed architecture

approved and not approved. Ideally, a 100% accurate model will fit a dataset in a manner that it exactly separates all the data points of one class from the other. Subsequently, a utility function (or fitness function) which measures how good the model is, or a cost function that measures how inadequate the model is, is defined. For linear regression problems, a cost function that measures the distance between the predictions of the linear model and the training examples is utilized to minimize this distance. Using this cost function, the ML model's parameters are tuned such that it reduces the cost function and thereby fits the data in an ideal manner. In case of M3, an ANN is utilized, for which the steps are delineated later.

Stage 3—Using the trained model to get the desired outputs. Once the training of the model is complete, it is necessary to comprehend how well a model can be generalized to new test cases. A suggested method is to deploy the model in production and monitor how well it performs. If the response is good, then no issues exist. But if the model underperforms, it will result in a huge loss to the credit card company. Hence, integrating the model directly into real-world applications is not a recommended practice. A preferred option is to split the data into two sets, namely a training set and a testing set.

The training set is first used to train the three models, and subsequently, the data that is hitherto unseen by the models are used as the test set. This enables detecting how well the proposed models M1, M2 and M3 can be generalized to solve new problems. This split varies depending on the dataset and the problem statement under consideration. For instance, in this project, 70% of the data is selected for training and 30% of the data is utilized for testing.

Stage 4—Grid searching to improve the model's accuracy. In this study, the accuracy of the model M2 is improved using a technique called grid search. Grid searching is the process of scanning a specified range of hyper-parameter combinations to obtain optimal hyper-parameters for a given model. In grid searching, the various hyper-parameters of the model are first, collectively specified. Next, distinct models are trained iteratively for every combination of the specified hyper-parameters. The best possible hyper-parameter combination is finally used to train and thereby obtain the best model.

The accuracy of the ML model defines the total percentage of predictions that are made correctly by the model. The confusion matrix specifies the performance of any classification model [8]. It includes the different combinations of actual and predicted values comprises of true positive (TP), true negative (TN), false positive (FP) and false negative (FN). Among these, TP indicates that the actual and the predicted cases are true, and TN represents the actual and the predicted cases are false. These two values should be high if the accuracy of the model is good. The next metric for evaluating accuracy of the model is area under the receiver operating characteristics curve (AUC-ROC). It is the plot of true positive rate (TPR) versus true negative rate (TNR).

2.2 Logistic Regression

Logistic regression is a powerful classification algorithm capable of fitting both linear and non-linear datasets [9]. Linear datasets are those that can be separated by a single straight line while nonlinear datasets are those that require a polynomial function to fit the data. In the case of logistic regression, the prediction function is the sigmoid function as shown in Eqs. (1) and (2).

$$\text{sigmoid}(z) = \frac{1}{1 + e^{-z}} \tag{1}$$

$$z = W_0 + W_1 * X_1 + W_2 * X_2 + W_3 * X_3 + \cdots + W_n * X_n \tag{2}$$

where z is the linear equation of the features, W_0 is the bias term, W_1 to W_n represent weight terms and X_1 to X_n refer to the 'n' training instances. The logistic regression cost function is represented in Eq. (3) along with the varying value of y in Eqs. (4) and (5).

$$J(\theta) = \frac{1}{m} \sum_{i=1}^{m} \left(\cos t \left(h_\theta \left(x^{(i)} \right) - y^{(i)} \right) \right) \tag{3}$$

$$\cos t \left(h_\theta \left(x^{(i)} \right) - y^{(i)} \right) = -\log(h_\theta(x)); \quad \text{if } y = 1 \tag{4}$$

$$\cos t \left(h_\theta \left(x^{(i)} \right) - y^{(i)} \right) = -\log(1 - h_\theta(x)); \quad \text{if } y = 0 \tag{5}$$

The cost function refers to the difference between the actual values of the target and the values predicted by the model, summed over all the training instances and divided by the number of instances. The goal is to minimize this function. This is performed using a process termed gradient descent.

In the gradient descent process first, our model's parameters (the weights and the bias of the model) are initialized randomly. Here, the gradients (i.e. the partial differentials) of each of the weights and bias are found according to Eqs. (6) and (7).

$$w = w - \alpha * d\omega \tag{6}$$

$$b = b - \alpha * db \tag{7}$$

where w and b represent the weight and bias terms. The α (alpha) term is a hyper-parameter termed as learning rate. This hyper-parameter determines how much w and b are altered at each step of the gradient descent. A large value of alpha will result in stochastic jumping of the parameters, and it will not converge to the minimum. A smaller value of alpha will consume a longer time for the parameters to reach the optimum value. In order to achieve a balance, the α value of 0.001 is chosen in this study [10].

2.3   Artificial Neural Network Model

ANNs are typically used for modelling and approximating a sophisticated relationship between input data (features) and the outputs (target). The ANN for model
M3 proposed in this work is two-layer model with 50 neurons in the first layer and
30 neurons in the second layer. Each neuron in the network has a set of associated
parameters termed as weights and biases. These values are altered during the
training phase so as to fit the dataset and obtain better results. The performance of
the neural network is measured with the help of a loss function [11]. For model M3,
a cross-entropy loss, also referred as the log loss function, is utilized as shown in
Eq. (8). The loss function depicts how close our model's predicted values are, to the
actual target values. A larger loss implies that M3 is not performing well and needs
more training. In contrast, a diminished loss signifies that M3 approximates the
input–output relationship very well.

$$\text{Cross Entropy loss, } l = -(y * \log(p) + (1 - y) * \log(1 - p)) \tag{8}$$

where y is the true target value, and p is the prediction of the model. If w_i and b_i are
current values of the weights and biases, then the update rule is as shown in Eq. (9).

$$w_{i+1} = w_i - \alpha(dl/dw); \quad b_{i+1} = b_i - \alpha(dl/db) \tag{9}$$

here α is a hyper-parameter called the learning rate. The learning rate determines the
step size of the update at each training epoch while moving towards a minimum of a
loss function. The derivatives 'dl/dw' and 'dl/db' represent the change of the loss
function with respect to the parameters 'w' and 'b', respectively. This loss function
is calculated, and the update is performed several times until the model's loss
converges at the optimum.

3   Results and Discussion

As stated earlier, the ML model used for M1 and M2 is logistic regression. The
features and the labels are passed as parameters to the ML model for training. The
accuracy score of this trained model is first determined without using any
hyper-parameter tuning techniques. Later, the accuracy is computed using grid
search, by tweaking the hyper-parameters of the model. The correct
hyper-parameters that give the best accuracy for the model are then chosen. The
model is then rerun without and with grid search technique termed as M1 and M2,
respectively. The five chosen hyper-parameters tol, max_iter, penalty and solver are
detailed in Table 1. After the grid search, the hyper-parameter combination which
provides the best accuracy is obtained as—dual: False, max_iter: 150, penalty: l1,
solver: 'liblinear' and tol: 0.0001. Figure 2a shows the plotted ROC curves.

Table 1 The hyper-parameters and values for grid searching

S. No.	Hyper-parameter	Description of the hyper-parameters	Values of the hyper-parameters
1	tol	Tolerance for stopping criteria	10, 1, 0.1, 0.01, 0.001, 0.0001
2	max_iter	Maximum number of iterations taken for the solvers to converge	50, 100, 150, 200, 250, 300
3	penalty	Used to specify the norm used in the penalization	l1, l2, elasticnet
4	solver	Algorithm used in the optimization problem	Newton-cg, lbfgs, liblinear, sag, saga

The blue and orange lines represent the ROC curves without (M1) and with (M2) grid search. From the two curves, it is evident that the M2 is better than M1. Specifically, the accuracy of the ML model without using grid search (M1) was computed as 82%, and after implementing grid search (M2), the accuracy increased to 86.69%. Figure 2b shows the plotted ROC curve for M3 using ANN. As evident from Fig. 2b, the accuracy of M3 using ANN was computed as 94% in testing. This value is comparatively better than models M1 and M2 signifying that ML model using ANN exhibits enhanced performance in credit card approval system.

The confusion matrix for models M1, M2 and M3 is shown in Figs. 3a, b and 4, respectively. As mentioned before, they consist of four quadrants, namely TP, TN, FP and FN. Comparing the figures, it can be deciphered that for model M2, the true negative obtained has improved by 0.73%, and correspondingly, the false positive has decreased by 0.73% when the grid search technique is utilized. Further, M3 exhibits the highest TN and TP values along with bare minimum values of FP and FN of all the three models M1, M2 and M3, thereby affirming that model M3 using ANN has the best performance. This observation combined with the performance improvement discerned from ROC curves confirms that the capability of the model M1 to reliably predict credit card approvals has improved by using grid search technique in model M2. This observation complies with prior studies [12]. Additionally, M3 exhibits the best overall performance among the three considered models.

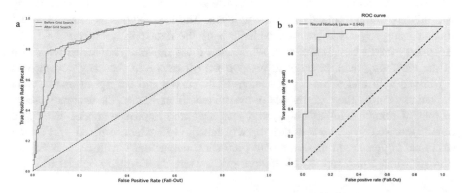

Fig. 2 ROC curve of **a** LR without and with grid search, **b** ANN model

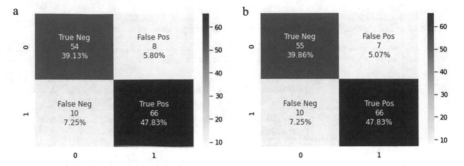

Fig. 3 Confusion matrix **a** without grid search (M1), **b** with grid search (M2)

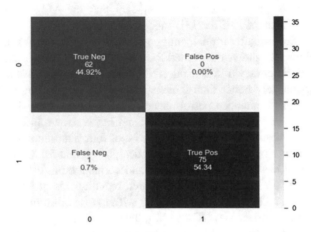

Fig. 4 Confusion matrix using ANN model (M3)

4 Conclusion

In this study, three ML models M1, M2 and M3 were built to predict credit card approvals. The commonly utilized processing steps such as scaling, label encoding and missing value imputation were employed. The ML models M1 and M2 were trained using logistic regression with 70% of the data retrieved from UCI ML Repository credit card approval dataset. Later, these models were tested with 30% of the remaining data. In order to observe the difference in accuracy, testing was performed with and without grid search. ROC curves and confusion matrix were plotted to quantify the performance of the models. In particular, the accuracy metric of the ML model without grid search was 82% and improved model using grid search was 86.69%. The results indicated that the M2 which utilized grid search exhibited higher credit card approval prediction competence with an accuracy of 86.69% when compared to M1 without grid search. Additionally, model M3 was built using ANN. The results indicated that the M3 performs comparatively better than M1 and M2 with an accuracy of 94% during testing. As a future enhancement, a larger dataset with an abundant of user entries and features can be used to train the model. This will help the model to give predictions with higher accuracies. Also, instead of using ML algorithms, more complex deep neural networks which requires a very large amount of data and processing power can be explored to obtain better prediction power with a high rate of accuracy.

References

1. Alzubi J, Nayyar A, Kumar A (2018) Machine learning from theory to algorithms: an overview. J Phys Conf Ser 1142(1):012012
2. Berry MW, Mohamed A, Yap BW (2019) Supervised and unsupervised learning for data science. Springer Nature
3. Bursztyn L, Fiorin S, Gottlieb D, Kanz M (2019) Moral incentives in credit card debt repayment: evidence from a field experiment. J Polit Econ 127(4):1641–1683
4. Ing E, Su W, Schonlau M, Torun N (2019) Support vector machines and logistic regression to predict temporal artery biopsy outcomes. Can J Ophthalmol 54(1):116–118
5. Ray A, Halder T, Jena S, Sahoo A, Ghosh B, Mohanty S, Nayak S (2020) Application of artificial neural network (ANN) model for prediction and optimization of coronarin D content in Hedychium coronarium. Ind Crops Prod 146:112186
6. Garud KS, Jayaraj S, Lee MY (2020) A review on modeling of solar photovoltaic systems using artificial neural networks, fuzzy logic, genetic algorithm and hybrid models. Int J Energy Res 1–30
7. Carneiro N, Figueira G, Costa M (2017) A data mining based system for credit-card fraud detection in e-tail. Decis Support Syst 95:91–101
8. Diaz GI, Fokoue-Nkoutche A, Nannicini G, Samulowitz H (2017) An effective algorithm for hyperparameter optimization of neural networks. IBM J Res Dev 61(4/5):9–1
9. Kahya MA (2019) Classification enhancement of breast cancer histopathological image using penalized logistic regression. Indonesian J Electr Eng Comput Sci 13(1):405–410
10. Syarif I, Prugel-Bennett A, Wills G (2016) SVM parameter optimization using grid search and genetic algorithm to improve classification performance. Telkomnika 14(4):1502
11. Yang X, Bian C, Yu L, Ni D, Heng PA (2017) Class-balanced deep neural network for automatic ventricular structure segmentation. In: International workshop on statistical atlases and computational models of the heart. Springer, Cham, pp 152–160
12. Zhang X, Wang D, Qian Y, Yang Y (2019) Prediction accuracy analysis with logistic regression and CART decision tree. In: Fourth international workshop on pattern recognition, vol 11198. International Society for Optics and Photonics, p 1119810

A MIMO-Based Compatible Fuzzy Logic Controller for DFIG-Based Wind Turbine Generator

K. Sudarsana Reddy◉ and R. Mahalakshmi◉

1 Introduction

Wind power is one of the most widely used and fastest growing renewable energies. According to the global reports, the capacity of generating electricity from wind has been increased to 75% in just two decades (1997–2018) both on offshore and onshore, respectively [1].

India being the fourth place across the globe in the installation of wind farm, and its installed wind energy capacity has been raised to 36.625GW as of 31 March 2019 [2]. Wind turbine generators (WTG) are used to generate electricity by capturing the kinetic energy of wind. There are three main different types of WTG utilizing induction generators. Type 1 WTG consisting of AC-DC-AC converter in addition to the induction generator before the grid. Type 2 WTG which is connected to grid directly utilizes wound rotor induction generator whose rotor speed is adjusted by using rheostat. The type 3 WTG consists of DFIG connected to the grid, where the speed of the rotor gets adjusted with the help of MSC and GSC converters [3]. DFIG has many advantages over other WTG types. Generation of the output signal with constant frequency irrespective of rotor speed and control of power factor is achieved easily in DFIG by maintaining it almost equal to unity. The controllers used in DFIG consist of low rated power electronic devices making whole control system cost to be effective [4]. Each MSC and GSC of DFIG is implemented with conventional PI controllers. This conventional control technique needs all the controllers to be tuned perfectly for achieving the desired output. The coordination between all the eight PI controllers at MSC and GSC is very difficult to achieve [5]. If in case any parameter is unknown, then these PI controllers do not work properly. So, many revolutionary control techniques have been introduced to

K. Sudarsana Reddy (✉) · R. Mahalakshmi
Department of Electrical and Electronics Engineering, Amrita School of Engineering,
Amrita Vishwa Vidyapeetham, Bengaluru, India
e-mail: d_mahalakshmi@blr.amrita.edu

© The Author(s), under exclusive license to Springer Nature Singapore Pte Ltd. 2022　　　11
R. R. Raje et al. (eds.), *Artificial Intelligence and Technologies*,
Lecture Notes in Electrical Engineering 806,
https://doi.org/10.1007/978-981-16-6448-9_2

overcome the issues created by conventional PI controllers such as fuzzy logic controllers (FLC) for RSC [6–9], deadbeat fuzzy controllers for GSC [9], PI and fuzzy hybrid RSC controllers [10] and self-tuning neural fuzzy controllers [11]. There are many adaptive controllers such as LQR [12, 13], LQG [14], particle swarm optimizing algorithm technique [15], etc. All existing control techniques proposed the separate control techniques at RSC and GSC. There were no reports on single unique control technique for both MSC and GSC together. This gap motivates to propose the single controller for RSC and GSC. The objective of this paper is to reduce the greater number of PI controllers of DFIG to a single MIMO-based controller so as to reduce the cost of the whole control system for DFIG and to replace the conventional PI controllers to FLC with introduction of compatible-based fuzzy algorithm to the FLC. The perfectly tuned PI controllers cannot respond to the dynamic changes whereas a perfectly defined knowledge base of a Fuzzy logic controller could mitigate the dynamic changes happening in the environment and stabilizes the requires output. The sole objective of this paper is to develop the MIMO-based compatible fuzzy logic controller for DFIG for extracting the maximum output power from the DFIG during different wind speed conditions and for the grid synchronization.

The novelty of parallel tuning of input and output scaling factors of the FLC and a MIMO-based system for DFIG has been achieved with the help of MATLAB/ Simulink, and its simulation results are being discussed. The paper has been organized into five sections. Section 2 describes the specification of the proposed system. In Sect. 3, the discussion of controllers of the proposed system along with its techniques is discussed. Section 4 dealt about the simulation results for the given system. Section 5 depicts the concluding remarks of the paper.

2   System and Specifications

The block diagram of the entire wind energy system using DFIG is depicted as shown in Fig. 1. The stator terminals of the DFIG are tied with grid and load. The extracted electrical power from wind energy through DFIG (P_s) is fed into the grid (P_g) and the three-phase constant load (P_L). The control pulses for the converters MSC and GSC are supplied from the single control unit which uses the compatible fuzzy algorithm which is an adaptive technique. The main purpose of these converters is to maintain constant DC link voltage (V_{dc}) and to extract the maximum amount of power from the varying wind velocity. The single controller is used for the purpose of gate pulse generation of the MSC-GSC converters. P_{slip} is the slip power injected into rotor by MSC in order to make the generator to produce maximum power. L_{RSC} and L_{GSC} act as a filter component. The specifications of the DFIG machine are shown in Table 1.

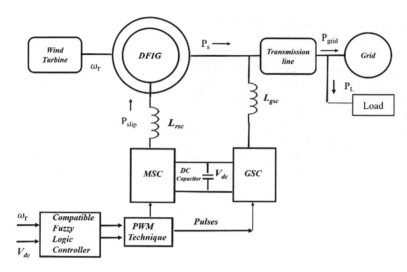

Fig. 1 Proposed system

Table 1 Specifications of the proposed system

Parameters	Values
Power (watts/Hp)	3730 W/5 Hp
Stator voltage (volts)	415 V
Grid frequency (Hz)	50 Hz
Stator impedance (Ω)	9.1411 Ω
No-load speed (rpm)	1500 rpm
Load	Star connected 1200 W (415 V, 50 Hz)

The plot in Fig. 2 describes the values of maximum power point (MPP) of stator power, rotor speed and torque of a 5 Hp DFIG-based WTG for different wind speeds (ranging from 7 to 11 m/s), respectively, which is obtained from the power versus wind speed characteristics of the wind turbine.

3 Control Technique

3.1 DFIG-Based WTG MSC and GSC Controller Techniques

The gate pulses for both MSC and GSC are obtained by each controller in different procedures. The gate pulses for MSC are obtained by sine pulse width modulation (SPWM) technique. The sinusoidal waves for the SPWM technique are obtained by

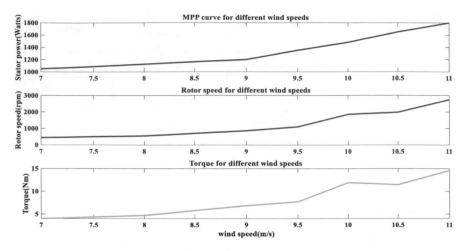

Fig. 2 Plots of maximum stator power, rotor speed and torque versus different wind speeds

giving the rotor speed in terms of wind speed as the input to a controller so as to provide the modulation index as the output and then multiplied with three sine waves of frequency (f_{slip}). This modulation index decides the magnitude of reference sine, and the rotor speed decides the slip frequency (f_{sl}) of the reference sine wave. Slip frequency is given by (1). Where N_s is the synchronous speed, N_r is the rotor speed, and f_{sys} is the stator (machine) frequency.

$$f_{sl} = \frac{N_s - N_r}{N_s} * f_{sys} \tag{1}$$

The pulse from SPWM decides the output slip voltage and current of the MSC. The output of MSC and its corresponding injected slip power to the rotor decides amount of power delivered from the DFIG for the particular wind speed as given Fig. 2. Table 2 represents the data of speed versus modulation index for generating MSC pulses. Using these data, the MSC control procedure is designed with the help of fuzzy logic control technique considering the input membership function as speed and the output membership function as modulation index [6]. These modulation indices for different wind speeds (rotor speed) are found by injecting slip voltage with the particular magnitude and slip frequency into the rotor circuit of DFIG and finding whether the maximum power is delivered or not. This procedure is followed for all speeds, and the data is consolidated in Table 2. This helps in forming input and output membership functions for the fuzzy controller at MSC to extract the maximum amount of power from DFIG.

The GSC is used to maintain the voltage across the capacitor C_{dc} (V_{dc}). C_{dc} is the capacitor which is interconnecting the GSC and MSC. The actual DC link voltage varies with the grid voltage variations. To maintain constant V_{dc}, conventionally PI controller is used. The error between $V_{dc,actual}$ and $V_{dc,ref}$ is sent to the tuned PI

Table 2 Speed versus modulation index

Speed (in rps)	Modulation index (m_a)
100	0.573
105	0.545
110	0.516
115	0.486
120	0.46
125	0.43
130	0.4
135	0.366
140	0.336
145	0.301
150	0.27
155	0.251
160	0.222
165	0.228
170	0.45

controller, and the controlled signal which is proportional to the modulation index is given to the comparator. The signal is compared with the ramp signal to produce the gate pulses. The gate pulses for GSC are obtained with the help of comparator mechanism whose inputs being a triangular wave and the modulation index that has been obtained by a PI controller. The PI controller of GSC is tuned for different grid voltage conditions, and this PI-based GSC is also transformed into CFLC and is discussed in next section. This control technique suits for the grid voltage variations between 415 and 370 V. Output of the CFLC will be the modulation indices of both MSC and GSC.

3.2 Proposed MIMO-Based CFLC for DFIG

The above-discussed fuzzy control techniques for MSC and PI-based GSC are integrated together through compatible FLC technique and proposed a one single MIMO-based CFLC controller. Figure 3 represents the proposed block diagram for the complete process of generating gate pulses to the MSC and GSC controller block, respectively. The modulation index for generating the gate pulses for both MSC and GSC controllers, respectively, has been produced by the CFLC whose inputs are the DC link voltage (GSC) and speed (MSC) of the rotor, respectively. So, the controllers that are helping to achieve the gate pulses for MSC and GSC are combined to form a single controller making it as MIMO-based system. The MIMO based CFLC is trained to provide modulation indices for MSC and GSC corresponding to different wind rotor speed conditions and various error in DC link

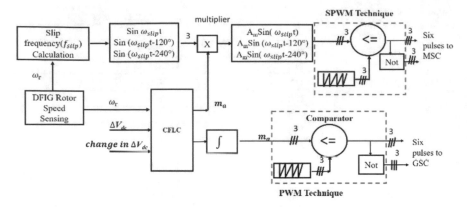

Fig. 3 Detailed block diagram for generating MSC and GSC pulses

voltage for the variations in the grid voltage. By using the proposed control, for different rotor speed conditions, the maximum power can be delivered, and the DC link voltage is maintained constant for varying grid voltage conditions.

3.3 Compatible Fuzzy Logic Controller

Disadvantages of the conventional controllers. The conventional controllers like PI and PID are used for as to generate gate pulses. So, to get desired output for the plant, these controllers need to be perfectly tuned, and if tuned, the output changes due to any dynamic change. This is the major drawback for the conventional controllers. The controllers like FLC, ANN-based controllers, machine learning technique-based controllers, etc., are used to overcome the drawback. But for any FLC to produce the desired response for any dynamic changes in the environment, one should properly define the rules or tune the scaling factors at both input and output sides. Generally, the rule base and scaling factors for a FLC are usually assumed manually for a system which is according to the performance of the plant. This becomes challenging for a human to continuously check for the system response which again becomes a challenge.

So, there are many algorithms and optimization techniques which automatically learn from the data and modify these rules or tune the scaling factors, like Genetic algorithm, Neural network, Queen bee colony algorithm or ant colony algorithm to name a few. These can modify the rule base and scaling factors according to the dynamic change. But these require a large amount of input data to be provided for proper tuning or modifying the rule base of FLC which becomes a challenge, and generally, these algorithms are used to modify the rule base. Similarly, there are few techniques which can properly tune the scaling factors alone. For example, consider

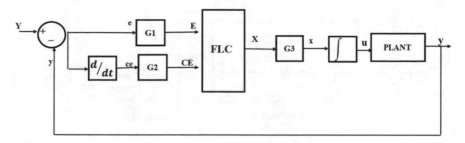

Fig. 4 First-order closed loop FLC for a plant

the first-order closed loop FLC plant shown in Fig. 4. whose open loop transfer function is given in (2). But those are best suited for only first order and possibly for a second-order control system.

The open loop transfer function (first order) of the plant is given by (2)

$$G(s) = \frac{Ke^{-ms}}{1 + Ts} \tag{2}$$

$$G1 = \frac{1}{Y(0) - y(0)} \tag{3}$$

$$G2 = minimum\left(T, \frac{m}{2}\right) * G1 \tag{4}$$

$$G3 = \frac{1}{K * G1\left(K_c + \frac{m}{2}\right)} \tag{5}$$

where K, m and T are the gain, time delay and time constant, respectively. Therefore, the input scaling factors $G1$, $G2$ and output scaling factor $G3$ can be defined by (3–5). Where $Y(0)$ represents the initial reference, and $y(0)$ represents the initial output. Where K_c is the closed loop gain of the plant.

Equations (3), (4) and (5) represent the input and output scaling factors for a first-order-based FLC, respectively. For any higher order, those equations do not work properly. To overcome all the drawbacks, a novel controller technique has been proposed. This controller can automatically scale its tuning factors with the help of FLCs whose inputs are same as the main FLC making it data efficient and perfectly suited for any dynamic change. Such a FLC can be known as compatible fuzzy algorithm-based fuzzy logic controller (CFLC), and its working has been explained in the next section.

Novel single CFLC controller for DFIG. The block diagram for the CFLC has been shown in Fig. 5. According to the application provided, the CFLC here is suited for the fuzzy controllers whose inputs are ω_r, error (E) and change in errors (CE) of V_{dc}. The control signal rotor speed is processed for the MSC converter, and

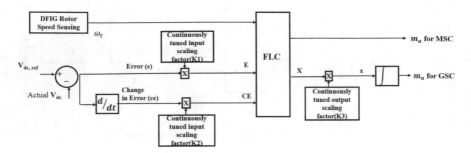

Fig. 5 Proposed novel compatible fuzzy logic controller

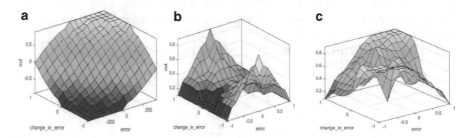

Fig. 6 a, b Surfaces for input scaling factors, **c** surfaces for output scaling factor

E, CE are processed by CFLC for GSC pulses. The outputs of CFLC are modulation indices for MSC and GSC.

Figure 5 depicts the proposed compatible fuzzy logic controller which is MIMO-based controller for the complete DFIG-based WTG. The lower part of Fig. 5 shows the control technique of CFLC only for GSC pulses generation in which the inputs are E and CE and are given by (6) and (7), respectively. Difference between actual V_{dc} and reference V_{dc} being the parameters for calculating the error and the change in error which are the inputs to the control technique for variations in the grid voltage.

$$E = e * K1; \tag{6}$$

$$CE = ce * K2; \tag{7}$$

e, ce are the unscaled or actual value of error and change in error, respectively. The input for the integrator x is given by (8).

$$x = X * K3; \tag{8}$$

where X is the output of the CFLC for GSC, and $K1$, $K2$, $K3$ are the continuously tuned scaling factors.

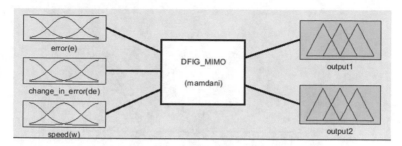

Fig. 7 Fuzzy diagram for the proposed MIMO controller

Figure 6 shows the surfaces of the continuously tuned scaling factors. Figure 6a, b shows the surfaces for error and change_in_error scaling factors which are input scaling factors for the main FLC. And the output scaling factor has been shown in Fig. 6c.

Figure 7 shows the fuzzy diagram for the modelled MIMO controller for the DFIG WTG. The description of the fuzzy variables is as follows:

Error: represents the deviation of the DC output of the GSC.

Change in Error: represents the deviation of present error to the previous error of the DC output of GSC controller.

Speed: represents the speed of the machine at the very instant.

Output1: gives the modulation index for the GSC controller.

Output2: gives the modulation index for the MSC controller.

Figure 8 shows the membership functions of the input variables of the FLC (error ΔV_{dc}, change_in_error (change in ΔV_{dc}), speed (ω_r) variables, respectively), and Fig. 9, respectively, shows the membership functions for the output variables (output1 and output2, respectively).

In the main FLC, the input crisp variables error, change in error and speed will be converted to linguistic variables with the help of respective input triangular or trapezoidal membership functions. The universe of discourse for error and change in error has been divided into seven fuzzy sets, namely NB, NM, NS, ZE, PS, PM, PB, respectively, where NB and PB are trapezoidal remaining being triangular membership functions, and every linguistic data is subset of these members with some degree of membership changing in the range of $[-1, 1]$. Similarly, the universe of discourse for speed has been divided into nine fuzzy sets, namely mf1, mf2, mf3, mf4, mf5, mf6, mf7, mf8, mf9, respectively, and each of the variable is subset of these members with finite degree of membership varying in the range $[80, 160]$. These fuzzy data get processed by inference mechanism. Here, the inference mechanism used is max–min. The rules for the inference mechanism are given by Table 3 for output1 variable, and Table 4 depicts the rule base for the output2 variable. The single fuzzy sets are then defuzzified which convert the fuzzy data (variable) to a crisp data. The range of values in the fuzzy set will be defuzzified to form a single number. The defuzzification process has been done by "centroid" method where the single number will be chosen from the centre of the area of the fuzzy set.

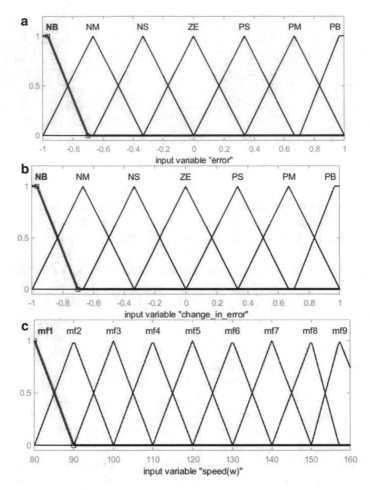

Fig. 8 Membership functions for the **a** error variable, **b** change_in_error variable, **c** speed variable

4 Results

The DFIG with MSC and GSC along with single CFLC has been simulated in MATLAB (Simulink), and the respective results have been discussed in this section. The aimed system has been simulated for different wind speeds in the range 7–12 m/s. The plots have been taken at the wind speed of 9.2 m/s with rotor speed being 1400 rpm. The comparison in the results discussed here are done by considering the system with CFLC and with un-tuned PI controllers. The DC link voltage between MSC and GSC, the stator power along with torque had been compared for two controllers that are mentioned above. The grid voltage is dynamically varied as 375 V, 380 V, 410 V, 415 V at regular intervals of 0.5 s, 1 s, 1.5 s, 2 s, respectively. Figure 10a, b depicts the DC link voltage for

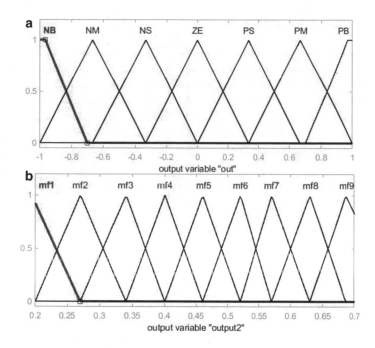

Fig. 9 Membership function for the **a** output1 variable, **b** output2 variable

Table 3 Rule base for Output1 (GSC)

Change in error	Error						
	PB	PM	PS	ZE	NS	NM	NB
PB	PB	PB	PB	PB	PM	PS	ZE
PM	PB	PB	PB	PM	PS	ZE	NS
PS	PB	PB	PM	PS	ZE	NS	NM
ZE	PB	PM	PS	ZE	NS	NM	NB
NS	PM	PS	ZE	NS	NM	NB	NB
NM	PS	ZE	NS	NM	NB	NB	NB
NB	ZE	NS	NM	NB	NB	NB	NB

MIMO-based CFLC and PI controllers, respectively. From Fig. 10, it can be observed that MIMO-based CFLC maintained the DC voltage constant (500 V), while the other has been reducing and getting little stable at 300 V whose grid voltages have dynamically changed at 0.2 s, 0.4 s, 0.6 s and 0.8 s, respectively, with being same dynamic voltages as with proposed controller. But the rise time taken by PI controllers is being less when compared to MIMO-based CFLC.

Figures 11, 12 represent the stator power and torque of both controllers, respectively. It can be inferred from Fig. 11a that the MIMO-based CFLC maintains the maximum stator power at 1480 W, and torque is being stabilized at 10 Nm as

Table 4 Rule base for Output2 (MSC)

Speed	Output
mf1	mf9
mf2	mf8
mf3	mf7
mf4	mf6
mf5	mf5
mf6	mf4
mf7	mf3
mf8	mf2
mf9	mf1

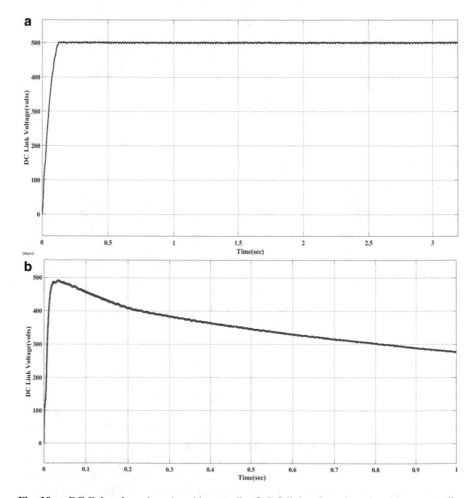

Fig. 10 **a** DC link voltage in volts with controller, **b** DC link voltage in volts without controller

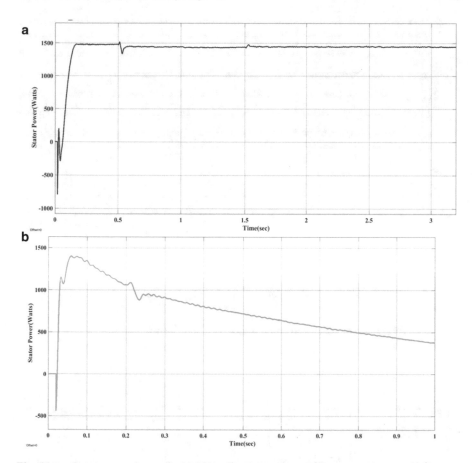

Fig. 11 **a** Stator power in watts with controller, **b** stator power in watts without controller

shown in Fig. 12a even with dynamic changes occurring at four (0.5, 1, 1.5, 2 s) instants, respectively. From the plot shown in Fig. 2, the stator power and torque at wind speed of 9.2 m/s are 1480 W and 10 Nm, respectively. Only at the instants t = 0.5 s and t = 1.5 s, the changes are predominately seen. The stator power and torque in Figs. 11b and 12b of PI controllers are continuously dropping from 1450 W and 9 Nm during dynamic change, and at $t = 0.2$ s, the changes are clearly seen in the figure and more noticeably when compared to MIMO-based CFLC results.

The respective plots from Figs. 13 and 14 correspond to the MIMO-based FLC at wind speed of 9.2 m/s (rotor speed of 1400 rpm). Figure 13a shows the stator voltage in d-q frame. It can be observed that V_d has been maintained at 415 V (rms line) [338 * 1.732/1.414] and V_q being 0 V at every instant even though dynamic change is occurring at four different instances. Figure 13b shows the stator current being maintained at 10.67 A rms with the frequency of 50 Hz. Figure 14 shows the

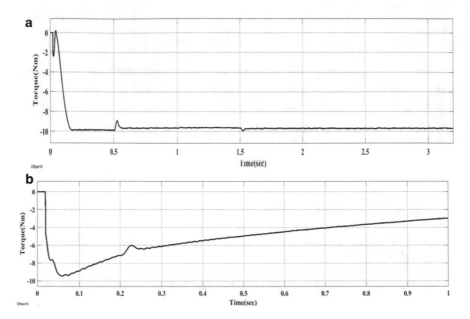

Fig. 12 **a** Torque in Nm with controller, **b** torque in Nm without controller

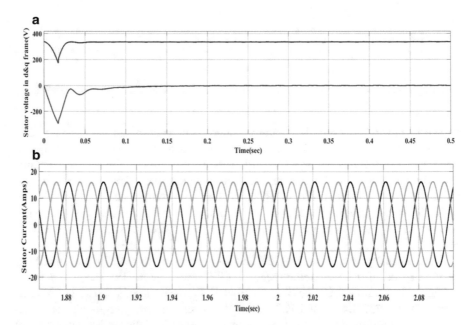

Fig. 13 **a** Stator voltage in volts in *d-q* frame, **b** stator current in amps

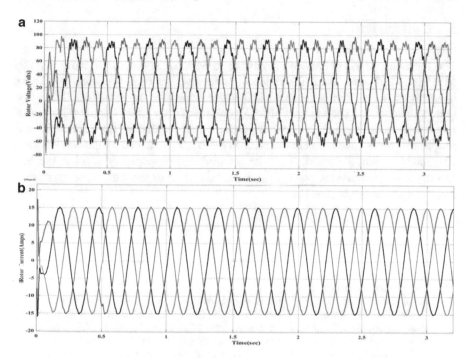

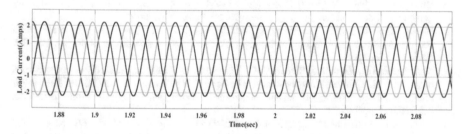

Fig. 14 **a** Rotor voltage in volts at 1400 rpm, **b** rotor current in amps at 1400 rpm

Fig. 15 Load current in amps

rotor voltage and rotor current at the slip frequency of 3.4 Hz. From (1), the slip frequency can be calculated as 3.33444 Hz at given slip of 0.66667, and rotor speed of 1400 rpm with stator frequency of 50 Hz respectively.

The stator of the DFIG is connected to the 1200 W constant load and to the grid. It is observed that the stator power is delivered to load power, and remaining power is injected to the grid. From Fig. 15, we can infer that the load specified has been drawing 2.16 A peak current (1.527 A rms) from the stator as to deliver 1200 W load which is connected to machine at 415 V, 50 Hz. The remaining 9.142 A rms has been drawn by the grid so as the overall 10.67 A rms from the stator has been utilized.

5   Conclusions

The DFIG-based WTG has been simulated in MATLAB (Simulink) along with its MSC and GSC controllers. The different controllers for each MSC and GSC have been integrated to form a single MIMO-based controller for the whole DFIG system. The controller used for the MIMO system is fuzzy adaptive controller whose input and output scaling factors have been tuned parallelly. The results for this proposed system have been compared with the results obtained by the conventional PI controllers and discussed. It can be inferred that proposed system has the best results, and the whole control system is cost effective as the number of controllers has been decreased to one.

References

1. www.irena.org/wind.com
2. Global Wind Energy Council. http://www.gwec.net/
3. Mahalakshmi R, Viknesh J, Thampatty KCS (2016) Mathematical modelling of grid connected doubly fed induction generator based wind farm. In: 2016 IEEE international conference on power electronics, drives and energy systems (PEDES), Trivandrum, pp 1–6
4. Raj KR, Mahalakshmi R (2017) Enhancement of oscillatory response of DFIG based wind energy conversion system. In: 2017 international conference on smart grids, power and advanced control engineering (ICSPACE), Bangalore, pp 93–98
5. Bhukya J, Mahajan V (2018) Fuzzy logic based control scheme for doubly fed induction generator based wind turbine. Int J Emerg Electr Power Syst 34(2)
6. Mahalakshmi R, Sindhu Thampatty KC (2020) Slip frequency control technique for DFIG based wind turbine generators. Advances in Electrical and Computer Technologies, vol 672. Springer, Singapore
7. Mahalakshmi R, Viknesh J, Ramesh MG, Vignesh MR, Thampatty KCS (2016) Fuzzy logic based rotor side converter for constant power control of grid connected DFIG. In: 2016 IEEE international conference on power electronics, drives and energy systems (PEDES), Trivandrum, pp 1–6
8. Elkhadiri S, Elmenzhi PL, Lyhyaoui PA (2018) Fuzzy logic control of DFIG-based wind turbine. In: 2018 international conference on intelligent systems and computer vision, pp 1–5
9. Bo Q, Xiao-yuan H, Wen-xi Y, Zheng-yu L, Guerrero JM (2019) An optimized deadbeat control scheme using fuzzy control in three-phase voltage source pwm rectifier. In: Twenty-fourth annual IEEE applied power electronics conference and exposition, pp 1215–1219
10. Poitiers F, Bouaouiche T, Machmoum M (2009) Advanced control of a doubly-fed induction generator for wind energy conversion. Elect Power Syst Res 79(7):1085–1096
11. Muneer A, Bilal-Kadri M (2013) Pitch angle control of DFIG using self-tuning neuro fuzzy controller. In: 2nd international conference on renewable energy research and applications (ICRERA), Madrid, Spain, pp 20–23
12. Phan DC, Trinh TH (2019) Application of linear quadratic regulator to control directly power for DFIG wind turbine. J Electr Syst 15(2):42–52
13. Azri HE, Essadki A, Nasser T (2018) LQR controller design for a nonlinear, doubly fed induction generator model. In: 2018 6th international renewable and sustainable energy conference (IRSEC), Rabat, Morocco, pp 1–6

14. Barrera-Cardenas R, Molinas M (2012) Optimal LQG controller for variable speed wind turbine based on genetic algorithms. Energy Procedia 20:207–216
15. Bounar N, Labdai S, Boulkroune A (2018) PSO-GSA based fuzzy sliding mode controller for DFIG-based wind turbine. ISA Trans

Robotic Process Automation

Raj Choudhary and A. Karmel

1 Introduction

The main foundational concept in any software testing company is the creation of the same base environment before testing the new update. This leads to a lot of time invested in creating the same base environment for testing the update. Moreover, in case of failure of services, companies prefer to deploy the same service in some other server while they think about whether they should have their employees debug the crashed service or not. This leads to two major benefits for the organization. They can continue to provide their services as it has been deployed in some other server which in turn leads to an interrupted flow of services and also revenue generation. Secondly, if the debugging of the crashed service would take a longer time to rectify the error and would, in turn, lead to loss of money to the organization, they can just delay the debugging or even discard it based on the decision of the managerial board. Even when creating a new environment and deploying the service again sounds easy, it requires a lot of setups to be made as well as files to be copied so that it is ready for the general public or even the testing team to use. This leads to significant investment of time from employee(s). If this setup could have been automated, then it would just have required a single employee to run the task, and after the task would have been finished, he would just have to once check it for its working before making it available to others. This leads to minimal involvement from the employee, and he can use the time saved for other productive tasks. This not only leads to saving the employee's time but also a better return on investment (ROI) for the organizations, as their employees can make better use of their time. This type of automation can be made for general tasks like installing and starting services, deleting them, copying files, and for even more specialized tasks like creating a whole base environment for testing a new increment in the software.

R. Choudhary (✉) · A. Karmel
Vellore Institute of Technology, Chennai, Tamil Nadu 600127, India
e-mail: karmel.a@vit.ac.in

R. R. Raje et al. (eds.), *Artificial Intelligence and Technologies*,
Lecture Notes in Electrical Engineering 806,
https://doi.org/10.1007/978-981-16-6448-9_3

2 Literature Survey

The manufacturing industry was the first to implement RPA as they used hardware components to automate their tasks that were tedious and/or required great precision. As in [1], the authors discuss how robotic arms were used to perform thermoplastic fiber placement in both the factory and offsite locations of ABB motor factory in Switzerland. Many other industries like the metallurgy as well as the automobile industry employ robotic components to automate tasks like carrying components from one location to another as well as painting, welding, etc., that are repetitive tasks and require great precision for quality results. In [2], the author discusses what is RPA, why is there a need for RPA as well as the difference between RPA and straight through processing (STP). To illustrate the need for RPA, the author discusses the example of a person transferring from one country to another. In such a case, the data of the user has to be changed in multiple locations manually, and this takes a lot of time. If this same process would have been automated, the human involvement would have been reduced significantly. Given this automation of tasks, the author discusses what type of tasks could be automated using RPA. The distribution between the type of tasks and their frequency showcases a Pareto distribution. In [3, 4], the authors summarize that for RPA production, traditional delivery methodologies like the waterfall model are over-engineered. Agile delivery methods are the best for the development of RPA in an iterative manner. This process is continued until all the cases and situations in the process definition document are covered. This development ideology is currently at the conceptual level and needs to be adapted for the corporate setting. The authors in [5] provide a four-step approach to classify if a process is RPA applicable or not. It classifies the process into the following categories—not suitable for RPA, less suitable for RPA, moderately suitable for RPA, or highly suitable. It uses six properties collected in the four-step approach by getting responses to a questionnaire from business experts and analysts. In [6], the authors try to address the problem of identifying whether a task could be automated from their process description. Since a company has a tremendous amount of these process descriptions, manually identifying whether a process could be automated or not is not feasible. So, they proposed a supervised machine learning workflow that would automatically identify a task as manual, user interaction, or automatic.

The paper [7] starts with the authors discussing how automating enterprise resourcing planning (ERP) and customer relationship management (CRM) tasks are helpful as they involve structured data with definite steps for creating the results. These tasks are the best for automating as they produce definite results and involve copying the human process using a set of rules. They also highlight the fact that RPA sits on top of the business models and all the other software of the company. Moreover, since RPA does not store any transactional data, no data model is needed to be employed. The author analyzes the case of RPA used in a BPO provider. In [8], the authors discuss how DT employed RPA for not only moving toward digital transformations but also use them to target the areas of most customer dissatisfaction. To find out potential areas for automating, they created teams of project leaders, 20 users, and 5 software developers who worked together for 2 weeks to find potential RPA ideas. In [9], the authors compare the top three RPA service providers in the world—UiPath, Blue Prism, and Automation Anywhere. The services are compared based on different factors that define the ease of use of the service like visual process designer, control through coding, script-based designer, etc.

3 Proposed Architecture of Web Interface

The front face of all the Web sites is HTML which is generated by Python CGI scripts. The main webpage sums up all the tasks that the user can perform on the managed nodes. In the current implementation, the user can configure YUM on Red Hat systems, configure HTTPD server, install and uninstall packages, as well as create and delete an existing user. Here, the user can input his/her choice in the text box or can input the field using voice recognition. The list of available IPs for different tasks is generated dynamically using Python helper functions created from scratch. This helper functions only to display the IPs of the systems on which the particular tasks could be performed. For example, YUM could only be configured on Red Hat systems; therefore, when the configuration of YUM is selected as the task to be performed, the system reads the hosts inventory file and displays the IP of only those managed nodes which were defined under the Red Hat group as shown in the figure. Whereas when the configuration of the HTTPD server was selected as the automation task which could be performed on all the managed nodes in the networks, the helper function displays all the possible IPs available in the inventory file in the drop-down menu. The helper function also creates subgroups in the drop-down menu as per the grouping present in the hosts inventory file. In the present contents of the inventory file, there are three IPs available which are grouped into two groups, namely—Red Hat with two nodes present under it and Ubuntu which has one node present under it. So, the helper function presents the user with the following options to run the automation task on the managed nodes.

4 Environmental Setup

The system configuration of managed node and controller node is given in Tables 1 and 2, respectively.

Table 1 System configuration of managed node

System configuration of managed node	Details
Operating systems	RHEL 7.5 cli, Ubuntu 19.04 GUI
Packages required	Python3, SSH
Tasks to be executed before automation	Static IP

Table 2 System configuration of controller node

System configuration of controller node	Details
Operating systems	RHEL 7.5 cli, Ubuntu 19.04 GUI
Packages required	SSH, HTTPD, FTP, Ansible
Tasks to be executed before automation	FTP and HTTP servers should be started. SSH connection with all the managed nodes

4.1 Requirements for Implementation

For the proper functioning of the system, the tasks that are to be implemented on the controller node are, the FTP server should be started and made permanent; HTTP server should be started and made permanent; Ansible package should be installed; SSH connection should be made with all the managed nodes with ssh-copy-id after generating the key using ssh-keygen so that future SSH requests do not prompt the user for a password for connecting with the servers and the firewall should be configured to allow access to a Web site hosted on the HTTP server. The IP of the managed node should be made static for the network so that it does not change on its own. Appropriate user should be created on the managed nodes to which the request could be made for SSH connection.

4.2 Configuration of Python CGI and Ansible

After the httpd package is installed, /var/www/ folder is created. This folder contains two subfolders—HTML and cgi-bin. The static HTML pages are stored in /var/www/HTML and can be accessed using http://IP_OF_SERVER/name_of_file. html, whereas the dynamic webpages generated by running a script are stored in /var/www/cgi. All the scripts used should contain the shebang of the command in the system as the first line in the file so that the server could execute them. After Ansible is installed, its inventory and configuration file can be accessed in /etc./ ansible. The inventory file is named hosts, and it contains the IPs of all the managed nodes in the network. The IPs can also be formed into groups that could be used to run a particular task on all the IPs in that group. Some changes have to be made in the Ansible configuration file named ansible.cfg. These changes are to be made for the following reasons: to disable SSH host key checking and to enable privilege escalation.

5 Results

In the first attempt, obviously network bandwidth for connection establishment, copying of files, and downloading packages from the FTP server would be used. Installing packages, creating files, starting services as well as making changes in their configuration would use computation power of the systems. But wasting these precious network bandwidth and computation power on servers in which the tasks of the concerned Ansible playbook are already done and are in the same consistent state as required by the playbook is not desired and leads to wastage of resources. To avoid this scenario, the consistent state of the servers should be checked with that of the executed Ansible playbook, and when there is some task that is not in the same state as that required by the playbook, then that task should be executed or else the consistent task should be skipped to preserve network bandwidth as well as the computing power of the server. Ansible ensures this by gathering facts about the servers involved in a task. It uses this returned JSON file from the servers to check if the tasks are performed on the server or not and if they are in the same state as that of the playbook or not. If the task is performed and in the same state as that of the playbook, Ansible would skip this task, whereas if the task needs to be reperformed, then Ansible would execute that task on the concerned server. The time analysis of Ansible playbooks and Python automation scripts is given in Tables 3 and 4, respectively.

Table 3 Time analysis of Ansible playbooks

Task performed	Time taken in first execution of the playbook on all the servers (in seconds)	Time taken in second execution of the playbook on all the servers (in seconds)
Configuring YUM on Red Hat servers	18.781	3.760
Configuring HTTPD server	17.23	8.17
Installing a package	7.924	2.508
Uninstalling a package	7.647	2.487
Adding a new user	2.052	1.959
Removing a user	7.213	1.858

Table 4 Time analysis of Python automation scripts

Task performed	Time taken in first execution of the python script on Red Hat servers (in seconds)	Time taken in second execution of the python script on Red Hat servers (in seconds)
Configuring YUM	59.854	7.227 (since failed in first command execution)
Configuring HTTPD server	46.143	43.611
Installing a package	23.491	11.294
Uninstalling a package	12.469	11.335

From the detailed analysis of execution time taken and for which different tasks within the playbook, how much time was taken, we can understand the importance of this checking of consistent states in the managed nodes well before execution of the playbook. The major time-consuming task in all the playbooks in all the scenario is that of gathering facts. This is the major time-consuming task as it has to send request to all the managed nodes requested by the playbook and get the system details as the response. After getting the JSON response, it has to decide which tasks in the playbook are to be executed and which tasks upon execution does not change the state of the managed node and can be ignored. This checking of tasks is done separately for all the managed nodes, and as the number of nodes on which the Ansible playbook is increased, the time required by gathering facts is also increased. Even though, gathering facts takes up lot of time in the playbook execution time, it leads to significant time saving in the overall execution of the Ansible playbook in the subsequent runs. This can be seen from the drastic reduction in the execution time in the second execution of the Ansible playbooks irrespective of the tasks and the fact that the playbooks were executed on all the managed nodes. This would prove beneficial where some or the other task is to be performed on some of the managed nodes requested by the playbook. As, it would lead to overall saving in the execution time of the Ansible playbook.

From the above time analysis, we can see that automation with Python, even though ran only on two servers, took more time in all the tasks than Ansible playbook which ran on all the available managed nodes in the implementation. The time difference between the first run and second run of the script is also not significant as Python cannot check for the consistent state of the managed nodes until and unless the output of each command is taken and processed to find if the tasks was successful or not. This only does add to the amount of code, but it also makes the script more and more OS and even OS version dependent as the output of the same command changes in different OS types and even in different versions of the same OS.

6 Conclusion

Robotic process automation system for the software development cycle has been proposed, in which time-consuming tasks are automated so that the developers can run these tasks on multiple servers at once without any bit of coding. The software robot will handle all the tasks and perform them in the remote servers and notify the developer whether the tasks were successful or not. Even if some subtasks are not performed, the developer would be notified of the same. In case of errors, the error as well as the remote servers on which the tasks were not performed, the IP of these systems would be displayed to the user. From time analysis, it is clear that in comparison with automation using Python, the implementation produces much more platform-independent results with very less execution time taken.

The work implemented could be extended by trying different types of controller and managed node hierarchy in the automation scenario; automating more complicated tasks that are needed to take into account the configuration of the systems and their specifications; collecting data on errors reported by the automation tasks so that they could be used to create models that could handle even errors on behalf of the users. These developments would make the automation process in the software development industry fully autonomous and would lead to saving more valuable time of FTE. This in turn would lead to greater ROI to the organizations that deploy these software robots.

References

1. Ahrens M, Mallick V, Parfrey K (1998) Robotic based thermoplastic fiber placement process. In: Proceedings of 1998 IEEE international conference on robotics and automation (Cat. No. 98CH36146), vol 2. IEEE, pp 1148–1153
2. van der Aalst WM, Bichler M, Heinzl A (2018) Robotic process automation
3. Cewe C, Koch D, Mertens R (2017) Minimal effort requirements engineering for robotic process automation with test driven development and screen recording. In: International conference on business process management. Springer, Cham, pp 642–648
4. Jimenez-Ramirez A, Reijers HA, Barba I, Del Valle C (2019) A method to improve the early stages of the robotic process automation lifecycle. In: International conference on advanced information systems engineering. Springer, Cham, pp 446–461
5. Leshob A, Bourgouin A, Renard L (2018) Towards a process analysis approach to adopt robotic process automation. In: 2018 IEEE 15th international conference on e-business engineering (ICEBE). IEEE, pp 46–53
6. Leopold H, van der Aa H, Reijers HA (2018) Identifying candidate tasks for robotic process automation in textual process descriptions. In: Enterprise, business-process, and information systems modeling. Springer, Cham, pp 67–81
7. Aguirre S, Rodriguez A (2017) Automation of a business process using robotic process automation (RPA): a case study. In: Workshop on engineering applications. Springer, Cham, pp 65–71

8. Schmitz M, Dietze C, Czarnecki C (2019) Enabling digital transformation through robotic process automation at Deutsche Telekom. In: Digitalization cases. Springer, Cham, pp 15–33
9. Issac R, Muni R, Desai K (2018) Delineated analysis of robotic process automation tools. In: 2018 second international conference on advances in electronics, computers, and communications (ICAECC). IEEE, pp 1–5

Automatic Road Surface Crack Detection Using Deep Learning Techniques

S. Aravindkumar, P. Varalakshmi, and Chindhu Alagappan

1 Introduction

Road plays a major role in the development and growth of our economy in day-to-day life. In India, as per the 2019 statistics, on an average, 16 deaths occur for an hour in road accidents. The major source of road crashes is due to the presence of damages. The reason for the occurrence of crack and damage can be identified, and some preventive must be undertaken to increase the durability of road surfaces. Cracks are due to overloading of vehicles in a thin basement; changing temperature cycle will result in shrinking or expansion of asphalt pavement and improper road joint parts. Consider a scenario of a National Highway which has cracks. In the next few days, there will be higher possibility of a crack to become damage which paves way for accidents to take place. In bridges, the issue is even more alarming. Repairing and maintenance of road surfaces from such distress is important task to be undertaken to save lives.

Deep learning algorithms are increasingly used in object detection mechanism which is helps in artificial intelligence. To assist the work of repairing and maintenance of road surfaces, various deep learning applications are deployed in detecting cracks and damages. Convolutional neural network (CNN) is the base network with significant number of layers used for extracting visual features.

In this paper, we propose a novel Faster R-CNN approach for detecting and classifying cracks and damages by plotting bounding box around them. Faster R-CNN approach uses pre-trained models like VGG16 and ResNet152. In VGG16, pre-trained model contains convolution, pooling, fully connected and a softmax layer. Convolution layer uses 3X3 kernel along with increasing number of filters at each level. Max pooling is performed based on the maximum value from each kernel, whereas pooling is used to reduce the feature map size. In fully connected

S. Aravindkumar (✉) · P. Varalakshmi · C. Alagappan
Department of Computer Technology, MIT Campus, Anna University, Chennai 600044, India

© The Author(s), under exclusive license to Springer Nature Singapore Pte Ltd. 2022   37
R. R. Raje et al. (eds.), *Artificial Intelligence and Technologies*,
Lecture Notes in Electrical Engineering 806,
https://doi.org/10.1007/978-981-16-6448-9_4

layer, one-hot vector values will be generated. However, VGG16 network contains 16 layer of deep neural functionalities. Residual networks were commonly referred as ResNet. ResNet has various versions like ResNet50, ResNet101 and ResNet152. ResNet152 will train the deep neural networks against 152 layers with a considerable amount of time to increase the crack detection performance. The dataset consists of 3,533 images. Those images were categorized into alligator crack, linear crack, nonlinear crack, damage and non-crack image. The training contains 70%, validation of 10% and testing uses 20% from the dataset.

2   Related Work

During inspection of road surface, crack and damage plays a major role in municipal roads, streets, highways, tunnels and flyovers. Edge detection and segmentation algorithm were used in [1]. Various edge detection and feature extraction methods are produced higher accuracy only if there occurs a good continuity of crack and high contrast input images. In [2], the author uses a deep convolutional neural network (DCNN) for edge detection, contour detection and boundary segmentation along with VGG16 model. Batch normalization is performed after each convolution layer. At each scale of the network, the convolved features from encoder and decoder are fused into a single-scaled feature map. The single-scaled feature maps are combined at all stages of the network to form a multi-scale fused feature map for detection of cracks. Segnet architecture is used in [3] for semantic segmentation. Various benchmark dataset like CrackTree260, CRKWH100, CrackLS315 and Stone331 are used for crack segmentation. The segmented region is evaluated using F-measure. A mobile measurement system (MSS) [4] is used to obtain high accuracy of geospatial information of a moving vehicle. Kitti dataset [5] is used, which consists of 15,435 instances of road surfaces. LG Nexus 5G app is used to collect the datasets from various streets of Japan. It detects crack regions and classifies its types into vertical, horizontal and crocodile. Asphalt 3D pixel-level crack detection is made in [6].

Naive Bayes CNN is used in [7] for crack detection using convolutional neural networks. The advantage of this scheme is to enhance the performance and robustness. In [8], pothole is detected using CNN model and virtualisation approach is used for feature selection. In [9], crack width transform algorithm is used for identifying the crack region of the image. Crack width transform initially computes the ratio filtering and then crack region search and then hole filling and relative thresholding. The crack width is measured accurately by extracting crack edge pixels. Multi-scale fusion crack detection (MFCD) algorithm is used in [10] to scan the cracks region of the image. Support vector machine (SVM), random forest, random structured forest are used for classification. In [11], genetic programming is used for detecting crack. The fitness function is used before the crossover and mutation process for the MSRA Dataset-1000. A Faster R-CNN [12] is built for segmentation and classification uses road damage dataset. Semi-automatic pavement crack labeling algorithm is developed in [13] used for accurate segmentation of pavement cracks. A multi-scale feature extraction mechanism is used to learn crack features from its background. Moreover, their proposed method also has the ability to overcome the noise produced by many environmental factors.

3 Proposed Work

Based on the above survey, many of the existing methods were used to identify and detect cracks and damages with less accuracy and performance. Our paper proposes Faster R-CNN approach along with two different pre-trained models like ResNet152 and VGG16 are used for crack detection and classification. The working procedure of our proposed model as shown in Fig. 1a, b shows the crack detection and classification using region proposal network.

3.1 Ground-Truth Identification and Dataset Pre-processing

Various traditional image processing techniques are used to identify ground-truth image and for dataset pre-processing. Image enhancement technique is applied to improve the quality of an image. Various image enhancement techniques are implemented and concluded that sharpening performs better. Image transformation

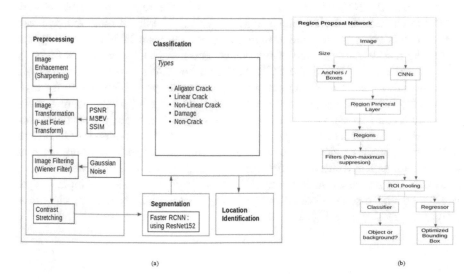

(a)   (b)

Fig. 1 **a** Overall proposed architecture; **b** crack detection and classification using region proposal network

is applied after sharpening; fast Fourier transform (FFT) is used to convert an image of a particular domain into another for extracting the features required. Wiener Filter is used in our system to reduce noise. The abovementioned pre-processing techniques are applied for dataset, whereas, additionally, Canny edge detection method is used to generate the ground-truth image. FFT and inverse FFT are computed using Eqs. (1) and (2), respectively.

Fast Fourier transform:

$$f(x,y) = \sum_{u=0}^{N-1}\sum_{v=0}^{M-1} f(u,v)e^{-j2\Pi\left(\frac{ux}{N}+\frac{vy}{M}\right)} \tag{1}$$

Inverse fast Fourier transform:

$$f(u,v) = \frac{1}{NM}\sum_{x=0}^{N-1}\sum_{y=0}^{M-1} f(x,y)e^{j2\Pi\left(\frac{ux}{N}+\frac{vy}{M}\right)} \tag{2}$$

where $f(x, y)$—indicates the pixel coordinates of u and v, $f(u, v)$—indicates frequency domain of u and v, N and M indicates the dimension.

3.2 Faster R-CNN

Faster R-CNN is the improved version of Fast R-CNN; selective search algorithm is used in Fast R-CNN, whereas it is replaced with the region proposal network (RPN) in Faster R-CNN. The input image is given to series of convolution layer, and then, it is given as the input to region proposal network. RPN uses anchors for proposing and detecting region of the object. Faster R-CNN uses anchor ratios of 1:1, 1:2 and 2:1. 1282, 256 and 5122 are the anchor sizes, respectively. The main objective is to predict crack regions by computing its probability value for each anchor. Non-maximum suppression is a function that is applied to ensure that the proposed regions have no overlaps. Overlapping bounding boxes are removed, and the maximum numbers of unique regions are identified in Faster R- CNN. The maximum overlaps of identified bounding box with ground truth were predicted as a crack region. The width and height of the bounding box will be generated from regressor, and softmax function is used for classification. The sample input images and the generated sample output images using faster—R-CNN with its classification types are shown in Fig. 2.

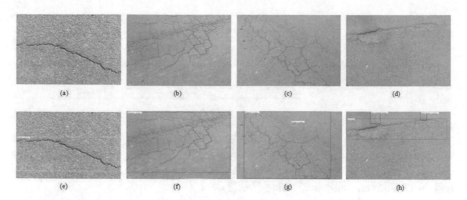

Fig. 2 a–d Sample input images; **e–h** sample output images with crack detection and classification

4   Experiments and Results

In this section, the experimental set up and results were analyzed for our proposed system using precision, recall and F-measure.

4.1   Implementation Details

The training dataset contains 6,358 bounding boxes labeled on five categories bundled in 3,200 images. Finally, a json file is obtained for the entire dataset which contains the image name, annotation of all bounding boxes within the image, image width and image height. Each annotation contains the following details—label of the bounding box, shape, four coordinates $[(x1, y1), (x2, y2), (x3, y3), (x4, y4)]$. Training process contains 300 numbers of epochs, and the batch size is 8.

In testing, the model weights are loaded along with their training weights. For each class, a color is allocated which helps in differentiation while obtaining the final output. The images for testing are loaded into the system. 333 images are used for testing the proposed system. The results of testing provide us with the predicted class label, coordinates of the bounding boxes and elapsed time for testing all possible regions in an image.

4.2   Evaluation Metrics

Pre-processing is evaluated using peak signal to noise ratio (PSNR), signal to noise ratio (SNR) and mean square error (MSE) values. Table 1 represents the evaluation metrics of pre-processing.

Table 1 Evaluation metrics of pre-processing

PSNR	SNR	MSE
27.544	23.8542	114.45559
27.51	23.9106	115.1912
27.882	24.3029	105.7700
27.7802	24.0521	108.4071

Peak signal to noise ratio (PSNR) is used to measure the quality between compressed and original image. Mean square error (MSE) is used to calculate the cumulative square error between original and compressed image. Precision, recall and F-measure are computed using Eq. (3).

$$Precision\_cls = \frac{TP}{TP + FP}$$
$$Recall = \frac{TP}{TP + FN} \tag{3}$$
$$F - measure = 2 \cdot \frac{Precision \cdot Recall}{Precision + Recall}$$

4.3 Comparison Results of ResNet and VGG16 Pre-trained Models

In this subsection, we compare the two different pre-trained models used in our proposed system. ResNet152 and VGG16 are the two pre-trained models used as a base for defining a faster R-CNN-based network. The overall accuracy for crack detection and classification using Faster R-CNN network with VGG16 has 0.8787. The results obtained for crack detection and classification using VGG16 model are shown in Table 2. Similarly, the results obtained for crack detection and classification using ResNet152 model are shown in Table 3. The overall accuracy of Faster R-CNN network with ResNet152 model is found out to be 0.9099. The results are represented as graphs for better visualization and understanding. Figure 3a represents the mean number of bounding boxes overlapping from RPN overlapping with the ground-truth boxes. The total class accuracy is 0.914 as shown in Fig. 3b. Figure 3c–g shows the loss for our proposed system. Figure 3h shows the time elapsed in processing the image in minutes.

Table 2 Results obtained for crack detection and classification using VGG16 model

Crack types	PSNR	SNR	MSE
Alligator crack	0.89	1.00	0.94
Damage	0.86	1.00	0.93
Linear crack	0.92	1.00	0.96
Nonlinear crack	0.85	0.80	0.91

Table 3 Results obtained for crack detection and classification using Resnet152 model

Crack types	PSNR	SNR	MSE
Alligator crack	0.90	1.00	0.94
Damage	0.91	1.00	0.95
Linear crack	0.93	0.96	0.94
Nonlinear crack	0.87	0.84	0.93

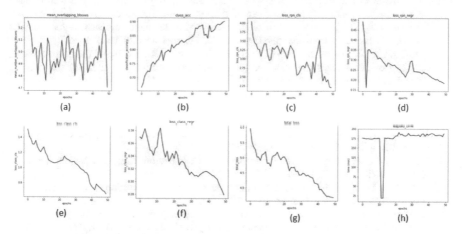

Fig. 3 Curves for our framework with ResNet152 model. **a** Mean overlapping boxes, **b** class accuracy, **c** loss RPN classification, **d** loss RPN regression, **e** loss class classification, **f** loss class regression, **g** total loss, **h** elapsed time

5 Conclusion

A novel Faster R-CNN network was proposed to identify crack, damage detection and classification. Our model is implemented with our own dataset, comprised of five class labels which are used for training and testing process. During the training process, the images are passed through a faster region convolutional neural network, and the regions were identified using region of interests (ROI). The ROI is used to detect the crack region, and it is indicated by regressor, whereas the classification is made using softmax function. Thus, the trained model is tested with the dataset. In order to convert it into an application, the locations of the images are stored in the database, and the authorities are notified with such occurrences of cracks in the road surfaces. Thus, proposed system can be used to identify and detect cracks and damages on surface and report to the nearest authority office. The obtained results were measured using precision, recall and F-measure and compared results with different pre-trained models. The result showcases that Faster R-CNN using Resnet152 performs better.

References

1. Song Q et al (2019) Real-time tunnel crack analysis system via deep learning. IEEE Access 7:64186–64197
2. Zou Q, Zhang Z, Li Q, Qi X, Wang Q, Wang S (2019) DeepCrack: learning hierarchical convolutional features for crack detection. IEEE Trans Image Process 28(3):1498–1512
3. Zhang Y et al (2018) A kinect-based approach for 3D pavement surface reconstruction and cracking recognition. IEEE Trans Intell Transport Syst 19(12):3935–3946
4. Maeda H, Sekimoto Y, Seto T, Kashiyama T, Omata H (2018) Road damage detection using deep neural networks with images captured through a smartphone. Arxiv
5. Basavaraju A, Du J, Zhou F, Ji J (2020) A machine learning approach to road surface anomaly assessment using smartphone sensors. IEEE Sensors J 20(5):2635–2647
6. Fei Y, Wang KC, Zhang A, Chen C, Li JQ, Liu Y, Yang G, Li B (2019) Pixel-level cracking detection on 3D asphalt pavement images through deep-learning based CrackNet-V. IEEE Trans Intell Transport Syst 1–12. https://doi.org/10.1109/TITS.2019.2891167
7. Chen F, Jahanshahi MR (2018) NB-CNN: deep learning-based crack detection using convolutional neural network and Naïve bayes data fusion. IEEE Trans Ind Electron 65(5): 4392–4400
8. Karmel A, Adhithiyan M, Kumar PS (2018) Machine learning based approach for pothole detection. Int J Civ Eng Technol 9:882–888
9. Cho H, Yoon H, Jung J (2018) Image-based crack detection using crack width transform (CWT) algorithm. IEEE Access 6:60100–60114
10. Li H, Song D, Liu Y, Li B (2019) Automatic pavement crack detection by multi-scale image fusion. IEEE Trans Intell Transport Syst 20(6):2025–2036
11. Qu Z, Chen Y, Liu L, Xie Y, Zhou Q (2019) The algorithm of concrete surface crack detection based on the genetic programming and percolation model. IEEE Access 7:57592–57603
12. Wang W, Wu B, Yang S, Wang Z (2018) Road damage detection and classification with faster R-CNN. In: 2018 IEEE International Conference on Big Data (Big Data). Seattle, WA, USA, pp 5220–5223
13. Jia G, Song W, Jia D, Zhu H (2019) Sample generation of semiautomatic pavement crack labelling and robustness in detection of pavement diseases. Electron Lett 55(23):1235–1238

Hand Signs Recognition from Cellphone Camera Captured Images for Deaf-Mute Persons

Asif Irfanullah Masum and Ayatullah Faruk Mollah

1 Introduction

Communication has been a vital point of human civilization and technological advancement. Hand gesture recognition is a form of communication through which a machine can understand and extract information from hand sign images or video frames. It may act as a powerful interface between deaf-mute persons and the world around them by enabling communication with other human beings and computers without conventional input devices. However, development of such human–computer interaction (HCI) systems, specifically for deaf-mute communities, is a non-trivial research problem. Many studies and research works in this direction have been reported in literature. A few articles on survey or review of the reported methods are also available [1–4]. Hand gesture recognition, as evident in literature, may be broadly categorized into two groups, (i) sensor-based methods [5–9] and (ii) vision-based methods [10–17]. The sensor-based methods employ different kind of sensors to acquire hand posture images or videos. Depth sensor is the most popular among them. Using depth sensors such as Microsoft Kinect, Asus Xtion Pro, Intel Real Sense, depth information for object pixels at various locations is acquired, which is often useful in segmenting the region of interest (ROI), i.e., the hand portion containing the palm and the fingers from the acquired images containing surroundings as well. However, such sensors are not popular and economic, and hence, its usability is limited in comparison to vision-based systems which work with RGB images that are pervasively used at population scale.

Vision-based gesture recognition has received significant attention of researchers for wide availability and usage of digital camera since the last two decades. Even cellphones, nowadays, have built-in high-resolution digital cameras which acquire RGB images. Moreover, web-cam integrated with laptop or attached with desktop

A. I. Masum (✉) · A. F. Mollah
Department of Computer Science and Engineering, Aliah University, IIA/27 New Town, Kolkata 700160, India

© The Author(s), under exclusive license to Springer Nature Singapore Pte Ltd. 2022   45
R. R. Raje et al. (eds.), *Artificial Intelligence and Technologies*,
Lecture Notes in Electrical Engineering 806,
https://doi.org/10.1007/978-981-16-6448-9_5

computer may also play a significant role in HCI. Therefore, research findings are being increasingly reported in this field. Islam et al. [10] have presented a method for hand gesture recognition by subtracting background, segmenting ROI, and recognizing the segmented symbols. As segmentation of ROI is one of the most challenging problems in gesture recognition, some methods provide a predetermined area (as a visible square on camera preview) to guide the user to position his/her hand [18]. Some approaches employ some kind of markers to segment hand ROI from surroundings of the acquired images. Other methods based on blob analysis [19], deep learning [20], skeleton analysis [21], etc. have been reported. Few works [22, 23] have also been found to focus on recognition of hand gestures in real time. Use of markers around the wrist for better segmentation of hand ROI is also adopted in some works, whereas marker-less approaches are desirable [24] to foster development of hand gesture recognition from unconstrained environment, which is necessary for practical applications.

In short, progress in hand gesture research is not adequate for real-life situations, and more research is needed to address real complexities of the problem. Particularly, development of automatic recognition of hand gesture images or video frames in an unconstrained environment is not yet reported. In this paper, we report (i) the development of a dataset of reasonably unconstrained smartphone camera captured hand gesture images of American Sign Language (ASL), (ii) a method to preprocess, segment and recognize the acquired gestures, and (iii) initial benchmark performance obtained with the developed method on this dataset. We also intend to release the dataset for academic/research purpose and invite research fraternity to develop methods to outperform on this dataset [25].

2   Dataset Development

Dataset is a crucial need for development of models and their quantitative assessment. Here, we report the development of a hand gesture dataset of reasonably complex images acquired with a cellphone built-in camera. In Sect. 2.1, we discuss about preparation of the dataset, and in Sect. 2.2, we mention the challenges associated with the images of this dataset.

2.1   Preparation

We have acquired RGB images of hand signs following ASL convention with the built-in camera of iPhone 7. The original resolution of the acquired images was 4032×3024 pixels. However, these images have been down-sampled to 1008×756 pixels. While most of the images are in portrait mode, some images are captured in landscape mode as well. There are a total of 1500 images constituting 150 images of every gesture type. Fifteen persons were involved in data collection, and each of them has volunteered to give 10 samples for 10 gesture types. Thus, every person has given 100 samples and for 15 persons, we have obtained 15 × 100, i.e., 1500 images.

2.2 Associated Challenges

The developed dataset reflects a number a challenges associated with hand gesture recognition (Fig. 1). In that respect, the dataset may be characterized as follows.

Free Acquisition. While capturing the images, utmost care is not taken to acquire the sample in the best way. Rather, reasonable care with free mode of acquisition is followed.

Presence of Shadow. Images of the dataset often contain shadow of palm or wrist portions, which makes the segmentation task more challenging.

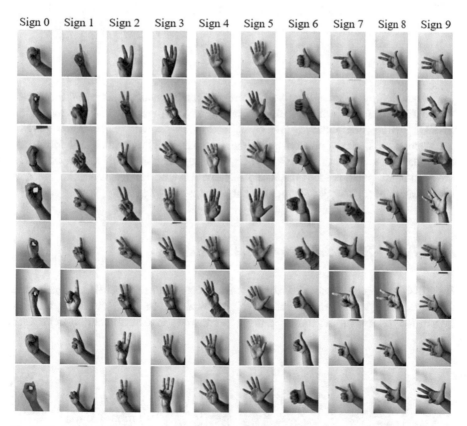

Fig. 1 Variations in hand gestures for different numerals. Besides differences in the finger patterns and background, varying amount of wrist portions has been included in the images

Background. Though the background apparently appears uniform in most images, many images contain objects, shades, and non-uniform backgrounds.

Arbitrary Hand Position and Orientation. Positions of palm with fingers are varying extensively. Moreover, orientation of the hand signs is not same. Due to free mode of acquisition, gestures have appeared in different orientation in acquired images.

Variable Wrist Presence. Presence of wrist in acquired images is of variable quantity, which poses challenges in proper segmentation and recognition

Apparels Near Wrist. To entertain unconstrained environment, no restrictions are imbibed with apparels near the wrist portion during image acquisition. As a result, some samples have empty wrists, and some others have different kinds of apparels.

Aspect Ratio. Though number of pixels in all images of the dataset is the same, number of row and column are not always the same because some images are captured in portrait orientation, whereas some others are captured in landscape orientation.

3 Methodology

The methodology adopted in this work for recognizing hand signs in HCI interface for deaf-mute persons employs four major stages, viz. preprocessing, segmentation of ROI, feature computation, and classification, as shown in Fig. 2. As part of preprocessing, K-means clustering is applied to partition the input image into some segments as discussed in Sect. 3.1. Then, ROI segmentation and normalization of orientation are performed as discussed in Sect. 3.2, and finally, Sect. 3.3 presents computation of feature descriptors for subsequent classification.

3.1 Image Partitioning by Pixel Clustering

Input RGB images are, at first, converted to gray-scale images. Then, pixels of a gray-scale image are clustered into three groups using K-means clustering method. It partitions an image into three segments as shown in Fig. 3c. It is an unsupervised

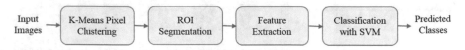

Fig. 2 Block diagram of the proposed pipeline for hand gesture recognition

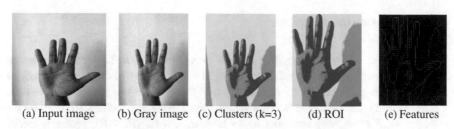

   (a) Input image   (b) Gray image (c) Clusters (k=3)   (d) ROI   (e) Features

Fig. 3 Visual illustration of different stages of the proposed hand gesture recognition pipeline

algorithm to classify unknown data to some labels which may deem fit. It is used to cluster the data points of a given dataset into K-clusters.

3.2 Segmentation of ROI and Orientation Normalization

Post-clustering, our goal is to segment the ROI, i.e., the hand portion containing the palm and the fingers. In order to do so, we binarized the clustered image with Otsu's method. On this binary image, we compute all the contour pixels, and the largest one among them is considered to supposedly identify the hand region in the binarized image. Using the pixels of the largest contour, we find out the coordinates of the bounding rectangle and crop the image using the coordinates of the normal bounding rectangle.

Now, in order to normalize the orientation of hand gesture samples, we compute the angle at which minimum bounding rectangle is stationed with respect to the coordinate axes and normalize the image by rotating the image to the opposite direction. Finally, the cropped images are resized to 300 × 300 pixels and used for feature extraction.

3.3 Feature Extraction

We employ HOG which is a popular feature descriptor used to characterize objects. Basically, it counts the number of occurrences of gradient orientation in localized portion of an image, i.e., ROI. In this work, we compute this descriptor by considering cell size of 16 × 16 pixels, block size of 4 × 4 cells, and orientation of 8 directions. Figure 3e shows visual representation of computed feature values for a sample hand gesture.

4 Experimental Results

We carried out experiments with multiple classifiers and n-fold cross-validation approach. Instead of experimenting on a single classifier, exhaustive experiments as adopted in this work reflect more insights into the recognition process. Moreover, instead of only classification accuracy, multiple evaluation metrics such as precision, recall, F-score, negative root mean squared error (NRMSE), and area under curve (AUC) have been taken into account to report the results obtained with multiple sets of experiments for validating the developed method.

Table 1 presents a summary of results obtained in the current experiments. In n-fold cross-validation, samples of the developed dataset are split into n folds, from where samples of $n - 1$ folds are used for training and that of the remaining fold is used for prediction. Thus, for n-fold cross-validation, n number experiments are eventually conducted, and mean values of evaluation metrics are reported. And, the same strategy is repeated for all the classifiers considered in this work.

It may be noted from Table 1 that SVM with RBF kernel has outperformed all other classifiers. Therefore, we select this SVM with RBF kernel as the classifier of the proposed method. It may also be noted that with the increase in the value of n in n-fold cross-validation, the performance has gradually improved to an extent. This

Table 1 Performance of the proposed method with various classifiers for threefold, fivefold, and tenfold cross-validation

CV	Classifier	Accuracy	Precision	Recall	F-score	NRMSE	AUC
3	Naïve Bayes	0.9220	0.9299	0.9220	0.9231	−0.7818	0.9636
	KNN	0.9700	0.9710	0.9700	0.9695	−0.5241	0.9991
	MLP	0.9867	0.9852	0.9753	0.9827	−0.3033	0.9984
	SVM (Lin)	0.9973	0.9974	0.9973	0.9973	−0.0682	0.9999
	SVM (Poly)	0.9940	0.9942	0.9940	0.9940	−0.2400	0.9999
	SVM (RBF)	0.9893	0.9900	0.9893	0.9893	−0.1456	0.9999
5	Naïve Bayes	0.9320	0.9406	0.9320	0.9330	−0.6883	0.9655
	KNN	0.9793	0.9799	0.9793	0.9791	−0.5165	0.9991
	MLP	0.9907	0.9871	0.9900	0.9867	−0.5187	0.9996
	SVM (Lin)	0.9980	0.9981	0.9980	0.9980	−0.0577	0.9999
	SVM (Poly)	0.9967	0.9968	0.9967	0.9967	−0.0515	0.9999
	SVM (RBF)	0.9920	0.9926	0.9920	0.9919	−0.1108	1.0000
10	Naïve Bayes	0.9367	0.9478	0.9367	0.9375	−0.6100	0.9701
	KNN	0.9813	0.9824	0.9813	0.9812	−0.3883	0.9992
	MLP	0.9927	0.9914	0.9900	0.9913	−0.2718	0.9998
	SVM (Lin)	0.9980	0.9981	0.9980	0.9980	−0.0408	1.0000
	SVM (Poly)	0.9980	0.9981	0.9980	0.9980	−0.0408	1.0000
	SVM (RBF)	0.9947	0.9951	0.9947	0.9946	−0.0527	1.0000

In all cases, SVM has found to outperform all other classifiers

may be realized from the fact that with increasing number of folds in cross-validation, the number of samples in the training set increases, which in turn empowers better learning of the classifier models, thereby resulting better prediction.

5 Conclusion

Hand gesture recognition is an important problem in HCI for deaf-mute persons. Due to pervasive availability and usage of digital camera, vision-based gesture recognition may play a key role in the said HCI. In this paper, we present a featured dataset of hand gesture images acquired with cellphone camera and initial benchmark results with multiple classifiers and multiple folds of experiments. We designed an effective segmentation strategy to segment the hand region of interest and computed HOG features for classification purpose. We have obtained highest performance of 99.47% accuracy with SVM classifier equipped with RBF kernel for tenfold cross-validation. We also release this dataset in a public repository [25] and make it available for academic, scientific, and non-commercial purposes. In future, we would like to apply the developed method on other datasets as well and analyze in terms of variations in cluster size, orientations, local and global feature combination, and deep features.

References

1. Xia Z, Lei Q, Yang Y, Zhang H, He Y, Wang W, Huang M (2019) Vision-based hand gesture recognition for human-robot collaboration: a survey. In: 5th international conference on control, automation and robotics, pp 198–205
2. Badi H (2016) Recent methods in vision-based hand gesture recognition. Int J Data Sci Analytics 1(2):77–87
3. Rautaray SS, Agrawal A (2015) Vision based hand gesture recognition for human computer interaction: a survey. Artif Intell Rev 43(1):1–54
4. Garg P, Aggarwal N, Sofat S (2009) Vision based hand gesture recognition. World Acad Sci Eng Technol 49(1):972–977
5. Paul S, Basu S, Nasipuri M (2015) Microsoft kinect in gesture recognition: a short review. Int J Control Theory Appl 8(5):2071–2076
6. Sahana T, Paul S, Basu S, Mollah AF (2020) Hand sign recognition from depth images with multi-scale density features for deaf mute persons. Procedia Comput Sci 167:2043–2050
7. Paul S, Bhattacharyya A, Mollah AF, Basu S, Nasipuri M (2020) Hand segmentation from complex background for gesture recognition. In: Emerging technology in modelling and graphics. Springer, pp 775–782
8. Skaria S, Al-Hourani A, Lech M, Evans RJ (2019) Hand-gesture recognition using two-antenna doppler radar with deep convolutional neural networks. IEEE Sens J 19(8):3041–3048
9. Chugunov I, Zakhor A (2019) Duodepth: static gesture recognition via dual depth sensors. In: IEEE international conference on image processing, pp 3467–3471

10. Islam MM, Islam MR, Islam MS (2020) An efficient human computer interaction through hand gesture using deep convolutional neural network. SN Comput Sci 1(4):1–9

11. Kamruzzaman MM (2020) Arabic sign language recognition and generating arabic speech using convolutional neural network. Wireless Commun Mob Comput. https://doi.org/10.1155/2020/3685614

12. Kılıboz NÇ, Güdükbay U (2015) A hand gesture recognition technique for human–computer interaction. J Vis Commun Image Representation 28:97–104

13. Rasel AAS, Yousuf MA (2019) An efficient framework for hand gesture recognition based on histogram of oriented gradients and support vector machine. Int J Inform Technol Comput Sci 12:50–56

14. Li C, Xie C, Zhang B, Chen C, Han J (2017) Deep fisher discriminant learning for mobile hand gesture recognition. Pattern Recogn. https://doi.org/10.1016/j.patcog.2017.12.023

15. Zhi D, de Oliveira TEA, da Fonseca VP, Petriu EM (2018) Teaching a robot sign language using vision-based hand gesture recognition. In: IEEE international conference on computational intelligence and virtual environments for measurement systems and applications (CIVEMSA). IEEE, pp 1–6

16. Ahmed W, Chanda K, Mitra S (2016) Vision based hand gesture recognition using dynamic time warping for Indian sign language. In: International conference on information science (ICIS). IEEE, pp 120–125

17. Ghosh DK, Ari S (2016) On an algorithm for vision-based hand gesture recognition. Signal Image Video Process 10(4):655–662

18. Tamiru HG, Yan RS, Long DH (2018) Vision-based hand gesture recognition for mobile service robot control. In: 8th international conference on manufacturing science and engineering (ICMSE). Atlantis Press

19. Ganokratanaa T, Pumrin S (2017) The vision-based hand gesture recognition using blob analysis. In: International conference on digital arts, media and technology (ICDAMT). IEEE, pp 336–341

20. Oyedotun OK, Khashman A (2017) Deep learning in vision-based static hand gesture recognition. Neural Comput Appl 28(12):3941–3951

21. De Smedt Q, Wannous H, Vandeborre J-P (2016) Skeleton-based dynamic hand gesture recognition. In: IEEE conference on computer vision and pattern recognition workshops, pp 1–9

22. Liu K, Kehtarnavaz N (2016) Real-time robust vision-based hand gesture recognition using stereo images. J Real-Time Image Process 11(1):201–209

23. Negi PS, Pawar R, Lal R (2020) Vision-based real-time human–computer interaction on hand gesture recognition. In: Micro-electronics and telecommunication engineering. Springer, Singapore, pp 499–507

24. Haria A, Archanasri S, Nivedhitha A, Shristi P, Nayak JS (2017) Hand gesture recognition for human computer interaction. Procedia Comput Sci 115:367–374

25. Aliah University Hand Gesture Dataset. https://github.com/iilabau/AUHGdataset

Prediction of In-Cylinder Swirl in a Compression Ignition Engine with Vortex Tube Using Artificial and Recurrent Neural Networks

Manimaran Renganathan

1 Introduction

Transportation sector is gaining importance nowadays due to mobility requirements for maintaining the business, human relationship, etc. However, due to alarming climate changes, the engine emissions from vehicles must be taken into consideration for reduction of pollutants as per the government standards in lowering the overall pollution. Combustion in engines plays a crucial part in pollutant formation. In order to achieve a clean combustion, the swirl ratio in a four-stroke compression ignition engine helps in the formation of air–fuel mixture [1]. Swirl ratio is defined as the ratio of rotational velocity imparted inside the cylinder with respect to the engine speed [2].

1.1 Machine Learning and Deep Learning Techniques

A numerical model for neural networks based on algorithms in mathematics was developed long back in the nineteenth century. The purpose is to solve problems similar to a human brain with large interconnected neural blocks. The main advantage is to simplify the complicated process to design and optimize the given scenario [3]. RNN operate on sequences of time series data and their weights are obtained by back-propagation through time. Algorithms that use long short-term memory (LSTM) involve the current instance of time with the appropriate gating mechanisms to deposit and relieve the data regarding the previous time instant data, thereby reducing the problems with the gradients and the passing of the memory for long term duration [4].

M. Renganathan (✉)
Thermal and Automotive Research Group, School of Mechanical Engineering,
Vellore Institute of Technology, Chennai, Tamil Nadu 600127, India

© The Author(s), under exclusive license to Springer Nature Singapore Pte Ltd. 2022
R. R. Raje et al. (eds.), *Artificial Intelligence and Technologies*,
Lecture Notes in Electrical Engineering 806,
https://doi.org/10.1007/978-981-16-6448-9_6

1.2   Computational Fluid Dynamics (CFD)

The evolution of new combustion models and flow field investigations to validate the experimental results are always associated with novel mathematical schemes and computational algorithms. The open-source architecture [5], for example, OpenFOAM, an open-source and cross-platform software, gives enough flexibility in modeling the engine in-cylinder phenomena due to complex flow field and expensive empirical constants obtained from measurements. CFD enables the fluid physicists or engineers to quickly model and solve the flow field with reliable, accurate calculations. But the time taken to obtain the results from CFD calculations vary depending on the number of control volumes (grid based on mesh size) and time step during iteration.

1.3   Objective

The motivation behind this study is to examine the prediction of time series swirl ratio data using the machine learning and deep learning algorithms that can be devoted in the reduction of cost and time of CFD simulations. Analysis of swirl ratio inside the engine cylinder could provide a valuable information for IC engine manufacturers. Reduction of time taken involved in the swirl analysis with the use of ANN/RNN techniques is the main consideration of this study.

2   Methodology

2.1   Computational Fluid Dynamics

The physical space and its computational domain of vortex tube and engine cylinder are shown in Fig. 1a, b, respectively. Experimental and computational studies on the vortex tube [6–8] are carried out with high pressure air supply from air compressor. Air enters the vortex tube in a tangential fashion, and a portion of fluid leaves the vortex tube past the cone placed at the annulus located farther end from inlet (hot exit flow), while the remaining portion leaves at the central orifice near inlet (cold exit flow). The engine cylinder is connected from the cold exit flow of the vortex tube (details and specifications in Table 1) so that the swirl in the engine cylinder can be enhanced [9] with the vortex motion of cold exit flow from vortex tube. As the flow entering the engine inlet manifold is at lower room temperature compared to outside conditions [10], dense air–fuel mixture distribution is possible for better combustion as similar to an intercooled–supercharger case [11]. SonicFOAM solver [5] is accompanied to solve the standard equations of conservation of mass, momentum and energy. 463,644 control volumes are chosen in computational domain under which finite volume method is applied with 0.5

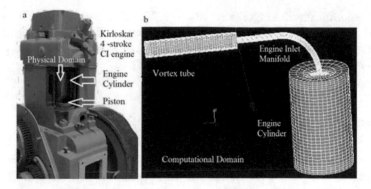

Fig. 1 **a** Engine cylinder; **b** vortex tube

Table 1 Vortex tube details [6] and engine specifications [10]

Vortex tube inlet (rectangular type)	2 mm height and 6 mm width	Kirloskar AV1 engine type	Four-stroke, compression ignition
Vortex tube orifice diameter	7 mm	Cylinder bore and stroke	80 mm, 110 mm
Vortex tube cone geometry	50° angle and 25 mm height	Cylinder compression ratio	16.5:1
Vortex tube length	120 mm	Engine speed and rated power	1500, 3.7 kW

micro-second time step. To model the turbulence effects, standard k-ε model is employed [10]. Inlet pressure (4 bar, abs) and hot end exit pressure (2 bar, abs) are set up to record average velocities at 18 locations in cylinder during the intake stroke, from which swirl ratio is found.

2.2 Artificial Neural Networks (ANN) and Recurrent Neural Networks (RNN)

The Levenberg–Marquardt (LM) method is chosen as the machine learning algorithm for the ANN and RNN models to in this work. Quick convergence of results

Fig. 2 **a** Feedforward ANN; **b** LSTM–RNN

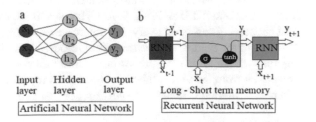

is obtained with the LM algorithm [12]. The LM algorithm is classified as feed-forward neural network (Fig. 2a) models in many studies [13, 14]. In the long short-term memory RNN (Fig. 2b), the hidden state is found as per the previous hidden states with the given current time data [15].

3 Performance Parameters

To have a measure between different techniques on the evaluation of swirl ratio performance, four parameters are chosen. Root mean square error (RMSE), coefficient of variation (CV) of RMSE, mean absolute error (MAE) and normalized mean bias error (NMBE) are the evaluation metrics for the prediction [16] defined as per the equations below. The variable y refers to outcome, swirl ratio as the ratio of tangential velocity to crank speed [2].

$$\text{Root mean square error is defined as RMSE} = \sqrt{\sum_{i=1}^{N} \frac{(y_{\text{TRGT}} - y_{\text{PRED}})^2}{N}} \quad (1)$$

$$\text{Coefficient of variation of RMSE is evaluated as CV (RMSE)} = \frac{\text{RMSE}}{y_{\text{TRGT}}} \times 100 \quad (2)$$

$$\text{Mean absolute error (MAE) is found as MAE} = \sum_{i=1}^{N} \frac{|(y_{\text{TRGT}} - y_{\text{PRED}})|}{N} \quad (3)$$

$$\text{Normalized mean bias error (NMBE) is evaluated as NMBE} = \frac{\text{MAE}}{y_{\text{TRGT}}} \times 100 \quad (4)$$

4 Results and Discussion

CFD calculations are performed with 0.5 micro-second time step size to arrive at an induction time period of 10 milli-seconds. The total elapsed CPU time is 30 h on Intel Core2Duo CPU E4600, 2.40 GHz, 4 GB RAM with nearly 1080 s for post-processing results in every 100 micro-seconds from serial CFD computations of swirl ratio. Figure 3a illustrates the velocity magnitude at three layers in the engine cylinder where the probes are marked as positive symbol (+). Streamlines emanating from the vortex tube inlet are found to spiral toward the engine cylinder where the tumbling squish and swirl motions are shown in Fig. 3b.

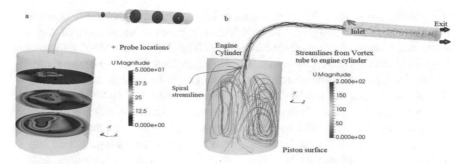

Fig. 3 a Probe locations in the engine cylinder; **b** boundary conditions in vortex tube and evolution of streamlines in cylinder

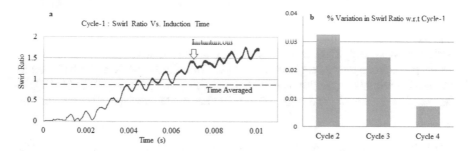

Fig. 4 a Time series of swirl ratio; **b** cyclic variations in swirl ratio

Overall size of the sampling data comes to 20,000, in which the time averaged swirl ratio is found to be 0.859. Time series data of swirl ratio is plotted in Fig. 4a. The cyclic variations of swirl ratio are found to be within 0.1% variation based on numerical error and shown in Fig. 4b. The data collected from 0 to 0.01 s are used for the training and testing of the machine and recurrent learning models, as shown in Fig. 4a. The data during the initial time period of 0.008 s were selected as the training data set to predict the swirl ratio, while the data for the duration 0.008–0.01 s were used for testing. There is only one input parameter (i.e., induction time of cycle) being used in the prediction. It is understood that the other in-cylinder engine variables in this work might influence the outcome. The ratio of data required for ANN training to testing is varied to understand the effect on the

Table 2 Performance evaluation of ANN model for training and testing

	70/30	80/20	90/10
RMSE	0.028	0.025	0.025
CV (RMSE) %	1.903	1.631	1.619
MAE	0.023	0.020	0.020
NMBE %	1.553	1.327	1.321

performance parameters. There arises an ambiguity in determining the quantity of data required for training and testing. Table 2 presents the results of the four performance indexes as introduced in Sect. 3. The RMSE and MAE decreases as the ratio of training/testing increases. This is due to the availability of more training data so that the model predicts the test data accurately. The RMSE and MAE for 80/20 and 90/10 cases are nearly closer to each other while CV (RMSE) and NMBE vary due to the factor (i.e., average of the target data). Hence, 80/20 ratio is selected and continued for other studies for comparison between ANN and RNN as per suggestion [15] of 80% data for training and 20% data for testing.

Though the ANN and RNN models are arrived at similar results (Fig. 5a), the RNN model underperforms as compared to the ANN model especially with the MAE index (Table 3). Table 4 shows that RMSE, CV (RMSE), MAE and NMBE increase with the decrease in sampling frequency. This indicates that a higher sampling frequency (Fig. 5b) could increase the prediction accuracy which means higher number of data points during the specified time period, so the accuracy of expectation should also be matched well between the target and training data sets. Both the ANN and RNN models show similar trend, while the RNN predicted results are quite higher as compared to ANN. For both the ANN and RNN models, the moving window algorithm is included and compared as shown in Table 5. With the moving window algorithm (Fig. 6a), the ANN performs better and the

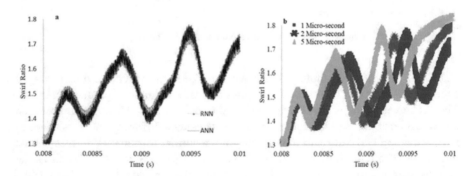

Fig. 5 a Results from ANN and RNN; **b** ANN predictions at different sampling frequencies

Table 3 Performance evaluation of the ANN and RNN models at 80/20 training/testing ratio

	ANN	RNN	% Improvement[a]
RMSE	0.025	0.026	4.0
CV (RMSE) %	1.631	1.671	2.45
MAE	0.020	0.021	5.0
NMBE %	1.327	1.367	3.01

[a]% Improvement |Metric_ANN–Metric_RNN|/Metric_ANN × 100%

Table 4 Performance of ANN and RNN models at different sampling frequencies

	ANN (1 micro-second)	ANN (2 micro-second)	ANN (5 micro-second)	RNN (1 micro-second)	RNN (2 micro-second)	RNN (5 micro-second)
RMSE	0.025	0.029	0.030	0.026	0.031	0.038
CV (RMSE) %	1.631	1.911	1.974	1.671	1.956	1.995
MAE	0.020	0.023	0.025	0.021	0.028	0.032
NMBE %	1.327	1.548	1.610	1.367	1.741	1.836

Table 5 Comparison of performance between simple and moving window algorithm

	ANN	ANN (moving window)	RNN	RNN (Moving window)
RMSE	0.025	0.021	0.026	0.022
CV (RMSE) %	1.631	1.365	1.671	1.436
MAE	0.020	0.017	0.021	0.018
NMBE %	1.327	1.128	1.367	1.169

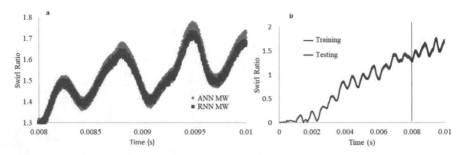

Fig. 6 a Results from moving window algorithm; **b** ANN test by moving window algorithm

evaluation parameters decrease as compared to case without the moving window. The coefficient of variation decreases from 1.631 to 1.365% in the ANN (without moving window) and the ANN (moving window), respectively. Between the cases of simple and moving window prediction, the normalized mean bias error for RNN is lowered from 1.367 to 1.169%. Figure 6b shows both the well-captured training and testing data of ANN moving window algorithm as against the validation CFD results in Fig. 4a.

5 Conclusion and Future Work

The prediction of swirl ratio from time series CFD observations in the engine cylinder is the main focus of the study. The usage of vortex tube at the inlet of engine cylinder reveals the swirling nature of streamlines inside the cylinder. The swirl ratio data are compared between the commonly used ANN and the RNN models. Subsequent results of data sampling frequency and the moving window algorithm on the expected outcome were analyzed. The swirl ratio prediction using the ANN model has greater accuracy as compared with the RNN model. Also, the CPU time for RNN is not appreciably higher as compared to ANN. In terms of statistical measure of comparisons, RMSE, CV (RMSE), MAE and NMBE can be improved by 4%, 2.5%, 5% and 3%, respectively. Based on the 1 micro-second time interval data set with the use of a moving window algorithm, higher sampling frequency can yield a better outcome in both ANN and RNN models. The training to testing data amount ratio is consistently maintained at 80/20 to acquire reliable accurate target values. Finally, the prediction results of ANN moving window algorithm match well with the CFD validation data set although a little over-fitting found with RNN model. The RNN model is able to predict the time series of swirl ratio better as compared to ANN model.

The limitations related to the computation of RNN and ANN are very minimal when compared to the full computational fluid dynamics simulations. The solution arrived from CFD runs requires a sufficiently longer period of more than 150 CPU hours. However, the results from ANN/RNN are obtained within ½ CPU hours to 1 h time period. The heat generation/maximum thermal dissipation from CPU in the CFD calculations is enormous on the order of several hundred kilowatts (~ 300 kW) as compared to RNN/ANN study (~ 2 kW). This study promises to encourage the mix of neural networks and CFD at suitable proportions to save the time and cost from further CFD calculations. The study could be extended in other directions such as full cycle simulation especially in the latest trend of homogeneous charge compression ignition (HCCI) engine strategies. As the adoption of new greener fuel (toward a lower pollution) is the key future, the fuel properties could be fed as input along with the combustion parameters to understand the effects on ANN output. The future scope of the study extends to prediction of mixture of gases for preparation of air–fuel mixture that is conducive for better combustion which leads to lower pollution. Other external variables such as intercooler boost pressure and exhaust gas recirculation level could be studied with neural network formulation along with the fuel injection studies.

Acknowledgements Author acknowledges the support received from the project grant (ECR/ 2018/000133) approved by the Science and Engineering Research Board (SERB), DST, India.

References

1. Manimaran R, Thundil Karuppa Raj R (2014) Computational studies of swirl ratio and injection timing on atomization in a direct injection diesel engine. Front Heat Mass Transfer (FHMT) 5:2:1–9. https://doi.org/10.5098/hmt.5.2
2. Tulwin T, Wendeker M, Czyz Z (2017) The swirl ratio influence on combustion process and heat transfer in the opposed piston compression-ignition engine. Comb Engines 170(3):3–7. https://doi.org/10.19206/CE-2017-301
3. LeCun Y, Bengio Y, Hinton G (2015) Deep learning. Nature 521(7553):436–444
4. Graves A, Fernandez S, Schmidhuber J (2007) Multi-dimensional recurrent neural networks. Artif Neural Networks-ICANN 549–558
5. OpenFOAM v2006 (2020). https://www.openfoam.com. OpenCFD Ltd
6. Manimaran R (2016) Computational analysis of energy separation in a counter-flow vortex tube based on inlet shape and aspect ratio. Energy 107:17–28
7. Manimaran R (2017) Computational analysis of flow features and energy separation in a counter-flow vortex tube based on number of inlets. Energy 123:564–578
8. Manimaran R, Ramakrishna PA, Ramakrishna M (2014) Experimental investigation of temperature separation in a counter-flow vortex tube. ASME J Heat Transfer 136:082801–1–6

9. Manimaran R, Thundil Karuppa Raj R (2018) CFD simulation of flow field and heat transfer in a single-cylinder HCCI engine at different boundary conditions. Comput Thermal Sci 10 (4):337–354
10. Manimaran R, Thundil Karuppa Raj R (2014) Comparative evaluation of the re-entrant and flat toroidal combustion chambers in a direct-injection diesel engine using computational fluid dynamics. Comput Thermal Sci 6(2):171–190
11. Renganathan M, Rajagopal T (2014) Computational study of HCCI-DI combustion at preheated and supercharged inlet air conditions. SAE Technical Paper 2014–01–1108
12. Beale HD, Demuth HB, Hagan M (1996) Neural network design. Pws Boston, Boston, MA
13. Halabi LM, Mekhilef S, Hossain M (2018) Performance evaluation of hybrid adaptive neuro-fuzzy inference system models for predicting monthly global solar radiation. Appl Energy 213:247–261
14. Hota H, Handa R, Shrivas A (2017) Time series data prediction using sliding window based RBF neural network. Int J Comput Intell Res 13(5):1145–1156
15. Connor JT, Martin RD, Atlas LE (1994) Recurrent neural networks and robust time series prediction. IEEE Trans Neural Network 5(2):240–254
16. Granderson J et al (2015) Assessment of automated measurement and verification (M&V) methods, LBNL report LBNL-187225 2015

Recent Trends and Study on Perspective Crowd Counting in Smart Environments

Vasupalli Jaswanth, Arun Reddy Yeduguru, Vura Seetha Manoj,
K. Deepak, and S. Chandrakala

1 Introduction

Crowd counting has become an innovative idea in smart environments. It has gained serious attention in recent years as there is rapid development in private and public places due to the increase in global population. With an increase in population, there is a large amount of hustle and bustle almost everywhere which can lead to massive chaos that can be life threatening and hence should always be prevented in critical environments. Single image-based crowd counting is still gaining attention and is one of the difficult topics due to the complex distribution of people, non-uniform illumination, low image resolution, and dense crowds that have excessive overlaps and occlusions within each other. Moreover, perspective effects can cause a huge contrast in human appearance. For example, in the regions of people close to the camera, the people heads are big and their respective density values are accordingly low, and in the regions of people farther from a camera, the heads are small and the density values are high.

In recent times, the crowd counting problem has been addressed by a huge number of methods such as SFANet [1] and SegNet [1], NAS [2], compact [3] convolutional neural network, and HYGNN [4]. The prevalent crowd counting methods can be broadly categorized into: Detection then counting, direct count regression, CNN-based methods, perspective-based methods. Detection then counting-based methods involve more computations, and these types of methods are only suitable for fewer crowd densities and fail if the density of crowd is high. Direct count regression methods reduce the computations, and it produces more accurate results compared to detection then counting methods but they are not efficient when there are excessive overlaps in an image. CNN-based methods pay attention to multi-scale and multi-column architecture that integrates features in

V. Jaswanth · A. R. Yeduguru · V. S. Manoj · K. Deepak · S. Chandrakala (✉)
School of Computing, SASTRA University, Thanjavur, Tamil Nadu, India

© The Author(s), under exclusive license to Springer Nature Singapore Pte Ltd. 2022   63
R. R. Raje et al. (eds.), *Artificial Intelligence and Technologies*,
Lecture Notes in Electrical Engineering 806,
https://doi.org/10.1007/978-981-16-6448-9_7

variable sizes of the respective fields and can detect the people in the case of excessive overlaps. But, CNN-based methods have not focused on perspective changes. Perspective-based methods focus on the continuous scale variations [5] of every single person and perspective information played a major role in the prediction. So far, perspective and CNN-based methods have achieved higher performance than other methods in terms of accuracy and robustness.

2 Categories of Approaches Perspective-Based Crowd Counting

As shown in Fig. 1, a generic crowd counting model learns the spatial and perspective features from input images.

2.1 Detection Then Counting

In most of the early approaches [6, 7], crowd counting estimation is done by first detecting and segmenting individual objects in the scene then followed by counting. Some of the challenges faced by these kinds of methods are, they are computationally expensive as they produce more accurate results than the overall count and are mostly suitable for scenes in which the crowd density is low, and they do not perform well on scenes having high crowd density. Another challenge is that a large amount of work is needed in scenes having high crowd density, which includes bounding box or instance mask to train the object detectors.

2.2 Regression-Based Methods

In this kind of method, the detection problem which is faced by detection-based methods is avoided, and image features are used to estimate crowd counts. Earlier methods [8, 9] gave poor performance because count prediction is done based on the information from the features, and the spatial awareness is completely ignored. In the later methods [10, 11], crowd count is obtained by first generating the density

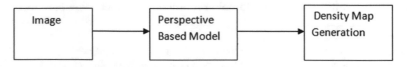

Fig. 1 Overview of perspective-based crowd counting

map and then combining all the information (pixel values) over the density map. Though spatial information is provided up to an extent by learning the density map, these methods lack in maintaining the high-frequency dissimilarities in the density map.

2.3 Graphical Methods

The concept of graphical neural network (GNN) was first introduced by Scarselli et al. in 2008 [12]. It extended recursive neural networks for processing graphical structure data. In 2016, Li et al. [13] proposed gated recurrent units to improve the representation capacity of GNN. To generalize the GNN, Gilmer et al. [14] in 2017 used message passing neural networks. The essential idea of GNN is to enhance the node representations by propagating information between nodes. Recently, GNN has been successfully applied in various applications like human-object interactions, attribute recognition, and crowd count estimation.

2.4 CNN-Based Methods

In recent years, there has been a rapid growth in CNN-based methods. Multi-scale, multi-task, and other techniques are usually carried out with the help of CNN-based approaches. Recently, there is an increase in the methods which incorporate handling of scale variation issues. Some of these include MCNN, which is a multi-column architecture proposed by Zhang et al. To obtain features with various scales, this architecture makes use of different filters on separate columns [15]. SANet is a novel encoder-decoder network proposed by Xinkhun et al. where multi-scale features are extracted by the encoder by using scale aggregation modules, and the high-resolution density maps are generated by decoder [16]. Also, there exist studies which focus on perspective maps [17] and region of interest (ROI) [18] to enhance the robustness and accuracy of the model.

3 Review on Recent Methods of PCC

3.1 Perspective Crowd Counting (PCC Net)

The entire PCC Net consists of three modules, which include density map estimation (DME), random high-level density classification (R-HDC), and fore-/background segmentation (FBS) [5]. Along with it, the Down Up Left Right (DULR) module is also present which takes the input from the full convolution network (FCN) and passes the features maps with encoded perspective changes to the DME and FBS module. Density map estimation (DME) module helps to generate density maps for crowd images. This module uses the FCN which can accept input of any dimension. In addition to that, upsampling is also done on the feature maps using the deconvolution layer present inside the full convolution network.

Random High-Level Density Classification (R-HDC). To learn global contextual information, the R-HDC module is used. In this, the entire density is divided into ten types of high-level labels. To perform this, an entire image is broken into many patches to cover the entire image. Then for each part, a random region of interest (ROI) is generated. This ROI generated must be as large as to cover more than 1/16 part of the image [5]. Then the pooling layer generates the feature maps for the ROI. Then the FCN layer classifies the feature maps as one among the ten high-level labels. In simple terms, the R-HDC model estimates the density of the image and then divides it into ten patches with labels.

Fore-/Background Segmentation (FBS). DME + R-HDC module neglects the contextual information in congested crowd scenes. To consider this, FBS is used. In this module, they generate a head segmentation map. This map helps us to cover the face region, its structure, and the semantic features. The last feature map in the FBS module is added with the last feature map of the DME module to obtain density map estimation [5].

Down Up Left Right (DULR) Module. By this point, the model can learn contextual features, global features, and local features. To translate perspective changes from four directions, this DULR module is used. The DULR module consists of four convolution layers, each handling four directions, namely top, down, left to right, and right to left, respectively. In each layer, the entire feature map is divided into h parts where h represents the height of the feature map [5]. Then each part is fed into the respective layer, and the output is then concatenated with the next part. For every convolution layer, this process is repeated iteratively for h parts. The output feature map of each layer is passed as the input feature map to the next layer. Note that the input feature map of the DULR module is of same shape as that of the output feature map of the DULR module.

3.2 Spatial Divide-and-Conquer (S-DC) Net

From Quantity to Interval. Rather than using regression to count the values in an open set, the local counts and classified count intervals are discretized. The interval partition of $[0, +\infty)$ is discretized as $\{0\}, (0, C1], (C2, C3], ..., (CM - 1, CM]$ and $(CM, +\infty)$. Here, $M + 1$ sub-intervals are present. The count value in $(C2, C3]$ is labeled as first class. This should not exceed the maximum local count present in the training set, which is obvious. The mid-value of every sub-interval is calculated dynamically when counting each interval; CM will be the last count value as the last sub-interval is (CM, $+ \infty$]. But this leads to error, and this error is reduced by S-DC Net.

S-DC Net consists of VGG16 [19] feature encoder, Unet [20] decoder, a count interval classifier, and a division decider [21]. In the classifier, the first average pooling layer will have a stride size 2 and the final prediction will be of stride size 64. The fully connected layers are removed by the feature encoder. Assume the input size of the as 64×64. The feature map $F0$ is obtained from the convolution layer 5. From, 1/32nd resolution of the input image and extracted feature map $F0$, the classifier predicts the class label of count interval CLS0. The local count $C0$ can be obtained from CLS0.

In the first stage of execution, the shared classifier gets the input from the fused feature map F1. The division count $C1$ is obtained from the shared classifier. Precisely, $F0$ is upsampled by $\times 2$ and attached to $F1$. The classifier extracts local features that related to spatially partitioned sub-regions. $C1$ is obtained from $F1$ and the classifier. Every 2×2 elements in $C1$ indicate the sub-count of the proportionate 32×32 sub-region. The division decider is used to divide among the obtained local counts $C0$ and $C1$. In the first stage of S-DC, the division decider produces a soft division mask $W1$ of similar size as $C1$ on F1 like for any w $W1$, w [0, 1]. No division is necessary at that position when w equals to zero. The division count $C1$ should be substituted in place of initial prediction when w equals to one. As $W1$ and $C1$ are double the count of $C0$, $C0$ is upsampled by $\times 2$. Initial stage division count is calculated as,

$$DIV1 = (1 - W1) \text{ o avg } (C0) + W1 \text{ o } C1 \tag{1}$$

Here, 1 represents the matrix packed with ones and has the same size as $W1$. "o" represents Hadamark product, and avg represents averaging redistribution operator. S-DC Net can also be implemented by dividing the feature map till the first convolution block output is obtained.

3.3 Spatial/Channel-Wise Attention Regression Networks (SCAR)

Overview. The SAM and CAM attention models are the two important modules of the SCAR [20] network. First, the image is fed into a local feature extractor [22], which consists of the VGG-16 as backbone (first ten convolutional layers) followed by the dilation module. Even though this output contains some spatial contextual information, it is not large enough, and also it does not encode attention features. To get rid of these drawbacks, two stream architectures (SAM and CAM) are designed to translate spatial attention features as well as channel-wise attention features. At last, the predicted density map is obtained by concatenating the two types of features maps (one from SAM and other from CAM) via convolution operation.

Spatial-Wise Attention Model (SAM). For global images, it can be observed that there is a certain uniformity in the density distribution locally and globally because of perspectives changes of crowded scenes. Also, there is a consistent gradual trend of density change. To encode these two observations, SAM is designed. SAM considers a large range of contextual information and identifies the density distribution change. The output from the VGG-16 backbone layer which is of size $C \times H \times W$ is fed into three different convolutional layers of kernel size 1×1 [22]. Then by applying reshape or transformations, three features maps $S1$, $S2$, and $S3$ are attained. The spatial attention map Sa [22] of size HW $\times$ HW is generated by performing matrix multiplication of $S1$ and $S2$ followed by applying softmax operation. The obtained Sa then undergoes matrix multiplication with $S3$, and then output is reshaped to $C \times H \times W$. Then the output is scaled by a learnable factor and undergoes sum operation with F (output from VGG-16 backbone) to give the final output of SAM.

Channel-Wise Attention Model (CAM). Channel-wise attention model (CAM) is similar in structure with SAM. The purpose of CAM is to translate large-range dependencies on channel dimension [22]. The similarity between the foreground and background textures can be addressed by using CAM. There are two main differences between SAM and CAM. SAM has three convolution layers of size 1×1, whereas CAM has only one, and the intermediate feature maps are of different dimensions in SAM and CAM.

3.4 S-DC Net + DULR

We explore S-DC Net + DULR architecture by integrating the S-DC Net and DULR module. The detailed architecture is given in Fig. 2.

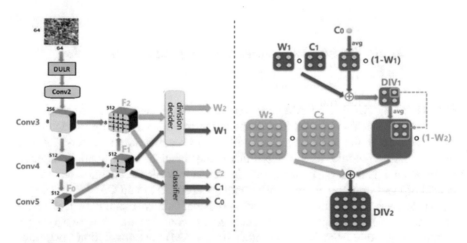

Fig. 2 Detailed architecture of S-DC Net with DULR module

Both the DULR module and the S-DC Net model are pre-trained with VGG16 NET. In the actual S-DC Net model, the image goes into the conv block which has two convolution layers. So, the first convolution layer from the conv block is replaced with the DULR module which encodes the perspective changes and passes the feature map to the second convolution layer in the conv block. From here, the whole functioning is the same as the S-DC Net model. The number of channels input to the DULR module is 3 and the number of output channels is 64 which is passed to the next convolution layer.

4 Experimental Studies

4.1 Dataset Description

The datasets used for the study are Shanghai part A and Shanghai part B. The Part A dataset has a total of 300 training images and a total of 182 testing images. The part B dataset consists of a total of 400 training images and 316 testing images. The datasets are trained and tested for the models, namely S-DC Net, PCC Net, SCAR, and S-DC Net + DULR, and the results were verified and recorded.

4.2 Experimental Results

The studied methods were evaluated on the Shanghai dataset, and the results are presented in Tables 1 and 2. The results clearly show that S-DC Net outperforms the other three models, namely SCAR, PCC Net, and S-DC Net + PCC Net.

Among the recent methods present, the S-DC Net is the best model for crowd counting, and it is proven by the results.

4.3 Analysis of the Studied Models

S-DC Net, PCCNet, and SCAR: The **S-DC Net** model gives the best accuracy as it is not affected by the perspective changes of the images. A divide-and-conquer

Table 1 Comparison of the four approaches over Shanghai Part A dataset	Methods	MAE	MSE
	S-DC Net [21]	58.3	95.0
	PCC Net [5]	73.5	102.7
	SCAR [22]	66.3	114.1
	S-DC Net + DULR	432.8	558.9

Table 2 Comparison of the four approaches over Shanghai Part B dataset

Methods	MAE	MSE
S-DC Net [21]	6.71	10.7
PCC Net [5]	11	19
SCAR [22]	9.5	15
S-DC Net + DULR	28.3	82.9

technique is used which causes the entire image to be compartmentalized. Thus, the whole process is repeated on the entire image by initially fixing the number of subparts. Each subpart contains the information of all its previous divided parts, thus making cumulative information to be available at each stage. In **PCC Net**, to prevent any loss from perspective changes, a DULR module is used. This module helps to get the spatial and contextual information from all the four directions. The information at a particular block of an image contains all the information of its previous blocks in all the four directions. Thus, the concatenated feature maps from the four directions form a resultant feature map that contains the perspective changes of the entire image. This model's accuracy is affected by occulted images. Hence, this architecture falls a little back of S-DC Net. **SCAR** has two modules that play a major role in the counting process. The spatial-wise attention module models the large contextual information and captures the changes in the density maps. The channel-wise attention model captures the contextual information between the three channels and also obtains the dependencies between the channels. This information helps to distinguish between the foreground and background. This model fills all possible gaps that are encountered in object counting and makes it a robust model for crowd counting.

Study on S-DC Net + DULR (Variant) Approach: Coming to the last model (S-DC Net + DULR), DULR module when added to S-DC Net made the model to be over-fit due to which the performance of the model was not good enough. Though conceptually the model looks perfect, the S-DC Net captures perspective changes along with spatial and contextual information. To this again, adding a DULR module which provides a feature map with encoded perspective changes makes the model to be over-fit due to which the model performance was moderate. As a future work, we are planning to propose a novel module to be integrated with S-DC Net for a better performance.

5   Direction for Further Research

- Focusing more on the spatial and contextual information to construct a more informative density maps that pushes the envelope further.

- Though the perspectives changes had been encoded well in the recent work, a little more focus toward the occlusion and background segmentation would yield better results.
- Integrating the recent work done on counting in images and applying it in real-time videos would have a good scope of exploring something innovative.

6 Conclusion

Crowd counting has become one of the most demanding tasks to be performed whether it be in security surveillance or in the advertisement sector. However, developing a model that fits the real-world environment is required. Among all the recent architectures that have been developed, S-DC Net appears to be the best model for crowd counting. The divide-and-conquer method used helps in the long run for better performance. Also, the PCC Net and SCAR works well as they have their perks. Based on the recent trends, new models are being developed to push the work done on crowd counting a little further.

Acknowledgements The authors would like to acknowledge Cognitive Science Research Initiative (CSRI)–(NO.DST/CSRI/2017/131 (G)) funding sanctioned by the Department of Science and Technology, Government of India and Council of Scientific and Industrial Research (CSIR) (09/1095(0043)/19-EMR-I).

References

1. Thanasutives P, Fukui K, Numao M, Kijsirikul B (2020) Encoder-decoder based convolutional neural networks with multi-scale-aware modules for crowd counting. arxiv: 2003.05586
2. Hu Y, Jiang X, Liu X, Zhang B, Han J, Cao X, Doermann D (2020) NAS-count: counting-by-density with neural architecture search. arxiv: 2003.00217
3. Shi X, Li X, Wu C, Kong S, Yang J, He L (2020) A real-time deep network for crowd counting. arxiv: 2002.06515
4. Luo A, Yang F, Li X, Nie D, Jiao Z, Zhou S, Cheng H (2020) Hybrid graph neural networks for crowd counting. arxiv: 2002.00092
5. Gao J, Wang Q, Li X (2019) PCC net: perspective crowd counting via spatial convolutional network. arXiv preprint arXiv: 1905.10085
6. Li M, Zhang Z, Huang K, Tan T (2008) Estimating the number of people in crowded scenes by MID based foreground segmentation and head-shoulder detection. In: ICPR, pp 1–4
7. Ge W, Collins RT (2009) Marked point processes for crowd counting. In: CVPR
8. Chan AB, Vasconcelos N (2012) Counting people with low-level features and bayesian regression. IEEE Trans Image Proces 21(4):2160–2177
9. Idrees H, Saleemi I, Seibert C, Shah M (2013) Multi-source multi-scale counting in extremely dense crowd images. In: Proceedings of IEEE conference on computer vision and pattern recognition, pp 2547–2554
10. Lempitsky VS, Zisserman A (2010) Learning to count objects in images. In: Proceedings of conference on neural information processing systems, pp 1324–1332

11. Pham V-Q, Kozakaya T, Yamaguchi O, Okada R (2015) COUNT forest: co-voting uncertain number of targets using random forest for crowd density estimation. In: Proceedings of international conference on computer vision, pp 3253–3261
12. Scarselli F, Gori M, Tsoi AC, Hagenbuchner M, Monfardini G (2008) The graph neural network model. TNNLS 20(1):61–80
13. Li Y, Tarlow D, Brockschmidt M, Zemel R (2016) Gated graph sequence neural networks. In: ICLR
14. Gilmer J, Schoenholz SS, Riley PF, Vinyals O, Dahl GE (2017) Neural message passing for quantum chemistry. CoRR abs/1704.01212
15. Zhang Y, Zhou D, Chen S, Gao S, Ma Y (2016) Single-image crowd counting via multi-column convolutional neural network. In Proceedings of the IEEE conference on computer vision and pattern recognition, pp 589–597
16. Cao X, Wang Z, Zhao Y, Su F (2018) Scale aggregation network for accurate and efficient crowd counting. In Proceedings of the European conference on computer vision (ECCV), pp 734–750
17. Shi M, Yang Z, Xu C, Chen Q (2018) Perspective-aware CNN for crowd counting. CoRR, abs/1807.01989
18. Liu W, Lis K, Salzmann M, Fua P (2018) Geometric and physical constraints for head plane crowd density estimation in videos. CoRR, abs/1803.08805
19. Simonyan K, Zisserman A (2014) Very deep convolutional networks for large-scale image recognition. Comput Sci
20. Ronneberger O, Fischer P, Brox T (2015) U-net: convolutional networks for biomedical image segmentation. In: International conference on medical image computing and computer-assisted intervention, pp 234–241
21. Xiong H, Lu H, Liu C, Liang L, Cao Z, Shen C (2019) From open set to closed set: counting objects by spatial divide-and-conquer. In: Proceedings of the IEEE/CVF international conference on computer vision (ICCV)
22. Gao J, Wang Q, Yuan Y (2019). SCAR: spatial-/channel-wise attention regression networks for crowd counting. arXiv preprint arXiv: 1908.03716

Short-Term Load Forecasting Using Random Forest with Entropy-Based Feature Selection

Siva Sankari Subbiah and Jayakumar Chinnappan

1 Introduction

The electricity load forecasting becomes an important research area in the power system to make proper decision on the planning and scheduling. It also gets an important role in the electricity market trading and the real-time dispatch. The nonlinearity, non-stationary and volatility nature of electricity becomes the hindrance for achieving an accurate forecasting. The load data is classified into short-term, medium-term and long-term load forecasting based on the hourly, monthly and yearly time horizon [1]. The weather factors such as temperature, humidity and other factors like holidays, festivals, past load data and tariff structure have major impact in load forecasting [2]. With an advent of artificial intelligence, the forecasting is achieved easily and efficiently [3]. The functioning of human brain stimulates the design and development of an artificial neural network (ANN) [4]. As the human brain, the ANN consists of number of neurons like units grouped in different layers, namely input, hidden and output layers [5]. ANN has an ability to model the linear and nonlinear complicated datasets and discovers easily the hidden patterns and correlation among them [6]. The optimization techniques also can be utilized for improving the performance of the network [7]. The neural network has designed to provide an improved performance by introducing deep feed-forward networks. Recurrent neural networks (RNNs) are considered as an alternative model [8]. ANN has proved that it has an ability for solving any problem such as disease prediction [9], text mining [10], toxity prediction [11] and crop

S. S. Subbiah (✉)
Department of Information Technology, Kingston Engineering College, Anna University, Vellore, India

J. Chinnappan
Department of Computer Science and Engineering, Sri Venkateswara College of Engineering, Pennalur, India

© The Author(s), under exclusive license to Springer Nature Singapore Pte Ltd. 2022   73
R. R. Raje et al. (eds.), *Artificial Intelligence and Technologies*,
Lecture Notes in Electrical Engineering 806,
https://doi.org/10.1007/978-981-16-6448-9_8

identification [12]. It has the ability for providing high accurate results for the problems that has an uncertain nature and usually can be solved only by human [13].

The real-world load data may have relevant and also irrelevant data. The dataset with relevant data produces accurate results [14]. The feature selection can be used to identify relevant data from the load dataset [15]. The relevant features can be identified using feature selection. It also prevents the curse of dimensionality issue. Sarhani et al. [16] developed a short-term load forecasting (STLF) model using CFS and support vector regression (SVR). The performance of the model has been improved by using feature selection. Ghiasi et al. [17] designed a hybrid forecasting system for predicting an electricity load where the preprocessing was done using the weighting feature selection technique, and the accurate forecasting was done by utilizing the backpropagation neural network (BPNN). Rana et al. [18] introduced an ANN-based load forecasting model for achieving an improved accuracy of 29%. Del Río et al. [19] developed a model for handling an imbalanced big dataset using machine learning. The model utilized RF and achieved an improved accuracy, versatility and robustness.

Rahman et al. [20] introduced a forecasting model for electricity generation of USA. The author utilized BPNN for forecasting and achieved high performance with the MAPE of 4.13% for all states and 4% to 9% MAPE for individual states as a result. Rana et al. [21] performed a forecasting on the UK and Australia electricity demand data for predicting an interval of electricity demand using BPNN, and the result is compared with and without MI, CFS and partial autocorrelation (AC) feature selection. As a result, the MI and CFS helped to produce better accuracy than AC. Huang et al. [22] introduced a MI-based forecasting model for Northest China electricity data. The model utilized simple forward selection and generalized minimum redundancy and maximum relevancy (G-mRmR) techniques for feature selection and RF and BPNN for learning purpose. The model with G-mRmR achieved better results. The feature selection discussed in the literature identifies the relevant features. But, the redundant features also need to be removed for improving the performance. So, in this paper, the EBFS is introduced to consider both feature to target and feature to feature relationships. It removes both irrelevant and redundant features from the list of input features provided for the learning process. It utilizes the information theoretic and correlation-based feature selection concepts for selecting relevant features and improving the performance of STLF.

2 Proposed Model

The proposed model consists of two stages, namely feature selection and prediction. The feature selection selects the relevant features from the load dataset using entropy-based feature selection. The information theoretic-based MI and the correlation-based symmetric uncertainty feature selections are utilized to select the relevant features. It helps to improve the performance of the forecasting by eliminating the irrelevant and redundant features from the dataset. The first stage uses entropy-based feature selection for dimensionality reduction and the second stage uses the RF for predicting the load. The performance of the RF is measured in terms of mean absolute percentage error (MAPE) and compared against BPNN.

2.1 Entropy-Based Feature Selection

The EBFS is an entropy-based filter feature selection that identifies the relevant and redundant features in two phases. In the first phase, it identifies the relevant features by calculating the mutual information between each feature and target. After that, it selects the relevant features that has high information gain value by plotting the graph and selecting the number of features from that. In the second phase, the redundant features are identified from the list of selected features using the symmetric uncertainty measures [23]. The mutual information between pair of features 'M' and 'N' is represented as $I(M; N)$. As the independent features do not have any common information, the mutual information $I(M; N)$ is 0. It is defined as follows,

$$I(M;N) = \int_a \int_b p(M,N)^{(m,n)} \log\left(\frac{p(M,N)^{(m,n)}}{p_M(m)p_N(n)}\right) dm dn \qquad (1)$$

The number of data in 'M' and 'N' is represented as 'm' and 'n,' respectively. The $p(A, B)$ represents the joint probability density function. The 'p_A' and 'p_B' are marginal probability density function. The symmetric uncertainty 'SU,' information gain 'IG' and entropy 'H' are calculated as follows,

$$SU(M,N) = 2\left[\frac{IG(M|N)}{H(M)+H(N)}\right] \qquad (2)$$

$$IG(M|N) = H(M) - H(M|N) \qquad (3)$$

$$H(M) = -\sum_i P(F_i) \log_2(P(F_i)) \qquad (4)$$

$$H(M|N) = -\sum_j P(F_j) \sum_i P(F_i|F_j) \log_2(P(F_i|F_j)) \qquad (5)$$

where 'F' is the feature, '$H(M)$' is an entropy of the feature 'M,' '$H(M|N)$' is an entropy of the feature 'M' after observing the value of feature N and '$IG(M|N)$' is the information gain value [24].

2.2 Backpropagation Neural Network

The most important phase of neural network started with the discovery of BPNN algorithm. It is applied to multilayer feed-forward networks that has a number of processing elements through which the data flows in forward direction from input layer to output layer. The BPNN calculates an error at the output and distributes it back through the hidden layers, and then updates the weight and bias at that layers. It is an algorithm for supervised ANN using gradient descent that calculates the error contributed by each of the neuron after a batch of data is processed [25]. The working of BPNN consists of four stages. First, initialization stage, where the biases, weights, inputs and outputs are initialized. Second, feed-forward stage in which the input is forwarded from input layer to output layer through number of hidden layers. At each layer from input to output, the net input for the layer is calculated, and the input is squashed using the activation function. Third, the backpropagation of error stage, where the error at each neuron and the net error for the layer are calculated using the squared error function and by summing them. Then, the error is propagated in backward direction, and the weight for each input is updated. Fourth, the biases and weights are updated in each layer [22].

2.3 Random Forest

The random forest can be utilized for both classification and regression problems. But in this paper, the random forest is utilized for regression problem of STLF. It is a category of an additive model which takes different subsamples from the given load dataset with replacement and construct tree for each set of subsamples as follows,

$$f(x) = f_0(x) + f_1(x) + f_2(x) + \cdots + f_n(x) \tag{6}$$

where 'f_i' represents base model and 'f' represents final model. It follows the model ensemble technique that consists of two stages, namely construction of random forest and prediction of load. In the first stage, 'p' number of features are taken from the 'A' number of samples, where $p < A$, then the best splitting criterion is identified by using information gain or Gini index or gain ratio, from the given list of 'p' features, and the base tree is constructed. This process is repeated till there is no more features in the p list of features. Similarly, the same set of processes are repeated for 'n' number of times and construct 'n' number of base trees. In the prediction stage, the test features are taken and applied on each base tree for calculating the outcome from them. Finally, the majority voting technique is introduced on the outcomes, and the outcome of the tree which gets highest vote is considered as the final prediction value.

3 Results and Discussion

The hourly load data of Australia electricity utility dataset from January 2004 to June 2008 is utilized for the experimental purpose. It consists of 27 features and 1586 samples. The data recorded from 2004 to 2005, 2006 and January 2007 to June 2008 is utilized as training, validation and testing datasets, respectively. Figure 1a shows the actual load from January 2007 to June 2008. As a sample, 24 h average load demand for the first week of January 2007 is shown in Fig. 1b. It shows that the average load peaks in most of the days during 7 to 11 h and also 18 to 22 h. So, in this paper, the relevant features for 8 h are identified as a sample by applying the EBFS feature selection. It identifies nine features as the relevant features. Then, the forecasting is done by using the RF and compared against BPNN. Figure 2a shows the comparison of the actual and predicted load using BPNN with all features and selected features. It shows that the BPNN with EBFS produces better results than BPNN without feature selection. Figure 2b shows the comparison of actual and predicted load using RF with all features and selected features. It shows that the RF with the selected features of EBFS achieves accurate result than RF with all features. The comparison of the actual and predicted load using RF and BPNN with selected features is shown in Fig. 3.

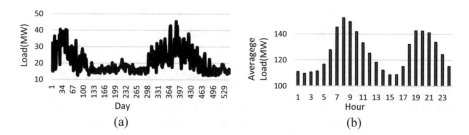

Fig. 1 **a** Actual load data from January 2007 to June 2008; **b** Sample average load demand for a week (1st week of January 2007)

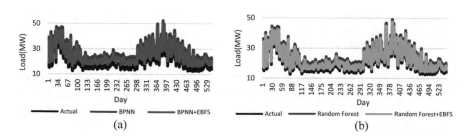

Fig. 2 **a** Comparison of actual and predicted load using BPNN; **b** Comparison of actual and predicted load using random forest

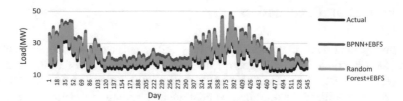

Fig. 3 Comparison of actual and predicted load using random forest and BPNN with selected features

Table 1 Comparison of actual and forecast load in terms of MAPE

Methodology	MAPE
Backpropagation neural network	6.582
Backpropagation neural network + EBFS	5.455
Random forest	3.974
Random forest + EBFS	2.601

It shows that the short-term load forecasted by random forest is more close to the actual load compared to the forecast load of BPNN. The performance of the forecasting is calculated in terms of the MAPE as follows,

$$\text{MAPE} = \frac{1}{n} \sum_{t=1}^{n} \left| \left(\frac{\text{ACTUAL}_t - \text{FORECAST}_t}{\text{ACTUAL}_t} \right) * 100 \right| \tag{7}$$

where 'n'—number of samples in dataset, 'ACTUAL_t'—actual load at time 't' and 'FORECAST_t'—forecast load at time 't.' The comparison of RF against BPNN with all features and selected features in terms of MAPE is shown in Table 1. It shows that the RF with EBFS produces more accurate result than others.

4 Conclusion

The feature selection plays a major role in STLF and helps to make proper decision in power systems. In this paper, entropy-based feature selection selects the relevant features by identifying the feature to target and feature to feature relationships. The mutual information is utilized for identifying the irrelevant features, and the correlation-based symmetric uncertainty is utilized for finding the redundant features. The result shows that the STLF with RF produces accurate result than BPNN, and also, the STLF with selected features produces accurate results than the STLF with all features.

References

1. Subbiah SS, Chinnappan J (2020) A review of short term load forecasting using deep learning. Int J Emerg Technol 11(2):378–384
2. Subbiah SS, Chinnappan J (2020) An improved short term load forecasting with ranker based feature selection technique. J Intell Fuzzy Syst (Preprint), pp 1–18
3. Senthil Kumar P (2017) A review of soft computing techniques in short-term load forecasting. Int J Appl Eng Res 12(18):7202–7206
4. Kiruthika VG, Arutchudar V, Senthil Kumar P (2014) Highest humidity prediction using data mining techniques. Int J Appl Eng Res 9(16):3259–3264
5. Senthil KP (2019) Improved prediction of wind speed using machine learning. EAI Endorsed Trans Energy Web 6(23):1–7
6. Senthil Kumar P, Lopez D (2015) Feature selection used for wind speed forecasting with data driven approaches. J Eng Sci Technol Rev 8(5):124–127
7. Karmel A, Jayakumar C (2014) Analysis on optimization of energy consumption in mobile ad hoc networks. In: Intelligent computing, networking, and informatics. Springer, New Delhi, pp 481–489
8. Karmel A, Adhithiyan M, Senthil KP (2018) Machine learning based approach for pothole detection. Int J Civil Eng Technol (IJCIET) 9(5):882–888
9. Diviya M, Malathi G, Karmel A (2019) Regression based model for prediction of heart disease recumbent. Int J Recent Technol Eng 8(4):6639–6642
10. Sivasankari S, Baggiya Lakshmi T (2016) Operational analysis of various text mining tools in bigdata. Int J Pharm Technol (IJPT) 8(2):4087–4091
11. Adhithiyan M, Karmel A (2019) Novel approach of deep learning in toxicity prediction. Int J Recent Technol Eng 7:698–704
12. Agila N, Senthil Kumar P (2020) An efficient crop identification using deep learning. Int J Sci Technol Res 9(1):2805–2808
13. Swaroop G, Senthil Kumar P, Muthamil Selvan T (2014) An efficient model for share market prediction using data mining techniques. Int J Appl Eng Res 9(17):3807–3812
14. Paramasivan SK, Lopez D (2016) Forecasting of wind speed using feature selection and neural networks. Int J Renew Energy Res (IJRER) 6(3):833–837
15. Senthil Kumar P, Lopez D (2016) A review on feature selection methods for high dimensional data. Int J Eng Technol 8(2):669–672
16. Sarhani M, Afia AE (2015) Electric load forecasting using hybrid machine learning approach incorporating feature selection. In: BDCA, pp 1–7
17. Ghiasi M, Jam MI, Teimourian M, Zarrabi H, Yousefi N (2019) A new prediction model of electricity load based on hybrid forecast engine. Int J Ambient Energy 40(2):179–186
18. Rana M, Koprinska I, Agelidis VG (2012) Feature selection for electricity load prediction. In: International conference on neural information processing. Springer, Berlin, Heidelberg, pp 526–534
19. Del Río S, López V, Benítez JM, Herrera F (2014) On the use of MapReduce for imbalanced big data using random forest. Inf Sci 285:112–1137
20. Rahman MN, Esmailpour A, Zhao J (2016) Machine learning with big data an efficient electricity generation forecasting system. Big Data Res 5:9–15
21. Rana M, Koprinska I, Khosravi A (2015) Feature selection for interval forecasting of electricity demand time series data. Artificial Neural Networks. Springer, Cham, pp 445–462
22. Huang N, Hu Z, Cai G, Yang D (2016) Short term electrical load forecasting using mutual information based feature selection with generalized minimum-redundancy and maximum-relevance criteria.Entropy 18(9):330
23. Yu L, Huan L (2003) Feature selection for high-dimensional data: a fast correlation-based filter solution. In: Proceedings of the 20th international conference on machine learning (ICML-03), pp 856–863

24. Piao M, Piao Y, Lee JY (2019) Symmetrical uncertainty-based feature subset generation and ensemble learning for electricity customer classification. Symmetry 11(4):498
25. Sivanandam SN, Deepa SN (2008) Principles of soft computing. Wiley India Edition

Depth Comparison of Objects in 2D Images Using Mask RCNN

Himanshu Singh and V. B. Kirubanand

1 Introduction

This paper suggests a unique way of comparing relative distances between objects from single 2D images, clicked from normal cameras. We are using mask RCNN for recognizing and marking the area of interest and then finding the depth of area of interest with a depth-estimation model. Mask RCNN is used to give the area of interest in which we have to apply depth model or DenseNet-201. Then, DenseNet is being used to give us the result such that depth of only ROI or region of interest is calculated and the mean of output of images from the model determines the difference between the distance of object from the camera.

This method is used for the camera which has similar focal point, if the focal length of the camera is fixed, this method accurately gives the output.

1.1 Literature Review

Limiting the number of objects in the images to be detected and sent to depth analysis is an important step of the result (Depth Estimation from Image Structure, Torraliba and Oliva) [1], it helps in reducing the background noise and hence have been implemented in this method by using the mask RCNN model. The model extracts the ROI or hand in our case and sends the ROI to the depth analysis, thus helping in better accuracy. The paper also concludes that mean depth is the key for the depth, and the conclusion of this paper also includes the mean depth for results.

H. Singh (✉) · V. B. Kirubanand
CHRIST (Deemed to be University), Bangalore, India
e-mail: Himanshu.singh@cs.christuniversity.in

V. B. Kirubanand
e-mail: Kirubanand.vb@christuniversity.in

© The Author(s), under exclusive license to Springer Nature Singapore Pte Ltd. 2022 81
R. R. Raje et al. (eds.), *Artificial Intelligence and Technologies*,
Lecture Notes in Electrical Engineering 806,
https://doi.org/10.1007/978-981-16-6448-9_10

In the paper, Object Localization and Size Estimation [2] from RGB-D Images, from Shree Ranjani Srirangam Sridharan, Oytun Ulutan, Shehzad Noor Taus Priyo, Swati Rallapalli, and Mudhakar Srivatsa, they use depth estimation for finding the size of the object. In proposed paper, we are using the same method of depth estimation to compare distance of an object in multiple images from same focal length. The proposed paper does not use RGB-D for the conclusion as it solely depends on the recognition of hand from the image or recognition of targeted object.

The use of CNN for depth analysis [3], proposed in Deeper Depth Prediction with Fully Convolutional Residual Networks by Iro Laina, Christian Rupprecht, Vasileios Belagiannis, Federico Tombari, and Nassir Navab, uses CNN model and uses map up sampling within the network for better accuracy.

For monocular depth estimation, CNN is used along with method of LeftRight consistency in Unsupervised Monocular Depth Estimation with LeftRight Consistency [4] proposed by Clément Godard, Oisin Mac Aodha, and Gabriel J. Brostow. The network estimates depth by inferring disparities that warp the left image to match the right one. The left input image is used to infer the left-to-right and right-to-left disparities. The network generates the predicted image with backward mapping using a bilinear sampler. This results in a fully differentiable image formation model (Table 1).

2 Dataset

A total of two datasets have been used for this implementation.

The first dataset contains images of hands with unknown distance from the camera, used to train mask RCNN. This is to recognize hands from images as shown in Fig. 1. Annotations have been done for proper hand recognition. The second dataset is divided into three categories based on the distance of hand from the camera. For our use, we have kept hand at 15, 20, and 25 cm distance from the camera and labelled the image with respect to this distance value and the models work on this dataset for the result as shown in Figs. 2 and 3.

Table 1 Summary of literature review

Paper	Conclusion
Depth Estimation from Image Structure, Antonio	Limits the number of target object which gives better result in depth analysis
Object Localization and Size Estimation from RGB-D Images	Use depth estimation for finding the size of the object
Deeper Depth Prediction with Fully Convolutional Residual Network	CNN model and uses map up sampling within the network for better accuracy
Unsupervised Monocular Depth Estimation with Left–Right Consistency proposed	CNN is used along with method of left–right consistency policies
Depth Comparison of objects in 2D images using Mask RCNN	Uses mask RCNN for extracting target and mean depth analysis for result technology

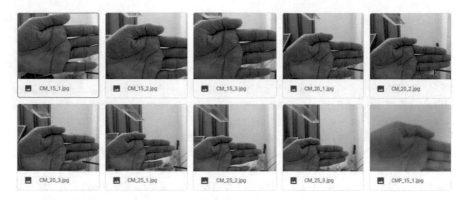

Fig. 1 Original hand images dataset

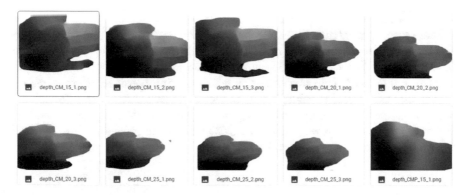

Fig. 2 Masked hand images dataset

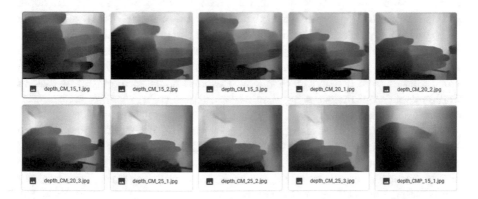

Fig. 3 Depth maps for the hands

3 Model Used

3.1 Mask RCNN

Mask RCNN is an extension of faster RCNN. Faster RCNN is widely used for object detection tasks. For a given image, it returns the class label and bounding box coordinates for each object in the image.

The masked RCNN model is built on top of faster RCNN. In addition to class labels and bounding boxes, it also returns object masks. It uses Resnet 101 architecture to extract feature maps from the images. These features act as input layer for the next layer. Now, it take the feature maps obtained in the previous step and applies a region proposal network (RPM) [5]. This basically predicts if an object is present in that region (or not). In this step, we get those regions or feature maps which contains some object as per the model prediction. After this, the region is converted into similar shape as we got them in different shapes.

Till now, the working is similar to faster RCNN. In addition, mask RCNN also generates the segmentation mask. Hence, it first compute the region of interest, so that the computation time can be reduced. For all the predicted regions, we compute the intersection over union (IoU) with the ground truth boxes.

Now, only if the IoU is greater than or equal to 0.5, it considers that as a region of interest. Otherwise, it neglects that particular region. It does this for all the regions and then selects only a set of regions for which the IoU is greater than 0.5. We have used pretrained weights from COCO dataset and trained the last layer for the detection and segmentation of hand.

3.2 DenseNet-201

Next comes the depth model in picture where we have used monocular depth estimation to produce depth maps required for our analysis. Monocular depth estimation is the task of estimating scene depth using single 2D images. We as humans are able to estimate depth from the surroundings because we have stereo vision, while 2D images do not preserve the same stereo information as a 3D world when clicked from a monocular camera. Vision systems accomplish this task of finding depth from images using depth cues such as color, shading, texture, perspective, and relative size.

For our work, we have used pretrained DenseNet-201 model, which contains 201 layers, previously trained on the NYU and KITTI datasets. We have used NYU weights for our purpose. In this work, our modifications mainly include scaling up of each image during processing to get a better view and hence a better prediction of depth for each object. The images are then scaled back to their original sizes. The depth maps that we obtain contain the distribution of color gradients and intensities in such a way that objects nearer to the camera appear darker, and those farther from the camera appear lighter in shade. Lack of a depth camera or a LIDAR camera leads us to rely on pretrained DenseNet-201 to obtain our depth maps [6].

4 Working

The project starts with data being sent to mask RCNN model to detect and segment the hands in the images. The mask RCNN model detects the hand from the images and gives an output with hand being detected and segmented as shown in Fig. 4. The area of interest we got from this model is then extracted from the image to increase accuracy of depth analyzer, so that background and other things do not alter the output of depth analyzer. For this, we obtain a binary mask from the output of mask RCNN with values of area of interest, i.e., hand pixels as 1 and the rest as 0.

We now run our test images through the depth model, upscale it, and get a depth map with darker shades showing nearer objects and lighter shades representing farther objects as shown in Fig. 2. The binary mask obtained from mask RCNN is then applied onto these depth maps to get only the original pixel values of hands and white pixel values for the background as shown in Fig. 3. The pixel values of these images are then plotted on a histogram which now shows us the distribution of depth values of the object of concern. The mean of all the histogram values for all images to be compared is then taken which tells us on the basis of the differences observed, that which hand (object) is closer and which is farther to the camera and by what factor exactly. Once we get to know this factor, it can be used for multiple other crucial purposes, like knowing an estimate of the actual size of the objects in the pictures with the help of certain other factors. A further implementation of a simple regression model could lead us to developing a complete model for actual size estimation of objects from 2D images which is by far one of the most difficult known existing problems.

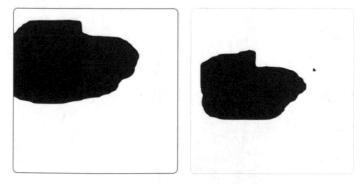

Fig. 4 Masked RCNN

As an outcome, we obtain that the mean values of 15 cm images are in the range 25–30, 20 cm mean is in the range 75–85, and 25 cm distant images from the lens of the camera lie in the range 100–110 as illustrated in table. Hence, this model was successful in comparing the depth of object in images taken from same focal length lenses cameras.

Distance	Mean (Range)
15 cm	35-55
20cm	65-80
25cm	85-100

5 Conclusion

Following are the results and discussion for the experiments conducted. The distances specified are the distances from the camera to the hand. Also, different images are also given to enable the readers have a view of the intermediate steps in the analysis.

Figure 5 represents intensity—frequency map of image in which hand was 15 cm away from the lens of the camera.

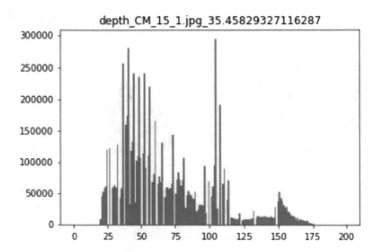

Fig. 5 15 cm distance

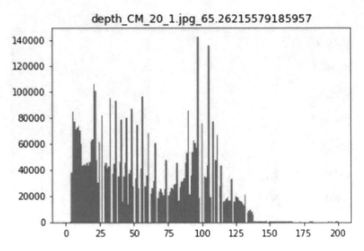

Fig. 6 20 cm distance

Figure 6 represents intensity—frequency map which was 20 cm away from the camera, and Fig. 6 represents intensity—frequency map for 25 cm distance image.

It can be seen in graph that mean for 15, 20, and 25 cm images are 35.46, 65.26, and 87.35 [accurate to two decimal points here]. As the distance of hand increases from the camera, the mean also increases for intensity–frequency graph. Thus, by comparing the means, the distance of objects can be compared.

Furthermore, the mean for a particular distance for similar lens focal length is always same, and hence, from mean, the distance of object from the image can also be predicted if not known.

5.1 Outputs

See Figs. 5, 6, and 7.

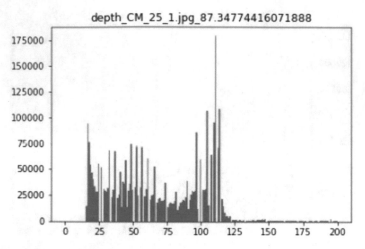

Fig. 7 25 cm distance

References

1. Toralba A, Oliva A (2002) Depth estimation from image structure. IEEE 24:13
2. SrirangamSridharan S, Ulutan O, Priyo SN, Rallapalli S, Srivatsa M. Object localization and size estimation from RGB-D images. arXiv:1808.00641
3. Laina I, Rupprecht C, Belagiannis V, Tombari F, Navab N (2016) Deeper depth prediction with fully convolutional residual networks
4. Godard C, Mac Aodha O, Brostow GJ (2016) Unsupervised monocular depth estimation with left-right consistency. arXiv:1609.03677
5. https://medium.com/@alittlepain833/simple-understanding-of-mask-rcnn134b5b330e95
6. https://towardsdatascience.com/depth-estimation-on-camera-images-using-densenetsac454caa893

Ensemble Methods with Bidirectional Feature Elimination for Prediction and Analysis of Employee Attrition Rate During COVID-19 Pandemic

Yash Mate, Atharva Potdar, and R. L. Priya

1 Introduction

Due to this pandemic, millions of people have been laid off from their jobs, and in developed countries like the USA, the national unemployment rate was at an astonishing 14.7% in the month of April 2020. Alongside the unfavorable situation and increasing loss in a business, various factors account for employee attrition. The factors affecting employee status are extensively studied, and the observations found are further explained in detail.

Attributes such as department, job role, and education have to be primarily considered for analyzing the trend. But considering only these factors is not sufficient to successfully comprehend this issue. Many other factors that might even appear to be trivial at a first glance have to be included to significantly improve the quality of this research. These are stated and explained further.

2 Background

The workforce of a nation largely determines its economic progression [1, 2]. The dissatisfaction of employees in an organization could be a potential warning that an organization needs to change its policies [3, 4]. The study done by Silpa et al. looks at statistical measures like the coefficient of correlation, Chi-square test, and mean of employee's data to understand the reasons behind them leaving the organization.

The researcher in his proposed study [5] has explained unique factors like turnovers in BPO sectors. But the individual employee factors mentioned are hardly two, and the data collection was restricted to only 100 employees.

Y. Mate (✉) · A. Potdar · R. L. Priya
Department of Computer Engineering, V.E.S Institute of Technology, Mumbai, India
e-mail: 2017.yash.mate@ves.ac.in

The study [6] done by Jaya Sharma et al. included various reasons for employee attrition. But in the study, it was concluded that compensation was the most important factor for employee attrition. There are many factors equally as important as compensation which can determine employee attrition.

In the research [7] done by Sunanda et al., the author has clearly explained features like working conditions, career growth, and organizational culture. But the scope of the study was limited to the employees from the age group 20–40 of only one organization.

3 Conceptual Design

The proposed system is designed for analysis of factors that have an impact on the attrition of employees. In addition to that, the system also makes a highly accurate prediction of whether an employee would be churned out from the organization or not based on the selected set of attributes from the dataset. The abstract representation of the entire data analysis and attrition prediction system is displayed below as shown:

According to Fig. 1, the dataset of the attrition rate of employees in the course of the COVID-19 pandemic is obtained from Kaggle [8] in collaboration with IBM. The raw data obtained in the form of a tabular representation is used for carrying out exploratory data analysis.

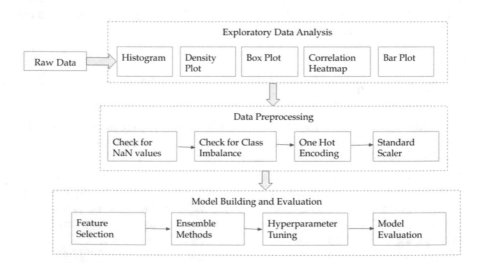

Fig. 1 Conceptual design of the proposed solution

4 Data Description

See Table 1.

5 Exploratory Data Analysis

Before building the model or carrying out any transformations in the data, it is a prime need to understand and summarize the data present in the dataset. Exploratory data analysis (EDA) helps in highlighting the relationships between attributes by plotting and reviewing the data collected (Figs. 2, 3, and 4).

Table 1 Attributes and their description

Data field	Description
Id	Anonymous ID for an employee
Age	Age of the employee
Attrition	Was the employee removed, 1—Yes, 0—No
Business travel	Frequency of traveling
Department	Working department
Distance from home	Distance between home and office
education field	Field of education
Employee number	Number of employees in the division of that employee
Environmental satisfaction	Satisfaction rating
Gender	Sex of the employee
Marital status	Status of marriage of employee
Monthly income	Income in USD
Num companies worked	Number of companies that employee had worked in before joining
Over time	Does the employee work over time
Percent salary hike	Average annual salary hike in percentage
Stock option level	Company stocks given to an employee
Total working years	Work experience
Training times last year	No of trainings taken by the employee in the past year
Years at company	Years worked for the company
Years in current role	Years spend at the current role in the company
Years since last promotion	Number of years since last being promoted
Years with curr manager	Number of years under the current manager

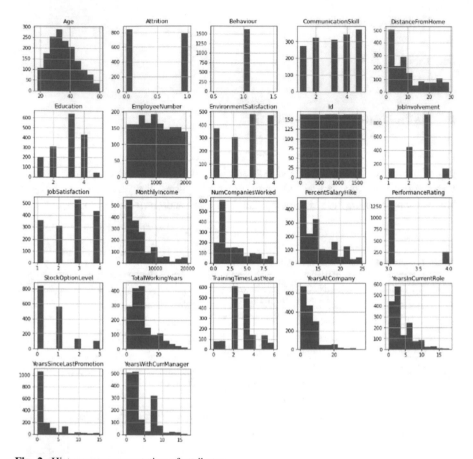

Fig. 2 Histogram representation of attributes

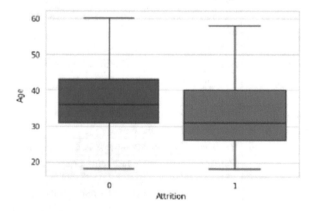

Fig. 3 Box plot representation of age

Fig. 4 Box plot
representation of monthly
income

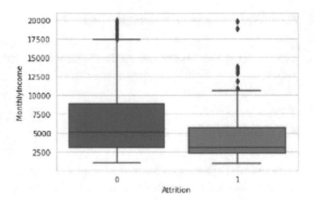

6 Data Preprocessing

Generally, machine learning algorithms make assumptions regarding the data that it
gets fed into. Different algorithms make different presumptions. Hence, we need to
apply certain transformations to make the data compatible with the model. There
are a total of 1628 records/instances in the dataset which are divided into 20%
validation and 80% training and evaluated using cross-validation.

6.1 Checking for Missing Values

The dataset into consideration has no missing values. Therefore, no extra imputation methods are required.

6.2 Class Imbalance

The attrition dataset does not have the class imbalance problem as both the classes
are fairly proportionate.

There are a total 843 instances of class with label '0' and 785 instances of class
with label 1.

6.3 One Hot Encoding

One hot encoding encodes each category to a separate column in the dataset having
a binary value. After one hot encoding, all the variables are converted to the data
type of int which is compatible with all machine learning models.

6.4 Splitting into Predictors and Target

The dataset in the form of tabular data is now separated into independent and target variables.

6.5 Standard Scaler

Data scaling [9] forms a vital part of preprocessing that reduces the training time by reducing the scale of variables to a single standard range.

$$\text{Standardized\_x} = (x - \text{x\_mean})/\text{standard\_deviation\_x}$$

7 Training and Testing

7.1 Stratified K-Fold Cross-Validation

For model building [10], the dataset should be split into training and testing sets. The training data would be used for developing the machine learning model which would be in turn used for prediction. A more advanced approach is using a special form of cross-validation called *stratified K-fold cross-validation.*

7.2 Logistic Regression

This algorithm [11] is used when the variable that needs to be predicted is categorical in nature. The cost function for logistic regression is the maximum likelihood estimation.

7.3 Linear Discriminant Analysis (LDA)

LDA requires the data to have a Gaussian distribution and works well for classification problems. LDA [12] calculates the statistical properties of data like mean and variance.

7.4   K-Nearest Neighbors (KNN)

KNN [13] works on the principle that data points belonging to a particular class lie closer to each other as compared to other classes. Proximity using Euclidean distance is what KNN looks at for making the decision regarding which class the new unseen data point belongs to.

7.5   Decision Trees

Decision trees are also referred to as classification and regression trees (CART). The system uses CART [14] for binary classification and forms the basis of more advanced classification models like random forests, extra tree classifier, adaptive gradient boosting, and many other ensemble techniques. Decision trees are highly interpretable and ensure transparency.

7.6   Gaussian Naive Bayes

This algorithm is fundamentally based on Bayes theorem which calculates the probability of an event A occurring given that another event B has already occurred.

7.7   Support Vector Machines

SVM is a supervised machine learning algorithm that is used for classification problems. SVM separates the classes by introducing a boundary known as a hyperplane.

7.8   Random Forest Classifier

Random forest [15] is a supervised machine learning algorithm that works on the principle of ensembling, and random forest comprises numerous decision trees. More trees make the model more robust.

7.9 Extra Tree Classifier

It is an ensemble method used primarily based on decision trees. Like random forests, extra tree classifier adds an element of randomness in decisions and overcomes overfitting. Extra tree has the lowest variance as compared to decision trees and random forests (Table 2).

8 Feature Selection

It reduces the dimensionality of the high-dimensional data and makes it more interpretable. Less relevant and redundant features are eliminated through the process of feature selection [16].

A brute force approach [17] for feature selection would involve trying out all possible combinations and evaluating the performance for each combination and selecting the one that gives the maximum accuracy.

8.1 Bidirectional Feature Elimination

It is also known as stepwise selection and falls in the category of wrapper-based feature selection approach:

1. Define two variables, S_IN (Significance level for selecting a feature) and S_OUT (Significance level for removing a feature).
   S_IN = 0.05 and S_OUT = 0.05 with a confidence of 95%
2. For forward selection, the new feature should have a p-value < S_IN.
3. For backward elimination, any previously added feature to the feature set should have p-values > S_OUT to be removed.
4. Repeat the above two steps, until the most optimal set of features are obtained.

Table 2 Comparison of performance metrics of classification algorithms without feature selection

Algorithm	Accuracy	Precision	Recall	F1-score
Logistic regression	0.714	0.716	0.714	0.714
Linear discriminant analysis	0.794	0.796	0.794	0.794
K-nearest neighbors	0.796	0.856	0.796	0.788
Decision trees	0.927	0.940	0.928	0.927
Gaussian–Naive Bayes	0.687	0.701	0.687	0.683
Support vector machine	0.636	0.636	0.636	0.634
Random forest	0.978	0.983	0.979	0.982
Extra tree classifier	0.989	0.992	0.990	0.988

The null-hypothesis says that the coefficient has no effect on the target output.

Lower p-value indicates that the attribute has a meaningful weightable to the model as variance in the independent variables has a corresponding change in the target variable. The p-value chose is 0.05. It means that if an attribute has a p-value greater than 0.05, then it is insignificant and the null-hypothesis holds true (Table 3).

8.2 Ensembling Methods

Ensembling is a machine learning technique that combines many base classifiers to produce an optimal predictive model.

8.3 Boosting

The basic building block of ensembles in the decision tree. Boosting [18] is a type of ensemble machine learning model where the model is built one tree at a time. The newly added trees aim at fitting and correcting the prediction errors made before the one.

8.4 Adaptive Boosting Classifier

It works on the principle of combining many weak classifiers together to formulate a stronger classifier. A special type of decision tree with a single split called a "decision stump" is the fundamental unit of the AdaBoost classifier.

Table 3 Comparison of performance metrics of classification algorithms with feature selection

Algorithm	Accuracy	Precision	Recall	F1-Score
Logistic regression	0.788	0.778	0.784	0.781
Linear discriminant analysis	0.789	0.772	0.798	0.784
K-nearest neighbors	0.863	0.780	1.0	0.876
Decision trees	0.933	0.875	1.0	0.934
Gaussian–Naive Bayes	0.741	0.739	0.722	0.73
Support vector machines	0.893	0.861	0.931	0.894
Random forest	0.979	0.955	1.0	0.978
Extra tree classifier	0.987	0.977	1.0	0.985

8.5 Gradient Boosting Classifier

It is one of the most effective classification algorithms which delivers models with extremely high accuracy. The model [19] constructed is using a random differentiable loss function with an optimization algorithm as gradient descent.

8.6 Extreme Gradient Boosting Classifier

XGB is an extremely powerful advancement of the gradient boosting classifier [20]. XGB uses regularization techniques to reduce overfitting and is alternatively known as "regularized boosting" technique.

8.7 Light Gradient Boosting Machine

LGBM is a high-performance boosting algorithm based on distributed computing [21]. As opposed to XGB, LGBM splits the tree leaf-wise and not depth-wise.

8.8 Bagging

Bagging is an ensemble approach that reduces overfitting by averaging out the predictions of the individual units through a process of voting. The base classifiers are constructed by randomly sampling out data from the training set with replacement.

8.9 Voting

Voting classifier takes into account the prediction generated by all the base classifiers to form its final prediction. There are two types of voting: Hard and Soft. Hard voting considers the count of each class that is outputted by each classifier (Table 4).

Table 4 Comparison of performance metrics of ensembling classifiers with feature selection

Algorithm	Accuracy	Precision	Recall	F1-Score
Adaptive gradient boosting	0.787	0.784	0.775	0.779
Gradient boosting classifier	0.872	0.854	0.918	0.884
Extreme gradient boosting	0.959	0.921	1.0	0.959
Light gradient boosting machine	0.954	0.919	1.0	0.957
Bagging classifier	0.962	0.923	1.0	0.991
Voting classifier	0.994	0.983	1.0	0.964

9 Hyperparameter Tuning

The goal of hyperparameter tuning is to identify the best combination of these hyperparameters that minimize the loss or maximize the accuracy.

The method used for hyperparameter tuning is called "randomized search."

The hyperparameters are selected from the following range of values:

Max depth: [3, 5, 10].
N estimators: [10, 100, 200, 300, 400, 500].
Max features: [1, 2, 3].
Criterion: Gini, Entropy.
Bootstrap: True, False.
Min samples leaf: [1, 2, 3, 4].
Search is carried out over 40 iterations.

10 Evaluation Metrics

The various evaluation metrics used here are accuracy, precision, recall, and F1-score.

- Accuracy is defined as total no. of correct predictions divided by total no. of predictions.
- Precision is calculated as the total no. of true positives divided by sum of true positives and false positives.
- Recall is calculated as total no. of true positives divided by sum of true positives and false negatives.
- F1-score is the weighted harmonic mean of precision and recall.

Random Forest

The hyperparameters after tuning had n_estimators equal to 400, minimum leaf samples as 1, maximum features as 1, and the criteria for splitting was Gini Index.

Accuracy attained by random forest classifier is 99.5%.

Extra Tree Classifier

The hyperparameters after tuning had $n\_$estimators equal to 400, minimum leaf samples as 1, maximum features as 1, and the criteria for splitting was Gini Index.

Using the hyperparameters obtained above, the model achieves an accuracy of 99.6%.

11   Conclusion

Employee attrition is certainly inevitable in the times of a pandemic. In this proposed system, various factors affecting employee attrition have been studied in detail. The most impactful features are then observed and selected eliminating the rest of them. Random forest and extra tree classifier were optimized with the help of randomized search hyperparameter tuning. This resulted in an accuracy of 99.6% on the extra tree classifier model.

References

1. IMF, World Economic Outlook, Deloitte India workforce and increment trends survey overview of results, March 2020
2. IMFBlog. COVID-19 crisis poses threat to financial stability. https://blogs.imf.org/2020/04/14/covid-19-crisis-poses-threat-to-financial-stability/
3. Silpa N et al (2015) A study on reasons of attrition and strategies for employee retention. Int J Eng Res Appl 5(12 Part I):57–60. ISSN: 2248-9622
4. Saranya S, Sharmila Devi J et al (2018) Predicting employee attrition using machine learning algorithms and analyzing reasons for attrition. Int J Adv Eng Res Technol (IJAERT) 6(9). ISSN No.: 2348-8190
5. Latha Lavanya B et al (2017) A study on employee attrition: inevitable yet manageable. Int J Bus Manage Invention 6(Sept):38–50. ISSN (Online): 2319-8028, ISSN (Print): 2319-801X
6. Sharma J et al (2015) Employee attrition and retention in a cut-throat competitive environment in India: a holistic approach. Indian J Res 4(2). ISSN 2250-1991
7. Sunanda K et al (2017) An empirical study on employee attrition in it industries—with specific reference to Wipro technologies. Int J Manage Stud September 2017. ISSN (Print) 2249-0302, ISSN (Online) 2231-2528
8. Kaggle Dataset [https://www.kaggle.com/c/summeranalytics2020/overview]
9. Wan X (2019) Influence of feature scaling on convergence of gradient iterative algorithm. J Phys: Conf Ser 1213:032021. https://doi.org/10.1088/1742-6596/1213/3/032021
10. Raschka S (2018) Model evaluation, model selection, and algorithm selection in machine learning
11. Peng J, Lee K, Ingersoll G (2002) An introduction to logistic regression analysis and reporting. J Educ Res 96:3–14. https://doi.org/10.1080/00220670209598786
12. Tharwat A, Gaber T, Ibrahim A, Hassanien AE (2017) Linear discriminant analysis: a detailed tutorial. AI Commun 30:169–190. https://doi.org/10.3233/AIC-170729
13. Cunningham P, Delany S (2007) k-nearest neighbour classifiers. Mult Classif Syst
14. Rokach L, Maimon O (2005) Decision trees. https://doi.org/10.1007/0-387-25465-X_9
15. Biau G (2010) Analysis of a random forests model. J Mach Learn Res 13

16. Miao J, Niu L (2016) A survey on feature selection. Procedia Comput Sci 91:919–926. https://doi.org/10.1016/j.procs.2016.07.111
17. Guyon I, Elisseef A et al (2003) An introduction to variable and feature selection. J Mach Learn Res 3:1157–1182
18. Bühlmann P, Hothorn T (2008) Boosting algorithms: regularization, prediction and model fitting. Stat Sci 22. https://doi.org/10.1214/07-STS242
19. Natekin A, Knoll A (2013) Gradient Boosting Machines, a tutorial. Front Neurorobotics 7:21. https://doi.org/10.3389/fnbot.2013.00021
20. Chen T, Guestrin C (2016) XGBoost: a scalable tree boosting system
21. Machado MR, Karray S, de Sousa IT (2019) LightGBM: an effective decision tree gradient boosting method to predict customer loyalty in the finance industry, pp 1111–1116. https://doi.org/10.1109/ICCSE.2019.8845529

Face Recognition with Mask Using MTCNN and FaceNet

Abhishek Sunil Tiwari, Prajul Gupta, Aanya Jain,
Hari Vilas Panjwani, and G. Malathi

1 Introduction

Face recognition systems are becoming an integral part of authentication at organizations nowadays. It is widely adopted due to its non-invasive and contactless
nature. These systems are capable of identifying and verifying a person uniquely
and are modern substitutes to traditional biometric systems based on finger
impression. Other applications include an automatic grouping of images, video
surveillance, and also important for better human–computer interaction. After going
through so many recognized and established works of researchers, there was a need
to design something that could accommodate the current situation and be able to
present a solution to the need of the hour. So this paper provides that much-needed
solution thereby helping to predict the identity of the person with the mask. But the
preexisting system of recognizing a person with face will not work as keeping
health guidelines it is important to wear the mask and at the same time. Facial
recognition is also needed for authentication purposes. So, we have proposed facial
recognition systems even for people with a mask based on features of eyes and
forehead and used the multi-task cascaded convolutional neural network (MTCNN)
[1] for getting the pixel coordinate values of the left eye, right eye, and bounding
box. Further, they are passed into the FaceNet [2] model to generate embeddings
and lastly passing these to the classifier. This research paper consists of seven
sections including the introduction section. Section 2 explains the various previous
approaches and models used for face recognition. Section 3 describes the dataset.
Section 4 proposes our novel architecture, Sect. 5 explains in detail about our

A. S. Tiwari (✉) · P. Gupta · A. Jain · H. V. Panjwani · G. Malathi
School of Computer Science and Engineering, Vellore Institute of Technology,
Chennai 600127, India
e-mail: abhisheksunil.tiwari2018@vitstudent.ac.in

© The Author(s), under exclusive license to Springer Nature Singapore Pte Ltd. 2022   103
R. R. Raje et al. (eds.), *Artificial Intelligence and Technologies*,
Lecture Notes in Electrical Engineering 806,
https://doi.org/10.1007/978-981-16-6448-9_12

proposed work, Sect. 6 discusses outcomes of several experiments conducted and results obtained, and finally, the last section concludes the research paper best architecture among all the experiments conducted.

2 Related Works

Face recognition is a very popular field among researchers, and over time, several approaches have been proposed. Principal component analysis (PCA) [3, 4] was used for facial recognition purposes, and it basically reduces large dimensional data space to small dimensional feature space, by computing a subspace vector also known as Eigenfaces. It does so by representing 2D images into 1D vectors of pixels also known as Eigenspace vectors. Linear discriminant analysis (LDA) method for face recognition purposes popularly termed as Fisherface [5] method. Although for detecting masked faces [6] used LLE-CNN, it only detects whether a face is masked or not, and that with 76.4% accuracy and also does not do face recognition. And further major recent benchmarks in facial recognition domain developed are as follows MTCNN [7] rained on the WiderFace [8] become a popular face detection framework due its proposal of adopting three stages carefully engineered deep convolutional networks that particularly predict in landmark and face location in coarse to fine manner, exploiting the inherent correlation between them to improve their performance. FaceNet [9] which gives better representational efficiency due to its novel algorithm in which they do training with triplets aligned roughly with matching/non-matching face patches generated using a method of online triplet mining, which helps in directly optimizing embedding itself. Deepface [1] which performed affine transformation in piecewise form then obtained face representation from a deep neural network consisting of nine layers. This network involves close to 120 million parameters without weight sharing of many locally connected layers. Huang et al. [2] defined distribution distillation loss which specifically gives a more generic method for soft and hard samples of image. Bendjillali et al. [10] proposed algorithm which does enhancement using modified contrast limited adaptive histogram equalization algorithm (M-CLAHE) [11] and deep CNN architectures like VGG16, ResNet50 which is majorly designed for making face recognition robust to illumination variations. As we all know, post-COVID masks will be a common thing at work places and that time another challenge of detecting a person in spite of wearing a mask, existing approaches would not be enough to overcome this challenge. Ejaz and Islam [12] have tried faces with masks, in which they used MTCNN [7] for face detection, FaceNet [9] for face embeddings, and SVM [13] for classification. Ejaz and Islam have done experimentation on standard datasets which have mixed images. We have compared the same with our approach in which we have used LDA [14] for classification, giving decent accuracy of 94% which outperforms all existing approaches in the domain of face recognition of masked faces. We have also made our own dataset for training purposes. Our architecture is novel because we only take the upper half of face for embedding purposes, so that forehead and eyes features are taken into consideration rather than mask type and color which may rather act as noise for actual recognition.

3 Dataset

Although it was hard to find the labeled dataset of images with people wearing masks, we have created our own dataset by taking mask images from our friends and relatives for training the model. The dataset has been divided to training set and testing set. Both have images of 49 people wearing face masks. The images looked quite alike, so to create variation and a diversified dataset, images have been augmented by changing angle, height, and width. Finally, a new dataset has been generated with better chances of training the model. The dataset consists of two subparts, training set and testing set. The training has 20 images per person with masks on their faces with different variations. The other is the testing set with 10 images per person with masks to actually test that the person wearing mask belongs to any one of the labels in training set.

4 Proposed Architecture

In the current paper, we developed a system which detects the face of the person with a mask. In our system architecture, the first stage is the augmentation of an image dataset using Image-DataGenerator class in Python. After the initial stage, the augmented image is passed through the MTCNN model for face detection purpose, and extracted face image will be the output of that model. Then, FaceNet is used for converting the detected image into an embedded vector. After that we have just preprocessed our dataset and splitted them into train and test images. For the classification, linear discriminant Analysis (LDA) is used. The overview of current system is demonstrated in Fig. 1.

5 Proposed Work

This work is an improved version of a preexisting recognition system which took full face as input. But as we know post-COVID era people will be wearing masks and as for recognizing person masks, we have come with architecture which would take features of eyes and forehead features and will generate encoding using FaceNet model architecture.

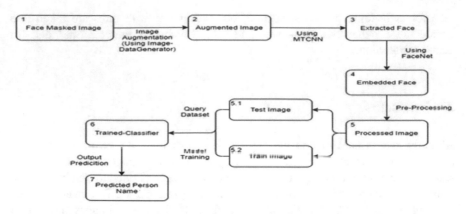

Fig. 1 Architecture diagram

5.1 MTCNN (Multi-task Cascaded Convolutional Neural Network)

The MTCNN works by extracting face from given filename and then converting to RGB, getting the pixel coordinates which include center coordinates, detected face height and width, obtained and then we have done height/2 for a portion of the face above mask. It is resized to (160, 160). After resizing, store the resized image into trainX, testX variable and store class name in trainY, testY and finally store all variables in a .npz file. So up till now, we have extracted the face upper part only without a mask consisting of eyes and forehead.

5.2 FaceNet Model

Now creating face embedding comes into play, and we had done this part using FaceNet. So in face embedding, features are extracted from the face and are represented in the form of a 128 element vector. Later on, with the help of these vectors distinguishing between 2 persons can be done. Suppose two-element vectors of the same person will be similar, whereas for different people, element vectors will vary a lot. The work of FaceNet in our project is simple, it will generate the embedding for a given face in the form of an element vector and will store them with the name of the person whose face is embedded. After MTCNN is done, load the detected faces .npz file using load function in NumPy library and again store values in train and test. Loading the pretrained FaceNet keras model is the easiest step as it helps to generate embeddings for face. Now scale the pixel values and then standardize them, so that they can meet the input expectations of the FaceNet model. Expand the dimensions and then use the "facenet_keras.h5" model for prediction and extraction of the resultant embedding/element vector. The function will receive the model as well as detected face as input and will return the embedded vector. New variable created after getting embeddings each embedding of (128X1). Now the most important steps, i.e., face classification, come. Under this, we will fit a different classification model using Pycaret on our dataset. So, firstly we loaded face-embeddings.npz file, and then, we developed our classification model. But before developing our classification model, we required to perform some data preprocessing.

5.3 Data Preprocessing

Under data preprocessing, do normalization by using normalizer class which scaled our values to unit length. This normalization is performed both on training dataset as well as testing dataset. After normalization, label encoding is done using the LabelEncoder class. Label encoding converts our string target variables, i.e., name of the person into some integers. This leads to our end of data preprocessing. Now comes the final part which is developing a model and fitting that model to our train dataset and then predicting the accuracy score.

5.4 Models Using Pycaret

Using Pycaret which is an open-source machine learning library, it is simple and easy to use. It allows us to compare different machine learning models and fine-tune them very easily. Initialize the models by creating them and tune them to give best results, as Pycaret itself does grid search to find the best parameters in case of getting maximum accuracy. Now just compare all the classification models to check which model did its job best by doing the most accurate prediction of the person wearing the mask just by inputting a masked-image of them.

6 Result and Discussion

As of now, our proposed architecture has outperformed all existing approaches with very good accuracy in spite of the mask on the face. We have experimented with several algorithms and some of our key observations are recorded in Table 1.

After successful experimentation on several algorithms and with various approaches, we have found that linear discriminant analysis is performing considerably well in comparison with others. The top-five models, linear discriminant analysis, SVM–Linear kernel, extra trees classifier, ridge classifier, K-neighbors classifier, gave accuracies of: 94.81, 84.99, 83.88, 83.34, and 81.23%, respectively in Fig. 2.

Table 1 Comparison of different classification algorithm

Model	Accuracy	Recall	Prec	F1	Kappa
LDA	0.9481	0.9492	0.9586	0.9451	0.9468
SVM–Linear	0.8499	0.8465	0.8742	0.8373	0.8462
Extra trees	0.8388	0.8345	0.8645	0.8282	0.8349
Ridge classifier	0.8334	0.8283	0.8479	0.8151	0.8293
K-neighbors	0.8123	0.8109	0.8257	0.7957	0.8077
Logistic regression	0.7958	0.7907	0.8091	0.7764	0.7907
Naive Bayes	0.7958	0.7965	0.8295	0.7828	0.7908

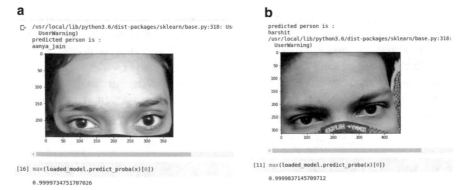

Fig. 2 **a** Tested image 1 **b** Tested image 2

Above are some results for the tested images with the name of the predicted person and the prediction probability.

7 Conclusion

After experimenting with several approaches as discussed above, the paper finally concludes that with the use of MTCNN for face extraction followed by FaceNet for getting the embeddings of the extracted face, and then, different classification algorithms were used, and the best accuracy was given by linear discriminant analysis (LDA) of 94%. The model is independent of the mask shape, size, color, design, etc. It will predict more accurately irrespective of the person having a mustache or beard. Future scope of our system is to work with more datasets and try for getting better accuracy.

References

1. Parkhi OM, Vedaldi A, Zisserman A (2020) Deep face recognition. Accessed: 22 July, 2020
2. Huang Y et al (2020) Improving face recognition from hard samples via distribution distillation loss. In: Vedaldi A, Bischof H, Brox T, Frahm JM (eds) Computer vision—ECCV 2020. ECCV 2020. Lecture notes in computer science, vol 12375. Springer, Cham
3. Wendy S. Yambor Bruce A. Draper J. Ross Beveridge, "Analyzing PCA based Face Recognition Algorithms: Eigenvector Selection and Distance Measures", July 1, 2000.
4. Yambor WS (2000) Analysis of PCA based and fisher discriminant-based image recognition algorithms. M.S. Thesis, July 2000
5. Belhumeur PN, Hespanha JP, Kriegman DJ (1997) Eigenfaces versus Fisherfaces: recognition using class specific linear projection. IEEE Trans Pattern Anal Mach Intell 19(7):711–720
6. Ge S, Li J, Ye Q, Luo Z (2017) Detecting masked faces in the wild with LLE-CNNs. In: 2017 IEEE conference on computer vision and pattern recognition (CVPR), Honolulu, HI, pp 426–434. https://doi.org/10.1109/CVPR.2017.53
7. Zhang K, Zhang Z, Li Z, Qiao Y (2016) Joint face detection and alignment using multitask cascaded convolutional networks. IEEE Signal Process Lett 23(10):1499–1503
8. Yang S, Luo P, Chen C, Tang X (2016) Wider face: a face detection benchmark. In: CVPR
9. Schroff F, Kalenichenko D, Philbin J (2015) FaceNet: a unified embedding for face recognition and clustering. In: 2015 IEEE conference on computer vision and pattern recognition (CVPR), pp 815–823, June 2015. https://doi.org/10.1109/CVPR.2015.7298682
10. Bendjillali RI, Beladgham M, Merit K, Taleb-Ahmed A (2019) Illumination-robust face recognition based on deep convolutional neural networks architectures, vol 18, pp 1015–1027. https://doi.org/10.11591/ijeecs.v18.i2.pp1015-1027
11. Mohan S, Ravishankar M (2013) Modified contrast limited adaptive histogram equalization based on local contrast enhancement for mammogram images. In: Das VV, Chaba Y (eds) Mobile communication and power engineering. AIM 2012. Communications in computer and information science, vol 296. Springer, Berlin, Heidelberg. https://doi.org/10.1007/978-3-642-35864-7_60
12. Ejaz MS, Islam MR (2019) Masked face recognition using convolutional neural network. In: 2019 International conference on sustainable technologies for industry 4.0 (STI), Dhaka, Bangladesh, pp 1–6. https://doi.org/10.1109/STI47673.2019.9068044
13. Evgeniou T, Pontil M (2001) Support vector machines: theory and applications, 2049, pp 249–257. https://doi.org/10.1007/3-540-44673-7_12
14. Prince SJD, Elder JH (2007) Probabilistic linear discriminant analysis for inferences about identity. In: 2007 IEEE 11th international conference on computer vision, Rio de Janeiro, pp 1–8. https://doi.org/10.1109/ICCV.2007.4409052

Speech Audio Cardinal Emotion Sentiment Detection and Prediction Using Deep Learning Approach

Sachit Bhardwaj and Akhilesh Kumar Sharma

1 Introduction

Sentiment analysis is the study of person's emotional state. Speech is one of the most important way by which a person can convey emotions, knowledge and most importantly can express himself. Human mood apprehension can be helpful in dealing many typical situations [1]. A person through this can introspect himself as whether his decisions are driven by emotions or the practicality of the world. Sentiment analysis through speech can be used in the medical field as a disease diagnosing method. For e.g., dyslexia can be diagnosed with sentiment analysis as it is related with various changes in behaviour and mood swings [2, 3].

Emotions combined with several factors can help in detecting or confirming possible theories. Emotions play a vital role in our community as we can have moments when emotional levels get very sensitive. So there sentiment analysis can be helpful in understanding the emotional aspects of a person. With the help of sentiment analysis, agents can be monitored in a company which can give company an insight of the interest level of the employees toward their job. Hence, it will increase the efficiency of the company by helping their emotionally sensitive employees so that they would do their work more happily and productively. By this sentiment analysis through speech, customer care services can predict their customer's mood and can handle the customer accordingly, giving the best smooth experience. Sentiment Analysis through speech have limitless application and can be used in number of fields and hence enhancing the comfort of our lives. There are seven cardinal emotions, i.e. angry, disgust, fear, happy, neutral, pleasant and sad, as shown in Fig. 1.

In this paper, deep learning models like deep neural network and a special case of recurrent neural network—long-term short memory is taken into consideration

S. Bhardwaj · A. K. Sharma (✉)
Department of Information Technology, Manipal University Jaipur, Jaipur, Rajasthan, India

© The Author(s), under exclusive license to Springer Nature Singapore Pte Ltd. 2022   111
R. R. Raje et al. (eds.), *Artificial Intelligence and Technologies*,
Lecture Notes in Electrical Engineering 806,
https://doi.org/10.1007/978-981-16-6448-9_13

Fig. 1 Cardinal emotions

for the experimentations. In addition to this, support vector machine's special cases like polynomial, Gaussian and sigmoid variants have been used to obtain the best result possible.

2 Literature Review

Volkmen et al. discussed in their research the measurement of the psychological magnitude pitch. The researchers predicted the scale for measurement of the pitch from the determinations of the half-value of pitches at different frequencies [4]. Deller et al. in their research proposed the methods and approaches toward the discrete time signals, mainly giving the insights of behind the mathematical portions of Mel frequency, DFT and various methods [5]. In this study, while comparing different implementations of MFCC, the authors predicted that the performance of Mel frequency cepstral coefficients (MFCC) by factors such as number of filters, cut off filters, orientation of the filters and the way power spectrum is developed has been considered [6]. This research work studies the performance of MFCC on speaker verification. The authors have proposed the filter bank design, approximation of the nonlinear pitch perception by human and the filter bank output compression [7].

   Maghilnan et al. in their research work analysed the performance of algorithms to predict the sentiment of the speakers indulged in the conversation [1]. Carolien et al. in their research work proposed a method by applying various techniques on dyslexia students to observe audio-support affecting learning similarity in students with and without dyslexia [2, 3].

   Kurpukdee et al. in their research work compared the performance between support vector machines (SVM) and binary support vector machines (BSVM) approaches in emotion recognition [8]. Abburi et al. in their research work used

Gaussian mixture models (GMM) and support vector machine (SVM) classifiers for the prediction of emotions with addition of extracting features from specific regions [9, 10].

Salah et al. in their research work proposed a debate graph extraction (DGE) framework to extract the features from the transcripts of political debates (positive and negative). Through these discussed features, the sentiment analysis is carried out with machine learning techniques [11].

Pan et al. in their research work proposed an emotion-detection model with overall mean recognition rate an error rate reduction [12]. Bertero et al. build a corpus from the punchlines and made a comparative study between a recurrent neural network (RNN) and a convolutional neural network (CNN) [13].

Zhao et al in their constructed a 1D CNN-LSTM network and a 2D CNN-LSTM network with global sentiment-related attributes from speech and log-mel spectrogram, respectively [10, 13]. A combination of useful techniques have been manoeuvred in this research work after scrupulously going through these relevant work. The later has been meticulously discussed in the following section, i.e. methodology.

3　Methodology

In the proposed methodology, a dataset produced by University of Toronto which available on Kaggle has been used. Dataset consisted over 1400 mini audio clips of a person expressing seven cardinal emotions, i.e. angry, disgust, fear, happy, neutral, pleasant and sad. The voice in this dataset is of a female actor aged 64 years expressing emotions over 200 target words. Detailed methodology is shown in Fig. 2.

This study is performed on Google Colaboratory environment which provides Intel(R) Xeon(R) CPU @2.20 GHz with 13 GB RAM.

3.1　Mel Frequency Cepstral Coefficients (MFCCs)

In this study, Mel frequency cepstral coefficients (MFCCs) has been used to extract features from the audio clips. MFCC is observed as compact description of spectral envelope's shape. MFCC feature extraction is performed in five steps as follows.

Pre-emphasis The higher frequencies are filtered in this step to ensure the balance of voiced sounds that have a steep roll off in high frequency region. The most often implemented pre-emphasis filter [$H(z)$] is as shown in Eq. 1. The waveform of an audio clip is shown in Fig. 3.

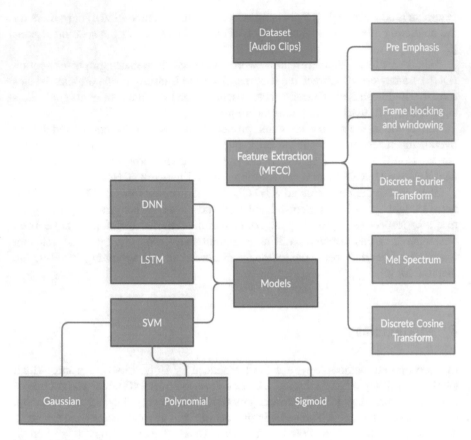

Fig. 2 Methodology

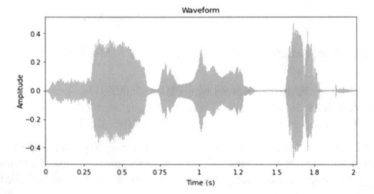

Fig. 3 Waveform

$$H(z) = 1 - bz^{-1} \tag{1}$$

where b is the slope of filter.

Frame Blocking and Windowing The Audio is divided into various segments in order to have a windowed section which helps in scrutinisation the slowly time varying or quasi-stationary signal. This helps in enhancing the harmonics of the audio signal. It also smoothers the edges and is observed to be reducing edge effect while taking Fourier transform of the signal. The window length is assigned as 512 in this study.

Discrete Fourier Transform Any sampled signal of length N can be represented uniquely and unambiguously by a finite series of sinusoids as shown in Eq. 2. This transformation is known as Fourier transform denoted as $X(k)$. The each windowed frame is converted into spectrum by the application of discrete Fourier transform. The power spectrum is shown in Fig. 4.

$$X(k) = \sum_{n=0}^{N-1} x(n) e^{\frac{-j2\pi nk}{N}}; \quad 0 \leq k \leq N - 1 \tag{2}$$

where N is number of points.

Mel Spectrum Processed DFT signal is passed through Mel filter bank (a set of band-pass filters). The applied filter in this study is shown in Fig. 5. Human ear frequency perceiving is the unit of measure for a Mel. The Mel scale is in the region of logarithmic spacing above 1 kHz and a linear frequency spacing below 1 kHz [4]. The scale as shown in Eq. 3.

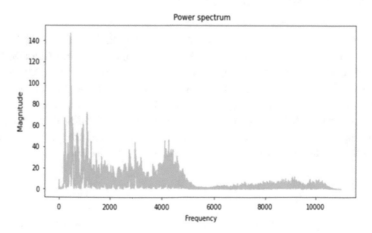

Fig. 4 Power spectrum

$$f_{\text{Mel}} = 2595 \log_{10}\left(1 + \frac{f}{700}\right) \tag{3}$$

where f denotes the physical frequency in Hz and f_{Mel} denotes the perceived frequency [5].

The Mel spectrum $[s(m)]$ is computed by multiplying the magnitude spectrum by each of the triangular Mel weighting filters which has been showed in Fig. 6.

$$s(m) = \sum_{k=0}^{N-1}\left[|X(k)|^2 H_m(k)\right]; \quad 0 \leq m \leq M - 1 \tag{4}$$

where M is total triangular Mel weighting filters and H_m is weight to kth energy spectrum bestowing to mth output band.

Discrete Cosine Transform The Mel spectrum is represented on logarithmic scale followed by the application of discrete cosine transform (DCT) which produces a set of cepstral coefficients as shown in Eq. 5.

$$c(n) = \sum_{m=0}^{M-1} \log_{10}(s(m)) \cos\left(\frac{\pi n(m - 0.5)}{M}\right) \tag{5}$$

where $n = 0, 1, 2, \ldots, C - 1, c(n)$ are the cepstral coefficients and C is the number of MFCCs. MFCCs obtained on performing it on an audio clip is shown in Fig. 7.

3.2 Deep Neural Network (DNN)

A three-layer deep neural network (DNN) is used with MFCCs as inputs. ReLU activation function is used in the all three layers while activation function used in the output layer is Softmax. A dropout layer is used in order to prevent overfitting of the model. Adam optimiser is used for optimising the model which uses momentum and adaptive learning to converge faster. Adam integrates the qualities of AdaGrad and RMSProp algorithms to provide an optimisation algorithm perform very well sparse gradients on noisy data. Sparse categorical crossentropy was used as loss function and update rule for AdaGrad as shown in Eq. 6.

$$v_t^w = v_{t-1}^w + (\delta w_t)^2 \tag{6}$$

It is clear from update_rule that gradient to be accumulated in v and w is the weight.

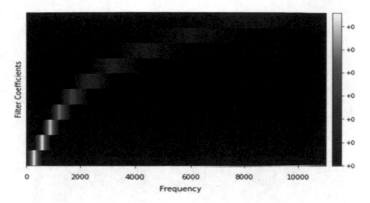

Fig. 5 Filter applied

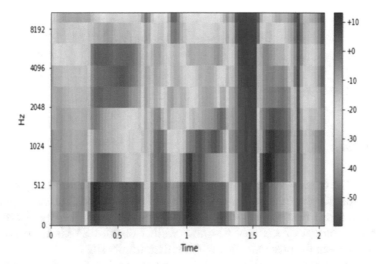

Fig. 6 Mel spectrum

3.3 Long-Term Short Memory-RNN (RNN-LSTM)

A two-layer RNN-LSTM was used with MFCCs as inputs. Activation function ReLU was used in the first layer followed by a dropout layer to prevent the overfitting of the model. The output layer uses activation function Softmax. Adam optimiser is used for optimising the model which uses momentum and adaptive learning to converge faster. Adam integrates the qualities of AdaGrad and RMSProp algorithms to provide an optimisation algorithm that perform very well sparse gradients on noisy data. Sparse categorical crossentropy was used as loss function as shown in Eq. 7.

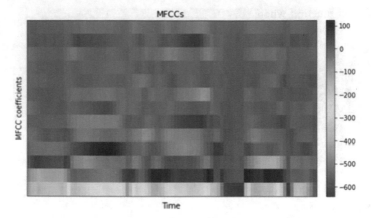

Fig. 7 MFCCs

$$\mathrm{CCE}(p,t) = -\sum_{c=1}^{C} t_{o,c} \log(p_{o,c}) \tag{7}$$

where p is prediction and t is target.

3.4 SVM

A linear SVM (that is, a model that implies that only the data is linearly separable) is to demonstrate the basic ideas. In order to illustrate the position of the mapping function, we will then extend the example to the nonlinear case and eventually explain the concept of a kernel and how well it enables SVMs to make use of high-dimensional features extracted while remaining feasible.

The support vector machine revolves around the concept of the hyperplane where the n-dimensional space can be categorised into various classes and the best decision boundary is to be identified. The data will be available in between the positive hyperplane and the negative hyperplane also known as the support vectors passing lines. Support vector machine can be categorised into linear as well as nonlinear SVM.

SVM uses the slightly right points for notifications along the judgement frontier. Notice that the SVM uses the $y(w \cdot x) < 1$ [instead of using the $y(w \cdot x) < 0$] condition, which is one of the primary aspects of this algorithm. The linear and nonlinear separation is handled differently in SVM. The transposition function works as a kernel function to separate the different classes apart from each other. Many known kernels were practical in use in various applications that served this purpose.

If the linear data is available, the linear classification can be done, but if the data items available are not linearly separable or nonlinear, we need to identify the transposition function where the data can be transferred into the third access using any transposition function. The benefit of this transposition function is that we can map the data points into the third axis in the 3D plane, and then, we can separate these transpose data using a straight line. And then, the true positive and false positive can be measured, and the accuracy can be calculated while identifying these data points into the confusion matrix-based performance evaluation where b is bias and w is the width.

$$h(x_i) = \left\{ \begin{array}{ll} +1 & \text{if } w \cdot x + b \geq 0 \\ -1 & \text{if } w \cdot x + b < 0 \end{array} \right\} \tag{8}$$

In terms of the SVM optimisation problem,

$$f(w) = \frac{1}{2} \|w\|^2 \tag{9}$$

$$g(w, b) = y_i(w \cdot x + b) - 1 \tag{10}$$

where $i = 1, \ldots, m$, the Lagrangian function is then

$$\mathcal{L}(w, b, \alpha) = \frac{1}{2} \|w\|^2 - \sum_{i=1}^{m} \alpha_i [y_i(w \cdot x + b) - 1] \tag{11}$$

4 Result

After training, deep neural network performed the best with 97.35% accuracy followed by LSTM with 94.08% accuracy. Gaussian-SVM performed the best among other kernels with 74% accuracy followed by polynomial-SVM with 67% accuracy and Sigmoid-SVM with 31% accuracy. Therefore, DNN performed the best, while Sigmoid-SVM performs the worst. Figure 8 shows the performance, and Tables 1 and 2 show the accuracy.

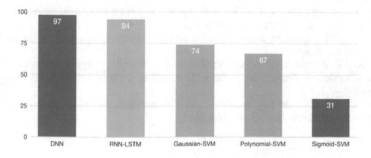

Fig. 8 Accuracy of models

Table 1 Comparison of models with parameters

Model	Adam learning rate	Batch size	Epochs	Accuracy (%)	Validation accuracy (%)
DNN	0.0001	32	200	97.33	89.29
	0.001	32	200	97.35	88.62
	0.0001	64	200	94.07	89.40
LSTM	0.0001	32	200	94.08	84.05
	0.001	32	200	91.94	84.57
	0.0001	64	200	91.50	84.43

Table 2 Accuracy of models

Model	Accuracy (%)
DNN	97.35
RNN-LSTM	94.08
Gaussian-SVM	74
Polynomial-SVM	67
Sigmoid-SVM	31

References

1. Maghilnan S, Kumar MR (2017) Sentiment analysis on speaker specific speech data. In: 2017 International conference on intelligent computing and control (I2C2), Coimbatore, pp 1–5
2. Campen CA, Segers E (2020) Effects of audio support on multimedia learning processes and outcomes in students with dyslexia. Comput Educ 150:2020
3. Knoop-van Campen CAN, Segers E, Verhoeven L (2020) Effects of audio support on multimedia learning processes and outcomes in students with dyslexia. Comput Educ 150
4. Stevens S, Newman EB (1937) A scale for the measurement of the psychological magnitude pitch. J Acoust Soc Am 8:185–190
5. Deller JR, Proakis JG, Hansen JH (1993) Discrete time processing of speech signals, 1st edn. Prentice Hall PTR, USA
6. Fang Z, Guoliang Z, Zhanjiang S (2000) Comparison of different implementations of MFCC. J Comput Sci Technol 16:582–589
7. Ganchev GKT, Fakotakis N (2005) Comparative evaluation of various MFCC implementations on the speaker verification task. In: Proceedings of international conference on speech and computer (SPECOM), pp 191–194
8. Abburi H, Gangashetty S et al (2017) Audio and text multimodal sentiment analysis using features extracted from selective regions deep neural networks
9. Salah Z (2014) Machine learning and sentiment analysis approaches for the analysis of parliamentary debates
10. Sharma AK, Panwar A, Chakrabarti P, Vishwakarma S (2015) Categorization of ICMR using feature extraction strategy and MIR with ensemble learning. Procedia Comput Sci 57 (201):686–694
11. Pan YC, Xu MX, Liu LQ, Jia PF (2006) Emotion-detecting based model selection for emotional speech recognition, pp 2169–2172

12. Bertero D, Fung P (2016) Deep learning of audio and language features for humor prediction. In: Proceedings of the tenth international conference on language resources and evaluation (LREC'16)
13. Zhao J, Mao X, Chen L (2019) Speech emotion recognition using deep 1D & 2D CNN LSTM networks. Biomed Signal Process Control 47:312–323

Comparative Investigation on Acoustic Attributes of Healthy Young Adults

V. Prarthana Karunaimathi, D. Gladis, and D. Balakrishnan

1 Introduction

Human voice, which is one's unique identification, is used for verbal communication, to create wonderful music through singing, and to express one-self. It is produced by a complicate mechanism which is subdivided into three main components. They are (1) the lungs, (2) the vocal folds within the larynx, and (3) the articulators. The voice can be affected by many disorders which are grouped into two classes such as organic and functional. Organic voice disorders are caused by the physical changes in the voice mechanism. It includes edema [1], vocal nodule which is caused due to the variation in vocal fold tissues [2], aging, which affects the structure of the larynx [3], and these impairments come under the category of structural voice disorder. On the other hand, neurogenic disorder includes vocal tremor [4], dysphonia [5], and paralysis of vocal fold [6], caused by the problems with the central nervous system. Functional voice disorder includes vocal fatigue [7], hyperkinetic dysphonia [8], diplophonia [9] and ventricular phonation [10]. Human emotions, moods, and inflection also cause voice variation [11]. In order to make a diagnosis of these impairments, a voice analysis application provides the pathologists to visualize and manipulate the voice of the patients. The speech signals, which are obtained from the subjects, are analyzed using these applications. This includes the extraction of the objective vocal features such as fundamental frequency, harmonics-to-noise ratio, and the percentage of jitter and shimmer. In

V. P. Karunaimathi (✉)
PG and Research Department of Computer Science, Presidency College,
University of Madras, Chennai, Tamil Nadu, India

D. Gladis
Bharathi Womens' College, University of Madras, Chennai, Tamil Nadu, India

D. Balakrishnan
Department of Audiology and Speech Pathology,
SRM Medical College, SRM University, Kancheepuram, Tamil Nadu, India

this study, a novel algorithm leading to an application named Ephphatha extracts the acoustic parameters, and the same were compared with the existing applications namely Dr. Speech and Praat.

2 Data Collection

The speech signals for analysis were collected from 19 female and 8 male volunteers without any voice impairment who are all in the age group of 18–25 years. The recordings were done in the speech laboratory with a microphone at a distance of 10 cm from the volunteers' lips. The tasks such as sustained vowel phonations //a//, //i//, and //u//, and rainbow passage reading were considered and recorded on their comfortable pitch without any interference. Speech signals were sampled at a rate of 44.1 kHz and 16 bits. Then, the collected voice signals were analyzed with Dr. Speech, Praat, and Ephphatha.

3 Feature Extraction

The assessment of vocal disorders is typically achieved using either sustained vowel phonations or running speech. Clinical practice has shown that the use of sustained vowels is very practical and sufficient for the assessment of many vocal disorders. In this study, the sustained vowel phonation /a/ is used at a comfortable pitch as long as possible, and as steady as possible for the parameter extraction.

3.1 Fundamental Frequency F0

F0 estimation, methods can be done in time and frequency domain or in the combination of both. Time-domain approaches include auto-correlation and cross-correlation. The frequency domain includes spectral [12] and cepstral approaches [13]. In this study, the correlation functions were analyzed. They are auto-correlation function (ACF), cross-correlation function (CCF), and normalized cross-correlation function (NCCF).

Auto-Correlation Function (ACF): Serial correlation is used for the correlation of a time-domain signal and its time delayed version of the same signal. Fundamental frequency can be extracted using this function. Consider s_p, $p = 0, 1, 2, \ldots,$ as a sampled speech signal with sampling interval $T = 1/F_s$. t and w are the frame interval and window size for analysis, respectively. Outside the window, s is assumed to be zero. The auto-correlation function is shown in Eq. (1).

$$R_{i,k} = \sum_{j=m}^{m+n-k-1} S_j S_{j+k} \tag{1}$$

where $k = 0, K - 1$; $m = iz$; $i = 0, M - 1$. i is the frame index for m frames, and k is the lag index or lag. ACF performs well, but still it has some notable drawbacks. In case of swift F0 changes, it loses the clear peaks, since it preludes the cycle-to-cycle variation in shorter periods. Praat [14] system uses this function as default for pitch extraction. In this system, the auto-correlation function of the windowed signal is divided by the auto-correlation function of the window, in order to overcome the shortcomings of the other existing methods. It is shown in the below Eq. (2).

$$r_x(\tau) \approx r_{wx}(\tau)/r_w(\tau) \tag{2}$$

It also includes cross-correlation method which can be chosen for the estimation of fundamental frequency. The dynamic programming namely shortest path is used as a post-processing algorithm in candidate generation in order to exclude the octave errors in F0 determination. The auto-correlation algorithm is moderately noise resistant, but it is very sensitive to sampling rate. Also the computational cost is relatively high. Vocal assessment module of Dr. Speech 3.0 (Tiger Electronics, Inc., USA) provides both acoustic and EGG feature analysis for sustained vowel phonations. It also contains phonetogram, which displays the dynamic range of voice in terms of pitch and volume. In this system, the signal is downsampled and clipped using three-level clipping followed by segmentation with fixed window size. Finally, auto-correlation is performed on the windowed clipped signal to estimate the F0. Post-processing method median smoothing is applied before the visualization of F0 contour [15].

Cross-Correlation Function (CCF): It is used to determine the correlation between two different signals. In this function, the analysis window size is based on the order of the single average glottal period. Therefore, the correlation interval is independent of the candidate search interval. It can be expressed as in Eq. (3).

$$X_{i,k} = \sum_{j=m}^{m+n-1} S_j S_{j+k} \tag{3}$$

In case of swift amplitude variation, normalization is insufficient to allow reliable candidate selection using sample threshold logic.

Normalized Cross-Correlation Function (NCCF): In Ephphatha, the F0 values expressed in Hz are estimated in robust algorithm for pitch tracking (RAPT) framework, which is based on NCCF [16, 17]. It solves the problem of candidate selection drawbacks occurred in the above correlation functions by increasing the computational cost slightly. Here, the window size is chosen to be in the neighborhood of the expected F0 period. The NCCF function is expressed as in Eq. (4).

$$\varphi_{i,k} = \frac{\sum_{j=m}^{m+n-1} S_j S_{j+k}}{\sqrt{e_m e_{m+k}}} \tag{4}$$

where

$$e_j = \sum_{l=j}^{j+n-1} s_l^2 \tag{5}$$

In this framework, the window size is also independent of the F0 range. It uses two versions of data signals, in which one is at the original sampled rate (44.1 kHz) and other at the reduced rate approximately 2 kHz. The peaks are refined by allowing two passes, in which the second pass searches the location which were already computed in the first pass. The improved peak is hence obtained in this framework and finally gives a candidate F0 for that frame.

Figure 1 shows the waveform representation and the F0 contour of sustained vowel phonation for the same subject using the three applications. It reveals that the pitch sigma is in control, since there are no much visible spikes or octave jumps in the contour.

3.2 Frequency and Amplitude Instability Measure

Jitter and shimmer are the two major features of quality of voice in acoustic analysis, which refers to the instability of frequency and amplitude, respectively.

Jitter% Estimation: It represents the average absolute difference between two successive periods [18] as shown in Eq. (6). Its percentage is calculated by dividing the JitterAbs by the mean period as shown in Eq. (7).

$$\text{JitterAbs} = \frac{1}{N-1} \sum_{i-1}^{N-1} |P_i - P_{i-1}| \tag{6}$$

$$\text{Jitter\%} = \frac{\text{JitterAbs}}{\frac{1}{N} \sum_{i-1}^{N} P_i} \times 100 \tag{7}$$

Higher jitter value is the result of uncontrolled vocal fold vibration. The vocal cord pathologies, including nodules, polyps, and weakness of the laryngeal muscles can be determined by the higher percentage of jitter [19].

Applications	Waveform and F0 Contour
Ephphatha	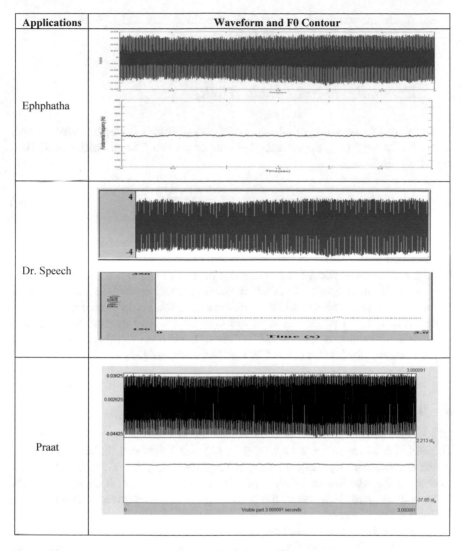
Dr. Speech	
Praat	

Fig. 1 Visualization of waveform and the F0 contour of sustained vowel //a// phonated by a female subject in Ephphatha, Dr. Speech, and Praat application

Shimmer% Estimation: It is defined as the average absolute difference between the amplitudes of successive periods, divided by the average amplitude [18]. It is expressed in percentage as shown in Eq. (8):

$$\text{Shimmer}\% = \frac{\frac{1}{N-1}\sum_{i=1}^{N-1}|A_i - A_{i+1}|}{\frac{1}{N}\sum_{i=1}^{N}A_i} \times 100 \qquad (8)$$

where A_i are the extracted peak-to-peak amplitude data and N is the number of extracted fundamental frequency periods.

3.3 Harmonics-to-Noise Ratio (HNR)

It represents the acoustic periodicity, which is expressed in decibel (dB). Higher level of noise represents the pathological state of a voice. The estimation of HNR is shown in Eq. (9).

$$\text{HNR} = 10 * \log_{10} \frac{\text{AC}_v(T)}{\text{AC}_v(0) - \text{AC}_v(T)} \tag{9}$$

$\text{AC}_v(T)$ and $\text{AC}_v(0)$ are the components of auto-correlation of all energy and the fundamental period energy of the given signal, respectively. The difference between them gives the noise energy of the signal. For a healthy subject, the value will be around 20 dB in the task of sustained vowel phonation //a//. In this task, both signal-to-noise ratio (SNR) and HNR are almost equal. Subjects with hoarse voice reports to have lower periodicity level less than 20 dB. Also its escalated value acts as an evident for elderly speakers [20]. This quantified noise value is hence useful in the treatment for hoarseness and vocal fold disorders [21].

4   Results and Discussion

The estimated parameters were stored in Comma Separated Value (CSV) files, which were statistically analyzed using R Programming [22] and t-test [23] in order to observe the significance level, which are given in Table 1. It also includes the mean and standard deviation of the parameters, t-values, degrees of freedom, and 95% confidence interval.

As seen in Table 1, the t-test results show that there is no significant difference in F0 parameters. The F0 extracted in both the applications are almost equal, since their significance levels are around 0.9 which is close to 1. In case of jitter and HNR, they are moderately correlated. But the amplitude perturbation values seem to be weakly correlated.

As seen in Table 2, the results obtained show no significant difference in fundamental frequency parameters except its standard deviation. The F0 values are almost equal in both the applications since their significance levels are around 0.9, which is close to1. In F0SD, the values are little deviated, but still there is good correlation between both the applications. The percentage of jitter and shimmer values are found to be moderately correlated with the p-value ranging from 0.04 to 0.06. HNR values are strongly correlated in both the applications, with the significance level greater than 0.05.

Table 1 Statistical results of acoustic parameters extracted using Ephphatha versus Dr. Speech

Parameters	Mean ± SD	t	Df	p-value	95% confidence interval	
EphMeanF0	197.01 ± 51.91	−0.0036	52	0.9971	−28.4039	28.3011
DrSMeanF0	197.06 ± 51.92			>0.05		
EphF0SD	1.42 ± 0.58	−0.1983	51.99	0.8436	−0.35050	0.2874
DrSF0SD	1.45 ± 0.59			>0.05		
EphMaxF0	200.73 ± 52.96	−0.0411	52	0.9673	−29.6426	28.4513
DrSMaxF0	201.32 ± 53.42			>0.05		
EphMinF0	193.29 ± 50.83	0.0372	52	0.9705	−27.1352	28.1607
DrSMinF0	192.78 ± 50.42			>0.05		
EphJitter%	0.303 ± 0.06	4.0608	52	0.000165	0.0360	0.1064
DrSJitter%	0.232 ± 0.065			<0.001		
EphShimmer%	5.986 ± 2.09	8.4986	34.73	5.32E−10	2.8181	4.5876
DrSShimmer%	2.283 ± 0.869			<0.001		
EphHNR	18.99 ± 2.21	−4.2546	44.14	0.000107	−4.9584	−1.7710
DrSHNR	22.36 ± 3.46			<0.001		

Table 2 Statistical results of acoustic parameters extracted using Ephphatha versus Praat

Parameters	Mean ± SD	t	Df	p-value	95% confidence interval	
EphMeanF0	197.01 ± 51.91	−0.0010	52	0.9992	−28.3628	28.3343
PraatMeanF0	197.03 ± 51.91			>0.05		
EphF0SD	1.42 ± 0.58	0.5520	51.99	0.5833	−0.2306	0.4056
PraatF0SD	1.33 ± 0.58			>0.05		
EphMaxF0	200.73 ± 52.96	0.0218	52	0.9827	−28.5706	29.1991
PraatMaxF0	200.41 ± 52.82			>0.05		
EphMinF0	193.29 ± 50.83	−0.0200	52	0.9841	−28.0372	27.4833
PraatMinF0	193.57 ± 50.83			>0.05		
EphJitter%	0.303 ± 0.06	−2.0782	42.67	0.0437	0.0972	−0.0015
PraatJitter%	0.353 ± 0.11			<0.05		
EphShimmer%	5.986 ± 2.09	1.9138	50.54	0.0613	−0.0588	2.4492
PraatShimmer%	4.791 ± 2.48			>0.05		
EphHNR	18.99 ± 2.21	−0.7433	49.39	0.4608	−1.8860	0.8673
PraatHNR	19.50 ± 2.79			>0.05		

Figure 2 shows the correlation between the acoustic parameters determined using Ephphatha, Dr. Speech, and Praat. The goodness of fit in the above regression model is shown by the coefficient of determination (R^2). It is given as in Eq. (10)

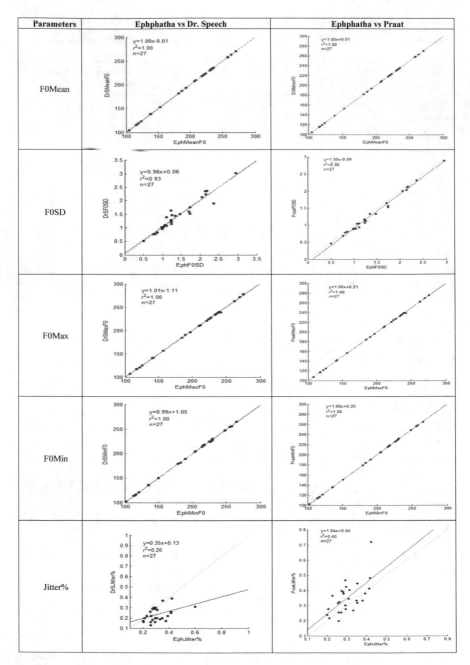

Fig. 2 Correlative visualization of acoustic characteristics between the newly developed application Ephphatha and the existing software Dr. Speech and Praat

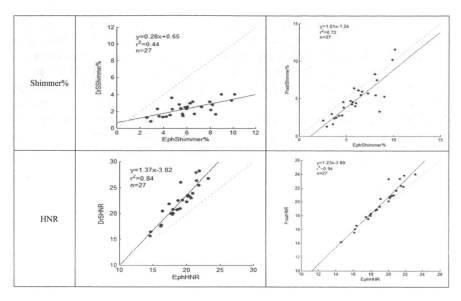

Fig. 2 (continued)

$$\text{R-Squared} = 1 - \frac{SS_{regression}}{SS_{total}} \tag{10}$$

where $SS_{regression}$ is the sum of squares due to regression and SS_{total} is the total sum of squares. The parameter F0 values are perfectly and positively correlated both in Dr. Speech and Praat, when compared with Ephphatha, since their R-Squared value is close to 1. Also the regression model reveals that the values of HNR are positively correlated in both Dr. Speech and Praat when compared with Ephphatha. The perturbation measures are weakly correlated between Ephphatha and Dr. Speech with the coefficient of determination ranging from 0.20 to 0.44. The correlation plot clearly shows that the Praat system is more comparable with the newly developed application Ephphatha than Dr. Speech.

5 Conclusion

The acoustic attributes which quantifies the voice quality and regularity plays a major role in the clinical analysis of voice. In this paper, the parameters were estimated using a newly developed algorithm viz. Ephphatha under MATLAB environment, and the same were compared with the existing software Dr. Speech and Praat which are currently being used in the speech laboratories. The sustained vowel phonation task //a// was taken into consideration for the signal processing technique applications, through which the attributes were extracted. The results of regression model and the statistical analysis *t*-test illustrate that in all the applications, the F0 values are satisfactorily correlated. There is no significant difference between all the acoustic characteristics of Ephphatha and Praat, except the feature of voice amplitude instability. The voice quality measures such as jitter%, shimmer % and HNR are weakly correlated between Ephphatha and Dr. Speech. Hence, the overall results depicts that the newly developed application's parameters are positively correlated with the existing software Praat than Dr. Speech.

References

1. Jackson-Menaldi CA, Dzul AI, Holland RW (1999) Allergies and vocal fold edema: a preliminary report. J Voice 13(1):113–122
2. Lancer JM, Syder D, Jones AS, Le Boutillier A (1988) Vocal cord nodules: a review. Clin Otolaryngol Allied Sci 13(1):43–51
3. Garvin HM (2008) Ossification of laryngeal structures as indicators of age. J Forensic Sci 53(5):1023–1027
4. Barkmeier-Kraemer J (2012) Updates on vocal tremor and its management. Persp Voice Voice Disorders 22(3):97–103
5. Hartman DE (1984) Neurogenic dysphonia. Ann Otology, Rhinology Laryngology 93(1): 57–64
6. Toutounchi SJS, Eydi M, Golzari SE, Ghaffari MR, Parvizian N (2014) Vocal cord paralysis and its etiologies: a prospective study. J Cardiovas Thoracic Res 6(1):47
7. Sander EK, Ripich DE (1983) Vocal fatigue. Ann Otology, Rhinology Laryngology 92(2): 141–145
8. Altman KW, Atkinson C, Lazarus C (2005) Current and emerging concepts in muscle tension dysphonia: a 30-month review. J Voice 19(2):261–267
9. Dejonckere PH, Lebacq J (1983) An analysis of the diplophonia phenomenon. Speech Commun 2(1):47–56
10. Maryn Y, De Bodt MS, Van Cauwenberge P (2003) Ventricular dysphonia: clinical aspects and therapeutic options. The Laryngoscope 113(5):859–866
11. Johar S (2015) Emotion, affect and personality in speech: the bias of language and paralanguage. In: Springer briefs in speech technology. Springer
12. Noll AM (1970) Pitch determination of human speech by the harmonic product spectrum, the harmonic sum spectrum and a maximum likelihood estimate. In: Proceedings of the symposium on computer processing in communications, vol XIX, Polytechnic Press, Brooklyn, New York, pp 779–797
13. Noll AM (1967) Cepstrum pitch determination. J Acous Soc Am 41(2):293–309
14. Boersma P (2001) Praat, a system for doing phonetics by computer. Glot International 5:341–345
15. Awan SN, Scarpino SE (2004) Measure of vocal F0 from continuous speech samples: an interprogram comparison. J Speech-Lang Pathol Audiol 28(3):122–131
16. Gladis D, Dalvi U (2015) A study of F0 estimation based on RAPT framework using sustained vowel. In: International conference on advances in computing, communications and informatics. IEEE, pp 2290–2295
17. Talkin D, Kleijn WB (1995) A robust algorithm for pitch tracking (RAPT). Speech coding and synthesis 495:518

18. Farrús M, Hernando J, Ejarque P (2007) Jitter and shimmer measurements for speaker recognition. In: Eighth annual conference of the international speech communication association
19. Teixeira JP, Oliveira C, Lopes C (2013) Vocal acoustic analysis–jitter, shimmer and HNR parameters Procedia Technol 9:1112–1122
20. Ferrand CT (2002) Harmonics-to-noise ratio: an index of vocal aging. J Voice 16(4):480–487
21. Yumoto E (1983) The quantitative evaluation of hoarseness: a new harmonics to noise ratio method. Arch Otolaryngol 109(1):48–52
22. R Core Team R (2013) A language and environment for statistical computing. In: R Foundation for Statistical Computing, Vienna, Austria. URL http://www.R-project.org/
23. Kim TK (2015) T test as a parametric statistic. Korean J Anesthesiol 68(6):540

Constraint-Based Parallel Clustering with Optimized Feature Selection for SDN-Enabled Traffic Anomaly Detection and Mitigation

T. Vadivu and B. Sumathi

1 Introduction

SDN is a model to support cloud and system engineers including administrators for making quick response to the many industry requirements by means of a unified control system. This helps to sustain the virtualized server and storage configuration of the latest data center. The network-based services through susceptible information are increased extensively with advanced information technologies. At once, end-systems and networks are attacked by different network attacks/anomalies. Those anomalies are detected. Nowadays, the most demanding task is detecting the anomalies since an amount of traffic sent via the networks is high. This problem is tackled by configuring such virtual networks in SDN.

Because of the characteristics of SDN like dynamic behavior, cost-effective, simple use and flexible, it is emerged in the innovated high-bandwidth purposes. It will execute the consistent centralized software-based regulator to facilitate the network's on-demand configurations and its resources [1–3]. Many investigators were experienced in anomaly identification by controlling the abilities of SDN. Among those, a sampled-DP scheme [4] has been developed depending on the sampling and unsupervised cluster-based feature selection (UCFS) approach to handle the huge high-dimensional and unlabeled data. Such approaches were adopted to extract the cluster centroids and outlier spots with the minimum storage demands. Using this scheme, the most relevant attributes were clustered according to their most idleness from every other and the unneeded attributes were eliminated. At last, an intermediate data was fused for obtaining the resultant decision.

Nonetheless, two fixed thresholds were adopted to select the most relevant attributes and eliminating the unneeded attributes. This may affect the accurate selection of the most relevant attributes and accurate outcome. For the real-time

T. Vadivu (✉) · B. Sumathi
Department of Computer Science, CMS College of Science and Commerce,
Coimbatore, India

data analysis, if this threshold was assigned to very low, then the most relevant attributes were ignored. If the threshold was assigned to very large, more noise attributes were decided as the most relevant attributes. So, an optimized feature selection with sampled-DP scheme [5] was suggested in which backward elimination iterative symmetric uncertainty-based FS (BE-ISU-FS) procedure was adopted to decide the minimum subset of attributes by eliminating the unneeded, i.e., less significant attributes from back. But, it does not able to handle the dataset with multiple relevant attributes/features. Thus, an OE-ISU-FS scheme [6] has been suggested using fractional-order Darwinian particle swarm optimization (FODPSO) that enhances the UCFS to detect the traffic anomalies in SDN. Conversely, quick search and discovering of clusters algorithm were applied for clustering that depends on the selection of variables: regional density, and distance. As well, it focuses on the previous data for deciding the appropriate cluster centroids only for static dataset.

Hence, in this article, a CP sampled-DP clustering scheme is developed using quick search and discovering of cluster centroids to handle the dynamic large-scale dataset without any previous data in parallel manner. In this algorithm, more probable cluster centroids are generated automatically via computing the regional density and distance values. Additionally, the structural data from the instance-level pairwise constraints is learned efficiently. Then, the decision graph is evaluated from multiple perceptions for supporting the complete clustering. Thus, CP sampled-DP can ensure no faulty-choice or miss-choice of variables encountered for clustering in the anomaly identification.

2 Literature Survey

Zhang [7] proposed an adaptive flow counting technique using linear prediction for controlling spatiotemporal counting function. In this technique, the prior samples were considered to predict the successive measurement granularity. Also, an adaptive algorithm was applied for allocating the flow counting processes among several switches. But it has less accurateness.

Wang et al. [8] introduced a FloodGuard scheme for SDN based on proactive flow principle estimator and data migration. Proactive flow principle estimator was applied for computing principles dynamically. Also, data migration was applied for temporarily caching the flooding data and including them to the SDN regulator by rate constraint and round-robin scheduling. But, the overhead during runtime was not reduced.

Sahri and Okamura [9] proposed an adaptive sampling strategy to accurately detect the anomalies in SDN through the most important traffic statistics. Additionally, a weighted K-means clustering was applied for categorizing the attacks via the network according to the determination of severity of monitored traffic. However, it requires an evaluation of training time, detection ratio, etc.

AlEroud and Alsmadi [10] developed an inference-based attack identification to recognize the cyber-attacks in SDN. In this technique, a packet aggregation scheme was applied to generate the attack signatures and detecting the attacks in SDN. But, the efficiency of identification was less.

Lee et al. [11] recommended an Athena model according to the completely distributed application hosting structure for scalable anomaly identification in. A common operation was carried out to synthesize the variety of anomaly identification activities and networking. But it has a high overhead during feature extraction.

Carvelho et al. [12] developed an SDB-based system to forecast the packet flow and proactively identify the anomalies which may degrade the appropriate network performance. The detected anomalies were neutralized to mitigate from the system. However, it requires relevant attributes to increase further efficiency.

3 Proposed Methodology

Initially, OE-ISU-FS scheme is applied to the actual dataset for reducing the data dimensionality via choosing the most relevant attributes. Then, the data are clustered by the CP sampled-DP which is differed from the clustering through quick search and discovering of density peaks according to the following assumptions: (i) cluster centroids involve the maximum densities amid its adjacent and the cluster centroids are remote from another maximum densities, (ii) the clusters with similar density distribution and without Cannot-Connect (CC) limits must possessed in similar class. The overall flow diagram of proposed traffic anomaly identification scheme is depicted in Fig. 1.

3.1 Definitions

Regional Density (ρ_i)

It is the inverse of the total of Euclidean ranges of data i from its top-k closest adjacent. The optimal of k is mostly effective compared to the optimal cutoff space. Despite of the variable k, for data in a compact area, distances from adjacent data are comparatively small and so ρ is large and vice versa. As k nearby data are essential to compute the density, larger k is more global data which is reflected by regional density.

$$\rho_i = \frac{1}{\sum_{j=1}^{k} d_{ij}} \tag{1}$$

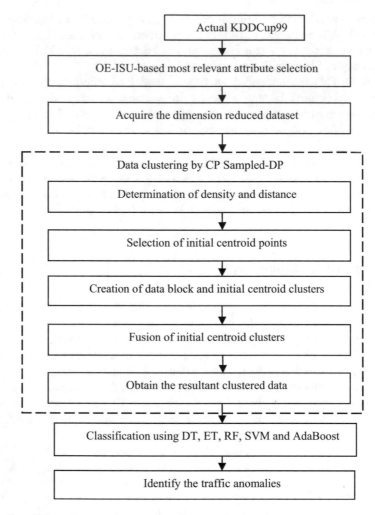

Fig. 1 Overall flow diagram of proposed traffic anomaly detection system

Must-Connect (MC) and Cannot-Connect (CC) Constraint

$MC(i,j)_{i\neq j} = 1$ specifies i and data j belongs to similar clusters.

$CC(i,j)_{i\neq j} = 1$ specifies i and j belongs to diverse clusters.

Transition of ML: If $MC(i,j)_{i\neq j} = 1$ and $MC(j,k)_{j\neq k} = 1$, then $MC(i,k)_{i\neq k} = 1$.

Inheritance of CL: If $CC(i,k)_{i\neq k} = 1$, then $CC(j,k)_{j\neq k} = 1$.

Data block (*dblock*)

$$dblock_k = \left\{ i | \forall j \subseteq dblock_k, \mathrm{CC}(i,j)_{i \neq j} \neq 1 \,\&\&\, \exists p \subseteq dblock_k, \mathrm{MC}(i,p)_{i \neq p} = 1 \right\} \quad (2)$$

All data in these *dblock* stand the ML data limits and not always interrupts CL data.

Primary Centroid Points (*icpoints*)

icpoints are generated after constraint-based clustering algorithm analyzes decision graph from different perceptions. Each *icpoints* have its unique label and may outnumber the valid types.

Initial Centroid Cluster (*icclust*)

$$icclust_k = \{ i | \exists j \subseteq icpoint, j \subseteq icclust_k \,\&\&\, i \subseteq icclust_k \} \quad (3)$$

The cluster created by the data comprising similar *icpoints* label is known as *icclust*.

3.2   Constraint-Based Parallel Sampled-DP Clustering Scheme

Computation of Density and Distance: To avoid more data having the maximum density simultaneously, the data requires small random perturbation that creates every data including varied ρ with no degrading of the data distribution.

   Selection of Initial Centroid Points: After determining the regional density and distance, the data is partitioned into four categories. These are situated at the core of the decision graph which are known as regional centroids imitating the hierarchical data with the maximum ρ and δ. Data with the average ρ and small δ situated nearby ρ axis are common data that have to be clustered. Initially, the logistic regression is used to set data on the decision graph.

   The highest ρ peaks of any cluster, few edge data, and partial regional centroids from dataset are secluded. Then, ρ is multiplied by δ for obtaining intermediate variable *tmult*. The data with larger *tmult* are normally the highest ρ peak of any cluster with the maximum ρ and δ, edge data with the minimum ρ and maximum δ, and regional centroids including the maximum ρ. Therefore, maximum ρ peaks of any cluster, few regional centroids, and many edge data are extracted via logistic regression.

   Further selection is required to eliminate edge data from the set of probable *icpoints*. A novel intermediate variable *clex* is obtained by splitting δ by ρ. The *clex* of edge data are unusually greater compared to *clex* of another data in *clex*-descending graph. Typically, *clex*-distribution relies the idea of long tail distribution.

As a result, the odd data with the minimum ρ are chosen and marked as edge data, because it is expected to lead uncertainty of *icpoints*.

Once the decision graph is examined from three perceptions, this parallel clustering can extract three types of special data. A *pot_1* consists of the maximum ρ peaks of any cluster, few edge data, and a partial regional centroid from the dataset.

A *pot_2* consists of the maximum ρ peaks of any cluster, few regional centroids, and edge data. A *pot_3* contains the most edge points and some regional centroids. But, the aim is obtaining *icpoints* indicating all maximum ρ peaks and most regional centroids. *icpoints* are equivalent to $(pot\_1 \mid \mid pot\_2) - pot\_3$. Regional centroids can aid for preventing miss-choice. But, the regional centroid of a cluster imitates hierarchical data to improve the clustering efficiency.

Creation of Block: At first, the nodes focus on the ML and CL are allocated to similar and multiple *dblocks*, respectively. After verifying all nodes subject to the MC and CC limits, it can further update. The *dblock* created is arranged for the following clustering. It decides the most descriptive data containing the maximum ρ in every *dblock* for denoting *dblock*'s another samples. If i suits to a *dblock*, the ρ of i is $\rho_{ibefore}$ before ρ reassignment and identical to ρ_{iafter} after reassigning its ρ. If j suits to no clusters, the ρ of j is ρ_j. If $\rho_{ibefore} < \rho_j < \rho_{after}$, then it indicates that i is not probable to be the closest data of j before ρ reassignment. However, after it updates ρ of ith data, it has likelihood to turn into the nearby data of jth data when no destruction of CL limits between them.

Creation of Initial Centroid Clusters: In clustering *dblocks*, *icclust* learns the data's fundamental details via incorporating *dblocks*. It has been observed that if a *dblock* disturbs CL limits including each *icclust*, a fresh *icclust* must be created according to *dblock*, because it infers the miss-choice of *icclusts*.

Fusion of Initial Centroid Clusters: As *icpoints* normally more than actual amount of labels, it is essential to merge *icclusts* according to the assumptions suggested by this algorithm. Initially, the limits between two clusters are verified. If no CL limits among clusters is ensured, *icclusts* are facilitated to fuse depending on the evaluation from the support vector machine (SVM) and distribution test. The joint intersection degree of clusters is evaluated by using linear SVM. If the precision is less compared to the threshold, then linear SVM will not able to differentiate clusters having the maximum precision and clusters must fuse. At last, the noise data are removed by the halo point method. These are fed to the different classifiers include decision tree (DT), extra trees (ET), random forest (RF), AdaBoost, and SVM for identifying the traffic anomalies.

Algorithm:

- Compute ρ and δ via its top-k closest adjacent.
- According to the decision graph's conversion and the outcome's permutation such that the confidence range is assigned to 0.05, *icpoints* are chosen.
- Allocate a set of data containing ML correlation into similar *dblock*. Simultaneously, a set of data belong to CL limits are isolated from *dblock*.
- Confirm that the complete *dblock* has such ML data.

- After that, every data are updated including the optimum relevant attributes.
- Update ρ and δ.
- Assign *dblocks* and distribute data into the cluster of its closest adjacent with greater ρ in parallel manner.

4 Results and Discussions

In this section, the efficiency of CP sampled-DP clustering scheme and existing schemes like Birch, K-means, DBSCAN, MeanShift, and sampled-DP is analyzed using MATLAB 2014a. In this experiment, KDDCup99 dataset is considered which comprises different categories of attacks: information gathering, Denial-of-Service (DoS), User-to-Root (U2R), and Remote-to-Local (R2L) attacks including the labeled and unlabeled files. Every labeled file has 41 attributes and one class, i.e., attack name. For discrete attributes, their entropy and similarity uncertainty (SU) are estimated with no preprocessing in the FS scheme. The comparison analysis is conducted for actual and reduced datasets based on the runtime, adjusted rand score (AdjRS), completeness score (CompltS), homogeneity score (HomoS), and detection accuracy (DA). Here, the reduced dataset is acquired via BE-ISU and OE-ISU procedures. Runtime is measured in seconds. AdjRS is a similarity between clusters via taking into account every sample couple. CompltS is the degree of every sample of a considered label assigned to the same cluster. Homo is the degree of clusters having only the samples of a single label. DA is the fraction of samples clustered to the exact cluster. The attributes in KDDCup99 dataset is given in Table 1 with either C (Continuous) or D (Discrete) category.

The efficiency of different clustering schemes on the reduced dataset acquired using BE-ISU and OE-ISU procedures is provided in Tables 2 and 3.

From this analysis, it is noticed that the CP sampled-DP clustering scheme can able to achieve enhanced efficiency compared to the other clustering schemes on the reduced dataset acquired from BE-ISU and OE-ISU procedures.

Figure 2 displays the comparison of the DA for the proposed and existing clustering on the both actual and reduced dataset resulting from the UCFS, BE-ISU, and OE-ISU based FS schemes. This analysis indicates that CP sampled-DP scheme maximizes the DA with the range of data. Since it does not need threshold for deciding the cluster data and cluster centroids, whereas 96.4% accuracy is taken as threshold only to fuse similar cluster centroids which have the maximum distance and densities from the nearby data. Thus, it is proved that the CP sampled-DP clustering scheme handles the dynamic dataset effectively and detects the traffic anomalies with higher accuracy.

Table 1 Summary of attributes/features in KDDCup99 dataset

No	Attribute name	Category	No	Attribute name	Category
1	Time	C	22	is_guest_login	D
2	protocol_category	D	23	Count	C
3	Activity	D	24	src_total	C
4	Flag	D	25	serror_value	C
5	src_bytes	C	26	srv_serror_value	C
6	dst_bytes	C	27	rerror_value	C
7	Land	D	28	srv_rerror_value	C
8	fake_fragment	C	29	similar_srv_value	C
9	imperative	C	30	var_srv_value	C
10	Hot	C	31	srv_var_host_value	C
11	amt_failed_logins	C	32	dst_host_total	C
12	logged_in	D	33	dst_host_srv_total	C
13	amt_compromised	C	34	dst_host_similar_srv_value	C
14	root_shell	D	35	dst_host_var_srv_value	C
15	su_attempted	D	36	dst_host_similar_src_port_value	C
16	amt_root	C	37	dst_host_srv_var_host_value	C
17	amt_doc_creations	C	38	dst_host_serror_value	C
18	amt_shells	C	39	dst_host_srv_serror_value	C
19	amt_access_docs	C	40	dst_host_rerror_value	C
20	amt_outbound_cmds	C	41	dst_host_srv_rerror_value	C
21	is_hot_login	D			

Table 2 Comparison of clustering schemes on reduced dataset using BE-ISU procedure

Schemes	Runtime (sec)	AdjRS	CompltS	HomoS	DA
Birch	4.02	0.51	0.712	0.84	0.901
K-means	1.24	0.51	0.712	0.851	0.913
DBSCAN	1.71	0.62	0.826	0.763	0.870
MeanShift	0.36	0.334	0.668	0.536	0.829
Sampled-DP	1.08	0.758	0.683	0.882	0.935
CP sampled-DP	0.75	0.793	0.725	0.904	0.942

Table 3 Comparison of clustering schemes on reduced dataset using OE-ISU procedure

Schemes	Runtime (sec)	AdjRS	CompltS	HomoS	DA
Birch	3.81	0.39	0.748	0.82	0.912
K-means	1.13	0.39	0.748	0.833	0.925
DBSCAN	1.45	0.51	0.851	0.741	0.881
MeanShift	0.21	0.307	0.693	0.514	0.843
Sampled-DP	0.79	0.724	0.714	0.859	0.950
CP sampled-DP	0.66	0.753	0.749	0.886	0.964

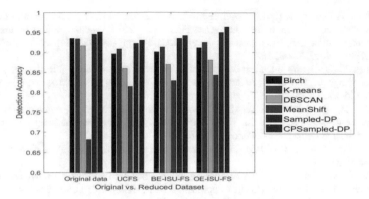

Fig. 2 Comparison of DA over different clustering schemes using both actual and reduced datasets

5 Conclusion

In this paper, the selection of regional density and distance of clustering schemes is resolved via developing the CP sampled-DP scheme. In this scheme, quick search and discovering of cluster centroids algorithm are proposed in parallel manner with three constraints to handle the dynamic large-scale dataset without any prior knowledge. By using such constraints, number of probable cluster centroids is produced automatically by determining the regional density and distance values. Moreover, the structural data from the instance-level pair wise constraints is learned efficiently. Further, the decision graph is analyzed from different perceptions for supporting the complete clustering. Finally, the experimental outcomes concluded that the CP sampled-DP scheme has the maximum efficiency than the existing sampled-DP scheme.

References

1. Xia W, Wen Y, Foh CH, Niyato D, Xie H (2015) A survey on software-defined networking. IEEE Commun Surv Tut 17(1):27–51
2. Kreutz D, Ramos FM, Verissimo PE, Rothenberg CE, Azodolmolky S, Uhlig S (2015) Software-defined networking: a comprehensive survey. Proc IEEE 103(1):14–76
3. Khan S, Gani A, Wahab AWA, Abdelaziz A, Ko K, Khan MK, Guizani M (2016) Software-defined network forensics: motivation, potential locations, requirements, and challenges. IEEE Netw 30(6):6–13
4. He D, Chan S, Ni X, Guizani M (2017) Software-defined-networking-enabled traffic anomaly detection and mitigation. IEEE Internet Things J 4(6):1890–1898
5. Vadivu T, Sumathi B (2019) A novel feature selection using optimized eliminated iterative distance correlation for SDN-enabled traffic anomaly detection and mitigation. Int J Adv Sci Technol 28(16):1380–1388

6. Vadivu T, Sumathi B (2019) Swarm optimization techniques using attribute assortment and clustering for machine learning classifiers. In: 2019 International Conference on Communication and Electronics Systems (ICCES), pp 834–837. https://doi.org/10.1109/ICCES45898.2019.9002279
7. Zhang Y (2013) An adaptive flow counting method for anomaly detection in SDN. In: Proceedings of the ninth ACM conference on emerging networking experiments and technologies, pp 25–30
8. Wang H, Xu L, Gu G (2015) Floodguard: a DoS attack prevention extension in software-defined networks. In: IEEE 45th annual IEEE/IFIP international conference on dependable systems and networks, pp 239–250
9. Sahri NM, Okamura K (2015) Adaptive anomaly detection for SDN. Proc Asia-Pacific Adv Netw 40:57–63
10. AlEroud A, Alsmadi I (2017) Identifying cyber-attacks on software defined networks: an inference-based intrusion detection approach. J Netw Comput Appl 80:152–164
11. Lee S, Kim J, Shin S, Porras P, Yegneswaran V (2017) Athena: a framework for scalable anomaly detection in software-defined networks. In: IEEE 47th annual IEEE/IFIP international conference on dependable systems and networks, pp 249–260
12. Carvalho LF, Abrão T, de Souza Mendes L, Proença ML Jr (2017) An ecosystem for anomaly detection and mitigation in software-defined networking. Expert Syst Appl 104:121–133

Predictive Policing—Are Ensemble Methods More Accurate Than Regression Methods?

Ronit Kathuria and Vinish Kathuria

1 Introduction

Policing characterized by its propensity to be proactive, smart, and effective is undeniably preferred to simple, often delayed, reactions to criminal acts. Predictive policing refers to proactive policing strategies that focus on predicting crimes before they occur [1]. In recent years, the focus has been on predicting the time and location of where crime is likely to take place [2, 3]. Across the USA, police and civic authorities have amalgamated multiple predictive policing approaches into the crime prevention and detection framework.

Predictive policing methods can be categorized into four umbrella sections based on end usage: predicting crimes, predicting future offenders, predicting perpetrators' identities, and predicting victims. This research focuses on predictive policing methods related to predicting crimes, which is identifying *times* and *places* that correspond to an increased risk of crime. Following most forecasting methods in operation, indicating future criminal events is established upon examining data on past crimes and using multiple techniques for analysis. This research aims to compare two widely used predictive policing methods with the concerted goal of predicting the time and location where crime is most likely to occur. *Individual regression methods* describe a mathematical relationship between the variable that it aims to predict and independent "expository" variables. *Ensemble methods* combine a myriad of predictive models by either deriving a weighted average of the individual models' outputs or by having them vote to yield a final overall prediction.

R. Kathuria
The Shriram, Gurgaon, India

V. Kathuria (✉)
Indian Institute of Management, Lucknow, India
e-mail: efpm07014@iiml.ac.in

The focus of said comparison hinges on the accuracy of the varied predictive policing models, which can be evaluated on various metrics that analyze the relationship between historical crime events and how often the models anticipate or miss these events. This research utilizes four primary metrics as measurements of accuracy:

- **True positive rate**: Percentage when the model correctly predicts the occurrence of a crime
- **True negative rate**: Percentage when the model correctly anticipates that a crime would not happen.
- **False positive rate**: Percentage when the model incorrectly anticipates the occurrence of a crime
- **False negative rate**: Percentage when the model misses the occurrence of a crime

This research compares an individual regression method and an ensemble method to predict the number of crime incidents in a given zip code in a particular hour, given the day of the week, the time of day, the month within the crime takes place, and the weather forecast. The research uses the datasets from two cities, San Francisco and Chicago. In summary, the number of crimes is regarded as the dependent variable, while the remaining features act as independent variables.

2 Literature

Existing research supports the theory that crime is predictable (in the statistical sense), mainly because criminals tend to operate in their comfort zone. This view is supported by major criminal behavior theories, such as routine activity theory, rational choice theory, and crime pattern theory [4]. Blended theory [3] states that like ordinary people, criminals and victims follow patterns, and overlap in those patterns indicates an increased probability of crime. As people move within the patterns, geographic and temporal features (area, target's suitability, risk of getting) caught influence the "rational" decisions criminals take about committing crimes.

Researchers have studied in-depth different proactive policing techniques for crime reduction [5]. Studies in the evidence-based policing field demonstrate the possibility of proactive prevention and reduction in crime, and examples of some studies are on the evidenced-based policing matrix. It is possible to integrate predictive analytics into most of these programs. Lot of research in predictive policing has focused on using geospatial modeling to predict future hot spots [3]. Crime control programs focusing on geographic targets have been met with the most success [6]. Models and methods to predict individuals' criminal behavior are less prominent in criminal justice crime prevention literature.

Place-based predictive policing is centered around: (a) mathematical forecasting methods that can be used to foresee future crime risk in narrowly prescribed geographic locations; and (b) the supply of police resources to those locations to disrupt the opportunity for crime [2]. Using a randomized controlled experimental approach of predictive policing conducted in Los Angeles, researchers showed that algorithmic methods predict twice as much crime as existing best practice and double the amount of crime prevented [2]. While this treatment effect can be measured in the field, predictive policing's specific mechanisms that deliver more significant crime reduction are not immediately apparent.

Predictive policing models may use various regression models [3] or ensemble methods [6]. Ensemble methods focus on getting better predictive performance using multiple learning algorithms compared to what could be derived from any constituent learning algorithms alone. The process involves starting with an initial simple classification model (which is probably weak, or not terribly accurate) and iteratively adding simple classification models. The final prediction can be a weighted average or majority vote of all the simple classification models. Random forest methods iteratively generate many simple decision trees (a "forest"). They "grow" a simple decision tree on randomly selected subsets of input variables and input data in each iteration. The final prediction is the plurality vote of all the simple trees. Bagging methods generate multiple classifiers by building a classification model in each iteration on a randomly selected subset of the input data. The classifiers are typically decision trees or neural nets. The final prediction is either the average or plurality vote.

3 Dataset and Research Study

This research utilizes crime statistics data from two major cities in the USA—San Francisco and Chicago. The dataset for the San Francisco dataset was taken from the open data initiative by DataSF. This government-sponsored platform seeks to transform the way the city works using data [7]. The Chicago dataset has been acquired from the public archives of the Chicago Police Department [8]. Preprocessing of data-included eliminating inconsistencies and noise in information and transforming the dataset into a form that can be processed. The dataset variables included address, location coordinates, date and time, day of the week, and description and resolution of the crime. Exploratory data analysis (EDA) was utilized to encapsulate and analyze the dataset's specific characteristics through visualization. A preliminary analysis of data was carried out to recognize determining features for developing a predictive model. Figure 1 showcases some relevant statistics for the San Francisco dataset.

The cyclical pattern of different crimes as per hour of the day based on data crime records from 7 years (2008–2015) shows that as the given hour of the day varies, certain types of crime see a fall in the number of incidents, while others rise. Interestingly, there also seems to be a considerable and consistent drop in crime early in the morning; it seems that the offense does sleep at roughly 05:00 a.m. Figure 1 also shows the number of crimes per day in San Francisco for consecutive

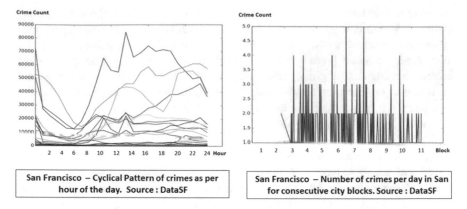

Fig. 1 Crime data from San Francisco, CA, USA

city blocks. The X-axis is a given block, and the Y-axis represents the rate of crime. The figure provides a visual representation of the crime rate in a set of 11 blocks in the city throughout 2013. The crime rate varies significantly across neighborhoods but does not reveal a distinct pattern. Analysis of the crime rate in Chicago for various categories of crime over 15 years (2000–2015) indicates that the year's month is a causal variable for crime rate.

Regression Methods. Regressions project future crime risk based on a wide range of data, such as the month, time of the day, and geographic location. The train-test split is carried out chronologically since the data itself is time-oriented—for San Francisco, statistics for 2016 are used as the training data, and the crime statistics from 2017 are used as the test data. For the Chicago dataset, datasets for 2013 and 2014 are used similarly. The process below outlines the steps that we followed to predict the number of crimes in an hour.

The least-squares method and tenfold cross-validation are utilized to estimate the model's mean squared-error (MSE) model. The MSE for the training data is calculated and reported to facilitate comparison with the estimate for determining the amount to which the results of the linear regression will generalize to an independent dataset. Ridge regression is leveraged to reduce overfitting, the adequate alpha for which is calculated to be 1 using RidgeCV. This function enables ridge regression, a technique for analyzing multiple regression data that suffer from multicollinearity. When multicollinearity occurs, least-squares estimates are unbiased, but their variances are large, so that they may be far from the true value. The four metrics for accuracy are then calculated for the Chicago and San Francisco dataset for the model's predictions for 2014 and 2017.

Figure 2 depicts the regression model's predictions for the number of crimes in San Francisco from 7 to 8 p.m. on July 27, 2017. A darker shade corresponds to a higher number of crimes.

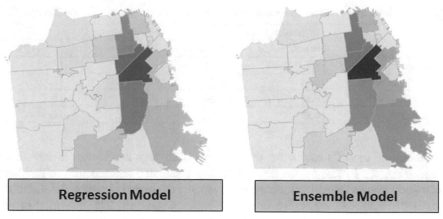

A darker shade corresponds to a higher number of crimes

Fig. 2 Predictions for the number of crimes in San Francisco from 7 to 8 p.m. on July 27, 2017

Ensemble Methods. They are centered on generating a large set of individual predictive models and combining them in a manner that aims to increase the accuracy level of predictions best. Most notable of these combination processes include having the unique models vote, averaging the individual models' predictions, or using the individual models' results as inputs for a broader and far more complex supermodel. The datasets are the same as for regression methods mentioned above. The process below outlines the steps that we followed to predict the number of crimes in an hour.

The foundational approach used to develop the ensemble method for this research is a bagging method. The models used include multiple versions of an multilayer perceptron (MLP) regressor, a class of feedforward artificial neural networks, and a clustering algorithm that leverages a K-nearest neighbors (KNN) model. The ensemble method used as part of this research consists of four varied MLP models compatible with the crime statistics dataset. The models are distinguished by the number of hidden layers and the nodes that comprise each layer, which vary between 1 and 4, and 50 and 200, respectively. Each MLP regressor is trained on a randomized sample from the training set dictated by the bagging method. For the KNN model, the optimal number of neighbors is selected to be 7. For each model, the learning rate is 0.3, the momentum is 0.2, and the number of epochs is 500. The implementation of the ensemble method uses a learning rate annealer. The learning rate annealer decreases the learning rate after a certain number of epochs if the error rate does not change.

The averaging ensemble method is leveraged to combine the results of the individual MLP regressors. Each network's outputs are passed through dense layers with a fixed number of neurons to equalize them. The result shows an output that shares each input's dimensionality, but that is the average of them. For this research, the ensemble size is set to 50, and 5× twofold cross-validation is used to

Table 1 Regression and ensemble model predictions for San Francisco, CA, and Chicago, IL

	Regression model		Ensemble model	
	San Francisco	Chicago	San Francisco	Chicago
True positive rate	0.8	0.81	0.88	0.68
True negative rate	0.74	0.73	0.81	0.61
False positive rate	0.16	0.16	0.11	0.21
False negative rate	0.13	0.13	0.09	0.20

estimate the model's root mean squared-error (RMSE). Analogous to individual regression, the RMSE for the training data is calculated and reported to facilitate comparison with the estimate above to determine the extent to which the ensemble method results will generalize to an independent data set. The four metrics for accuracy are then calculated for the Chicago and San Francisco dataset for the model's predictions for 2014 and 2017, respectively.

4   Findings

The ensemble method leveraged for this research reports an average true rate (average of true positive and true negative rates) of 0.76 and 0.75 for San Francisco and Chicago, respectively. The same method reports an average false rate of 0.18 and 0.16 for the two cities. On the other hand, the individual regression method reports an average true rate of 0.68 and 0.63 and an average false rate of 0.17 and 0.13 for San Francisco and Chicago, respectively.

In the case of accurately anticipating when crimes will and will not happen, the ensemble method outdoes the individual regression method, as showcased by the 11.7% and 19.1% higher average true rate for the ensemble method relative to the individual regression model for the San Francisco and Chicago datasets, respectively. On the contrary, when it comes to anticipating when crimes will and will not happen inaccurately, the ensemble method is worse off than the individual regression method, as showcased by the 5.9% and 23.1% higher average false rate for the ensemble method for the two cities in consideration. Outlined below are three features that aim to explain this disparity inaccuracy.

Crime Distribution. In general, the risk of suffering a crime is not uniformly distributed over a region, and because of this risk heterogeneity, crime is highly concentrated in certain population groups. Therefore, to test the impact of crime distribution on the accuracy of methods, we train and test the two models used earlier in this research on a subsample of the San Francisco dataset with an equal and uniform crime distribution. Table 1 showcases those results. The 9.6% increase in the average true rate for the individual regression methods, as compared to the 1.7% increase in the average true rate for the ensemble method, sheds light on the

Table 2 Regression and ensemble model predictions for San Francisco, CA, and Chicago, IL, with a sample of uniform crime distribution and spatiotemporal information

San Francisco	With a sample of uniform crime distribution		With spatiotemporal information	
	Ensemble	Regression	Ensemble	Regression
True positive rate	0.83	0.81	0.88	0.68
True negative rate	0.74	0.73	0.81	0.61
False positive rate	0.16	0.16	0.11	0.21
False negative rate	0.13	0.13	0.09	0.20

notion *that ensemble methods adapt their predictions more effectively to a higher crime distribution relative to individual regression methods.*

Spatiotemporal effects. An influential factor in the accuracy of predictions made by predictive policing methods is the relationship between crime and the crime's environment. Therefore, it is imperative to expand the prediction problem to include various environmental and temporal features of the crime location. Accordingly, additional determining characteristics such as temporal proximity to other events, type of place, and economic data from the crime area are added to a subsample of the Chicago dataset. Figure 3 showcases the accuracy metrics for the training and testing process performed with a subsample with additional features, which effectively investigates the effects of the environment on the crime.

As underscored by the 11.2% increase in the average true rate and a 7.8% decrease in the average false rate, the ensemble method effectively leverages the spatiotemporal information and offers more accurate predictions. On the other hand, the individual regression model sees a measly 1.4% increase in the true positive rate and a considerably detrimental 6.4% increase in the average false rate. The amalgamation of these results shed light on the notion that *ensemble methods adapt their predictions more effectively in tandem with spatiotemporal analysis relative to individual regression methods.*

5 Discussions and Implications

This research contributes to the evolving body of academic literature that analyzes the intricacies of predictive policing methods, and the broader use of the machine and deep learning techniques to facilitate crime detection and prevention. The results of the comparison between the accuracies of the two distinct predictive policing techniques reveal the superiority of ensemble methods in terms of accurately anticipating the occurrence or lack of crimes in each region, which is reflected by its significantly higher average true rate. An investigation into the determining features behind the disparity in accuracy levels reveals ensemble methods' ability to deal with higher rates of crime distribution and account more effectively for spatiotemporal effects. Individual regression methods' incapacity to perform accurately in regions with a considerably distributed crime can be traced back to the phenomenon of runaway feedback loops [9]. When leveraged for predictive policing tasks, individual regression models showcase the tendency to repeatedly send police forces to the same neighborhoods, regardless of the real crime rate, because predictions made by the model influence the data that it treats as inputs in the future. Ensemble methods mitigate the effects of runaway feedback loops due to the process of taking a weighted average of the individual models' outputs, a process that is intrinsic to ensemble methods.

A critical assumption that this research rests on is that reported crime rates correspond to actual crime rates. There are a few aspects that this research does not explore in detail. While the importance of crime distribution can be determined to be rooted in runaway feedback loops, the reasons behind the characterization of spatiotemporal effects as determining features of accuracy levels can be investigated through the creation of hypothetical crime statistics dataset focused on examining the relationship between reported crimes and the environment in which they take place. When it comes to considering practices, further discussion on predictive policing could revolve around biases within predictive policing methods.

References

1. Bordua DJ, Reiss AJ Jr (1966) Command, control, and charisma: reflections on police bureaucracy. Am J Sociol 72:68–76
2. Mohler GO, Short MB, Malinowski S, Johnson M, Tita GE, Bertozzi AL, Brantingham PJ (2015) Randomized controlled field trials of predictive policing. J Am Stat Asso
3. Perry W, McInnis B, Price C, Smith S, Hollywood J (2013) Predictive policing: the role of crime forecasting in law enforcement operations. RAND Corporation
4. Clarke RVG, Felson M (eds) (1993) Routine activity and rational choice. Transaction Publishers, New Brunswick, NJ
5. Braga A, Papachristos AV, Hureau DM (2012) The effects of hot spots policing on crime: an updated systematic review and meta-analysis. Justice Q:1–31
6. Rummens A, Hardyns W, Pauwels L (2017) The use of predictive analysis in spatiotemporal crime forecasting: building and testing a model in an urban context. Appl Geogr 86:255–261
7. DataSF (2020) Office of the chief data officer: city and county of San Francisco (2020, June 24). Retrieved 12 Sept 2020, from https://datasf.org/
8. Chicago PD (2018) Year-end summary crime statistics (May 4, 2018). Retrieved 4 Oct 2020, from https://home.chicagopolice.org/statistics-data/crime-statistics/
9. Ensign D, Friedler SA, Neville S, Scheidegger C, Venkatasubramanian S (2018) Runaway feedback loops in predictive policing. FAT

A Fast Method for Retinal Disease Classification from OCT Images Using Depthwise Separable Convolution

S. Meenu Mohan and S. Aji

1 Introduction

The retina, which is a thin layer of tissue, situated inside the back wall of the eye, near the optic nerve. They preprocess light through a layer of photoreceptor cells. The macula present in the midpoint of the retina is responsible for accurate and highly sensible information. Typical retinal diseases include choroidal neovascularization (CNV), diabetic macular edema (DME) [1] and Drusen [2]. CNV is the growth of the blood vessel starting from the choroid and passes through the brunch membrane into the sub retinal space. CNV is one of the major causes of vision loss. DME is a complication of diabetics. These diseases affect the center of the macula known as the fovea. DME is responsible for the blurred or wavy vision and faulty perception of colors. Drusen are yellow deposit under the retina. They occur between the brunch membrane and retinal pigment epithelium (RPE). They are like tiny pebbles of debris which can also affect in the optic nerve. It is associated with aging and macular degeneration. It may affect the risk of developing age-related macular degeneration (AMD).

One of the standard imaging technologies used for eye imaging is optical coherence tomography (OCT). OCT is a non-contact, non-invasive imaging technique that can identify the layers of the retina by looking at the inference pattern of the reflected laser light. Tomography is the reconstruction of cross-sectional images of an object by using its projection. OCT uses near-infrared light for providing the cross-sectional view of the retina which is similar to a histopathological specimen. The time domain OCT (TD-OCT) was hindered by its comparatively slow scanning speed, which restricted the number of retinal images that could be sampled during image acquisition. It provides a 2D view of the retina. The recent technology used for the OCT scan is SD-OCT, which provides a 3D view of the retinal morphology

S. Meenu Mohan (✉) · S. Aji
Department of Computer Science, University of Kerala, Thiruvananthapuram, Kerala, India

© The Author(s), under exclusive license to Springer Nature Singapore Pte Ltd. 2022   153
R. R. Raje et al. (eds.), *Artificial Intelligence and Technologies*,
Lecture Notes in Electrical Engineering 806,
https://doi.org/10.1007/978-981-16-6448-9_18

and also increased the scanning speeds. It also provides more accurate information. Figure 1 shows the images of the OCT scan for each retinal disease and normal scan image.

For the past few years, several retinal OCT classification techniques were introduced. Recently, deep learning techniques were introduced, which classifies with high accuracy and more concise time. Among the deep learning models is the convolutional neural network [3–8], which has shown a great impact on image classification and recognition. Recently, CNN frameworks are also used for retinal OCT classification and feature extraction. CNN architecture includes several layers that are used for feature extraction, dimensionality reduction and finally classification. But traditional CNN model requires lots of mathematical computation which increases the time and space complexity of the model. They also increase the number of parameters used for classification. In this paper, we use a variation of the CNN architecture which is depthwise separable convolution (DSC). DSC decomposes standard convolution into 2D spatial convolution and 1×1 pointwise convolutions. DSC reduces the number of computations required than that of the traditional CNN model. It reduces the complexity of the network by reducing trainable parameters and by reducing time and space complexity.

In recent years, several methods were introduced for the classification of retinal diseases. Fang et al. [9] propose an iterative fusion method in a convolution neural network for the classification of retinal diseases. IFCNN introduces an iterative layer fusion strategy that incessantly integrates features of the current convolutional layer with those of all previous layers. The system tries to categorizes four different classes (CNV, DME, Drusen and normal) and offers an overall accuracy of 87.3%. Kamble et al. [6] introduced a fusion of residual connections along with Inception v3 known as Inception-Resnet-v2. The model uses BM3D for the filter stage and obtained 100% in specificity, sensitivity and accuracy.

Li et al. [10] developed a method that integrates handcrafted and deep features for the classification of retinal diseases from OCT images. The paper proposes three different feature integration methods for classification; they are early, late and full. The proposed RC-Net uses dense block and sum operation to replace the convolutional block and concatenation, which reduces the parameter and improves the

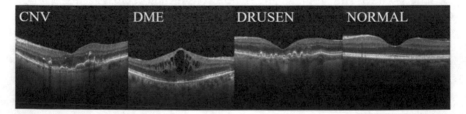

Fig. 1 OCT scan images, the left-most figure shows a scan image of ChoroidalNeovascularization (CNV), middle left shows diabetic macular edema (DME), middle right shows Drusen, right-most shows a normal scan image. Image *Source* http://www.cell.com/cell/fulltext/S0092-8674(18) 30154-5

performance of the network. The method is tested on three different groups of data in which each group obtained an accuracy above 97%.

Alqudah [11] proposed an automated classification method for multiclass retinal disease classification. The paper presents a CNN architecture for the classification of four retinal diseases CNV, DME, AMD, Drusen and normal retinal scan images. The model was trained using Adam optimizer, which gives more accurate and lowers time cost results were obtained. The model obtains an accuracy of 95.30%. Li et al. [12] proposed a novel ensemble of four classification models for automatic detection of retinal diseases from OCT images. The models were based on improved Resnet50 architecture. The model was evaluated using both pretrained weights and by not using pretrained weights which obtained an accuracy of 86.7% and 86.5%, respectively.

Kermany et al. [13] proposed a transfer learning method for the classification of four retinal disease. The paper presents a pretrained Inception v3 from ImageNet for classification. They could attain 96.6% accuracy, 97.8% sensitivity and 97.4% of specificity. Srinivasan et al. [14] proposed a fully automated system for the detection of diabetic macular edema and age related macular degeneration. They have used histograms of oriented gradients (HOG) as features that are fed to support vector machine (SVM) classifier. The system mainly uses three separate stages before feature extraction; they are image denoising, flatting retina curvature and cropping ROI.

Perdomo et al. [15] proposed a new deep learning approach for the classification of retinal diseases. The new approach was named OCT-NET which classify three retinal diseases DR, DME, AMD and normal scan images. They evaluated the model with SERI-CUHK and A2A SD-OCT dataset and obtained an accuracy of 93%. The model outperforms from conventional deep learning approaches. Ji et al. [16] proposed a transfer learning approach that uses pretrained Inception v3 model for classification of AMD, DME and normal retinal images. The model tries to improve classification performance and shorten the training time.

The proposed work is trying to classify three retinal diseases CNV, DME, Drusen including normal retinal images. The models use depthwise separable CNN for the classification which gives notable accuracy with better classification time. It also reduces the number of parameters used for classification. The method also provides discriminative localization of scan images which improves the classification of the model. The method was evaluated on a publically available OCT2017 dataset, which was developed from the Heidelberg Spectralis OCT device. There are 108,312 images for training and the test set contains 250 images for each class. The experimental evaluation of the model was done by the performance metrics accuracy, precision, recall and f1-score. The model was also evaluated based on some of the baseline models. The experimental results show that the model could reduce the training time considerably with an accuracy of 93.5%.

The remaining section of the paper is organized as: in Sect. 2 concepts and methods of the proposed model are explained. Section 3 discusses the proposed methodology. Section 4 shows the experimental evaluation, dataset used and the result analysis. Finally, Sect. 5 concludes the paper.

2   Concepts and Methods

This module gives details of SD-OCT image classification based on a deep neural networks which consist of four stages. First stage receives raw SD-OCT scan volume which contains abnormalities and then the preprocessing stage where the layers of the retina are identified. The preprocessed images are inputted to the proposed CNN model for feature extraction. These images are also passed through the class activation map stage to underline the relevant zones of the image. Finally, the classification is performed by the proposed model.

2.1   SD-OCT Image Preprocessing

SD-OCT [17–19] scan images are volumetric array which provides the cross-sectional view of the retina. During image acquisition, these scan images gets affected by speckle noises and it has to be reduced for better classification. The RoI was detected using a median filter with a 3×3 kernel, and a threshold value of 0.3 is used to highlight the internal limiting membrane (top layer) and retinal pigment epithelium (bottom layer).

2.2   Model Architecture

The proposed model was derived from the model proposed by Kermany et al. [13] and Perdomo et al. [9]. Kermany et al. [13] proposed a transfer learning method which uses pretrained Inception v3 from image net for the classification of retinal diseases. In the pretrained network, the CNN model is updated using pretrained weights which reduce the overfitting in the model. The method also performs occlusion testing to identify the area contributing most of the neural network. The model gains an accuracy of 96.6%. Since the Inception v3 network uses more layers gradient vanishing or gradient divergence may happen. Perdomo et al. [9] propose a new CNN model, OCT-NET for classifying diabetics related retinal diseases. The method consists of a preprocessing stage for the removal of speckle noise and a class activation map stage is also used for discriminative localization of the image. But the time and space complexity of the model is high which increases the computational cost. The proposed model tries to overcome the disadvantages of these methods; by using separable convolution, the number of parameters and time required for training is reduced. The number of layers of the CNN model is lesser than that of pretrained Inception v3 (Fig. 2).

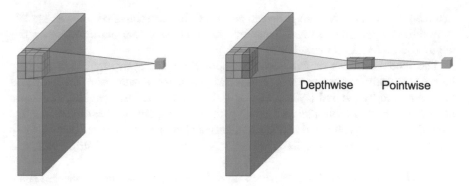

Fig. 2 Normal convolution and Depthwise separable convolution. *Image source* https://arxiv.org/abs/1610.02357

2.3 Depthwise Separable Convolution

Depthwise separable convolution [17, 20, 21] decomposes standard convolution into spatial convolution and pointwise convolution layers. Spatial convolution performs a 2D filter to every input channel that considerably reduces the number of computation than that of normal convolution; however, solely input channel corresponding to each input channel is calculated. A pointwise convolution is employed to mix the outcome of spatial convolution and to make new features.

2.4 Class Activation Map

The class activation map is intended to validate the CNN model for image region identification for a selected category. The network mainly consists of a large number of CNN layers and a pooling layer just before the output layer. The features obtained are given to a fully connected layer having softmax activation that produces the specified output. We were able to identify the required image regions by returning the weight of the output layer on the convolution feature map obtained from the last convolution layer.

3 Proposed Methodology

The proposed model uses depthwise separable convolution [22, 23] that is a combination of 2D spatial convolution. It applies a 2D filter to each input channel and pointwise convolution that combines the resulting output. In this model, the combination of separable convolution and max-pooling will take care of the feature extraction, and the remaining layers are used for classification of the model. A nonlinear operation rectified linear unit (RELU) was also performed after the depthwise separable convolution layer.

The model consists of five separable convolutional layers, four max-pooling layers, two fully connected layers and a dropout layer as shown in Table 1. The input received by the first layer is of size $224 \times 224 \times 1$. The convolutional layer outputs the feature map and max-pooling layer is responsible for dimensionality reduction. The convolutional layers use a kernel size of 3×3 and stride of 1×1, and for dispensing non-maximal qualities, a pool size of 2×2 and stride of 2×2 are used.

The classification sub-block carries with it three layers: one fully connected layer with 4096 neurons, a dropout layer with a rate of 0.5 and at last a fully connected layer with a number of classes(4). The dropout layer is answerable for avoiding overfitting, and it permits learning with various neurons the similar information improving the generality of the model.

By using separable convolution instead of standard convolution to the proposed model reduces the time requires for training the model. Separable convolution is performed in two stages (depthwise and pointwise convolution), where the convolution operation is performed on each channel, and results are combined. So, the number of computation required is less compared to standard convolution which reduces the training time, and the number of features used is also reduced so space required is also reduced. In Kermany et al. [13] model, when the Inception v3 network with more layers is trained, gradient vanishing or gradient divergence was easier to happen. So, the amount of layers employed by the proposed model is additionally reduced compared to the previous model which avoids the matter of gradient vanishing or gradient divergence.

Table 1 Structure of the proposed CNN model

Layer	Output shape	Parameters
Input	224,224,1	0
SeparableConv2d	224,224,16	91
SeparableConv2D	224,224,16	416
SeparableCon2D	112,112,16	416
Maxpooling2D	112,112,16	0
Add (Add)	112,112,16	0
Maxpooling2D	56,56,16	0
SeparableConv2D	54,54,32	688
Maxpooling2D	27,27,32	0
SeparableConv2D	25,25,64	2400
Globalaveragepooling2D	25	0
Dense	4096	106,496
Dropout (rate = 0.5)	4096	0
Dense	Classes-4	16,388

4 Experimental Evaluation

4.1 Dataset

The experiments in the proposed model were carried out with the help of publically available 'OCT2017' dataset. The dataset consists of SD-OCT images which are validated and analyzed in a recent work of Kermany et al. [13]. Retinal OCT is an imaging technology used for obtaining high resolution cross-sectional view of a patient's retina. Optical coherence tomography (OCT) images (Spectralis OCT, Heidelberg Engineering, Germany) were selected from patients of Shiley Eye Institute of the University of California San Diego, the California Retinal Research Foundation, Medical Center Ophthalmology Associates, the Shanghai First People's Hospital, and Beijing Tongren Eye Center between July 1, 2013 and March 1, 2017. The dataset is mainly divided into two folders train and test folders. Each folder contains four sub-folders CNV, Drusen, DME and normal. The dataset consists of 108,312 images, in which the train sub-folder consists of 37,206 images in CNV, 11,349 images in DME, 8617 images in Druse and 51,140 images in normal. The test folder consists of 1000 images in total where each sub-folder consists of 250 images, respectively.

4.2 Experiment and Results

4.2.1 Qualitative Analysis of Preprocessing

The preprocessing stage was used for the deletion of speckle noise from the OCT scan images. These noises are occurred during image acquisition and cause abnormalities that affect the classification of images. After performing preprocessing, the noises are removed; it also highlights the upper and lower layers of the retina for classification using threshold function. Figure 3 represents the scan images before and after preprocessing.

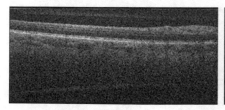

Fig. 3 Raw SD-OCT scan image (left) contains noises and not visible and scan images after preprocessing (right) which highlights the upper and lower retinal layers

4.2.2 Model Evaluation

The model was trained using Nadam (Nesterov-accelerated Adaptive Moment Estimation) optimizer which is Adam with Nesterov momentum (NAG). Nesterov-accelerated gradient (NAG) is a way to give momentum a kind of pre-science and Adam is a genre of stochastic gradient descent with an adaptive learning rate for each parameter with momentum. The batch size, learning rate and number of epochs are set to 200, 0.001 and 15, respectively.

The model was evaluated on the test set of OCT2017 data using accuracy, recall, precision and f1-score as performance metrics which are defined by Eqs. (1)–(4). Table 2 shows the performance metrics value obtained by each class in the test set of the dataset

$$Accuracy = \frac{TP + TN}{TP + TN + FP + FN} \tag{1}$$

$$Recall = \frac{True\ Positive}{Total\ Actual\ Poitive} \tag{2}$$

$$Precision = \frac{True\ Positive}{Total\ Prdicted\ Positive} \tag{3}$$

$$F1\text{-}Score = \frac{2 * Precision * Recall}{Precision + Recall} \tag{4}$$

where *TP*—true positive, *TN*—true negative, *FP*—false positive and *FN*—false negative numbers obtained in the experiments.

Table 2 gives the value obtained by each class for the corresponding performance metrics. From the table, it is noted that the DME class obtains a better value for all the three matrices. Drusen class obtains the lowest value for all matrices. Even though each class obtains a different value for performance matrices, the model achieves an overall better accuracy.

To evaluate the proposed CNN model for the classification of four retinal diseases, the model was tested on the test dataset of the OCT2017 dataset. The model thus obtains an accuracy of 93.5%. The AUC curve of the proposed model is shown in Fig. 4.

Figure 4 shows the performance of the model on categorical cross-entropy loss on both the train and validation datasets. The right side of the figure shows the

Table 2 Performance metrics obtained by each class

Class	Precision	Recall	F1-Score
CNV	0.95	0.93	0.94
DME	0.94	0.98	0.96
Drusen	0.92	0.90	0.91
Normal	0.96	0.96	0.96

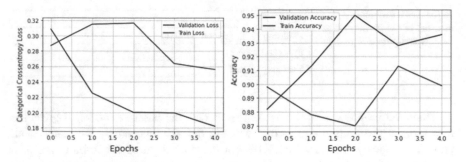

Fig. 4 Loss (left) and accuracy (right) were obtained by the proposed model on the testing

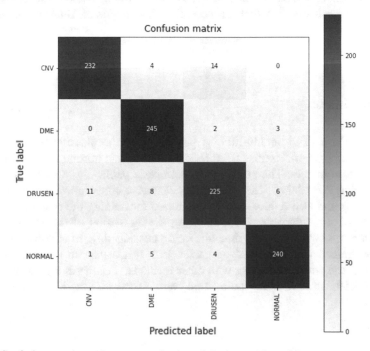

Fig. 5 Confusion matrix-performance evaluation of the proposed model

accuracy obtained by the train and validation dataset. The model obtains an accuracy of 95% during validation and 91.5% during training. Figure 5 shows the confusion matrix obtained by the model for the test set.

In the confusion matrix in Fig. 5, the DME class has a high classification accuracy; the number of images misclassified is less compared to other classes. The images from Drusen and CNV classes are misclassified mostly compared to others, but the images misclassified are less compared to correctly classified images.

The work proposed by Kermany et al. [13] and Perodomo et al. [9] was chosen as baseline models for the proposed model. The best accuracy obtained during the test set of each model is evaluated and presented in Table 3.

Table 3 Performance measure of the baseline models and proposed model

Model	Accuracy (%)
Perdomo et al. [9] (OCT-NET)	93.0
Proposed model	**93.5**
Kermany et al. [13]	96.6

When the proposed model was compared with the baseline model, it gives the best score. Even though comparing with the model proposed by Kermany et al. [13], the accuracy is reduced by 3.1% but the time required for training the proposed model is reduced by 50%. Since the proposed model uses depthwise separable convolution instead of standard convolution which reduces the computation required in each epoch which reduces the time required. But the accuracy is improved from OCT-NET, and the dimension of the features required by the model is also reduced.

5 Conclusion

In this work, a CNN architecture is proposed for the classification the retinal diseases. The system classifies three retinal diseases and normal scan images from SD-OCT scan images. The proposed model uses a publically available dataset for both training and testing. The model uses separable convolution which improved the accuracy and also reduces the time required for training the model. The model also reduced the complexity by decreasing the dimension of the feature set and hence improved the space and time utilization considerably. In addition, the model was also trained with NADAM optimizer. The system has an overall better performance,, and when comparing with other methods, it achieves a greater accuracy rate of 93.5%. In the future, the method can be used for classifying more retinal diseases.

References

1. Rong Y, Xiang D, Zhu W, Yu K, Shi F, Fan Z, Chen X (2019) Surrogate-assisted retinal OCT image classification based on convolutional neural networks. In: IEEE J Biomed Health Inform 23(1):253–263
2. Awais M, Müller H, Tang TB, Meriaudeau F (2017) Classification of sd-oct images using a deep learning approach. In: IEEE international conference on signal and image processing applications (ICSIPA). IEEE, pp 489–492
3. Simonyan K, Zisserman A (2014) Very deep convolutional networks for large-scale image recognition. arXiv:1409.1556
4. Gu J et al (2018) Recent advances in convolutional neural networks. Pattern Recog 77:354–377

5. Venhuizen FG, van Ginneken B, Bloemen B, van Grinsven MJ, Philipsen R, Hoyng C, Theelen T, Sánchez CI (2015) Automated age-related macular degeneration classification in OCT using unsupervised feature learning. In: Medical imaging 2015: computer-aided diagnosis, vol. 9414, international society for optics and photonics, p 94141I

6. Kamble RM et al (2018) Automated diabetic macular edema(DME) analysis using fine tuning with inception-resnet-v2 on OCT images. In: Annual conference of the IEEE engineering in medicine and biology society, vol 2018, pp 2717–27198

7. Indolia S, Goswami AK, Mishra SP, Asopa P (2018) Conceptual understanding of convolutional neural network—a deep learning approach. Proc Comput Sci

8. Zeiler MD, Fergus R (2014) Visualizing and understanding convolutional networks. In: 2014 13th European conference on computer vision (ECCV), pp 818–833

9. Fang L, Jin Y, Huang L, Guo S, Zhao G, Chen X (2019) Iterative fusion convolutional neural networks for classification of optical coherence tomography images. J Vis Commun Image Represent 59:327–333

10. Li X et al (2019) Integrating handcrafted and deep features for OCT based retinal disease classification. https://doi.org/10.1109/ACCESS.2019.2891975

11. Alqudah A (2019) A OCT-NET a convolutional network automated classification of multiclass retinal disease classification using SD-OCT images. In: International federation for medical and biological engineering

12. Li F, Chen H, Liu Z, Zhang X-d, Jiang M-s, Wu Z-z, Zhou K (2019) Deep learning based automated detection of retinal diseases using OCT images. R.59:327–33

13. Kermany DS, Goldbaum M, Cai W, Valentim CC, Liang H, Baxter SL et al (2018) Identifying medical diagnosis and treatable diseases by image based deep learning. Cell 172(5):1122–1131

14. Srinivasan PP et al (2014) Fully automated detection of diabetic macular edema and dry age-related macular degeneration from optical coherence tomography images. Biomed Opt Express 5(10):3568–3577

15. Perdomo O, Otálora S, González FA, Meriaudeau F, Müller H (2018) Oct-net: a convolutional network for automatic classification of normal and diabetic macular edema using SD-OCT volumes. In: 2018 IEEE 15th international symposium on biomedical imaging (ISBI 2018). IEEE, pp 1423–1426

16. Ji Q, He W, Huang J, Sun Y (2018) Efficient deep learning-based automated pathology identification in retinal optical coherence tomography images. Algorithms 11(6)

17. Pekala M, Joshi N, Freund DE, Bressler NM, DeBuc DC, Burlina PM (2018) Deep learning based retinal OCT segmentation. arXiv:1801.09749

18. He Y, Carass A, Jedynak BM, Solomon SD, Saidha S, Calabresi PA, Prince JL (2018) Topology guaranteed segmentation of the human retina from OCT using convolutional neural networks. arXiv:1803.05120

19. Karri SP, Chakraborty D, Chatterjee J (2017) Transfer learning based classification of optical coherence tomography images with diabetic macular edema and dry age-related macular degeneration. In: Biomed Opt Express 8(2):579–592

20. Albawi S, Mohammed TA, Al-Zawi S (2017) Understanding of a convolutional neural network. In: International conference on engineering and technology (ICET)

21. De Fauw J, Ledsam JR, Romera-Paredes B, Nikolov S, Tomasev N, Black-well S, Askham H, Glorot X, O'Donoghue B, Visentin D et al (2018) Clinically applicable deep learning for diagnosis and referral in retinal disease. Nat Med 24(9):1342

22. Krizhevsky A, Sutskever I, Hinton GE (2012) ImageNet classification with deep convolutional neural networks. In: Proc Adv Neural Inform Process Syst 1097–1105

23. Chollet F (2016) Xception: deep learning with depthwise separable convolutions [Online]. Available: https://arxiv.org/abs/1610.02357

Machine Learning-Based Smart Surveillance and Intrusion Detection System for National Geographic Borders

Mrinal Sharma and C. R. S. Kumar

1 Introduction

Artificial intelligence applications are rapidly growing and have taken entire world's attention towards the military applications. The future military systems are likely to be dominated by unmanned systems. The defence architecture of border areas which are prone to terrorists infiltration consists of high mesh of concertina wire followed by search lights, hooters, thermal imagers, minefields and patrols for recce. To ensure elimination of terrorists, fencing was kept under strict surveillance and various surveillance equipment were extensively used to cover the complete fence [1]. The efficiency of the intrusion detection system depends upon the ability to carry out effective surveillance of complete fence. Following are the challenges/ problems in ensuring continuous surveillance of border fence:

1. Requirement of large number of troops since the entire fence has to be physically manned 24×7.
2. No existing surveillance CCTV cameras all along the border to keep entire area under observation remotely.
3. Intrusion detection using the surveillance equipment takes place manually. Hence, plethora of manpower and assets is employed to guard the complete border.

The need is to design a smart system which can perform the surveillance task of detecting terrorist/UAV intrusion automatically without requiring human assistance [2]. Though humans are highly skilled at recognizing and detecting small movements, the strenuous nature of continuous monitoring several camera footages simultaneously is the main cause of committing errors. But a smart and autonomous system can perform the task all the time, and chances of human errors are reduced

M. Sharma (✉) · C. R. S. Kumar
Defence Institute of Advanced Technology, Girinagar, Pune, India
e-mail: suthikshnkumar@diat.ac.in

© The Author(s), under exclusive license to Springer Nature Singapore Pte Ltd. 2022
R. R. Raje et al. (eds.), *Artificial Intelligence and Technologies*,
Lecture Notes in Electrical Engineering 806,
https://doi.org/10.1007/978-981-16-6448-9_19

to minimum. Such systems if deployed successfully, can not only save resources but extensive manpower deployed for patrolling the fence, hence reducing the risk to life and save the existing surveillance equipment for other important tasks.

Solution to the problem lies in **smart surveillance and intrusion detection system (SSIDS)** which utilizes cameras and sensors incorporating artificial intelligence, suitably integrated, to provide a coherent picture which will contribute immensely in further enhancing the effectiveness of the border areas [2]. The solution involves installation of high-resolution cameras interfaced with AI-based intrusion detection module (IDM), incorporating machine learning/deep learning/computer vision algorithms, all along the fence except places which are less susceptible to intrusion. Since border fence has been erected to prevent terrorist infiltration, the following tasks can be executed automatically and intelligently by intrusion detection modules (Fig. 1):

1. **Automatic detection of terrorists (human)** during day and night.
2. **Automatic detection of weapons** carried by the humans.
3. **GPS module**. Suitably integrated with IDM to give location of the intrusion.
4. **Bluetooth scanner**. These scanners can detect the signature of any Bluetooth device like mobiles carried by the terrorists (in case the Bluetooth is active) while infiltrating, augmenting the detection possibility.

In addition to detection of terrorists' intrusion, the IDM can be employed to detect any UAV (UAV Detection) in the field of view of the cameras [3]. UAVs have been used widely in reconnaissance and observation from the low flying heights and supply weapons and cargo to terrorist near border fence by air dropping. Hence, detection of UAVs flying in the vicinity to fence with the objective of carrying out surveillance is an essential task. As the signature of human/weapon/Bluetooth/UAV is detected by the system, an alert signal along with the location

Fig. 1 Human, weapon and UAV detection

(Lat–Long) is sent to the surveillance centre and immediate reaction teams can be activated to respond to the intrusion with exact coordinates. On detection of UAV intrusion, higher Head Quarters can be informed and standard operating procedure can be followed.

2 Literature Review

In the field of computer vision and image processing, object detection plays a vital role in detecting objects (such as humans and weapons) in images and videos. Machine learning-based algorithms or deep learning-based algorithms are widely used for object detection. Machine learning algorithms require extraction of features from videos/images and then carrying out the classification using techniques like support vector machine [4]. Deep learning algorithms, typically based on convolutional neural networks (CNN), does not specifically require defining input features and execute end-to-end object detection.

Greeshma and Viji Gripsy [5] have presented an effective system of object classification and recognition using HOG and SVM as classifier. Vishwanath and Perumal Sankar [6] have proposed a methodology for border security surveillance using various sensors like IR sensors for human detection, gas sensors for poisonous gas detection and flame sensors for fire detection. All the sensors are integrated with Raspberry Pi for seamless operations. Setjo and Faridah [7] have implemented the Haar cascade classifier to detect humans using thermal images from IR cameras. The methodology used is much faster than HOG-based feature extraction methods with support vector machine as classifier.

Wang et al. [8] used HOG for feature extraction of training samples and classification using SVM classifier. The output of SVM classifier is preclassification results which are used as training samples to train the transfer network of CNN. This gave a new transfer learning model which is used to classify preclassification samples. Dwivedi et al. [9] propose CNN architectures by considering VGG16 as a base model for weapon classification. To train the proposed networks, weights of convolutional layer are initialized with the weights of pretrained VGG16 model while weights of fully connected layer are randomly initialized. Weights of the proposed network are fine-tuned by training this network with the images of guns. Sommer et al. [3] have proposed automatic UAV detection systems based on image differencing for UAV detection and classify each detection by a convolutional neural network (CNN) into the classes UAV.

Arreol et al. [10] have explained the use of Haar feature-based cascade classifier used for fast object detection using quadrotors. The data set requires positive and negative images to train the model and carry out detection. Redmon et al. [11] have described You Only Look Once (YOLO), a new technique towards object detection. The algorithm carries out object detection by predicting class probability and bounding box using neural network. Blizzard et al. [12] have presented approach for real-time human detection and tracking using Bayesian filtering framework and convolutional neural network on IP PTZ camera. Lakshmi Devasena et al. [13] have described an automatic video surveillance and object tracking system using background foreground subtraction technique to detect objects and track the objects by template matching to activate the Pan Tilt mechanism.

Though plethora of research has been conducted in the field of object detection like humans and weapons using known algorithms of deep learning and machine learning which have been applied in field of border surveillance in research papers, an integrated and holistic approach covering possible aspects of surveillance and intrusion detection is missing and hence needs to be explored.

3 Proposed Methodology

The proposed methodology involves an overall framework for intrusion detection and surveillance which include detection of humans with weapons during day and night, detection of UAVs/flying objects and detection of Bluetooth signature of intruder (using mobile phone or any Bluetooth-enabled device), combined together and integrated with a small microcomputer like Raspberry Pi for smooth and efficient operations. It will ensure a seamless and robust surveillance grid for the protection and security of border areas. The architecture of the proposed solution is shown in Fig. 2. The architecture has following components:

3.1 Intrusion Detection Module (AI-Based Module)

It consists of:

1. **High-resolution camera with IR capability for day and night vision**. Camera works well in the environment that has enough illumination or light intensity like daytime. In the environment with low light intensity or during the night, IR cameras can perform better since object (like human and weapons) that has contrast temperature with the surrounding environment is much easier to distinguish in the infrared camera.
2. **GPS module**. A GPS module can be interfaced with Raspberry Pi which can be coded for activating the GPS and extracting the exact location using Python coding.
3. **Bluetooth scanner**. Raspberry Pi has inbuilt Bluetooth which can be used as scanner. Python coding for activating the Bluetooth scanner, detecting the Bluetooth signature and sending alert signal, can be carried out.
4. Each module, housed in a climate protected box, can be mounted on an elevated platform at a distance of approx 100 m from adjacent camera with each camera

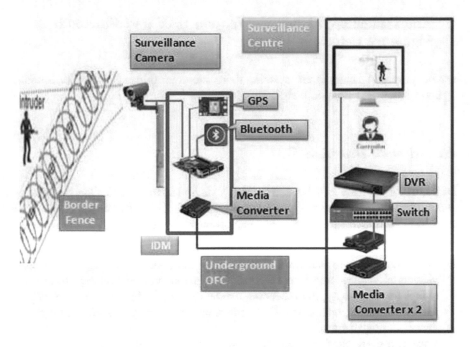

Fig. 2 Architecture of the proposed machine learning-based surveillance and intrusion detection system

focus overlapping with adjacent cameras. The power requirement of the cameras can be fulfilled with the help of generators and electric connectivity provided for illumination of the fence lights. Each IDM can be connected to an OFC-based media converter.

3.2 Surveillance Centre

Each surveillance centre can be connected to approx 15 IDM. The other pair of media converters, connected at IDM locations, will be present at surveillance centre. All media converters are connected to a DVR via switch. The video feeds of all the location can be viewed on a central monitoring system comprising of a high-end computer with a big screen connected to DVR to view 15 video feeds. Connectivity between surveillance centre and each IDM location will be over underground OFC. The communication could not be planned over wireless network keeping in mind the security of the data and possible interception or disruption by enemy. Communication over underground OFC is secure and sustainable in harsh climatic condition with minimum disruption and maintenance.

4 The Technology: Human, Weapon UAV and Bluetooth Signature Detection

In the proposed autonomous system, there are four primary task, i.e. human, weapon, UAV and Bluetooth signature detection.

4.1 Human Detection

Detection of humans can be carried out using machine learning/deep learning techniques. In the process of human detection, the video stream received from camera is processed in the form of video frames (images) which is followed by various stages:

1. **Coloured image to grayscale conversion**. The concept of carrying out conversion of coloured image to grayscale conversion is to process each frame and reduce its complexity, thus speeding up the subsequent background subtraction and segmentation operations. Grayscale images are generally processed much faster in comparison to RGB images.
2. **Foreground–Background Subtraction**. It is one of the fundamental methods used in the domain of video surveillance. The goal is to separate the human (foreground) and the environment (background) [14] as shown in Fig. 3. The general equation of subtraction is given by

$$(F_c(x, y) - F_p(x, y)) > T_h \qquad (1)$$

where $F_c(x, y)$ represents current frame, $F_p(x, y)$ represents previous frame and T_h represented threshold.

Fig. 3 Foreground–background subtraction

3. **Segmentation**. Segmentation divides the complete frame into small segments in which presence of human is detected. The process of segmentation is carried out by sliding window mechanism in which a fixed rectangular region of fixed width and height slides across an image [15]. If the object of interest is not found in the segment, sliding window is moved horizontally by certain number of pixels. For a video frame of 640×480 pixels, a sliding window of size of 200×100 pixel is selected for detection of human.

4. **Feature extraction**. It is the process of extracting features like shape, colour and pattern of object of interest. Various feature detection algorithms used are Harris keypoint detection (corner detection), SIFT or histogram of gradient (HOG). HOG is one of the most fundamental and robust feature descriptors in computer vision since it describes the complete shape of the object of concern by using intensity gradient and edge direction gradient, hence have a natural invariance to changes in lighting conditions and colour variation [16]. Pixelwise histograms of gradient directions are generated, and descriptor is obtained by concatenating them. In this, first the gradients of all pixels are computed. For an image T, the gradient filters used are

$$B(x) = [-1, 0, 1]B(y) = [-1, 0, 1]T \tag{2}$$

Let D_x and D_y be the gradient matrices generated by

$$D_x = I * B(x)D_y = I * B(y) \tag{3}$$

Here, $*$ represents the convolution. For each pixel, the gradient value is

$$d(i,j) = \sqrt{(D_x(i,j)2 + D_y(i,j)2)} \tag{4}$$

After this, a cell histogram is created. Based on the gradient values computed earlier, each point in a cell casts a weighted vote for the histogram channel. To provide an invariance in changes in illumination and contrast, the cells are grouped into spatially connected larger blocks and normalized locally. Finally, HOG feature vector is created by concatenation of these normalized cell histograms.

5. **Classification**. Human classification can be done using fundamental methods with machine learning classifier like support vector machine. The feature vector created by HOG algorithm is used as input. Quality of classification greatly depends upon the quality of input vector [17]. Support vector machine is a classifier that constructs a hyperplane or set of hyperplanes (decision boundaries), which can be used for classification of the data points. Data points lying on either side of the hyperplane can be attributed to different classes. Hence, given labelled training data, the algorithm outputs an optimal hyperplane that distinctly classifies new examples into one of the class or output. The number of

features decides the dimension of the hyperplane. The hyperplane is just a line when the number of input features is 2 and a two-dimensional plane when the number of input features is 3.

4.2 Weapon and UAV Detection

Weapons and UAV may be classified either by using machine learning algorithms or deep learning-based techniques. However, varying lighting conditions, fluctuating background contrast and rapidly changing object dimensions pose numerous challenges in detecting UAVs and weapons in video data [18]. Weapon/UAV detection has following stages:

1. **Coloured image to grayscale conversion**. It is the most fundamental method used to reduce the complexity of each frame and speed up the process as carried out in human detection.
2. **Background subtraction**. It is performed to distinguish the object or moving object from foreground and the environment (background) as carried out in human detection.
3. **Edge detection**. Edge detection is an important prerequisite for the next stages of the process. Canny edge detection is widely used algorithm which takes the filtered foreground object as input and gives the edges of the required object as output as shown in (Fig. 4).

Fig. 4 Canny edge detection

4. **Sliding window**. Sliding window technique is used to detect the UAV or weapon which can be at any location in the foreground frame. As seen in human detection, the area under examination by the learning algorithm is significantly reduced by sliding window technique. Hence, in the video frames of resolution 640 × 480, size of the weapon or UAV is significantly small. Hence, window size of 100 × 100 is optimal.

5. **Classification**. Classification of a weapon is done by a TensorFlow-based convolutional neural network (CNN). CNNs are highly layered structural neural network that have learnable weights and biases [19]. The input to the CNN is a feature vector that is generated in the previous stage (sliding window operation). It is passed through a set of a batch normalization layer, activation and max pooling layer. After passing the input features through them, flattening is executed and then fed to fully connected layers to retrieve an output of classification.

4.3   Training and Testing

Training of a machine learning/deep learning model involves feeding the training data set to the model to learn. Firstly, adequate number of video/image data set needs to be collected to train the model [20]. To train our model, three different data sets are required—human data set, weapons data set and UAV data set. Machine learning repository like kaggle or open-source TensorFlow data sets can be used to obtain the data sets and train the model. Next, we label the human, weapon and UAVs. For training of deep learning networks, a large number of positive and negative video frame samples need to be labelled. Positive samples refer to video frames/images which contain required object, and negative samples refer to video frames/images which do not contain required object [21]. Training of model is carried using TensorFlow or MATLAB. In the testing phase, detection is performed with sliding window method to extract the features using HOG and carry out the classification/detection.

4.4   Bluetooth Smart Scanners

Low-power, lightweight, easily upgradeable, secure and robust Bluetooth technology has wide range of practical application which can be exploited for military operation like detecting the Bluetooth signatures of infiltrating terrorist. Raspberry Pi platform has BCM43438 highly integrated single chip integrated Bluetooth and Wi-Fi chipset which support BLE (Bluetooth Low Energy, also known as Bluetooth Smart). So, there is no need to for external Bluetooth dongle. This frees up a USB port for other uses, which would otherwise be used up by the Bluetooth dongle. On-board Bluetooth of Raspberry Pi needs to be programmed in Python for activation, initiate scanning and sending alert signal on detection of any Bluetooth signature. Once enabled, Bluetooth scanning will look for Bluetooth signatures, respond quickly and send alert signal.

5 Operations on Border Areas

The sequence of action of the events are as follows (described in Fig. 5).

1. IDM continuously monitors for human/weapon/Bluetooth/UAV detection.
2. The operator in the surveillance centre receives the video feed of all the cameras under his area of responsibility on screen.
3. Any terrorist/UAV intrusion is detected by the IDM with Python codes and algorithms running continuously behind the system.
4. On detection, the IDM transmits an alert signal to the surveillance centre. This highlights the feed of the corresponding camera to red and alarms an alert.
5. The location of the surveillance camera can be retrieved from the GPS module connected to the IDM from surveillance centre using Python codes.
6. The operator can inform the immediate reaction teams about the location of intrusion and counter action can be performed. On detection of UAV intrusion, higher HQ can be informed and standard operating procedure can be followed.

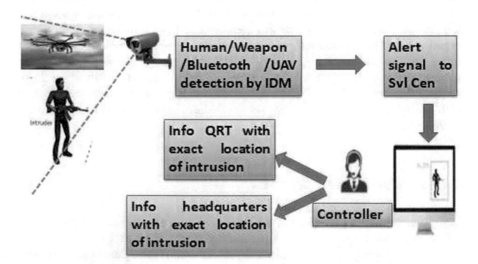

Fig. 5 Sequence of actions

6 Conclusion

Off lately, there has been a dynamic shift to replace the traditional and conventional equipment by smart and autonomous machines that trains and learns itself by observation, trial and error to enhance operational efficiency. In this paper, general overview of autonomous surveillance systems has been presented. The proposed system will enhance the security of border areas prone to cross-border terrorist movements, minimizing the requirement of manpower and judicious utilization of existing surveillance equipment and resources.

Intuitively, proposed surveillance system can be integrated with existing sensor and surveillance network at border areas which will provide more accurate information about the targeted object. Sensor fusion may happen at the proposed surveillance centres where each sensor/surveillance equipment/human, weapon and UAV detection module sends its original data which can be integrated and collated centrally augmenting the overall design of SSIDS with enhanced decision-making capability.

References

1. Forty Seventh report, Standing committee of Defence (2018–2019), Ministry of Defence
2. Problem definition statement-39 (2020) Compendium of problem definition statements, Ministry of Defence
3. Sommer L, Schumann A, Muller T, Schuchert T, Beyerer J (2017) Flying object detection for automatic UAV recognition. IEEE AVSS 2017. Lecce, Italy
4. Object Detection (2019) Learn computer vision using OpenCV: with deep learning CNNs and RNNs
5. Greeshma KV, Viji Gripsy J (2020) Image classification using HOG and LBP feature descriptors with SVM and CNN. Int J Eng Res Technol (IJERT). NSDARM-2020 conference proceedings
6. Vishwanath N, Perumal Sankar S (2019) Multisensor smart robot for border security surveillance with human action prediction. International J Innovative Technol Exploring Eng (IJITEE) 8(10S). ISSN: 2278-3075
7. Setjo CH, Faridah BA (2017) Thermal image human detection using HaarCascade classifier. In: 7th international annual engineering seminar (InAES). IEEE, Indonesia
8. Wang Y, Wang H, Luo L, Zhou Z (2019) Image classification based on transfer learning of convolutional neural network. In: Proceedings of the 38th Chinese control conference, Guangzhou, China, 27–30 July 2019
9. Dwivedi N, Singh DK, Singh Kushwaha D (2019) Weapon classification using deep convolutional neural network. In: 2019 IEEE conference on information and communication technology (CICT)
10. Arreol L, Gudi G, Flores G (2019) Object recognition and tracking using Haar-like features cascade classifiers: application to a quad-rotor UAV. The international conference on unmanned aircraft systems, Cornell University
11. Redmon J, Divvala S, Girshick R, Farhadi A (2016) You only look once: unified, real-time object detection. Conference on computer vision and pattern recognition (CVPR), IEEE

12. Blizzard D, Davar S, Mohammadi A (2017) Real-time and event-triggered object detection, recognition, and tracking. In: 60th international mid-west symposium on circuits and systems (MWSCAS). IEEE
13. Lakshmi Devasena C, Revathi R, Hemalatha M (2011) Video surveillance systems—a survey. Int J Comput Sci 8(4)
14. Singh P, Deepak TS, Murthy MDP (2015) Real-time object detection and tracking using color feature and motion. IEEE ICCSP 2015 conference
15. Yilmaz A, Javed O, Shah M (2006) Object tracking: a survey. ACM computing surveys, vol 38, No 4, Article 13
16. Hou Y-Y, Chiou S Y, Lin M H (2017) Real-time detection and tracking for moving objects based on computer vision method. In: 2017 2nd international conference on control and robotics engineering
17. Bhagyalakshmi I, Lakshmi B (2019) Real time video surveillance for automated weapon detection. Int J Trend Sci Res Dev 3(3)
18. Yazdi M, Bouwmans T (2018) New trends on moving object detection in video images captured by a moving camera: a survey. Comput Sci Rev
19. Rohith Sri Sai M, Veeravalli S (2019) Object detection and identification. Thesis paper, KL University, Department of Computer Science and Engineering
20. Borji A, Cheng M-M, Hou Q, Jiang H, Li J (2019) Salient object detection: a survey, vol 5, no 2. Computational Visual Media
21. Grega M, Matiolanski A, Guzik P, Leszczuk M (2016) Automated detection of firearms and Knives in a CCTV image. Sensors J

Real-Time Big Data Analysis Using Web Scraping in Apache Spark Environment: Case Study—Mobile Data Analysis from Flipkart

Pushpita Ganguly, Giriraj Parihar, and M. Sivagami

1 Introduction

Currently we are living in an era that is greeting "Big Data Generation." A newly defined word "Big Data" which refers to a huge volume of data obtained from various types of data sources and extreme frequent renovation of data. Businesses and enterprises can learn more about their operations and do analysis with the big data, so that they can convert that knowledge into upgraded decision-making processes and can perform better too. The obtainability of big data can be utilized in some special sort of big data applications that require real-time decisions to enhance the execution of their services or for the growth in profits. Examples of these applications are weather forecasting, smart transportation, share market analysis, surveillance, making decisions for military operations, trading market analysis, smart grids, and emergency response. These applications require working with stream and historical both kinds of data to carry out quick analysis aiding in instantaneous decision making. Untimely collection of data and processing leading to a delay in making decisions can remarkably reduce the execution of the applications. Real-time data analysis is kind of similar to enlarged data mining techniques. The reason is real-time data mining techniques that group data as a transaction and detect the frequent patterns from various transactions can also be applied to real-time big data analytics. However, there are few challenges in case of applying directly the real-time data mining techniques as it can deal with only structured data and as it is known that big data characterizes as both structured and unstructured data. Also, changing the scale of the data rapidly for the requirements of new knowledge is another challenge for real-time data analysis. So with these challenges applying real-time analysis will be an important challenge.

P. Ganguly (✉) · G. Parihar · M. Sivagami
School of Computer Science and Engineering, VIT University, Chennai, Tamil Nadu, India

© The Author(s), under exclusive license to Springer Nature Singapore Pte Ltd. 2022   177
R. R. Raje et al. (eds.), *Artificial Intelligence and Technologies*,
Lecture Notes in Electrical Engineering 806,
https://doi.org/10.1007/978-981-16-6448-9_20

In this paper, we proposed a model for real-time big data analysis that will contain major advantages for an enterprise to use or implement. The proposed model will eliminate all the challenges occurring for the real-time data analysis and then perform the analysis with the help of big data analytics tool apache Spark cluster. The paper surveys a real-time application of big data analytics, and then it discusses the model implemented by the proposed application and a short descriptive overview of the entire architecture of the model. In addition to that, it provides the analysis performed by the model and represents the importance of real-time data analysis for the current era of data science.

2 Related Work

Many research works have been done on big data analytics either in offline or real time. Some of them are highlighted here. Kim et al [1] proposed a model called real-time unstructured big data analysis (RUBA) framework. Mohamed and Al-Jaroodi [2] analyzed the challenges in real-time big data analytics with several applications. Akter and Fosso [3] represented a systematic review in e-commerce analytics of big data. Sivarajah et al. [4] produced a brief study on big data analytics with the state of the art review on challenges of big data analytics and its methods. Kolajo et al. [5] presented a brief view about big data real-time analysis using continuous queries and achieved analysis on the fly within the stream. Mittal and Sangwan [6] have discussed several challenges that may take place while using traditional machine learning tools for big data analytics and its possible solutions. Yadranjiaghdam et al. [7] constituted a brief survey on different methods in real-time analytics or stream analysis of the sheer volume data or big data in some specific domains of application. Ibtissame et al. [8] presented an overview of the already existing solutions for real-time big data processing and conducted an approximate comparison of the most favorable frameworks. Lv et al. [9] reviewed recent exploration in storage models, privacy, analysis methods, data security, data types and applications that are associated with the network of big data. Also, at the end of the paper, they have summarized the most obvious challenges and progress of big data to predict the current and future movement. L'heureux et al. [10] compiled, organized and summarized challenges occurring in machine learning with big data. Darwish and Bakar [11] discuss the ITS big data characteristics, a complete details of IoV environment and challenges that come across the implementation real-time big data analytics and fog computing and several computing technologies in the IoV environment and also discuss the issues should be considered to implement the proposed architecture more efficiently in the future work. Habeeb et al. [12] have surveyed anomaly detection-related real-time big data analytics technologies and described the issues faced in real-time anomaly detection. Qiu et al. [13] surveyed on advance machine learning approaches for big data processing. Thomas and Mathur [14] described the different types of scrapping the online data and its legal aspects. Al Walid et al. [15] have collected Twitter data by scrapping and analyzed on cancer-related tweets. Mikalef et al. [16] added a literature on the aspects of how real-time big data should be treated. Peng [17] proposed an approach to analyze data for any type and also any scale, using Robust Cloudera Hadoop pipeline. Haghighati and Sedig [18] describe Visual Analytics for RealTime Twitter datA (VARTTA), a visual analytics system to do analysis in real time with Twitter dataset. The proposed work basically highlights the data analysis in Apache Spark for real-time processing for online dataset.

3 Proposed Approach for Real-Time Big Data Analysis in Spark Environment

The proposed model considered with real-time big data analysis can analyze and predict more efficiently with online data. For example, it can predict based on previous data but with this model we can combine the real-time data like news and resources and predict it more correctly. The proposed architecture is unique and showed some great results. For scraping data in real time from the Web and all the processing, we have used Python language and spark.

Figure 1 shows the core architecture of our proposed model. The first procedure to analyze real-time data is extracting meaningful data from the Web. The model scrapes all the required information for creating the datasets using Python libraries and proceeds with the next level. Scraping the data can be considered as the initial stage of the model. Next, there are three steps that have been incorporated which will filter the data and will preprocess it more accurately for further analysis.

3.1 Preprocessing of Data

This second step is to clean the extracted data which is considered transforming raw data into an understandable format. As known, scraped data contains null values

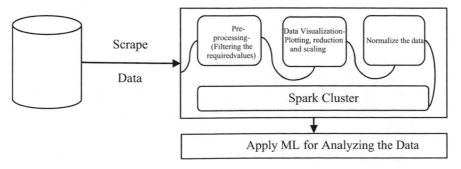

Fig. 1 Model architecture

and noise in it. So, this mechanism is applied for eliminating those null values and all the noise from the data. Now, the data is filtered, the null values and all kinds of noises are removed from the data. The mechanism also implies that all the peripheral columns are removed from the data.

3.2 Exploratory Data Analysis (EDA) Techniques

The model used EDA techniques to display the data in such a way that important and meaningful data become more apparent by plotting graphs using the data which have been already filtered with required attributes. Some of the EDA techniques are listed below.

Data plotting: Data plotting is a graphical representation of data that aids in more outward by constituting the relationship between two or more features. The aim of plotting the datasets is to gain insights into the underlying structure of the datasets.

Dimensionality Reduction: This is the technique used for decreasing the number of random variables by picking up some set of principle variables. Ideally the aim is to use more filtration to make the dataset more accurate as it transforms the data from a high-dimensional space to a low-dimensional space to retain only the meaningful properties that will be represented by low-dimensional space.

Multidimensional scaling: Multidimensional scaling can be referred to showing the homogeneous level of a dataset. Basically it is a set of techniques used for translating the information among individuals about pairwise distances. This technique also can be referred to as a nonlinear dimensionality reduction technique.

3.3 Normalization of Data

In the next step, the model used normalization of data to model it more correctly. Normalization can be referred to as a part of data preparation to analyze and that is why this step is needed to execute for further analysis. The main goal is to use a common scale in the dataset and so that it will change the columns based on a common scale. This step prepared the data completely to store and execute the analysis techniques on it. Normalization helped the model to perform analysis on the data very fast as it has already set a common scale on the dataset. Whichever analysis model needs to implement the model can easily perform without worrying about the data model. Linear scaling normalization technique is applied in this research work.

After the filtration and normalization of the extracted data which is obtained by scraping from the Web, now the model will store the clean and normalized data into Apache Spark Cluster. The model used Apache Spark tool for this sheer volume of data extracted for the purpose of storing. Using the required Spark library, the model stored all the important features that are needed for analysis in the Spark cluster that can be aided in real-time processing.

Now, in the last level, the model applied a couple of required machine learning models for analyzing the data as shown in the architecture using Python. After extracting the data from the Web, the inappropriate and uncalled data first filtered through required techniques to make it accurate and then scaled based on a common scale and made the data ready for analysis. Using machine learning tools for prediction and classification, the model now will analyze the data which will be a helpful insight for any business organization or any service enterprises. The main goal of the model is to make use of the real-time sheer amount of data which is greeted by the current era and help the enterprises to make decisions more correctly on the basis of these analyses which can encourage enterprises toward growth.

3.4 Applied Machine Learning Tools

Linear regression: Perhaps, it is the well-known tool for machine learning which is based on supervised learning. Linear regression performs regression tasks with a target prediction value based on the independent variables. The main use of this tool is to find out the relationship between two variables. In our proposed model, we have used linear regression to analyze the scraped information about mobile phones. Using this tool, we have made mobile rating predictions based on mobile price, RAM, and ROM. The other feature was also analyzed by the model as further analysis of the data. This analysis can make more profit and can give more valuable insights to business organizations.

Classification: Classification is a big part of machine learning which is used to know in what class an observation belongs to. Classifications can be either a binary classification problem or a multiclass problem both. It identifies the class in which data features belong to and is finest used when the outcome has restricted and discrete values. Data science comes up with a plethora of classification algorithms such as support vector machines, Naive Bayes classifier, random forest, and decision trees. The potentiality to very precisely classify various observations is exceedingly valuable for business fields and their applications.

In our proposed model, various classifiers such as random forest classifier with 71% accuracy, decision tree classifier with 71% accuracy, MLP or multilayer perceptron with 65% accuracy, SVM or support vector machine classifier with 60% accuracy, and KNN classifier with 71% accuracy are used for classification based on mobile rating.

Clustering: Clustering represents the grouping of all the data points. Using a clustering algorithm from a given set of data points, we can classify each data point into a particular specific group. In data science, we have several clustering algorithms and our model has used agglomerative clustering.

The proposed model using agglomerative clustering grouped all the features into clusters based on their similarities. It then represents the DENDROGRAM, the graphical representation of the hierarchy and follows the bottom-up manner. The model used clustering for visualizing and analyzing the data more efficiently taking each feature as a singleton cluster. In this research work, clustering has been done to view the dataset in a batter way, not toward improving the accuracy, which directly does not impact the quality of the result.

4 Results and Analysis

The applied architecture executed on a sample dataset and performed the analysis. For this analysis, we used Flipkart Web site for extracting all the cell phone data. In the mobile category in Flipkart Web site, there are lots of pages and every page contains more than 20 items. As our requirement, we need mobile phone names including their rating, price, RAM, and ROM. For scraping data, we used "Beautifulsoup" library.

After applying preprocessing techniques to remove the null values and convert the columns into required format, then all the data directly loaded into Apache Spark cluster. The graph has been drawn to show the correlation in between features so we can easily apply the analysis on the data and graph has been shown in Fig. 3. Lastly, we apply normalization (Fig. 4) on the data so it will be computationally low. In Fig. 2, sample data has shown.

We have created a local copy of data from spark cluster in which there are four attributes are there "Rating," "Price," "RAM," and "ROM" and changed the mobile rating column from continuous values to discrete values as S, A, B, C, D (rating 4.5 to 5 as A, 4 to 4.5 as B, 3 to 4 as C and values <3 are as D) and save it to another

	Unnamed: 0	mobileName	mobileRating	mobilePrice	mobileRam	mobileRom
0	0	Realme C11 (Rich Grey, 32 GB)	4.6	7499	2.0	32.0
1	1	Realme C11 (Rich Green, 32 GB)	4.6	7499	2.0	32.0
2	2	Motorola G8 Power Lite (Royal Blue, 64 GB)	4.4	9499	4.0	64.0
3	3	Motorola G8 Power Lite (Arctic Blue, 64 GB)	4.4	9499	4.0	64.0
4	4	POCO X2 (Matrix Purple, 64 GB)	4.5	17499	6.0	64.0
5	5	Realme 6 (Comet White, 64 GB)	4.4	15999	6.0	64.0
6	6	Realme 6 (Comet Blue, 64 GB)	4.4	15999	6.0	64.0
7	7	POCO X2 (Atlantis Blue, 64 GB)	4.5	17499	6.0	64.0
8	8	Realme 6 Pro (Lightning Orange, 64 GB)	4.5	17999	6.0	64.0
9	9	Realme 6 Pro (Lightning Blue, 64 GB)	4.5	17999	6.0	64.0

Fig. 2 Sample scraped dataset

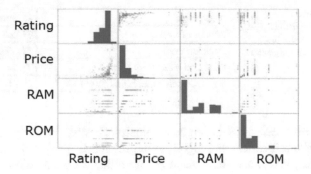

Fig. 3 Correlation between attributes of dataset

	A					B		
	mobileRating	mobilePrice	mobileRam	mobileRom		mobileRating	mobilePrice	mobileRam	mobileRom
0	4.6	7499	2.0	32.0	0	S	7499	2.0	32.0
1	4.6	7499	2.0	32.0	1	S	7499	2.0	32.0
2	4.4	9499	4.0	64.0	2	A	9499	4.0	64.0
3	4.4	9499	4.0	64.0	3	A	9499	4.0	64.0
4	4.5	17499	6.0	64.0	4	A	17499	6.0	64.0
5	4.4	15999	6.0	64.0	5	A	15999	6.0	64.0
6	4.4	15999	6.0	64.0	6	A	15999	6.0	64.0
7	4.5	17499	6.0	64.0	7	A	17499	6.0	64.0
8	4.5	17999	6.0	64.0	8	A	17999	6.0	64.0
9	4.5	17999	6.0	64.0	9	A	17999	6.0	64.0

Fig. 4 Data normalization

spark cluster. Now with two different spark datasets, one with continues value (Fig. 4a) for applying prediction and one with discrete value (Fig. 4b) for classification. In Fig. 3, the correlation between all the attributes of the dataset has shown.

Then processed data is saved in the Spark cluster and at the same time it is accessing the data from the Spark cluster also. In Fig. 5, you can see the structure of the data.

```
1  company_df.cache()
2  company_df.printSchema()

root
 |-- mobileRating: double (nullable = true)
 |-- mobilePrice: integer (nullable = true)
 |-- mobileRam: double (nullable = true)
 |-- mobileRom: double (nullable = true)
```

Fig. 5 Dataset schema

Table 1 Applied classification algorithms and their accuracy on mobile dataset—rating classification

Algorithms	Accuracy (%)
Random forest	71.06
Decision tree	71.06
K-nearest neighbors (KNN)	71.52
Support vector machine (SVM)	60.91

Now we can access the data and do the analysis easily. For analysis on the data, we are using machine learning algorithm-regression to predict the mobile rating based on price, RAM, and ROM features. With this we get 38.68% accuracy.

So as we know that for better accuracy we need more features like launching year, all the features, all the positive and negative feedbacks, and their reasons for mobile phones. Then we can predict ratings more accurately. Then the attribute ROM of the mobile is predicted based on RAM, price, and rating attributes of the dataset with 83.53% accuracy. It shows that the ROM attribute is correlated to RAM, price, and rating. On the other hand, classification algorithms are applied on the dataset to classify the rating. As mentioned above, the continuous values are converted as discrete values (Table 1).

5 Conclusion

In this paper, we have tried to represent a holistic observation of real-time big data analysis or stream analysis with a literature survey to discuss and understand the tools and technologies. Also, the challenges can occur for streaming data analysis discussed with the help of the model architecture and overview of the model. The proposed model showed that how it has handled the big data analytics in real time from data scraping, preprocessing, visualization, prediction, classification, and clustering with the help of sample dataset of this case study. The adequate of these challenges will effectively improve the development of streaming big data applications and that will provide a number of merits such as scaling down risks, improvement in quality of life, and strengthening profitability in turn.

Acknowledgements This paper would like to thank "Flipkart" Web site as the proposed model worked with the dataset which is scraped from this Web site.

References

1. Kim J, Kim N, Lee B, Park J, Seo K, Park H (2013) RUBA: real-time unstructured big data analysis framework. International conference on ICT convergence, pp 518-522. https://doi.org/10.1109/ICTC.2013.6675410
2. Mohamed N, Al-Jaroodi J (2014) Real-time big data analytics: applications and challenges. Proceedings of the 2014 international conference on high performance computing and simulation, HPCS 2014. https://doi.org/10.1109/HPCSim.2014.6903700

3. Akter S, Fosso WS (2016) Big data analytics in e-commerce: a systematic review and agenda for future research. Electr Markets 26:173–194
4. Sivarajah U, Kamal MM, Irani Z, Weerakkody V (2016) Critical analysis of big data challenges and analytical methods. J Bus Res 70:263–286
5. Kolajo T, Daramola O, Adebiyi A (2019) Big data stream analysis: a systematic literature review. J Big Data 6:47. https://doi.org/10.1186/s40537-019-0210-7
6. Mittal S, Sangwan OP (2019) Big data analytics using machine learning techniques. In: 2019 9th international conference on cloud computing, data science & engineering (confluence). IEEE, pp 203–207
7. Yadranjiaghdam B, Pool N, Tabrizi N (2016) A survey on real-time big data analytics: applications and tools. In: 2016 international conference on computational science and computational intelligence (CSCI). IEEE, pp 404–409
8. Ibtissame K, Yassine R, Habiba C (2017) Real time processing technologies in big data: comparative study. In: 2017 IEEE international conference on power, control, signals and instrumentation engineering (ICPCSI). IEEE, pp 256–262
9. Lv Z, Song H, Basanta-Val P, Steed A, Jo M (2017) Next-generation big data analytics: State of the art, challenges, and future research topics. IEEE Trans Industr Inf 13(4):1891–1899
10. L'heureux A, Grolinger K, Elyamany HF, Capretz MA (2017) Machine learning with big data: challenges and approaches. IEEE Access 5, 7776–7797
11. Darwish TS, Bakar KA (2018) Fog based intelligent transportation big data analytics in the internet of vehicles environment: motivations, architecture, challenges, and critical issues. IEEE Access 6:15679–15701
12. Habeeb RAA, Nasaruddin F, Gani A, Hashem IAT, Ahmed E, Imran M (2019) Real-time big data processing for anomaly detection: a survey. Int J Inf Manage 45:289–307
13. Qiu J, Wu Q, Ding G, Xu Y, Feng S (2016) A survey of machine learning for big data processing. EURASIP J Adv Signal Process 2016(1):67
14. Thomas DM, Mathur S (2019) Data analysis by web scraping using python. In: 2019 3rd international conference on electronics, communication and aerospace technology (ICECA). IEEE, pp 450–454
15. Al Walid M, Anisuzzaman DM, Saifuddin Saif AFM (2019) Data analysis and visualization of continental cancer situation by Twitter scraping. Int J Modern Educ Comput Sci 11(7)
16. Mikalef P, Pappas IO, Krogstie J, Giannakos M (2018) Big data analytics capabilities: a systematic literature review and research agenda. Inf Syst eBusiness Manag 16(3):547–578
17. Peng Z (2019) Stocks analysis and prediction using big data analytics. In: 2019 international conference on intelligent transportation, big data and smart City (ICITBS). IEEE, pp 309–312
18. Haghighati A, Sedig K (2020) VARTTA: a visual analytics system for making sense of real-time Twitter data. Data 5(1):20

Future Frame Prediction Using Deep Learning

Siddharth Itagi, Sinchana Gowda, Tanmaya Udupa, and S. S. Shylaja

1 Introduction

Videos are basically a set of continuous images that are captured throughout the duration. A term called frames per second (FPS) is used to determine the number of frames—which are single, still images—that are captured in a second. Higher the value of the FPS, smoother is the transition of the frames which results in a much better video output. Since a given video is a set of frames captured in a particular duration, the future frame would be the next frame present in the sequence if the video had continued beyond its specified duration. Given a set of images or frames, the aim is to predict the future frame of the given sequence. For the prediction of this future frame, deep learning methods provide a solution. Deep learning is based on similar lines of machine learning but with added benefits. Deep learning models consist of neural networks that are much deeper than artificial neural networks because they contain more layers and myriad weights to learn. These models are generally the first choice when there is a need to capture and learn lots of features especially in image or video processing. This is because there are a lot of attributes that need to be captured in images as each and every minute detail can aid in better learning of the model. Deep learning models, specifically those related to genera-tion of images, GANs, CNNs, LSTMs, and variational autoencoders (VAEs) are the ones that have been proven to generate better results. Furthermore, out of all these techniques, GANs are considered to be the best.

Autoencoders are unsupervised artificial neural networks that learn how to compress and encode the input data followed by reconstruction of the data from the encoded representation. Basically, they reduce the dimensions of data by learning how to ignore noisy data so as to capture only the essential features. Some appli-cations of autoencoders include dimensionality reduction, image denoising, and

S. Itagi (✉) · S. Gowda · T. Udupa · S. S. Shylaja
CSE Department, PES University, Bangalore, India
e-mail: shylaja.sharath@pes.edu

© The Author(s), under exclusive license to Springer Nature Singapore Pte Ltd. 2022 187
R. R. Raje et al. (eds.), *Artificial Intelligence and Technologies*,
Lecture Notes in Electrical Engineering 806,
https://doi.org/10.1007/978-981-16-6448-9_21

anomaly detection [1]. An autoencoder consists of four main parts—encoder, bottleneck, decoder, and reconstruction loss. The encoder learns how to compress the input data into an encoded or compressed form known as the bottleneck. The decoder, on the other hand, learns how to reconstruct the data back from this encoded form while trying to keep the output as close to the input as possible. The reconstruction loss is a measurement of similarity between the decoder's output relative to the original input. In order to train the model, backpropagation is used to minimize the reconstruction loss. A few different types of autoencoders include vanilla autoencoder, multilayer autoencoder, and convolutional autoencoder. A vanilla autoencoder is the simplest form of an autoencoder and contains one input layer, one hidden layer, and one output layer. In order to extend the autoencoder, multiple hidden layers can be added, resulting in a multilayer autoencoder. Instead of using fully connected layers, convolutional layers can be used as well where three-dimensional vectors are substituted for one-dimensional vectors. This paper makes use of convolutional autoencoders followed by GANs to further improve the results produced by the autoencoder and make the model generate better images.

Generative adversarial networks (GANs) are generative models that are capable of producing new images or content. Here the objective is formulated as a supervised learning problem with two submodels, mainly the generator and the discriminator. The generator is responsible for taking in a random input distribution and generating a sample in that domain. This sample that is generated should be similar to the original frame. A discriminator, on the other hand, is responsible for correctly distinguishing between an actual image and the one that is produced by the generator. Hence, the purpose of the generator is to fool the discriminator by producing a sample/image almost identical to the original image. A GAN can be treated as a two-player game. Consider the following example: the generator can be thought of as a fake money fraudster trying to make fake currency notes, while the discriminator is assumed to be the police whose job is to allow only legitimate money by identifying fraudulent money. In order to succeed in this game, the generator network must learn to create samples that are drawn from the same distribution as that of the training data, in the same way a fraudster tries to learn how to make currency notes that are indistinguishable from genuine currency notes. The objective of the GAN is to achieve equilibrium, often termed as the Nash equilibrium, where under no circumstance can one perform better than the other by hampering the opponent's performance. It is the most stable condition in which both the generator and the discriminator are trained to perform to the best of their abilities, i.e., the discriminator is 50% sure about an image being fake or real. Usually, the final winner in a GAN scenario is the generator as it is the one producing new images almost similar to the original input image. Recent studies have shown that GANs are considered the best in the class of generative models due to its inherent concept of training and competition between the generator and the discriminator. They also are capable of producing newer combinations of outputs that are of great help in certain problem statements which deal with generation of new variants of a particular product type.

2 Related Work

Multiple studies related to video prediction have been reviewed, and their studies have been analyzed. Video predictions involve the problem of image transformation. Johnson et al. [2] consider image transformation as a problem statement. This deals with the transformation of an input image into the output image. The usual methods or solutions to deal with these problems are typically done by training feed-forward convolutional neural networks. They take into consideration per pixel loss between the output image and the input image or the original image. They focus on extracting specific features from the image in order to make this regeneration of images efficient and obtain good accuracy for the same.

The proposed model [2] incorporates defining loss functions that use pretrained networks for the extraction of high-level features. Here, they try to utilize the benefits of both the approaches, i.e., trained feed-forward networks for image transformation tasks but rather than the per pixel loss functions for training, they focus on pretrained models which are capable of capturing the important aspects of an image that are also known as high-level features. These methods try to target the concept that the already learned networks encode and capture the perceptual information that would be used in loss functions.

The main emphasis and essence of the paper [2] lies in the unique loss functions taken into consideration. These are called the perceptual loss functions that are meant for capturing semantic and high-level perceptual differences between images. This is done by making use of a loss network that is pretrained for image classification. Here, with per pixel loss which is pixel-to-pixel comparison of the input and output image, these additional perceptual loss functions are taken into consideration to determine the cumulative loss. The reason for not just solely considering per pixel loss is because there may be two images that may slightly differ in their pixel orientation but would have all the features in great detail that are there in the input image too. However, since there are pixel differences, this would account for a good amount of error. After loss functions, we reviewed training deep neural networks. When deep networks are trained, once they start converging, a significant issue is faced. A degradation problem is imminent. Here, as the network depth goes on increasing, the accuracy takes a step back, i.e., it gets saturated and then starts to fall or degrade rapidly. This dramatic fall or degradation is not associated with overfitting or the presence of more layers which leads to the model adding a higher training error. The reason for this degradation is that not all systems are easier to optimize. There exists a model that is much deeper where identity mapping is used in added layers and the other layers are a mimic of the learned shallower model.

He et al. [3] talk about the existence of this constructed solution to convey that higher layers in a deep learning model are not supposed to produce higher training error than its shallower counterpart. They also talk about introducing a deep residual learning framework. Here, instead of assuming that every stacked layer matches and fits the corresponding stacked layer directly, they explicitly fit residual mapping to have these layers. This mapping is shown in Fig. 1.

Fig. 1 Formulation of
feed-forward network

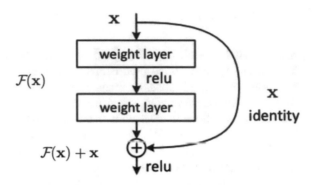

The formulation given in Fig. 1 can be understood as a feed-forward neural network with "shortcut connections." These connections are basically skipping one or more layers. They consist of identity mapping, and their outputs are concatenated to the outputs of stacked layers. The entire network can be trained end-to-end by using stochastic gradient descent [4] with backpropagation. On the ImageNet classification dataset, they have obtained excellent results with the usage of large deep residual networks. Their proposed network of 152 layers is the deepest network used on ImageNet and has a much lower complexity than other VGG networks. This strong evidence shows that the residual learning principle talked about in the paper and implemented is generic and hence can be used for other problems that deal with both vision and non-vision problems.

In specific, video prediction is understood as a problem of generating frames for a continuous set of frames. The solution to this problem can be used in various fields of video processing like anomaly detection, autonomous driving, and video coding. Extensive studies [2, 3, 5] have confirmed that deep neural networks have the ability to predict realistic future images. Most of the methods talk about using CNN for this frame prediction. However, the problem with these models is that they tend to give a blurry output, i.e., a blurry frame.

Kwon et al. [5] propose a deep network architecture for the generation of future frames or images. The model proposed talks about using GANs as the architecture. They firstly train a generator capable of predicting both future and past frames. They have two discriminators called frame and sequence discriminators. The frame discriminator is capable of distinguishing between fake frames individually while the sequence discriminator is responsible to decide whether the sequence contains fake frames or not. The generator is responsible for taking sequential images as input and then using those frames to generate the future frame. The generator is made up of a neural network consisting of four convolutional layers, four convolutional transpose layers, and nine residual blocks connected in the form of a UNet architecture. Five convolution layers make up the architecture of the discriminator with leaky rectified linear units as the activation function. The architecture for both sequence and frame discriminator is the same except for the number of input images it accepts. They introduce the idea of retrospective prediction which states that, if the predicted frame is realistic then it can be used in the sequence of images to predict a past frame as well. This would allow the model to predict the frames with high accuracy and can also be used to generate multiple future frames.

With forward and backward sequences given as input to the generator for training, and with an additional retrospective cyclic consistency, good training and results are ensured. They stated that it was experimentally verified that the proposed method was better than the previous methods tried in future frame prediction and is definitely one of the state-of-the-art methods.

3 Problem Statement

3.1 Overview

Our problem statement involves predicting the future frames of a given video which is represented as a series of consecutive frames. In the recent past, the introduction of deep learning techniques has successfully improved the performance of video prediction. However, most of the algorithms used generate blurry images, which is the drawback this paper aims to overcome.

In order to accurately predict future frames, the model aims to capture the visual representation of the input frames and predict the visual representation of the future frame, rather than predicting individual pixels. Through the help of generative models and different architectures for the encoder, future frames are predicted.

3.2 Dataset

The data being used for building the model is the KITTI dataset. KITTI dataset [6] was made of videos recorded by car-mounted cameras. It consists of images and videos captured from a Station Wagon car and are used in autonomous driving and mobile robotics. The raw dataset is divided into "City," "Residential," "Campus," and "Person" and is 180 GB in size. The one we are using is the "City" category of the dataset. The dataset consists of black-and-white frames captured while driving, of which we have used 1000 images, to train our model.

4 Our Approach

For the purpose of frame prediction, we propose to use GANs as our model. Here, the input to our model will be a set of frames, and the model will finally predict the next frame in the sequence. The discriminator is a vanilla CNN that is capable of distinguishing between the frame given by the generator and the actual frame at any specific time step. The training of the generator and the discriminator is done separately by keeping each frozen at the time of training of the other.

The objective of the GAN is to obtain equilibrium, i.e., a state where the generator and the discriminator perform optimally. Such a state is achieved by a gradual increase in the performance of one when there is a decrease in performance of the other. The overall system architecture is shown in Fig. 2.

Let us consider the generator first. The sole purpose of the generator is to generate a future frame that is similar to or the same as the actual frame. Hence, we have used an autoencoder to build our generator. It is built using UNet [7] Architecture helps the generator to learn more features of the image. In this autoencoder is capable of taking multiple frames as input. This input is given to the encoder which takes each frame at a time and generates a hidden/coded representation of the same. Now in order to take into consideration the sequence of frames, i.e., frames with respect to time, we have experimented with the following four approaches:

4.1 Approach-1 (Code Layer Overlapped Architecture)

As shown in Fig. 3, we pass each frame to the encoder and get the hidden/coded representation of that frame. Once the encoded layers of all the frames in a set are obtained, they are overlapped with each other and then the pixel values are

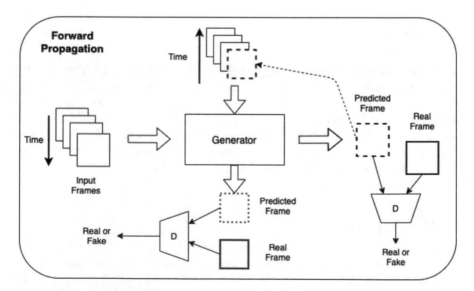

Fig. 2 System architecture

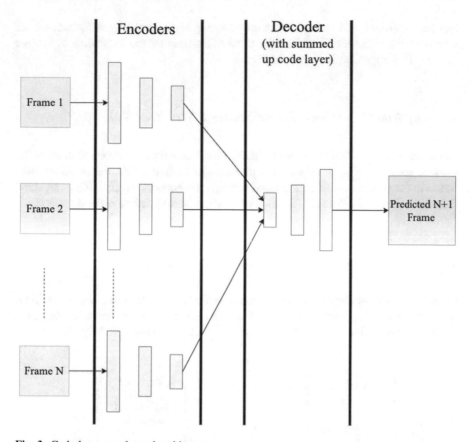

Fig. 3 Code layer overlapped architecture

normalized. This process is repeated for all the sets of N frames. Finally, the list of encoded layers of N different sets is then fed to the decoder to generate the predicted image.

4.2 Approach-2 (Code Layer Concatenated Architecture)

In this approach, we pass each frame to the encoder and get the hidden/code representation of that frame. Now we concatenate the code representations of each frame's code layer. The concatenated code layer is then sent to the decoder to predict the next frame. For example, suppose the code layer shape of a single frame is $32 \times 87 \times 128$, we concatenate the layers of n frames to form $(32 \times 87 \times 128) * n$. The way the code layer representations are concatenated is very similar to the kind of input given to a fully connected layer. Therefore, the way this can be visualized is a single vector containing all the code layer representations where the filters are modified in order to take into account the time sequence of the frames. This approach is shown in Fig. 4.

4.3 Approach-3 (Code Layer Stacked Architecture)

Here, we pass each frame to an encoder which gives a code representation of the same as shown in Fig. 5. This code representation is then overlapped with the next frame in the sequence and then passed through the encoder again. This is repeated for all the frames in the set. The final code layer is then fed to a decoder.

4.4 Approach-4 (Input Concatenated Architecture)

In this approach, the input images are concatenated in the beginning before giving it to the encoder. This can be visualized as a single vector containing all the input image representations one behind the other. This concatenation helps in making

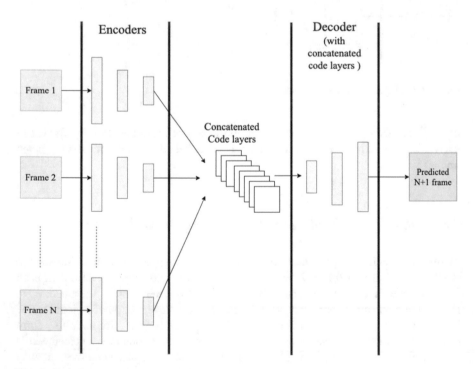

Fig. 4 Code layer concatenated architecture

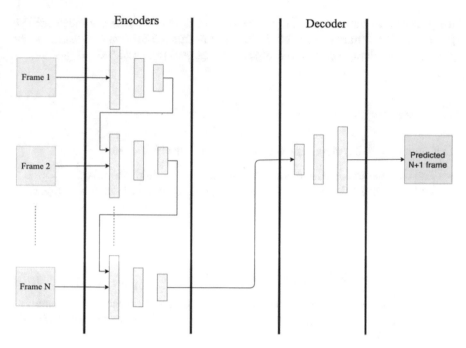

Fig. 5 Code layer stacked architecture

sure that the relative sequence of frames over a particular time frame are taken into consideration. This is shown in Fig. 6.

The above approaches explain how the generator is built and designed in order to take into consideration the time sequence of frames for it to be able to predict the future frame. We picked the input concatenated architecture for our generator model based on certain performance metrics used to determine the quality of the predicted

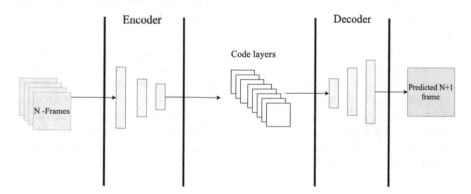

Fig. 6 Input concatenated architecture

images, such as mean squared error (MSE) and structural similarity index (SSIM) [8] obtained after training each of the four architectures. SSIM values determine the amount of similarity between the original image and the predicted image.

4.5 Algorithm

The following is the algorithm we used to train our GAN model.

- Train the generator with reconstruction loss for the initial training.
- Consider $D(x)$ to be the output of the discriminator when a frame x is passed to it.
- Consider $G(z^{m:n})$ to be the output of the generator when a sequence of frames, z is given to it.
- For each epoch of GAN training:

  – For each batch:

    Sample a set of real frames from the dataset and train the discriminator. Sample a set of generated frames from the generator and train the discriminator.

    $$\text{maximize } \log(D(x)) + \log(1 - D(G(z^{m:n}))) \tag{1}$$

    Sample a set of n frames from the dataset and pass it as input to the generator to predict the (n + 1)th frame and train it by keeping the discriminator constant.

    $$\text{minimize } \log(1 - D(G(z^{m:n}))) \tag{2}$$

The predicted frame obtained from the generator is subjected to the discriminator along with the original frame. The discriminator is then responsible for distinguishing between these two frames. The output of the discriminator is backpropagated to the generator which is used to improve the predictions by learning from the error measures obtained. We have used PatchGAN [9] for stable and better training of the entire GAN. This is more specifically implemented in what is known as the patch discriminator capable of producing multiple error values for different patches of the input image given to the discriminator. This helps in better capture of features and error propagation. The generator and discriminator work in sync to make sure that if either one of them is overperforming, the other one tries to train better to nullify this effect and hence reach equilibrium.

5 Results and Discussion

The autoencoder with the four different approaches mentioned previously was trained using a learning rate of 0.0002 and Adam [10] as the optimizer. We used mean square error (MSE) as the error metric. Also, according to the literature referred to previously, we picked $n = 4$ as the number of input frames being considered at a time to predict the future frame. This was run for a varied set of epochs ranging from 50 to 1000 epochs and the SSIM metric was used along with MSE in order to decide the best approach for the model. The first architecture, i.e., code layer overlapped architecture did not generate good future frames as all the code layers of the input frames were overlapped one above the other, and the sequence of the frames was not taken into consideration. The code layer stacked architecture, being one of the hardest to train, did not give promising results as well. The third and fourth architecture, i.e., code layer concatenated and input concate nated architecture considered the sequence of input frames and worked well in generating a future frame, thus giving good results. The best SSIM score was obtained for input concatenated approach with a value of 0.85 for training data and 0.84 for testing data. A summary of the SSIM scores can be seen in Table 1.

We then went ahead and used the input concatenated autoencoder model that produced the best result as the generator in the GAN model training in order to obtain better predictions of images. We chose $n = 4$ with binary cross-entropy as the error measure and used batch normalization to make the network more stable and speed up the entire training process. We trained the GAN on the KITTI dataset for a range of 100–300 epochs. The use of PatchGAN helped in the stability of GAN training and made sure that all the weights of the generator are updated with the error produced by the discriminator's classification. The best results obtained for GAN had an SSIM score of 0.88 for training and 0.875 for testing. A summary of the SSIM scores for the GAN training can be seen in Table 2.

Table 1 Autoencoder results

	Training		Testing	
	500 epochs	1000 epochs	500 epochs	1000 epochs
Code layer overlapped	0.64	–	–	–
Code layer concatenated	0.80	0.83	0.70	0.74
Input concatenated	0.78	0.85	0.75	0.84

Table 2 GAN results

	Training			Testing		
	100 epochs	200 epochs	300 epochs	100 epochs	200 epochs	300 epochs
SSIM	0.87	0.88	0.88	0.85	0.87	0.875
MSE	0.0043	0.0040	0.0039	0.0045	0.0041	0.0039

The results obtained by the autoencoders were enhanced by training of the GAN and led to an increase in SSIM values of the predicted frame by 0.035. The predictions can be seen on one of the sets of four frames in Figs. 7 and 8.

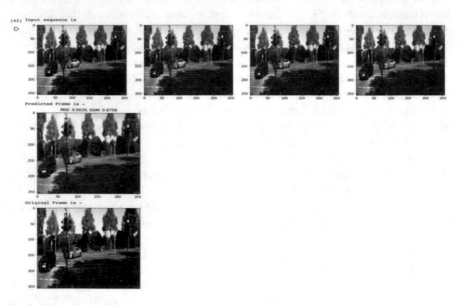

Fig. 7 Prediction results-1

Fig. 8 Prediction results-2

6 Conclusion

We built four different autoencoder architectures and trained them with different parameters in order to obtain the best results. This was satisfied by the input concatenated approach which is why we implemented it as the generator for our GAN. We also verified that GAN models predict better future frames than autoencoders by comparing their SSIM values, which is evident from Figs. 7 and 8. The best obtained SSIM scores for training and testing data were 0.88 and 0.875, respectively. Finally, the GAN model also helped in reducing the blurriness of the frames predicted. For future work, we aim to run the entire model for a much larger dataset since the results for our smaller dataset are promising, and we aim to achieve better results than the current existing model.

References

1. Liu W, Luo W, Lian D, Gao S (2018) Future frame prediction for anomaly detection—a new baseline. In: Proceedings of the IEEE conference on computer vision and pattern recognition, pp 6536–6545
2. Johnson J, Alahi A, Li F-F (2016) Perceptual losses for real-time style transfer and super-resolution. In: European conference on computer vision. Springer, Cham, pp 694–711
3. He K, Zhang X, Ren S, Sun J (2016) Deep residual learning for image recognition. In: Proceedings of the IEEE conference on computer vision and pattern recognition, pp 770–778
4. Ruder S (2016) An overview of gradient descent optimization algorithms. arXiv:1609.04747
5. Kwon Y-H, Park M-G (2019) Predicting future frames using retrospective cycle gan. In: Proceedings of the IEEE conference on computer vision and pattern recognition, pp 1811–1820
6. Geiger A, Lenz P, Stiller C, Urtasun R (2013) Vision meets robotics: the kitti dataset. Int J Robotic Res 32(11):1231–1237
7. Ronneberger O, Fischer P, Brox T (2015) U-net: Convolutional networks for biomedical image segmentation. In: International conference on medical image computing and computer-assisted intervention. Springer, Cham, pp 234–241
8. Wang Z, Bovik AC, Sheikh HR, Simoncelli EP (2004) Image quality assessment: from error visibility to structural similarity. IEEE Trans Image Process 13(4):600–612
9. Isola P, Zhu J-Y, Zhou T, Efros AA (2017) Image-to-image translation with conditional adversarial networks. In: Proceedings of the IEEE conference on computer vision and pattern recognition, pp 1125–1134
10. Kingma DP, Ba J (2014) Adam: a method for stochastic optimization. arXiv:1412.6980

Evaluation of Propofol General Anesthesia Intravenous Algorithm for Closed-Loop Drug Delivery System

Shola UshaRani

1 Introduction

Patient safety places an important role in all aspects of anesthesia. During the past few years, human factors contribute many mistakes or errors [1] on patient safety of anesthesia and from the lessons of aviation, safety should be a training component to be considered of the anesthesia. In the past decade, many critical incidents in anesthesia have been compared to the aviation disasters, leading to human factors affecting the performance of anesthetist to be looked into at a greater [2] depth. Research in anesthesia delivery has become as important factor to improve the patient safety.

It has been given that human error contributes a factor of 83% of anesthetic events occurred for every 2000 incident reports of Australia. This ensures the analysis and use of human factors to be embedded into this system and this will governance the healthcare setting. Closed-loop anesthesia feedback system will measure the quantified output from the patient. This will automatically adjust the anesthesia drug levels into the patient based on feedback from the patient. This will avoid the involvement of human to manually controlling the levels. So that the anesthesian will look after the patient toward other aspects. Designing the closed-loop system needs several steps to complete all scenarios of the anesthesia practice. An algorithm that consists of the rules necessary for drug infusion should be implemented. This paper defines the closed-loop delivery system with automatic delivery model with infusion algorithm, and how this algorithm can be evaluated is discussed here.

Many of the automated systems are available and designed but none of those are not up to the mark to meet the clinical approval. To convince regulatory authorities of the benefit of automated systems, both patient safety and improved outcome will have [3] to be demonstrated. The control algorithm used for the system should be

S. UshaRani (✉)
SCOPE, Vellore Institute of Technology, Chennai, India

© The Author(s), under exclusive license to Springer Nature Singapore Pte Ltd. 2022
R. R. Raje et al. (eds.), *Artificial Intelligence and Technologies*,
Lecture Notes in Electrical Engineering 806,
https://doi.org/10.1007/978-981-16-6448-9_22

robust enough, that will ensure the realistic patient variability and should be safe in all situations including failure modes of anesthetic drug typically used to maintain anesthesia.

2 Background

Initially the drug infusion is done manually where the physician will monitor the vital parameters of the patients. Then based on the observed values, the next rate of infusions can be decided and injects the drug by him physically. Two randomly enrolled groups are considered to compare the clinical performance of propofol, remifentanil, and rocuronium using closed-loop feedback control [4] with manually controlled group infusions. The performance is checked by calculating the BIS offset with four different categories of 'Excellent,' 'Good,' 'Poor,' and 'Inadequate.' The target BIS is fixed as 45. The results show that closed-loop group produces very significant results than manually controlled group.

After that the automatic drug infusion system is computerized and is done based on pharmacokinetic model-driven [5] approaches known as target controlled infusion (TCI) device. In TCI-based approach the target level of concentration is given, and the drug is infused based on this. The target level of the drug to be infused should be adjusted by the physician continuously based on the vital parameters. As the adjustment of the drug infusion is continuously should be checked in TCI, the research of automatic drug infusion is moved toward the closed-loop systems. A closed-loop system continuously monitors the controlling [6] parameters [7, 8] and automatically adjusts the infusion levels. These systems are well developed to perform computer trained sedation system (SEDASYS). This system integrates all vital parameter devices like pulse oximeter, capnometer, ECG, etc. Basically SEDASYS is used for safe administration of propofol to healthy adult patients those are undergoing elective colonoscopy. One of the important application areas of closed-loop anesthesia system is remote anesthesia monitoring [9, 10] where the physician uses telemedicine and telemonitoring technologies for general surgery and local anesthesia procedures to patients who are connected remotely. Closed-loop systems are still used for research purposes and have not yet been used in clinical practice. Decision support system has been another research area in anesthesia. These systems are program-based packages that are designed to assist trained and untrained doctors in hospitals.

Closed-loop anesthesia delivery system is used to maintain general anesthesia for maintaining hypnosis, analgesia, and muscle relaxation. Different drugs are used for automated delivery system. For example, propofol is used for hypnosis control, remifentanil for pain control, and rocuronium is used to inactivate the body movements of the patient. The system consists of computing device with closed-loop control algorithms, vital parameter monitoring device, and different infusion pumps. The basic block diagram representation of closed-loop automatic drug delivery system is shown in Fig. 1. The system has actual state and target or

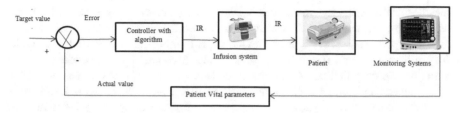

Fig. 1 Block diagram for closed-loop automatic drug delivery system, IR = infusion rate

reference values of patient vital parameters; here in this case the target value is bispectral index scale (BIS). The error is measured based on the difference between the two values. This error is used as input to the controller algorithm to calculate the next infusion levels [11]. The feedback output from the system is measured using sensors that are connected to the computing system. Despite the difficulties involved with the closed-loop system, most research analysis ensures that an automated systems is always more effective than the human control systems. These systems will ensure the delivery of doses within the specified range and reduces the recovery time of patients. In order to make this system to be very robust, the controller algorithm to be used should be strong enough to meet all the anesthesia practice constraints.

3   Closed-Loop General Anesthesia Algorithm

Closed-loop anesthesia is used as a tool to automate anesthesia care, reducing the anesthesiologist's workload and improving patient care [4]. It even improves the quality of drug delivery. Closed-loop systems consist of a computing system with built-in algorithms that compares the actual control variable with target control variable and produces the infusion rate to be delivered to an 'actuator,' such as a syringe pump of the infusion system as shown in Fig. 1. All these are combined with a feedback system to automate the control of drug delivery in order to maintain a preset target value. As regular sampling of the control variable and more frequent changes of drug delivery rates than with physically delivered anaesthesia, greater stability of the control variable would be achievable. The robustness or strength of the closed-loop system depends on the control variable used. So target parameter fixed for the general anesthesia decides the wellness of the drug delivery. The anesthesia levels of the patient are based on the levels of hypnosis levels of the brain. These levels are well defined by using EEG signal. The bispectral index (BIS) is a dimensionless number derived from processing component of the frequencies of the EEG which is used for control variable in regulating the drug levels. It is also used to identify the levels of hypnosis levels of the brain during anesthesia drug delivery. It ranges from 0 (NO_Signal) to 98 (Consciousness). A value from 40 to 60 is considered as representing an adequate state of hypnosis. During the anesthesia drug delivery procedures if BIS value reaches between 60 and 40, the surgery will be performed. In reality, BIS can be monitored using Vista monitor such as aspect medical systems and a vital signs monitor like CASMED 740, while drug infusion pumps (Graseby 3400, Graseby Medical, UK) functioned as the actuator. The drug delivery of anesthesia is three phases. The first one is induction phase, where here initial drug dosage will be given the BIS values reduced from conscious levels (98 or 100). After some time for successful drug infusions the values reach to 60. When this value is in between 60 and 40, the physician will perform the surgery. This phase is known as maintenance phase. Then once the surgery over the physician will stop the infusion, then the BIS values retains back to conscious levels. This is known as recovery phase.

The phases of both induction and maintenance phases are shown in Fig. 2. During the induction and maintenance period, BIS values are measured using Schinder-based pharmacokinetic pharmacodynamics (PK-PD) [12] equations defined in [11, 13, 14]. The PK-PD characteristics will define the levels of drugs infused in all the parts of the body. The flowchart for finding the drug rate levels during induction and maintenance phase is shown in Fig. 3. As shown from the flowchart, initially all the demographic data of the patient are entered through the user interface. Then on clicking the start button, based on patient demographic data the PK-PD characteristics are calculated. The PD equations produce the results of

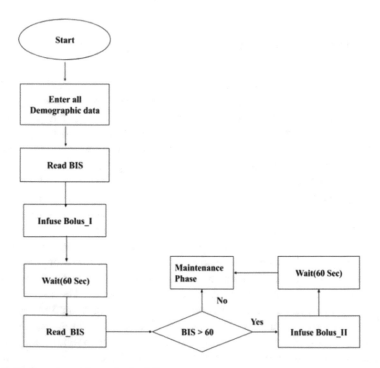

Fig. 2 Induction phase of anesthesia delivery

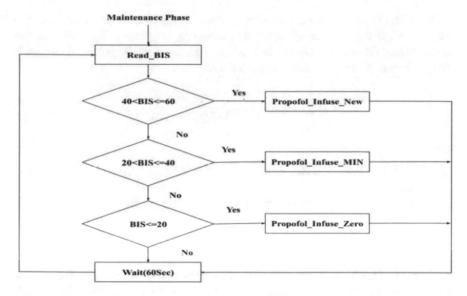

Fig. 3 Maintenance phase of anesthesia delivery

effective site compartment. The pharmacokinetic characteristics define the drug characteristics of the patient. Basically in standardized drug delivery pumps like TCI, these characteristics are calculated based on the average parameter set. But the interpatient [15] and interpatient [16] variations lead these parameters to depend on various parameters of the patients [17]. These characteristics act as a tool to predict the dosage levels at different parts of the body. The PK-PD, 4 compartmental model [13, 18] equations are shown below:

$$\frac{dX_1}{dt} = I_s - (K_{12} + K_{13} + K_{10})X_1 + K_{21} * X_2 + K_{31} * X_3 \tag{1}$$

$$\frac{dX_2}{dt} = (K_{12} * X_1) - (K_{21} * X_2) \tag{2}$$

$$\frac{dX_3}{dt} = K_{13} * X_1 - K_{31} * X_3 \tag{3}$$

$$\frac{dC_e}{dt} = K_{1e} * X_1 - K_{e0} * C_e \tag{4}$$

where

X_i represents the amount of drug present in ith compartment, where $i = 1, 2, 3$.
C_e represents the amount of drug present in effective site compartment.
K_{ij} is the drug transfer rates between the compartments i to compartment j.
I_s Drug infusion.

Equation (1) uses infusion rate of the primary compartment (or organ) and that rate will be distributed to all other compartments (or parts). The effective site compartment value from Eq. (4) is used as a measure of bispectral index scale (BIS) [18] value using the following equations.

$$BIS(t) = E_0 - E_{max}\left(\frac{C_e(t)^\gamma}{C_e(t)^\gamma + C_e 50^\gamma}\right) \tag{5}$$

where

E_0 represents the initial BIS value of patient without or before the drug infusion, i.e., at $C_e(t) = 0$.

C_{e50} is the effective site concentration when the drug effect is 50%.

γ is the steepness of BIS with respect to effective site concentration.

E_{max} maximum effective site concentration.

The entire computerized part of the closed-loop drug delivery system is developed in MATLAB–Simulink [19, 20]. The system will record the BIS values continuously for every 60 s (delay). In a particular time, the new dose is [4] calculated based on previous dose by using the following equation:

$$Dose_{New} = Dose_{Prev} * K_m * K_h. \tag{6}$$

where

K_m BIS error = |BIS act-BIS tar|

K_h |BISmean-BIStar|.

As discussed the maintenance phase is started from the induction phase, it will be continued in the phase until the end of the procedure. Figure 3 represents the operation procedure for induction phase, once all the demographic parameters are given. From the induction phase automatically it enters into maintenance phase when the value of BIS reduces lesser than or equal to 60. The corresponding functions in maintenance phase are shown in Fig. 4. There are three possible dosage levels that are maintained as injecting new drug dosage values or minimum drug dosage values and finally zero dosage values based on the variation of BIS between 60 and 20. The required data for the model will be given as GUI input to the Simulink.

4   Results and Discussion

As discussed the model is simulated in MATLAB [20], and the inputs of patient or demographic data [21] as age, weight, height, gender, minimum drug levels, and others are entered in the UI developed in the MATLAB. Table 1 is the sample data considered for evaluating the closed-loop controlled algorithm for propofol

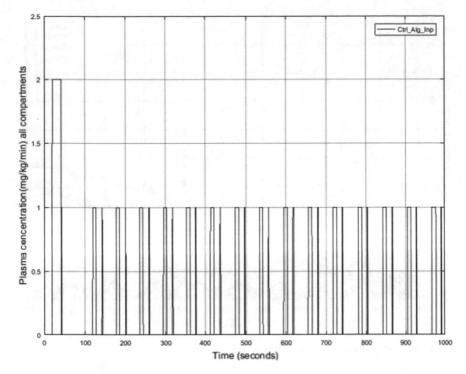

Fig. 4 Infusion rate for healthy patient

Table 1 Patient data taken for infusion algorithm

Age	Height	Weight	Gender	C_{50}	E_0	E_{max}	γ
40	163	54	F	6.33	98.8	94.1	2.24

infusion. The BIS measure provided from Eq. (1) is used for results of different cases from the above data. The following are two scenarios considered for the results. Figure 4 shows an example of a typical case of general anesthesia implemented for closed-loop propofol infusion for healthy patients using Table 1. Based on the BIS value till 120 s it is with bolus infusion of 2 mg/kg/min. After 120 s it enters the maintenance phase that will be the end of simulation. Based on this drug infusion, the four compartments results are shown in Fig. 5.

As discussed from [11], the primary compartment (X1) initially produces peak based on the bolus infusion later it distributes the drug to other compartments. Similarly the drug level of effective compartment (Xe) follows the primary compartment. Peripheral compartments (X2 and X3) slowly get increased on distribution of drug from primary compartment. Bolus infusion of 2 mg/kg/min was given below 100 s, the BIS reduces from E0 (from Table 1) to the target BIS 50, then its value continued from 60 to 40 with the infusion levels of 1 mg/kg/min. From the

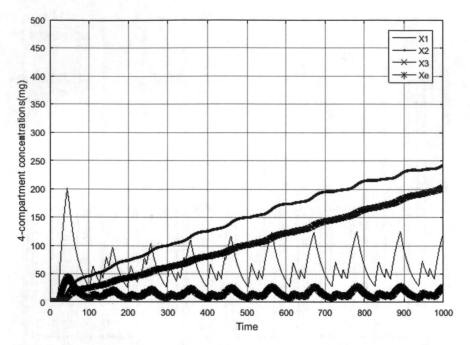

Fig. 5 Four compartments results based on Fig. 4

results of Fig. 6, the model infuses drug in induction phase till 120 s and after that it remains in maintenance phase until end of the surgery or end of the simulation period, i.e., 1000 s.

Figure 7 shows an example of a typical case of general anesthesia implemented for unhealthy patients. Propofol infusion to unhealthy patients or ASA III or IV or anesthesia-sensitive patients using Table 1. As these patients are sensitive to anesthesia, less bolus infusion is given to reduce the drug side effects during induction phase. This value is less as compared to the bolus given for healthy patients. Based on the BIS value before 100 s, the bolus infusion is given as 1 mg/ kg/min and it reaches to target value of 50. Later after 100 s it completes the maintenance phase until end of the surgery. The four compartments results are shown in Fig. 8. As discussed from [11] the primary compartment (X1) initially produces peak based on the bolus infusion later it distributes the drug to other compartments. Similarly the drug level of effective compartment (Xe) follows the primary compartment. Peripheral compartments (X2 and X3) slowly get increased on distribution of drug from other compartments. From the give bolus infusion of 1 mg/kg/min during induction phase below 100 s, the BIS reduces from E0 (from Table 1) to the target BIS 50, then its values between 60 to 40 with the maintenance infusion levels of 0.5 mg/kg/min. From the results of Fig. 9, this scenario infuses induction phase by 100 s and remains in maintenance phase until end of the surgery.

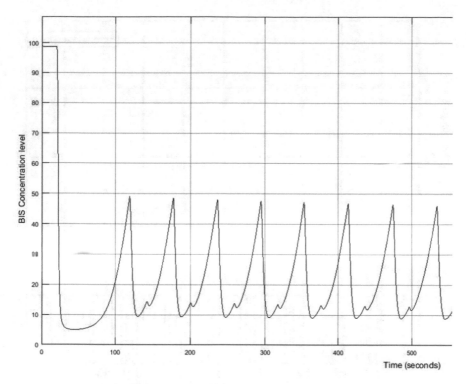

Fig. 6 Healthy patient BIS output based on Fig. 3

From the BIS equation to understand the results of healthy and unhealthy case scenarios, the γ, steepness is evaluated by varying it with different values. If the value is less than 1 the steepness of the BIS curve is very rear to the y-axis means the time to reach the maintenance phase is more. But if this value is greater than 1 and more the steepness is nearer to the y-axis. And the time to reach the maintenance phase is very fast. In Simulink both the healthy and unhealthy case scenario are verified based on γ. The variations of time to reach the maintenance phase with BIS value 50 and time to reach the deep conscious state with BIS among 40 to 20 are observed. The results are shown in Table 2. As shown in this table as the γ is more, the time to reach the both states increases. From the above results the induction phase can be seen more quickly in unhealthy patients than healthy patients. The elderly or anesthesia-sensitive patients staying more time in anesthesia than others and they requires more recovery time than the healthy patients.

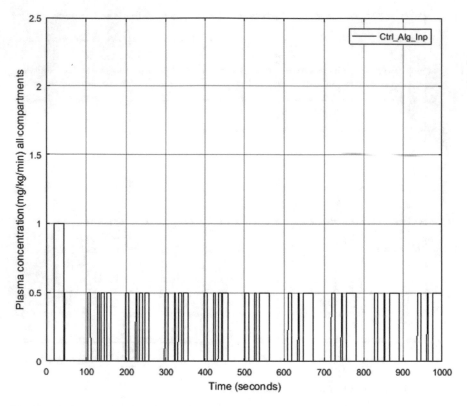

Fig. 7 Infusion rate for unhealthy patient

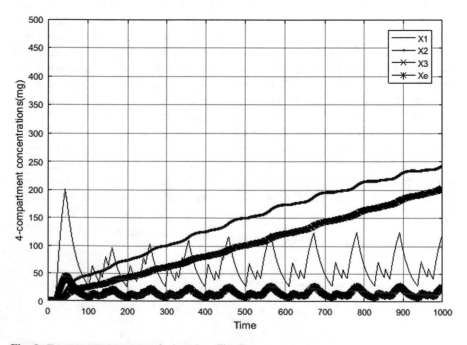

Fig. 8 Four compartments results based on Fig. 7

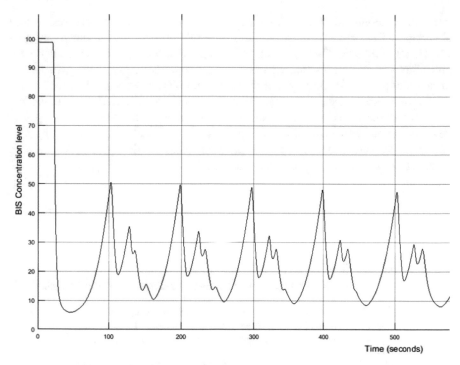

Fig. 9 Unhealthy patient BIS output based on see Fig. 8

Table 2 Comparison of BIS values for different values of steepness of the curve

SI no.	γ Value	Time to reach the target BIS value to 50 (s)	Time to reach BIS to deep consciousness state $20 < BIS \leq 40$ (s)
1	2.24	19	34
2	5	118	875
3	7	720	1623

5 Conclusion

The proposed algorithm for closed-loop general anesthesia propofol infusion is evaluated to Schinder-based PK-PD model. In existing system the anesthesia will be delivered manually or through open-loop computerized infusion systems. In the first case possibility of human errors or operational errors will happen and in open-loop computerized system the infusion is not delivered based on the patient feedback. If the feedback is not included into the system, the possibility of overdose or under dose will occur. This leads to abnormal or adverse event situations to the patients. The features of automatic drug delivery system will avoid the manual drug infusion procedure and feedback included will avoid the overdose and under dose situations. The characteristics of the algorithms are implemented through MATLAB–Simulink. The algorithm performs good in both the scenarios but the for unhealthy scenario shows that the induction phase comes early to the patients and these set of patients staying most of the time in maintenances stage. So much of the time they are in unconscious state. From this we can conclude that unhealthy patients take more recovery time than the healthy patient. Thus the closed-loop drug delivery system is developed and the performance of healthy and unhealthy scenarios is evaluated. But to increase the robustness of the system, it needs to demonstrate with more patient safety measures.

References

1. Shola UR, NNV (2017) A review on patient-controlled analgesia infusion system. Asian J Pharm Clin Res 10(13):117–121
2. Chandran R, DeSousa KA (2014) Human factors in anaesthetic crisis. World J Anesthesiol 27 (3):203–212
3. van Heusden K, West N, Umedaly A, Ansermino JM, Merchant RN, Dumont GA (2014) Safety, constraints and anti-windup in closed-loop anesthesia. In: Proceedings of the 19th world congress the international federation of automatic control Cape Town, South Africa, August 24–29
4. Hemmerling TM, Arbeid E, Wehbe M, Cyr S, Taddei R, Zaouter C (2013) Evaluation of a novel closed-loop total intravenous anaethesia drug delivery system: a randomized controlled trial. British J Anaesth 110(6):1031–1039
5. Alexander JC, Joshi GP (2018) Anesthesiology, automation, and artificial intelligence. Baylor University Medical Center Proceedings, vol 31, pp 117–119
6. Shola U, Neelanarayanan V (2019) Control of anesthesia concentration using PID controller. Int J Innov Technol Exploring Eng (IJITEE) 8(6S). ISSN: 2278-3075
7. UshaRani S, Neelanarayanan V (2019) Comparative study of 4-compartmental PK-PD model with effective site compartment for different parameter set. Int J Reliable Qual E-Healthcare 8(1)
8. Pambianco DJ, Vargo JJ, Pruitt RE, Hardi R, Martin JF (2011) Computer-assisted personalized sedation for upper endoscopy and colonoscopy: a comparative, multicenter randomized study. Gastrointest Endosc 73(4):765–772
9. Pambianco DJ, Whitten CJ, Moerman A, Struys MM, Martin JF (2008) An assessment of computer-assisted personalized sedation: a sedation delivery system to administer propofol for gastrointestinal endoscopy. GastrointestEndosc 68(3):542–547
10. Cone SW, Gehr L, Hummel R, Raq A, Doarn CR, Merrell RC (2004) Case report of remote anesthetic monitoring using telemedicine. Anesth Analg 98(2):386–388
11. Wehbe M (2014) Robotic anesthesia: a novel pharmacological robot for general anesthesia and robot-assisted intubation. A thesis submitted to McGill University
12. Yousefi M, van Heusden K, West N, Mitchell IM, Ansermino JM, Dumont GA (2019) A formalized safety system for closed-loop anesthesia with pharmacokinetic and pharmacodynamics constraints. Control Eng Pract 84:23–31
13. Usha Rani S, Neela Narayanan V (2015) Interoperable framework solution to ICU health care monitoring. Ictact J Commun Technol 06(01)
14. Usha Rani S, Maheshwari A (2020) Estimating depth of anesthesia for drug infusion using human EEG signals. Int J Adv Sci Technol, IJAST 29(3):9774–9785. ISSN: 2005-4238

15. Ilyas M (2019) Interpatient and intra-patient variability compensation in regulation of hypnosis based on modern control of propofol anaesthesia administration, PhD thesis, COMSATS University, Islamabad
16. Nas‚cu I, Krieger A, Ionescu CM, Pistikopoulos EN (2014) Advanced modelbased control studies for the induction and maintenance of intravenous anaesthesia. IEEE Trans Biomed Eng 62(3):832–841
17. Doyle DJ, Garmon EH (2019) American society of anesthesiologists classification (asa class). In: StatPearls [Internet]. StatPearls Publishing
18. Fiadjoe J, Gurnaney H, Muralidhar K, Mohanty S, Kumar J, Viswanath R, Rehman M (2009) Telemedicine consultation and monitoring for pediatric liver transplant. Anesth Analg 108 (4):1212–1214
19. Cascella M (2016) Mechanisms underlying brain monitoring during anesthesia: limitations, possible improvements, and perspectives. Korean J Anesthesiol 69(2):113
20. De Castro NC (2008) A Simulink compartmental model for depth of anesthesia. Technical report 37, INESC-ID
21. Schnider TW, Minto CF, Shafer SL, Gambus PL, Andresen C, Goodale DB, Youngs EJ (1999) The influence of age on propofol pharmacodynamics. Anesthesiol J Am Soc Anesthesiol 90(6):1502–1516

A Study on the Repercussions of the COVID-19 Pandemic in the Mental Health of the Common Public: Machine Learning Approach

Anusha Jayasimhan, Preetiha Jayashanker, S. K. Charanya, and K. Krithika

1 Introduction

The coronavirus pandemic is an ongoing global pandemic caused by SARS-CoV-2. It was first identified in 2019 in China, and it quickly spread worldwide as it is an infectious severe acute respiratory disease. Till date, it has affected over 36.5 million people, and killed over 1 million people worldwide. During this time, to reduce the risk of infection, many countries have implemented a lockdown which can be defined as an obligation for citizens to stay where they are, usually to avoid severe risk to themselves or to others if they move freely. This pandemic has caused a large upset to people of all ages and walks of life, as it has led to job loss, school and college closures, postponement or cancelation of sporting, religious, political, and cultural events, as well as widespread supply shortages exacerbated by panic buying [1, 2]. One of the most pressing concerns during these times is the mental health of the general population; therefore, this study mainly focuses on the anxiety levels people face due to the abrupt change in their day-to-day lifestyle. Anxiety was chosen as a measure of mental health, as it corresponds to one's body's natural response to stress. It is defined as a feeling of fear or apprehension about what's to come, and therefore, one of the earliest indicators of deteriorating mental health.

2 Literature Review

A study was conducted to analyze the impact of COVID-19 on learning styles, activities, and mental health of young Indian students. The online survey consisted of 22 questions which were split into five categories namely, digital connectivity, social life, online learning experience, overall moods, and thoughts during the

A. Jayasimhan · P. Jayashanker (✉) · S. K. Charanya · K. Krithika
Madras Institute of Technology, Chennai, India

© The Author(s), under exclusive license to Springer Nature Singapore Pte Ltd. 2022   215
R. R. Raje et al. (eds.), *Artificial Intelligence and Technologies*,
Lecture Notes in Electrical Engineering 806,
https://doi.org/10.1007/978-981-16-6448-9_23

pandemic. It was concluded that more than half of the participants terribly missed social life were dismayed due to missed events and opportunities, and students were in need of mental support from their faculties [3]. Another study was conducted on college students in China in May, assessing their anxiety levels using GAD-7 scale, and working with 7143 responses. This study found that 0.9% of its participants were experiencing severe anxiety, 2.7% moderate anxiety, and 21.3% mild anxiety. They also found that living in urban areas, the stability of a steady source of income (usually from parents), forms of social support and otherwise living with parents acted as mental health protection for many of these students. It also found that changes in daily life, economic stress factors, and delays in academics were proportional to levels of anxiety in these students [4]. The authors conclude by indicating that monitoring mental health of college students during pandemic situations in future would greatly improve their quality of life during these trying times [5].

3 Experimental Setup

The survey was conducted from June 19, 2020 to June 23, 2020 by circulating Google forms, and had a total of 32 questions. We selected the age groups based on 'Age Differences in Work Stress, Exhaustion, Well-Being, and Related Factors From an Ecological Perspective,' [6] and it was inspired by findings in [7–10]. The majority of responses came from India, notably from states Tamilnadu, Maharashtra, and Karnataka. To assess the anxiety levels, this survey gathered important information about the respondent, including factual information such as location, age, gender, health issues, and other visceral information based on large stress factors such as 'Do you feel like you have wasted your time during the lockdown?' [11] and 'Have you felt anxious to go to the hospital?' [12, 13]. The last seven questions of the survey were based on a standardized scale GAD, or Generalized Anxiety Disorder Scale, which accurately determines anxiety of the person and segregates it into one of the four basic categories, 0—Minimal Anxiety, 1—Mild Anxiety, 2—Moderate Anxiety, 3—Severe Anxiety. The last seven questions of the survey were based on a standardized scale GAD, or Generalized Anxiety Disorder Scale, which accurately determines anxiety of the person and segregates it into one of the four basic categories, 0—Minimal Anxiety, 1—Mild Anxiety, 2—Moderate Anxiety, 3—Severe Anxiety. We garnered a total of 832 responses, from 25 different locations worldwide. The majority of the responses, 441, came from the age group 18–25. On the whole, there were 428 (51.4%) responses that reported minimal anxiety, 267 (32.09%) responses that reported mild anxiety, 105 (12.6%) responses that reported moderate anxiety, and 36 (4.32%) responses that reported severe anxiety (Table 1).

Table 1 Important survey questions

Question number	Question
1	Is it difficult to get essentials (food/medicine)?
2	What is your gender?
3	During the lockdown have you been doing more household work?
4	How often do you use masks/hand sanitizers (per day)?
5	Do you feel like you have wasted your time during the lock down or have you ever felt pressured by your peer's accomplishments during the lockdown?
6	Are you stuck somewhere alone without support from friends or family?
7	Do you have children less than 3 years, or elders 65+ at your house?
8	Have you ever felt afraid or anxious to go to hospital during this pandemic?
9	Has your sleep cycle changed drastically?
10	Have you felt frustrated by having your activities, major life events or opportunities affected by the virus?
11	Is someone from your family in the medical field every day?
12	Do you have online classes/assignments/assessments/tests?
13	How often do you leave your house? (per week)
14	How often do you interact with people outside your house (friends, neighbors, etc.)?
15	Have you taken any medication to prevent corona (Chloroquine or kabasurakudineer) or any medication to increase your immunity?
16	Have you been affected by not being able to visit religious places?
17	Do you miss hanging out with people outside of your house?
18	Did you face a pay cut/job loss/take another job to cover expenses?

4 Methodology

Once the survey data had been converted into a majority of numerical values, we decided to split the data by age groups, which is a fundamental factor in determining anxiety levels, as seen through previous studies, and we conducted individual analysis on each of the age groups. Since there were no missing values, we began encoding labeled data to numerical data; we converted majority of the values to 1 if the response is *yes* and 0 if the response is *no*. For the question about gender, we converted male to 0 and female to 1. The anxiety levels were converted as follows: 0—Zero Anxiety, 1—Mild Anxiety, 2—Moderate Anxiety, and 3—Severe Anxiety.

4.1 Age Group: Less Than 18

4.1.1 Data Analysis Using EDA

We found 65 samples belonged to the age group less than 18, out of which 43 (64.6%) belonged to the female gender and 23 (35.3%) belonged to the male gender. Then, we further split up the data into 4 anxiety groups based on the GAD 7 scale. We conducted further analysis on each anxiety group and analyzed the different questions that could have contributed to a higher anxiety. We took each question and found the number of people who said Yes/Frequently to the question and then calculated the percentage of these people in each anxiety group. From Fig. 1, it was found that as the level of anxiety increases, the percentages also increase for a few questions (Highlighted in bold in Table 2). Also, it can be observed that for a few questions the percentages of people are higher in the case of moderate and severe anxiety and lower in the cases of minimal and mild anxiety (Highlighted in italics in Table 2).

When we summed up the people who answered:

1. Yes/Frequently for a question and have moderate and severe anxiety
2. No/Rarely for a question and have moderate and severe anxiety

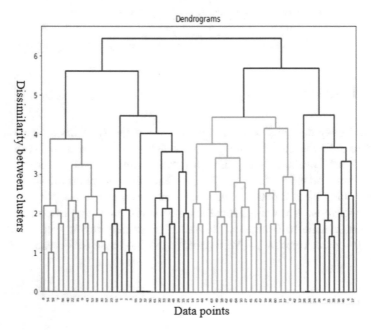

Fig. 1 Clusters formed

Table 2 Important features

Questions	0 anxiety (%)	Mild anxiety (%)	Moderate anxiety (%)	Severe anxiety (%)
Miss hanging out with people outside of your house?	**73**	**76**	**100**	**100**
Is it difficult to get essentials?	**0**	**11**	**11**	**50**
More household work?	**61**	**69**	**77**	**100**
Anyone you know has COVID/been tested/ recovered from COVID?	*38*	*34*	*55*	*100*
How often have you felt like you wasted your time or were pressured by peers' accomplishments during the lockdown?	*11*	*42*	*77*	*50*
How often do you eat outside food?	*11*	*7*	*11*	*25*
Do you have children less than 3 yrs. or elders 65+ at your house?	**26**	**23**	**44**	**75**
Have you ever felt afraid or anxious to go to hospital during this pandemic?	**38**	**42**	**77**	**100**
Has your sleep cycle changed drastically?	*38*	*68*	*88*	*75*
Have you felt frustrated by having major life events or opportunities affected by the virus?	**53**	**68**	**77**	**100**
Is someone from your family in the medical field everyday?	**15**	**19**	**22**	**50**
Do you have online classes/assignments/ assessments tests?	**96**	**96**	**100**	**100**

We found that if the sum of the latter is greater than the former, it does not contribute significantly to anxiety 1.

4.1.2 Feature Selection

There are 9 questions which satisfy the above condition which become trivial in their contribution to anxiety. For the questions that were selected as important from the data analysis, it is found that 3 of the questions satisfy the condition mentioned at 1 which makes them less significant in their contribution to anxiety. The rest of the 9 features has notable correlation with the anxiety of a person. On plotting several relationship plots, it was found that it is not possible to differentiate features based on gender which makes it an important parameter. On applying a feature selection method—Extra Tree Classifier, we got the top 10 features of importance. On comparing the same with the previously selected features 8 out of 10 overlapped. The most important 7 features are 2, 3, 5, 8, 9, 10, and 17 from Table 1.

4.2 Age Group: 18 to 25

4.2.1 Data Analysis Using EDA

We used python to perform Exploratory Data Analysis (EDA). After collecting some basic intuitions, we plotted histograms to find the frequency of type of response for each feature. There are totally 441 samples out of which 207 (47.1%) were male, and 232 (52.8%) were female. We found the following: Nearly, 91.79% have responded 'No' for the question asking if they find it difficult to get essentials. For the question, have you felt symptoms of COVID, 95.67% have said No. Almost 92.25% of the respondents have said that they rarely consume outside food. About 95.44% have reported that they are with family and 20 respondents said they are alone without the support from their family. Nearly, 75% of the people have said they terribly miss hanging out with their friends, and about 88.6% have said that they leave their home rarely. Those who have said they rarely feel frustrated for their career/major events being affected by the pandemic didn't have severe anxiety. Almost all those who didn't miss hanging out with friends didn't face severe anxiety. Only 3 people facing severe anxiety have said no to 'missing hanging out with friends.' Females were found to be more susceptible to severe anxiety than men as 13 out of 24 respondents were females who faced severe anxiety. Below is the composition of anxiety level for the age group 18 to 25. It can be inferred from Table 3 that most people experienced minimal or mild anxiety.

4.2.2 Feature Selection

To find out the important features that affect the anxiety level, we performed feature selection techniques. Since the output is categorical, we used 'Mutual Information' feature selection. To cross verify the features of high importance, we used the correlation function in pandas, and found that 6 out of 10 features overlapped: The important features are questions 2, 3, 5, 8, 9, 10 from Table 1.

4.3 Age Group: 26 to 55

4.3.1 Data Analysis Using EDA

Before analyzing the data, the location attribute was removed as more than 90% of all responses were from the same locale in India. Of the participants under this age

Table 3 Composition of anxiety level

Age group	Minimal anxiety (%)	Mild anxiety (%)	Moderate anxiety (%)	Severe anxiety (%)
18–25	45.50	33.02	15.94	5.46

group, 39.3% were male and 60.7% were female. To figure out which independent variables would affect the dependent variable the most, the present data had to be analyzed to determine how much each factor varied with the levels of anxiety. First, the four levels of anxiety were split up, then a 60% margin was taken for the number of negative responses, and a 75% margin was taken for the number of positive responses, for each feature in the data.

Many attributes maintained the same percentage of positive or negative responses regardless of anxiety levels, so these common attributes were removed as they could not help us differentiate between various levels of anxiety. The common attributes that were removed were questions 1, 3, 4, 11, and 18 in Table 1. After having eliminated the common columns, we were only left with 14 of the 25 original attributes. The next important step was to find the correlation between factors, a method similar to a regression graph, to show how dependent they were on one another. So, using the correlation function, we found the features that were the most highly correlated.

4.3.2   Feature Selection

To find which of these factors would prove valuable to our experiment, we selected those factors with the highest correlation with the target. All values below 0.1 had negligible correlation to the target, and negative correlation indicated that they were inversely proportional to the target. Of these, the questions that proved to have the greatest correlation with the target were questions 5, 9, 10, and 16 from Table 1. To verify our findings up to this point, we applied feature selection algorithms, to help determine which of these questions would be the most valuable in forming our clusters. We found that the most important features were 10, 5, 9, 16, 17, 13, 8, 15, 12, 14, 2, 7 from Table 1. Four of the five questions that were pegged as having the most importance based on correlation were also selected for feature selection, which indicates that those features will help us classify anxiety levels within the 26–55 age group accurately.

5   Implementation

5.1   Age Group Less Than 18

After analyzing the data and figuring out the intuitions and patterns among data, we tried to cluster the data using hierarchical clustering to observe clusters and similarities between data. We took the 23 features and constructed a dendrogram to work out ways in which samples can be allocated to clusters. We can observe that we a horizontal line can be drawn at $y = 6$, we have 4 clusters formed which is

Table 4 Tabulation of percentages of each anxiety level in each cluster

	Minimal anxiety (%)	Mild anxiety (%)	Moderate anxiety (%)	Severe anxiety (%)
Cluster 1	60	33	17	0
Cluster 2	36	43	21	0
Cluster 3	59	41	0	0
Cluster 4	23	41	18	18

equal to the four anxiety groups we have. On further analysis, it was found that the four clusters consisted of 12, 14, 17, 22 samples each (Table 4).

It is clear that Cluster 4 contains all the severe anxiety samples, and Cluster 3 contains only mild and minimal anxiety. Clusters 2 and 4 have the greatest percentage of moderate anxiety samples. On examining each cluster separately, we found the number of people who answered similarly for each question. For example, in the cluster with 12 samples, all the samples belonged to the male gender, and hence, the percentage of similarity within answers is 100%. And it is also important to note the majority response for each question. We tabulated the percentage of similarity and the majority response for each question under each cluster and counted the number of questions which have the majority response to be Yes/Frequently (Tables 6 and 7).

In cluster 1, there were 8 questions where the majority response was Yes/Frequently; in cluster 2, there were 4; in cluster 3, there were 7; in cluster 4 there were 9. It is interesting to note that cluster 4 has the highest number of questions with Yes/Frequently as the majority response and cluster 4 also houses all the severe anxiety samples. It was also found that for a few questions, the majority response was found to be the same for all the four clusters. Those questions have either a majority Yes/majority No response, in general, which minimizes its impact in the anxiety of a person (Highlighted in Italics in Tables 5, 6, 7 and 8).

5.2 Age Group: 18 to 25

1. *Perform Clustering*: After performing basic data analysis, we proceeded to perform clustering for the input data to find the patterns within them. We used dendrogram to find the number of clusters possible. From the below dendrogram (Fig. 2), we decided the number of clusters to be 4. We then applied Agglomerative Clustering (a type of hierarchical clustering) and the algorithm formed four clusters. Elucidation of each cluster is given below.
2. *Cluster Analysis*: We performed Agglomerative Clustering in Python, and we got 4 clusters. The composition of each cluster is given in Table 9.

It is clear from Table 9 that among the 4 clusters, cluster 2 contains more number of people with severe anxiety. Cluster 0 contains a smaller number of individuals

Table 5 Percentage of similarity and majority responses (Clusters 1 and 2)

Questions	Cluster 1		Cluster 2	
Is it difficult to get essentials?	*75%*	*No*	*100%*	*No*
Gender?	100%	Male	86%	Female
Felt symptoms of COVID?	*100%*	*No*	*100%*	*No*
Have any of the following conditions?	*100%*	*No*	93%	*No*
Doing more household work?	100%	**Yes**	79%	**Yes**
Anyone you know has COVID/been tested/ recovered from COVID?	92%	No	93%	No
How often do you use masks/hand sanitizers?	92%	**Frequently**	64%	Rarely
How often you felt like wasted your time or pressured by peers?	67%	**Frequently**	86%	Rarely
How often do you eat outside food?	92%	Rarely	100%	Rarely
Are you in a containment zone?	58%	No	100%	No

Table 6 Percentage of similarity and majority responses (Clusters 1 and 2)

Questions	Cluster 1		Cluster 2	
Are you stuck somewhere alone without support?	*100%*	*No*	93%	*No*
Do you have children less than 3 years, or elders 65+ at your house?	*92%*	*No*	71%	*No*
Have you ever felt anxious to go to hospital during this pandemic?	50%	**Yes**	86%	No
Has your sleep cycle changed drastically?	75%	No	86%	**Yes**
Do you felt frustrated by having major life opportunities affected by the virus?	58%	**Yes**	64%	**Yes**
Is someone from your family in the medical field every day?	*67%*	*No*	93%	*No*
Do you have online classes/assignments/assessments/ tests?	92%	**Yes**	100%	**Yes**
Did you face a pay cut/job loss/take another job to cover expenses?	92%	No	93%	No
How often do you leave your house?	*58%*	*Rarely*	93%	*Rarely*
How often do you interact with people outside your house?	67%	*Rarely*	93%	*Rarely*
Have you taken any medication to prevent corona or to increase your immunity?	58%	**Yes**	64%	No
Have you been affected by not being able to visit religious places?	*58%*	*No*	93%	*No*
Do you miss hanging out with people outside of your house?	75%	**Yes**	79%	**Yes**

Table 7 Percentage of similarity and majority responses (Clusters 3 and 4)

Questions	Cluster 3		Cluster 4	
Is it difficult to get essentials?	100%	No	86%	No
Gender?	76%	Female	77%	Female
Felt symptoms of COVID?	94%	No	100%	No
Have any of the following conditions?	94%	No	100%	No
Doing more household work?	88%	No	91%	**Yes**
Anyone you know has COVID/been tested/ recovered from COVID?	59%	**Yes**	73%	**Yes**
How often do you use masks/hand sanitizers?	94%	Frequently	77%	**Frequently**
How often you felt like wasted your time or pressured by peers?	76%	Rarely	59%	Rarely
How often do you eat outside food?	82%	Rarely	86%	Rarely
Are you in a containment zone?	88%	No	64%	**Yes**
Are you stuck somewhere alone without support?	100%	No	95%	No
Do you have children less than 3 years, or elders 65+ at your house?	71%	No	55%	No

Table 8 Percentage of similarity and majority responses (Clusters 3 and 4)

Question	Cluster 3		Cluster 4	
Have you ever felt anxious to go to hospital during this pandemic?	76%	No	91%	**Yes**
Has your sleep cycle changed drastically?	65%	**Yes**	82%	**Yes**
Do you felt frustrated by having major life opportunities affected by the virus?	76%	Yes	82%	**Yes**
Is someone from your family in the medical field every day?	94%	No	68%	No
Do you have online classes/assignments/assessments/ tests?	94%	**Yes**	100%	**Yes**
Did you face a pay cut/job loss/take another job to cover expenses?	88%	No	86%	No
How often do you leave your house?	100%	Rarely	95%	Rarely
How often do you interact with people outside your house?	65%	Rarely	77%	Rarely
Have you taken any medication to prevent corona or to increase your immunity?	94%	**Yes**	59%	No
Have you been affected by not being able to visit religious places?	71%	No	77%	No
Do you miss hanging out with people outside of your house?	88%	**Yes**	77%	**Yes**

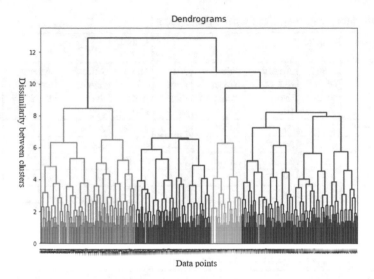

Fig. 2 Cluster in age group: 18 to 25

Table 9 Anxiety level composition in each cluster

Cluster	Minimal anxiety (%)	Mild anxiety (%)	Moderate anxiety (%)	Severe anxiety (%)
Cluster 0	*40*	37.33	17.33	5.33
Cluster 1	**66.40**	**23.20**	9.60	**0.80**
Cluster 2	23.76	42.57	*24.75*	*8.91*
Cluster 3	39.74	37.09	17.22	5.30

having minimal anxiety when compared to other clusters. Cluster 1 has more people with minimal anxiety. Since there is a myriad of data, we decided to group questions and analyze how people have responded to each group.

The questionnaire was segregated into 7 groups. They are as follows: Isolation—The way people feel without support from their family. The questions that come under this group are: Questions 6, 3, 17 from Table 1.

Economy—The way people feel without support from their family. The questions that come under this group are: Questions 1, 18 from Table 1.

Career—This category explains how people are affected by loss of opportunities and inability to use the lock down productively. Question numbers are 10, 5 (from Table 1).

Restlessness—This group explains how far the relaxing cycle has changed. The questions under this group are: Questions 3, 9, 12 (from Table 1).

Table 10 Group percentage for Cluster 0 and Cluster 1

Question group	Cluster 0				Cluster 1			
	None of them (%)	Atleast one of them (%)	Atleast two of them (%)	All three (%)	None of them (%)	Atleast one of them (%)	Atleast two of them (%)	All three (%)
Isolation	2.67	97.33	69.33	2	27.42	24.19	72.58	0
Economy	**61.33**	**38.67**	**7.33**	**1.33**	**63.70**	**3.22**	**36.29**	**0**
Career	18.67	81.33	28	0	54.03	11.29	45.97	0
Restlessness	0.67	99.33	84	42.67	*12.09*	*88*	*43.54*	*6.45*
Lockdown	2	98	73.33	14	25	75	31	4.03
Precautious	*0*	*100*	*94*	*67*	*0.80*	*99.19*	*71.77*	*17.74*

Table 11 Group percentage for Cluster 2 and Cluster 3

Question Group	Cluster 2				Cluster 3			
	None of them (%)	Atleast one of them (%)	Atleast two of them (%)	All of them (%)	None of them (%)	Atleast one of them (%)	Atleast two of them (%)	All of them (%)
Isolation	0.99	99.09	57.42	5.94	**4.96**	**95.24**	**52.38**	**3**
Economy	**67.33**	**32.67**	**6.93**	**0.99**	50.79	49.21	20.63	6.35
Career	*1.98*	*98.02*	*56.43*	*0*	12.70	87.30	30.16	0.00
Restlessness	*2.97*	*97.03*	*77.23*	*32.67*	*4.76*	*95.24*	*76*	*28.57*
Lockdown	0.99	99.01	74.26	26.73	*3*	*97*	*57.14*	*17*
Precautious	0.99	99.01	78.22	29.70	6	94	63	10

Lock down—These questions explain how they feel about lock down despite being together with their families. The questions under this group are: Questions 3, 16, 17 (from Table 1).

Precautions—This group explains about the precautionary measures people have taken though they rarely leave their home. The questions 4, 13, 15 comes under this group (from Table 1).

We analyzed the number of respondents who answered frequently/yes to none of the questions, at least one of the questions, at least two of them, all of them and calculated the percentage for each of these categories.

It can be deduced from Tables 10 and 11 that the individuals who come under cluster 0 are more concerned about the precautionary measures because about 67% have said that they frequently use masks, hand sanitizers, and medications though they go out rarely. Similarly in cluster 1, many people have taken precautionary measures, and also they have been doing more household work and have experienced a drastic change in their sleep cycle. In cluster 2, people have responded affirmatively to 'at least' one of them (around 99%) for the question groups isolation, career, restlessness, lock down, and being precautious. Around 33% have answered 'frequently' to 'all of them' for the group 'restlessness.' And a similar number of people have answered 'frequently' for the groups 'Lock down' and 'Precautions.' The individuals who belong to cluster 3 have answered 'frequently' to the group 'restlessness.'

5.3 Age Group: 26 to 55

Now that we have identified the most important features, it is time to classify our samples into their respective categories by using an unsupervised machine learning algorithm. We implemented a hierarchical clustering algorithm on the aforementioned attributes (in Fig. 3), and hence, we were able to classify our data into four separate clusters, as seen below.

From our chart, we can easily identify that the data was divided into four separate clusters, with a total of 303 data samples. Here, we have two tables to help us compare the actual results, and the results predicted through hierarchical clustering. In the predicted values table (Table 12), we have found the percentage of the number of responses from each anxiety level within each cluster. In the actual values table (Table 13), we've found how much of the total percentage of each

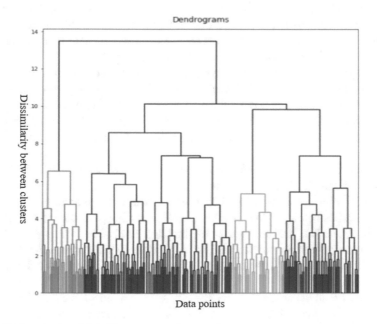

Fig. 3 Clusters in Age 26 to 55

Table 12 Predicted values

	Min (%)	Mild (%)	Mod (%)	Sev (%)	Prediction
Cluster 0	71.83	27.46	0.70	0	142
Cluster 1	0	28.50	53.57	17.86	28
Cluster 2	42.47	43.84	12.33	0	73
Cluster 3	98	2	0	0	50

Table 13 Actual values

	Min	Mild	Mod	Sev
Cluster 0 (%)	56.04	42.39	3.85	0
Cluster 1 (%)	0	8.69	57.69	100
Cluster 2 (%)	17.03	34.78	34.62	0
Cluster 3 (%)	26.92	0.11	0	0
Actual	182	92	26	5

anxiety group is within each cluster. No cluster consists purely of any one anxiety group, and hence, we must analyze what each cluster tells us.

The minimal anxiety classification group is strongly seen in both clusters 0 and 3, with 71.83% and 98%, respectively. This indicates that people with minimal anxiety exhibit more than one pattern of traits that can be used to identify them, and the algorithm has identified them separately. If we compare Table 14 to Table 13, we see that although both clusters 0 and 3 classify minimal anxiety, the majority of the samples, about 56%, come from cluster 0. Ideally, this would indicate that the pattern of traits in cluster 0 is more common or prevalent than those in cluster 3, even though 98% of samples in cluster 3 come from this group.

The mild anxiety classification group dominates cluster 2, with more than 40% of the samples within the cluster coming from this group. This group also identifies with the second cluster, as about a third of its samples belong under it. The minimal anxiety group holds almost the same number of samples in cluster 2, which indicates that there are many similarities, and maybe the features need to be made more specific.

The moderate anxiety classification group contains 26 samples, of which greater than 50% of samples belong in cluster

1. What is interesting to note is that 100% of the severe anxiety classification group's samples belong to this cluster as well, even though it comprises only 17% of the cluster. This is probably due to the fact that it only has 5 samples, and hence, a clear unique pattern cannot be identified for it alone. This indicates that the pattern for classification among these clusters is very similar, with the limited number of samples available.

In Table 14, all the answers with above 70% yes's or no's have been highlighted.

Now that we know that cluster 1 best identifies moderate and severe anxiety levels, we can see that people who have felt frustrated by having major life events

Table 14 Important survey questions

Important survey questions	Cluster 0		Cluster 1		Cluster 2		Cluster 3	
	No (%)	Yes (%)	No (%)	Yes (%)	No (%)	Yes (%)	No (%)	Yes (%)
Have you felt frustrated by having your activities, major life events or opportunities affected by the virus?	44	56	5	95	38	62	80	20
Has your sleep cycle changed drastically?	63	27	28	72	49	51	96	4
Have you been affected by not being able to visit religious places?	53	47	26	74	89	11	92	8
Do you have children less than 3 years, or elders 65+ at your house?	52	48	49	51	49	51	46	54
Do you have online classes/assignments/assessments/tests?	56	44	56	44	78	22	76	24
Have you ever felt afraid or anxious to go to hospital during this pandemic?	42	58	28	72	53	47	70	30
What is your gender?	25	75	41	59	69	31	36	64
Have you or anyone you know have corona virus/recovered from COVID/been tested for COVID?	49	51	54	46	75	25	94	6
Do you miss banging out with people outside of your house?	23	77	8	92	15	85	98	2
How often do you interact with people outside your house (friends, neighbors, etc.)?	73	27	77	23	74	26	92	8
Have you taken any medication to prevent corona or to increase your immunity?	31	69	36	64	89	11	64	36
Wasted your time during the lock down/pressurized by your peer's accomplishments during the lock down?	89	11	31	69	90	10	92	8
How often do you leave your house? (per week)	89	11	79	21	72	28	76	24

and opportunities affected by the virus, people anxious to go to the hospital during the pandemic, people who have experienced drastic changes in their sleep cycles, and people unable to visit their religious places, are those most likely to have moderate to severe anxiety during a pandemic. On the other hand, we can see how people often leave their house, how many people they interact with outside of their house, and if they miss hanging out with people outside of their house, have negligible effect on anxiety as most clusters indicate similar responses.

6 Conclusion

The COVID-19 pandemic has caused a lot of upheaval in the lives of millions of people across the globe. This has clearly resulted in an increase in mental health issues across the population as a whole. More specifically, we found that in the age group less than 18, factors such as drastic change in sleep cycle, frustration due to loss of opportunities, being in a containment zone, feeling anxious to go to a hospital have a greater impact on the anxiety of a person. In ages 18–25, we found that factors such as feeling like they wasted their time, drastic change in sleep cycle, frustration due to impediments in their career, and using hand sanitizers, taking self-medications has a significant impact on the level of anxiety but the economic related questions have comparatively less impact on the anxiety level. In ages 26–55, we found that change in sleep cycle, frustration due to changes in major life plans or regular activities, people anxious to go to the hospital during the pandemic, and people unable to visit their religious places, are those most likely to have moderate to severe anxiety during this pandemic. Since, the number of samples were too few for the age group greater than 55, and we refrained from doing an in-depth analysis of the data. However, these results are based on a survey that had 832 responses, and hence, we would require a comprehensive survey, with a larger volume of responses which would help to corroborate our results.

References

1. Nandini C The COVID-19 pandemic and its impact on mental health. https://doi.org/10.1002/pnp.666
2. Khattar A, Jain PR, Quadri SMK (2020) Effects of the disastrous pandemic COVID 19 on learning styles, activities and mental health of young Indian students—a machine learning approach. In: 2020 4th international conference on intelligent computing and control systems (ICICCS), Madurai, India, pp 1190–1195. https://doi.org/10.1109/ICICCS48265.2020.9120955
3. Khattar A, Jain PR, Quadri SMK (2020) Effects of the disastrous pandemic COVID 19 on learning styles, activities and mental health of young Indian students—a machine learning approach. In: 2020 4th international conference on intelligent computing and control systems (ICICCS), Madurai, India, pp 1190–1195

4. Ahmad A, Rahman I, Agarwal M (2020) Factors influencing mental health during Covid-19 outbreak: an exploratory survey among Indian population. medRxiv 2020.05.03.20081380
5. Zhang Y, FeeiMa Z (2020) Impact of the COVID-19 pandemic on mental health and quality of life among local residents in Liaoning Province, China: a crosssectional study", Lbsn. Int J Environ Res Public Health 17(7):2381
6. Hsu H-C (2018) Age differences in work stress, exhaustion, well-being, and related factors from an ecological perspective. Int J Environ Res Public Health 16(1):50. https://doi.org/10. 3390/ijerph16010050
7. Varshney M, Thomas Parel J, Raizada N, Sarin SK (2020) Initial psychological impact of COVID19 and its correlates in Indian community: an online (FEELCOVID) survey. PLoS ONE 15(5):e0233874
8. Xing J, Sun N, Xu J, Geng S, Li Y (2020) Study of the mental health status of medical personnel dealing with new coronavirus pneumonia. PLoS ONE 15(5):e02331452019
9. Kresimirŭosiu SP, Sarlija M, Kesedziŭ I (2020) Impact of human disasters and covid-19 pandemic on mental health: potential of digital psychiatry. Psychiatria Danubina 32(1):25–31
10. Cao W, Fang Z, Hou G, Han M, Xu X, Dong J, Zheng J (2020) The psychological impact of the COVID-19 epidemic on college students in China RSS. Psychiatry Res 287(112934)
11. Bhat BA, Khan S, Manzoor S, Niyaz A, Tak HJ, MuntahaAnees S-U, Gull S, Ahmad I (2020) A study on impact of COVID-19 lockdown on psychological health, economy and social life of people in Kashmir. Int J Sci Healthcare Res 5(2)
12. Mertens G, Lotte Gerritsen S, Duijndam ES, Engelhardb IM (2020) Fear of the coronavirus (COVID-19): predictors in an online study conducted in March 2020. J Anxiety Disorders 74:102258. ISSN: 0887-6185
13. Lee SA, Jobe MC, Mathis AA, Gibbons JA (2020) Incremental validity of coronaphobia: Coronavirus anxiety explains depression, generalized anxiety, and death anxiety. J Anxiety Disorders 74:102268. ISSN: 0887-6185. https://doi.org/10.1016/j.janxdis.2020.102268

Machine Learning-Based Categorization of Brain Tumor Using Image Processing

Muralidhar Appalaraju, Arun Kumar Sivaraman, Rajiv Vincent, N. Ilakiyaselvan, M. Rajesh, and Uma Maheshwari

1 Introduction

Image processing is a methodology that is capable of converting an image into discrete form, and it performs certain operations on an image, to achieve an enhanced image or to extract some vital information from it, identical to DSP. In image processing, the input is a picture (maybe a video frame and a photograph in any format) and the output is a picture or the characteristics of the input image. Tumor segmentation usually considers a picture 2D wave while processing [13]. It is one among the emerging technologies, with its branches of application widespread into several domains of business. The prediction method is a core research area in engineering, and it also acts as a thrust area in other disciplines of computer science. Researchers need image processing, as it offers real-world applications and the results derived from image processing algorithms are also made available to the hands of its user. Generally, medical images obtained from hospitals are in DICOS

M. Appalaraju · A. K. Sivaraman (✉) · R. Vincent · N. Ilakiyaselvan · M. Rajesh
School of Computer Science and Engineering, VIT University, Chennai, Tamil Nadu, India
e-mail: arunkumar.sivaraman@vit.ac.in

M. Appalaraju
e-mail: muralidhar.a@vit.ac.in

R. Vincent
e-mail: rajiv.vincent@vit.ac.in

N. Ilakiyaselvan
e-mail: ilakiyaselvan.n@vit.ac.in

M. Rajesh
e-mail: rajesh.m@vit.ac.in

U. Maheshwari
School of Computer Science and Engineering, Centre for Cyber Physical Systems, VIT University, Chennai, Tamil Nadu, India

© The Author(s), under exclusive license to Springer Nature Singapore Pte Ltd. 2022
R. R. Raje et al. (eds.), *Artificial Intelligence and Technologies*,
Lecture Notes in Electrical Engineering 806,
https://doi.org/10.1007/978-981-16-6448-9_24

communications and digital imaging in medicine format. These picture layouts are quite huge in size and require higher memory space for storage. For portability of these data, they are converted into GIF, BMP, JPEG, TIF, PNG file formats. Analysis of images in DICOS pattern is a tedious process, and the pictures are converted into any of the file formats listed above and are used worldwide. Image processing is extended to such medical image diagnosis to identify the pathologies present in our body, especially the pathologies present in the human brain that are complicated to diagnose. In [14], this work focused on this particular issue and is organized to resolve it. Meningioma using slices scan images is popularly used in the biomedical field for identification and visualization of finer details of the interior parts of the body. This technique is significantly used for detecting the variations between tissues and resulted to be a better technique when correlated with computed tomography (CT). Thus, slices technique is specially used for the detection and identification of cancer and meningioma imaging. Ionizing radiations are used for CT scan, and magnetization is used for MRI scan; here, a strong magnetic field is produced to adjust the nuclear magnetization and then uses radio frequencies for the coordinate of the magnetization to be identified by the image scanner [15].

That signal generated is further handled to derive the information from the body. By comparing both the types of images, MR image is safer than CT scan image because it is harmless for the human body. In the previous methods, radiologists used meningioma identification to manually discover the MRI image/pictures and attempt to identify and point out the abnormalities available in the MRI image. Figure 1 presents a generic flow diagram in MRI tumor detection and analysis to locate the abnormalities which consume time and efforts. Therefore, an assistant tool is needed to help accurately detect the presence of tumor in the slices image of the brain. Thus, detection of tumors in the brain plays a curial and tough job in the range of medical picture processing. The separation of a damaged or infected part from the brain along with its shape, size, and boundary is known as the identification of meningioma.

Rest of the paper is described in detail about the review of the literature in Sect. 2. The theoretical design and the phases of the scheme are described in Sect. 3, and the experimental results of the system are performed in Sect. 4. In Sect. 5, the study is concluded and has discussed the future work.

2 Literature Review

Segmentation for a brain tumor is described by various researchers [1–7]. Likewise, Arivoli et al. proposed a concept of super-pixel classification-based optic disk and optic cup. In the early stages, the brain tumor leads to serious effects and faces a lot of difficulties for treatment. They discussed about the bed fort and its effects on various people variations between age, sex, and different dates. The prevalence of brain tumors varies with the region and race advised that there is no specific cutoff

Fig. 1 Generic flow diagram
of brain tumor detection

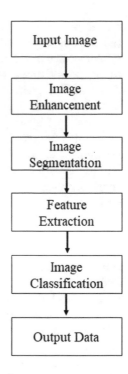

point for intraocular pressure (IOP) which brain tumor creates, although expanding IOP was a critical hazard factor for a brain tumor.

Primary hazard components of brain tumor are hoisted IOP applied by watery diversion, family history of brain tumor (genetic), astigmatism or partial blindness, brain tumor in the other eye, retinal separation, injury to the eye, diabetes, pigmentary scattering disorder, slender points, low fundamental circulatory strain, headache migraines or visual headaches, Raynaud's disorder, blood thickening, irregular visual field tests, undesirable optic nerve, corneal dystrophy, and pseudo-shedding.

Dionysiou and Stamatakos (2012) mentioned that primary hazard components of brain tumor are hoisted IOP applied by watery diversion, family history of brain tumor (genetic), astigmatism or partial blindness, brain tumor in the other eye, retinal separation, injury to the eye, diabetes, pigmentary scattering disorder, slender points, low fundamental circulatory strain, headache migraines or visual headaches, Raynaud's disorder, blood thickening, irregular visual field tests, undesirable optic nerve, corneal dystrophy, and pseudo shedding.

Shwakshar et al. stated brain tumor effects on all aged people as per the survey around Tamil Nadu, the southern part of India. They conducted the survey based on the Aravind Comprehensive Eye camp around the rural population in the state. India is the second most populated nation in the world. The effect of visual failure and visual deficiency from a brain tumor is in all likelihood expensive. In spite of its general well-being essentialness, there has been constrained information accessible on the pervasiveness of brain tumors and conceivable hazard factors for a brain tumor in India. Past populace-based investigations from India have announced the predominance of a brain tumor in urban populaces. There has been no investigation on the pervasiveness of brain tumors in provincial populaces from India. Also, in these earlier examinations, perimetry was restricted to the individuals who satisfied certain conditions like IOP or brain plate measuring.

Ansari and Singh portrayed diabetic meningioma as an endless disease, caused by inconveniences of diabetes mellitus, and established the essential driver of visual impairment among individuals of working age in created nations. As diabetic retinopathy is a difficult sickness, Nandpuru et al. recommended that laser photo-coagulation can forestall significant vision misfortune if identified in the beginning times.

Parveen and Singh brought up early recognition and consequent treatment which are fundamental to anticipate visual harm. With the new advances in advanced modalities [8–13] for retinal imaging, there is a dynamic need for picture handling instruments that give the quick and solid division of retinal anatomical structures. Sudharani et al. (2010) initiated a model for timolol maleate-controlled release ocular drug delivery system for the treatment of brain tumor. Likewise, most of the researchers focused on applying the retinal image and processed it for diagnosing. It minimizes the concept of complexity and helps the doctor to predict the disease in an early stage.

In the early stages, the classical segmentation algorithms [14–17] like thresholding, edge detection, and region growing techniques have resulted in some limitations. It is difficult to find the concept of boundary detection correctly because it gets varied during edge detection. Hence, the smoothness is mismatched while segmenting the image. Hence, various researchers focused on applying the concept of segmentation retinal images like local entropy thresholding-based fast retinal vessel segmentation by modifying matched filter (Perona and Malik), active contour model based on extended feature projection (Murthy and Sadashivappa), adaptive threshold-based algorithm (Demirhan and Guler), active contour techniques for automatic detection of brain tumor (Vijayarajan and Multan), etc.

3 Proposed System

The block diagram for brain tumor detection and classification consists of a set of input data called MR images; these images are fed to the preprocessing block where the images are smoothened and noise will be removed by the use of different algorithms. Next, the images undergo the process of segmentation mentioned in Fig. 2 where the precise boundary is obtained depending on the area of interest with human interaction or by automatic segmentation without human interaction. The segmented region's pixels are extracted by feature extraction algorithms to compare the extracted pixel with the victim case for detecting whether the brain region is affected or not to classify it into severe or benign.

Fig. 2 Brain tumor
prediction method

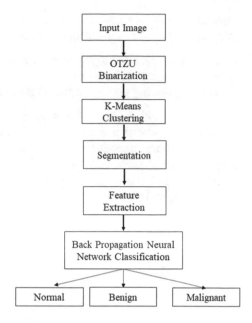

There are two major types of image classification methods. One is unsupervised which is programmed by software, and the other is supervised which is known as human-guided classification. Unsupervised classification is the measure of the software-based parameter analysis without any external affair or from the sample used to guide the classifier. Here, the classifier uses methods to determine which parameters or pixels of an image are similar and group them into particular classes. In this type of classifier, the user can provide the classification algorithms as inbuilt software which will be used for grouping the upcoming images into particular classes.

The input image to the registration phase will be the images produced by the noise reduction phase. The purpose of registration in this framework is to allow the use of multiple imaging modalities (co-registration) and the use of information derived from the alignment of a template in a standard coordinate system (template registration). Therefore, the output of the registration phase will be images in the different modalities that have been aligned with each other and have been additionally aligned with a template in a standard coordinate system. The registration methods should also not rely on segmentation or even the automatic recognition of specific landmarks, since the presence of large tumors can interfere significantly with these operations. These restrictions indicate that intensity or information-based registration algorithms are most appropriate, and fortunately, research is abundant into these methods for medical image registration.

Segmentation of a medical image is the process of dividing an image into different clusters, regions, or categories, which are equivalent to different objects or parts of objects. An image that contains the pixels is allocated to one of these categories. The best segmentation is typically one in which:

- Pixels with similar properties have a similar gray scale of multivariate values and form a region boundary,
- The neighboring pixels in different categories have dissimilar gray-level values.

The segmentation process involves image preprocessing to image analysis. As the image segmentation is carried out, the target information is based on segmentation, the feature extraction and other parameters measured to convert the original image into the discrete and compact form, which provides a high-level image to analyze and understand for diagnosis process here Otsu binarization and k means clustering algorithms complete the segmentation.

To reach a minimum of dissimilarity function, there are two conditions. These are given in Eqs. 1 and 2.

$$c_i = \frac{\sum_{j=1}^{n} u_{ij}^m x_j}{\sum_{j=1}^{n} u_{ij}^m} \tag{1}$$

$$u_{ij} = \frac{1}{\sum_{k=1}^{c} \left(\frac{d_{ij}}{d_{kj}}\right)^{2/(m-1)}} \tag{2}$$

Detailed algorithm of k-means proposed by Bezdek in 1978 is as follows.

$$\frac{\partial C}{\partial w_{jk}^l} = a_k^{l-1} \delta_j^l \text{ and } \frac{\partial C}{\partial b_j^l} = \delta_j^l. \tag{3}$$

In the backpropagation classification technique, that has been proposed, the supervised classifier works as a trained classifier which will compare the input data with the stored knowledge database and the unsupervised classifier stores the important features of the respective dataset for comparing with upcoming input data feature. Here, we have supervised classifier trained with victim dataset features to compare with feature extracted from the segmented MR image detailed in Eq. 3.

4 Results and Discussion

Initially, the input image is processed with the super pixel generation process (Fig. 3a), and then the task to identify the Glare area (Fig. 3b) is done. This is not visible in the framed pixel that is analyzed. It is important to detect the glare image because there are relatively constant aspects of digital cameras. The traditional methods of glare detection generally rely on one or two simplistic image properties. But this research utilizes various concepts to adapt to

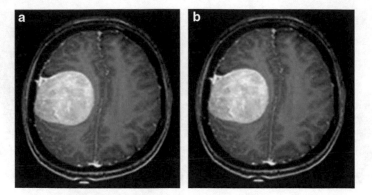

Fig. 3 **a** Source image. **b** Grayscale image

the segmentation process. Adaptive and threshold concepts are combined with the features. It is displayed in Fig. 3a.

The examination of proposed adaptively regularized kernel k-means clustering is tested with MATLAB simulation environment and has compared some other existing methodologies like Hough transform, gradient vector flow model, and Chan–Vese (CV) active contour model. Digital Retinal Images for Optic Nerve Segmentation Database (DCIOS) consists of optic nerve head segmentation from digital retinal images. It is considered to verify the concept of segmentation based on proposed fuzzy k-based clustering. The input samples are represented in Figs. 4 and 5.

The experiment is carried out for 15 different input sample images to verify and validate the brain tumor detection and classification. Figure 6a, b shows different sets of images; the first row (a) represents the registered image for selected source and destination images; the next step will be segmentation for brain tumor region

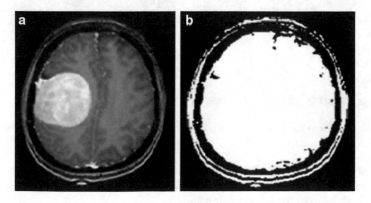

Fig. 4 **a** Filtered image. **b** Otsu binarization image

Fig. 5 MRI filtered segmentation images

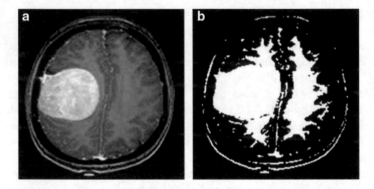

Fig. 6 a Cluster image. **b** Segmented tumor

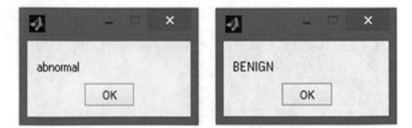

Fig. 7 Classification result (try to create a table as we discussed earlier and cite all the figures)

detection, as shown in the second column. Finally, the classification result is displayed using a message box as shown in Fig. 7, with two samples; one image is classified as benign, the other malignant.

5 Conclusion

Initially, we analyzed various brain tumor-based detection methodologies. The medical domain demands recent advancements to interact with the segmentation. With this motivation, this research adapted the segmentation process of clustering the MRI image downloaded from the Digital Brain Images for Brain Tumor Segmentation Database. Some traditional methods like threshold techniques and gradient-based methods are not effective when the intensity levels of the tumor cub are out of the region. In such cases, there is a need to design an exact algorithm to avoid blur and also to detect the glare area. Here, the regions are partitioned and segmented into various RGB-level histograms to achieve equalization. The role of Adaptively Regularized backpropagation gives high accuracy of each pre-stage of segmentation. Finally, the backpropagation-based clustering is applied to cluster the cub region exactly. From the results, it is noticed that the proposed methodology has maximum accuracy, average F-score, and expected average boundary distance. At last, the average correlation is achieved as expected. Hence, it is adaptable to any type of brain tumor disease. In the future, these concepts are adaptable for each real-time image sequence and verify the test results.

References

1. Selvakumar J, Lakshmi A, Arivoli T (2012) Brain tumor segmentation and its area calculation in brain MR images using K-mean clustering and Fuzzy C-mean algorithm. Int Conference Adv Eng Sci Manag 43(3):186–190
2. Reza SMS, Islam A, Iftekharuddin KM (2013) Multifractal texture estimation for detection and segmentation of meningiomas. IEEE Trans Biomed Eng 60(11):3204–3215
3. May C, Bauer S, Dionysiou D, Buchler P, Stamatakos G, Reyes M (2012) Multiscale modeling for image analysis of meningioma studies. IEEE Trans Biomed Eng 59(1):25–29
4. Shwakshar AS, Ahmmed R, Hossain MF, Rafiq MA (2017) Classification of tumors and it stages in brain MRI using support vector machine and artificial neural network. International conference on electrical, computer and communication engineering, pp 229–234
5. Singh G, Ansari MA (2016) Efficient detection of meningioma from MRIs using k-means segmentation and normalized histogram. 1st India international conference on information processing (IICIP), pp 1–6
6. Machhale K, Kapur V, Nandpuru HB, Kosta L (2015) MRI brain cancer classification using hybrid classifier (SVN-K-NN). International conference on industrial instrumentation and control (ICIC), pp 60–65
7. Parveen AS (2015) Detection of brain tumor in MRI images, using combination of fuzzy c-means and SVN. In: 2nd International conference on signal processing and integrated networks (SPIN), pp 98–102
8. Sarma TC, Sudharani K, Rasad KS (2015) Intelligent brain tumor lesion classification and identification from MRI images using k-NN technique (ICCICCT), pp 777–780
9. Perona P, Malik J (1990) Scale-space and edge detection using anisotropic diffusion. IEEE transactions on pattern analysis and machine intelligence, vol 12(7), pp 629–639

10. Sadashivappa G, Murthy TSD (2012) Meningioma segmentation using thresholding, morphological operations and extraction of features of tumor. International conference on advances in electronics computers and communications, pp 1–6
11. Guler I, Demirhan A (2011) Combining stationary wavelet transform and self-organizing maps for brain MR image segmentation. Eng Appl Artif Intell 24(2):358–367
12. Vijayarajan R, Muttan S (2015) Discrete wavelet transform based principal component averaging fusion for medical images. AEU 69:896–902
13. Zhang N (2011) Feature selection based segmentation of multisource images: application to meningioma segmentation in multi-sequence MRI. PhD thesis, L'Institut National des Sciences Appliquées de Lyon
14. John P (2012) Meningioma classification using wavelet and texture based neural network. Int J Sci Eng Res 3(10):1–7
15. Bian ZQ, Zhang XG (2000) Pattern recognition. Tsinghua University Press, Beijing
16. Schölkopf B, Smola AJ (2002) Learning with Kernels supervised learning models: regularization, optimization and beyond. MIT, Cambridge, MA
17. Zhou J, Chan KL, Chong VFH, Krishnan SM (2005) Extraction of meningioma from MR images using one-class support vector machine. IEEE, engineering in medicine and biology 27th annual conference, pp 6411–6414

Feature Explanation Algorithms for Outliers

Deepak Kumar Rakesh and Prasanta K. Jana

1 Introduction

Outliers are anomalous data points that differ from standard data in a dataset. They are usually removed from the dataset for classification or prediction. However, outliers can provide some information that can help solve some real-time applications. For example, suppose we have a dataset collected over some sensors which are deployed to measure various factors like temperature, humidity, wind speed, and air pollutants. In the dataset collected, the sensors represent the instances, and the environmental factors represent features. Assume a certain number of data points are identified as outliers on this dataset, using an outlier detection technique. The data point or the instance identified as an outlier can be because of two possible reasons: (1) The sensor is not working, thus producing random feature values, and (2) there is an abnormal environmental situation, like high temperature and wind speed. In this scenario, the system should identify the factors (features) whose values show much deviation from their normal range, creating abnormal behavior (outliers). This extracted information can provide useful information about the sensors' working conditions and the environmental conditions. In this paper, we are interested in finding the subset of features of a dataset containing the most abnormal feature values compared to others and hence responsible for the outlying behavior of the dataset. We term it as a feature explanation for outliers.

Now, we present a mathematical illustration to explain the research objective of this article. Consider an unsupervised dataset D, features $F = \{f_1, f_2, \ldots, f_{|F|}\}$, and sample size N. Suppose D forms one cluster C with mean m. Then,

$$m = C\{f_{1m}, f_{2m}, \ldots, f_{|F|m}\} \tag{1}$$

D. K. Rakesh (✉) · P. K. Jana
Department of Computer Science and Engineering, IIT (ISM) Dhanbad, Dhanbad, India

© The Author(s), under exclusive license to Springer Nature Singapore Pte Ltd. 2022   243
R. R. Raje et al. (eds.), *Artificial Intelligence and Technologies*,
Lecture Notes in Electrical Engineering 806,
https://doi.org/10.1007/978-981-16-6448-9_25

where $f_{1m}, f_{2m}, \ldots, f_{|F|m}$ are the coordinates of mean C. The range of feature values belonging to C is shown as

$$\delta_1 f_{1m} - f_{1m} + \delta_2 f_{1m}, \delta_1 f_{2m} - f_{2m} + \delta_2 f_{2m}, \ldots, \delta_1 f_{|F|m} - F_{|F|m} + \delta_2 F_{|F|m}$$
$$= \delta_1 f_{im} - f_{im} + \delta_2 f_{im}, f_i \in F, i \in \{1, 2, \ldots, |F|\} \tag{2}$$

where $\delta_1 f_{im}$ and $\delta_2 f_{im}$ are ranges of values corresponding to feature f_i that lies in cluster C. For any feature, $f \in F, I_f(C)$ represents the number of samples that belong to cluster C and $O_f(C) = N - I_f(C)$ represents the number of samples acting as outliers. Then, the outlier nature for feature f is measured as $\frac{O_f(C)}{N} = \frac{N - I_f(C)}{N} = 1 - \frac{I_f(C)}{N}$. Consider two subsets of features, such that $S, S' \subset F, \{f_{1S}, \ldots, f_{|S|S}\} = S$, and $\{f_{1S'}, \ldots, f_{|S'|S'}\} = S'$. The difference of outlier nature for these subsets S and S' is defined as

$$\Delta x = \left| \left\{ \frac{O_{f_{1S}}(C)}{N} + \cdots + \frac{O_{f_{|S|S}}(C)}{N} \right\} - \left\{ \frac{O_{f_{1S'}}(C)}{N} + \cdots + \frac{O_{f_{|S'|S'}}(C)}{N} \right\} \right|$$
$$= \left| \left\{ \frac{N - I_{f_{1S}}(C)}{N} + \cdots + \frac{N - I_{f_{|S|S}}(C)}{N} \right\} - \left\{ \frac{N - I_{f_{1S'}}(C)}{N} + \cdots + \frac{N - I_{f_{|S'|S'}}(C)}{N} \right\} \right|$$
$$= \left| \left\{ |S| - \left\{ \frac{I_{f_{1S}}(C)}{N} + \cdots + \frac{I_{f_{|S|S}}(C)}{N} \right\} \right\} - \left\{ |S'| - \left\{ \frac{I_{f_{1S'}}(C)}{N} + \cdots + \frac{I_{f_{|S'|S'}}(C)}{N} \right\} \right\} \right| \tag{3}$$

So, the main objective of this article is to find the set of subsets, where

$$\Delta x \text{ is maximum} \tag{4}$$

That is, the difference in the number of outliers generated on the selected subset of features and the number of outliers generated on the rest of the features is maximum. The selected subset of features will give us the subset of features with maximum outlying behavior compared to the rest of the features, thus most responsible for the dataset's outlier nature.

In this paper, we propose two algorithms for finding feature explanation for outliers. The proposed algorithms are based on isolation forest (IF) [1]. The first algorithm employs binary particle swarm optimization (BPSO) [2], and the second one uses greedy recursive feature selection (GRFS) approach to select an optimal subset of features. Accordingly, they are referred as IF-BPSO and IF-GRFS. The algorithms are experimented on four synthetic datasets, benchmark AQI dataset [3] and eight time-stamp of AQI dataset. The results show that the IF-BPSO is more effective than the IF-GRFS on stability and the proposed performance metric called feature consistency.

Feature selection and extraction are well-known research areas in various domains like pattern recognition, image processing, data mining, and machine learning. Many studies have been made with classification and regression [4–6] and also with outlier detection [7–10]. However, very few works have been done for feature explanation for outliers. Some researchers refer it as "outlying aspect mining" [11, 12]. Duan et al. [11] proposed a method called OAMiner that finds the subspace where a concerned object is ranked higher in terms of probability density measure. In [12], Vinh et al. developed a method called OARank. This method provides a ranking of features based on their prospective contribution to the externality of a query point. Micenková et al. [13] identified a subset of features that show the maximum deviation of an outlier object when projected along with the elements. The approaches mentioned above have not considered the state-of-the-art outlier detector method that are used to detect instance acting as an outlier. Instead, they have generated the synthetic data and trained a model different from the original outlier detector. Moreover, outlying aspect mining is different from the context, in which features' explanation proposed here. In the previous work, the authors are interested in finding the features which separate an instance from the rest of the instances of dataset, while the proposed article is interested in finding the features responsible for the outlier nature of the complete dataset. Siddiqui et al. [14] have acknowledged the importance of this problem but have not addressed it. This paper provides a primary solution to the aforementioned problem.

The rest of the paper is organized as follows. Section 2 discusses the proposed work. Section 3 discusses the performance metrics and results. At last, we conclude this the paper in Sect. 4.

2 Proposed Work

The objective of this work is to find the optimal subset of features responsible for outlier nature of the dataset. We use IF as outlier detection technique in our work. As IF has linear time complexity and low memory requirement. It has the ability to handle large dataset with high dimension and is thus more popular than other anomaly detection techniques [1]. Next, we describe the two proposed algorithms for the same.

2.1 IF-BPSO

We use BPSO [2], a population-based metaheuristic to select optimal set of features with incorporation of IF. Each particle in the BPSO is made of certain elements that indicate a potential solution. A fitness function is used to assess the suitability of a solution. We formulate the fitness function as follows:

Problem formulation (fitness function): For a dataset D having features Y, $2^{|Y|} - 1$ number of feature subsets are possible. Let S_i be the ith subset of features, and

$S_i' = Y - S_i$ includes rest of the features where $S_i, S_i' \subset Y$ and $1 \leq i \leq (2^{|Y|} - 1)$. Then, using IF

$$\text{Maximize CS} = \text{IF}(S_i) - \text{IF}\left(S_i'\right) \tag{5}$$

where IF(S) refers to outlier generated using IF on subset S.

Implementation detail of IF-BPSO: First, we initialize each particle (P_i) as represented by Eq. 6. It is generated randomly by selecting the value of X_{ij} from the binary set $\{0, 1\}$. Note that X_{ij} denotes the jth feature of ith particle. $X_{ij} = 1(0)$ indicates that the corresponding feature is selected (unselected). Line 1 calls function Initialization_of_Population() to generate initial population of the particles.

$$P_i = \left[X_{i,1}, X_{i,2}, \ldots, X_{i,D}\right] \tag{6}$$

Line 2 sets the condition for algorithm to stop as maximum number of iterations. From line 3, we start evaluating the fitness value of each particle and update their velocity and position vector, as discussed in paper [2], *Sect. 2*. We construct objective function as per Eq. 7 to evaluate particle of the population shown in line 4.

$$\text{Maximize CS} \tag{7}$$

This helps us to periodically update the personal best and global best of the particles. Our objective is to maximize the CS. In each iteration, the velocity and position of each particle P_i are updated as per their equations, paper [2], *Sect. 2*. After receiving the new position, the objective function of the particle is evaluated by Eq. 7. Next the updation of Pbest$_i$ and Gbest$_i$ is done as per Eqs. 8 and 9, respectively.

$$\text{Pbest}_i = \begin{cases} P_i, & \text{if fitness}(P_i) > \text{fitness}(\text{Pbest}_i) \\ \text{Pbest}_i, & \text{Otherwise} \end{cases} \tag{8}$$

$$\text{Gbest} = \begin{cases} P_i, & \text{if fitness}(P_i) > \text{fitness}(\text{Gbest}) \\ \text{Gbest}, & \text{Otherwise} \end{cases} \tag{9}$$

After termination, the particle Gbest provides the final solution.

Algorithm 1 : IF-BPSO
Input: Dataset D, Particles population P_p, IF and No_of_iterations (Max)
Output: Gbest (features_selected)
1 Initialization: Initialization_of_Population()
2 **while** *No. of iterations != Max* **do**
3   **foreach** $P_i, i = 1, ..., P_p$ **do**
4          $fitness = IF(P_i) - (\bar{P_i})$;
5   **if** $fitness(P_i) > fitness(Pbest_i)$ **then**
6              $Pbest_i = P_i$;
7   **if** $fitness(Pbest_i) > fitness(Gbest)$ **then**
8              $Gbest = Pbest_i$;
9   **foreach** *Particle* $P_i, i = 1, ..., P_p$ **do**
10   The Velocity and the Position vector is updated.
11 return Gbest.

2.2 IF-GRFS

Here, we recursively select features that decrease the outlier nature of the dataset. Our objective is to select the most outlying behaving features from the dataset in each phase.

Problem formulation: For a dataset (D) having Y features, we define data quality (DQ) as the average outliers generated using IF over iterations N.

$$DQ = \frac{\sum_{i=1}^{N} IF(D)}{N} \tag{10}$$

A feature $f \in Y$ is selected in subset $S, S \subset Y$ if it satisfies Eq. 11.

$$IF(Y - f) < IF(Y) \tag{11}$$

Implementation detail of IF-GRFS: We initialize the DQ using Eq. 10 (line 1). Lines 2 to 4 are the initialization part, as shown in Algorithm 2. Line 5 sets the stopping condition for the algorithm. We randomly select feature f from available features and calculate DQ' in lines 6 and 7 subsequently. Line 8 checks the conditions using Eq. 11. If it satisfies, then f is selected feature, DQ, and available features updated (lines 9 to 11). If not, feature f is discarded from available features (line 13). At each iteration, the counter is updated. Finally, the subset of features S is selected which is having maximum outlying behavior (line 15).

Algorithm 2 : IF-GRFS
Input: Dataset D (YxI)
Output: S (features_selected)
1   DQ ← DQ; // Calculating DQ using Eq. 10 with Y features.
2   Counter ← 0; // No. of features is traced.
3   Avail_feat ← Y ; // Total available features.
4   S ← ψ, // No feature selected.
5   **while** *counter < len (Y)* **do**
6   f ← random (Avail_feat); // Randomly selecting features
7      $DQ' \leftarrow IF\ [:, [Avail\_feat - f]]$;
8   **if** $DQ \leftarrow DQ'$ **then**
9        $S \leftarrow S \cup f$; // Appending features
10       $DQ \leftarrow DQ'$;
11   Avail_feat ←Avail_feat – f;
12   **else**
13   Avail_feat ← Avail_feat – f ;
14   Counter ← Counter + 1;
15 return S;

2.3   *Proposed Work*

Now, we discuss the experimental results of the proposed algorithms. We used scikit-learn library [13] for our experiment, as and when required. We employed 10-fold cross-validation for validation. We applied our algorithms on four synthetic datasets, AQI dataset [2], and eight time intervals of AQI dataset. Each synthetic dataset consists of 11,000 instances.

These datasets are imputed with global outliers so that the outliers generated on the specific subset of features are maximum as compared to the rest of the features. To test the performance of the proposed algorithms on synthetic datasets, we used two performance metrics: stability and the proposed feature consistency. Stability measures the robustness of preferences of outlier features produced by reduced datasets, as discussed in [14]. The proposed feature consistency's objective is to measure the consistency of the outlier features produced by the algorithm concerning the known outlier features of.

3   Results and Discussion

Now, we discuss the experimental results of the proposed algorithms. We used scikit-learn library [15] for our experiment, as and when required. We employed 10-fold cross-validation for validation. We applied our algorithms on four synthetic datasets, AQI dataset [2], and eight time intervals of AQI dataset. Each synthetic dataset consists of 11,000 instances. These datasets are imputed with global outliers so that the outliers generated on the specific subset of features are maximum as compared to the rest of the features. To test the performance of the proposed algorithms on synthetic datasets, we used two performance metrics: stability and the proposed feature consistency. Stability measures the robustness of preferences of outlier features produced by reduced datasets, as discussed in [16]. The proposed feature consistency's objective is to measure the consistency of the outlier features produced by the algorithm concerning the known outlier features of the synthetic dataset.

Feature consistency (*fc*): Suppose a synthetic dataset D having features F, then $\vec{F} = 0(1)$ where 0 represents normal features, 1 represents the outlier features, and $\vec{F_i} = 0(1)$ is the feature vector generated using an algorithm A at ith iteration. The feature consistency fc is defined as:

$$fc = \frac{\sum \vec{F} \cdot \vec{F_i}}{t} \qquad (12)$$

where t is the total number of iterations for which algorithm runs. The value of fc will range between $(0, 1]$, and 1 represents that an algorithm can generate similar outlier features to known outlier features $\vec{F} = 0(1)$ in each iteration. Table 1 shows the comparison of stability and feature consistency for synthetic datasets. Here, we can observe that IF-BPSO performs better than IF-GRFS. In other words, IF-BPSO generates more or less same and equivalent outlier features as imputed in original synthetic datasets in comparison with IF-GRFS.

As we do not know the nature of the real-time environment, we choose air quality data [2] to cross-verify whether the outlier features generated resemble a real-time scenario or not. We show only those outlier features in Table 2, consistent

Table 1 Stability and feature consistency comparison on synthetic datasets

	Outlier_features	Stability	Feature consistency
		IF-GRFS IF-BPSO	IF-GRFS IF-BPSO
Syn_data1	f6, f8, f1, f5	0.65 0.71	0.52 0.62
Syn_data2	f11, f9, f7, f8	0.68 0.75	0.59 0.68
Syn_data3	f1, f3, f5, f10, f14	0.55 0.67	0.49 0.53
Syn_data4	f5, f20, f18, f14, f6, f3, f2	0.60 0.69	0.42 0.51

Table 2 Outlier features on AQI and different time intervals of AQI dataset

	AQI	Outlier_features		IF-GRFS	IF-BPSO
		PT08.S1, C6H6, PT08. S4, NO$_x$, PT08.S2		0.60	0.69
		Least	Most		
Time intervals	7:00	PT08.S1	Temp	0.67	0.71
	9:00	Temp	PT08.S1	0.63	0.64
	12:00	PT08.S4	NO$_2$	0.71	0.75
	14:00	RH	NO$_x$	0.72	0.78
	16:00	C6H6	NO$_x$	0.65	0.68
	18:00	PT08.S1	NO$_x$	0.61	0.63
	20:00	NO$_2$	C$_6$H$_6$	0.58	0.64
	22:00	PT08.S3	NO$_2$	0.73	0.77

in different iterations for both the algorithms. On the AQI dataset, the outlier features generated for AQI are NO$_x$, PT08.S2, PT08.S5, C6H6, and PT08. The result shows that the vehicle emission is the main reason for the outlier nature of air. However, each time interval shows the different sets of outlier features generated due to various social, industrial, and environmental effects. In the early afternoon, as vehicular movement starts, PT08.S1 shows the most outlying nature. Similarly, in the evening, PT08.S1, PT08.S4, and C6H6 show high outlying nature. Nitrogen oxide is the vehicle emitted gas, which takes the various forms of oxides in the presence of sunlight [17] and remains in the environment. So, it shows outlying nature for most of the day. Here, we can observe that the proposed algorithms' outlier features are relevant to real-time scenarios present in the environment. From the results, we can observe that IF-BPSO generates relatively stable outlier features as compared to IF-GRFS.

4 Conclusion

We have proposed two algorithms to identify a subset of features that are responsible for the outlying behavior of the dataset. We have used isolation forest to find outliers in the dataset, whereas binary particle swarm optimization and greedy recursive feature selection technique are used to select features. The algorithms were tested on four synthetics, the complete air quality dataset, and different time stamps of the AQI dataset. We have compared the results with respect to stability and the proposed metric, i.e., feature consistency. Through analysis of the results generated by the algorithms, we have shown that the IF-BPSO demonstrates better performance than the IF-GRFS.

References

1. Liu FT, Ting KM, Zhou ZH (2012) Isolation-based anomaly detection. ACM Trans Knowl Disc Data (TKDD) 6(1):1–39
2. Kennedy J, Eberhart RC (1997) A discrete binary version of the particle swarm algorithm. In: 1997 IEEE International conference on systems, man, and cybernetics. Computational cybernetics and simulation, vol 5. IEEE, pp 4104–4108
3. De Vito S, Massera E, Piga M, Martinotto L, Di Francia G (2008) On field calibration of an electronic nose for benzene estimation in an urban pollution monitoring scenario. Sens Actuators B Chem 129(2):750–757
4. Kohavi R, John GH et al (1997) Wrappers for feature subset selection. Artif Intell 97(1–2):273–324
5. Guyon I, Eliseeff A (2003) An introduction to variable and feature selection. J Mach Learn Res 3:1157–1182
6. Aggarwal CC, Philip SY (2005) An effective and efficient algorithm for high-dimensional outlier detection. VLDB J 14(2):211–221
7. Song L, Bedo J, Borgwardt KM, Gretton A, Smola A (2007) Gene selection via the bahsic family of algorithms. Bioinformatics 23(13):i490–i498
8. Lazarevic A, Kumar V (2005) Feature bagging for outlier detection, In: Proceedings of the eleventh ACM SIGKDD international conference on knowledge discovery in data mining, ACM, pp 157–166
9. Azamandian F, Yilmazer A, Dy JG, Aslam JA, Kaeli DR (2012) Gpu-accelerated feature selection for outlier detection using the local kernel density ratio. In: 2012 IEEE 12th international conference on data mining. IEEE, pp 51–60
10. Chen Xw, Wasikowski M (2008) Fast a roc-based feature selection metric for small samples and imbalanced data classification problems. In: Proceedings of the 14th ACM SIGKDD international conference on knowledge discovery and data mining, pp 124–132
11. Duan L, Tang G, Pei J, Campell A, Tang C (2015) Mining outlying aspects on numeric data. Data Min Knowl Disc 29(5):1116–1151
12. Vinh NX, Chan J, Bailey J, Leckie C, Ramamohanarao K, Pei J (2015) Scalable outlying-inlying aspects discovery via feature ranking. In: Pacific-Asia conference on knowledge discovery and data mining. Springer, pp 422–434
13. Micenková B, Ng RT, Dang XH, Assent I (2013) Explaining outliers by subspace separability. In: 2013 IEEE 13th international conference on data mining. IEEE, pp 518–527
14. Siddiqui MA, Fern A, Dietterich TG, Wong WK (2019) Sequential feature explanation for anomaly detection. ACM Trans Knowl Discov Data (TKDD) 13(1):1–22
15. Pedregosa F, Varoquaux G, Gramfort A, Michel V, Thirion B, Grisel O, Blondel M, Prettenhofer P, Weiss R, Duborg V et al (2011) Scikit-learn: machine learning in python. J Mach Learn Res 12:2825–2830
16. Kalousis A, Prados J, Hilario M (2007) Stability of feature selection algorithms: a study on high-dimensional spaces. Knowl Inf Syst 12(1):95–116
17. Pandey B, Agrawal M, Singh S (2014) Assessment of air pollution around coal mining area: emphasizing on spatial distributions, seasonal variations and heavy metals, using cluster and principal component analysis. Atmos Pollut Res 5(1):79–86

Recognition and Classification of Stone Inscription Character Using Artificial Neural Network

K. Durga Devi and P. Uma Maheswari

1 Introduction

Stone inscriptions have been found in substantial numbers, holding huge volumes of authentic information from the ancient to the medieval period, beginning from 3rd millennium BCE. Presently, epigraphical efforts are focused on locating places where inscriptions are found, labor-intensive copying called estampaging, and the manual transliteration and scanning of estampage that is indecipherable. Further, the man power required for transliteration is very scarce, leading to a great loss of information. Durga and Maheswari [1] presented a survey about optical character recognition (OCR) system that has been developed for handwritten and scanned document images for international languages like Arabic, Chinese, Thai, and English, as well as for national languages like Hoysala, Bangla, Marathi, Telugu, and Tamil. However, very few studies have been carried out on stone inscription images. In today's fast-moving world with its incredible technology, character recognition is a demanding area of study. In recent decades, the most demanding research area has been the design of an OCR system.

K. Durga Devi (✉)
Department of ECE, SRMIST Ramapuram, Chennai, India

P. Uma Maheswari
Department of CSE, CEG, Anna University, Chennai, India

2 Literature Survey

Emerging technologies today in the field of image processing train machines to receive input, interpret information, and describe pictorial information in text form. In the field of character recognition, several methodologies that have been implemented are highlighted below.

Beulah and Sahana [2] developed an efficient system for translating Tamil Grantham characters into modern Tamil characters. The characters are transliterated by constructing a Grantham font and using a mapping technique. Elakkiya et al. [3] developed a character recognition system for handwritten Tamil script images. Here, the K-nearest neighbor (KNN) classifier is used to transliterate Tamil scripts into an editable format. Mari and Raju [4] recognized handwritten Tamil characters by new feature extraction method. Here, 80 feature values are extracted and classified using the SVM and multilayer perceptron (MLP) classifiers. Manigandan et al. [5] handled Tamil characters from the ninth to eleventh centuries. The SIFT algorithm is used to extract feature vectors from the character, followed by training with an SVM classifier. Shanthi and Duraiswamy [6] developed a recognition system for classifying Tamil handwritten scanned images using an SVM classifier.

Mahalakshmi and Sharavanan [7] transliterated Tamil Brahmi inscriptions into twenty-first century characters using Laboratory Virtual Instrument Engineering Workbench (LabVIEW). Vellingiriraj et al. [8] handled Tamil Brahmi and Vattaezhuttu characters. The Brahmi translation demonstrated higher accuracy. Seethalakshmi et al. [9] handled printed Tamil documents, and the characters were classified using SVM. Bharath and Sriganesh [10] developed an online character recognition system based on the HMM, for handwritten Devanagari and Tamil scripts. Karunarathne et al. [11] developed an OCR system for Sinhala inscriptions. Characters are classified using the ANN and backpropagation algorithm and implemented with the CNN. Ruwanmini et al. [12] developed an OCR system to transliterate inscriptions into modern Sinhala characters. The characters are classified using the K-means algorithm. Rajithkumar and Mohana [13] developed a system for recognizing characters from Kannada stone inscriptions with the scale-invariant feature transform (SIFT). Pal et al. [14] implemented an SVM for classifying Bangla and Devanagari texts. Hence, there is very little ongoing research on recognizing characters from real-time images. Based on the insights drawn from the discussion above, it is evident that no efficient techniques have been developed that recognize characters from stone inscriptions.

3 Digital Acquisition and Preprocessing

The proposed work is twofold and incorporates (1) digital acquisition and (2) character recognition and classification. A total of full script 100 stone images, captured from Tanjore Brihadeshwara temple using a DSLR camera, the sample inscription image is shown in Fig. 1. Once the acquisition is done, a sequence of methods is used to clean the image. Durga and Maheswari [15] proposed preprocessing algorithm in which first step is the point processing method termed linear contrast adjustment (LCA). Next, to completely eliminate background, binarization is carried out using a technique called the modified fuzzy entropy-based adaptive thresholding (MFEAT) to separate the inscribed characters with the iterative bilateral filter (IBF) to preserve edges and denoise the image, and the output image is shown in Fig. 2. The characters were cropped and stored in the database for training and testing purpose.

4   Feature Extraction

Extracting meaningful feature is an essential part in character recognition system. The local features such as area, moment, entropy, and loops are extracted from stone inscription images. Entropy provides standard information on an image in terms of luminance, contrast, and pixel value. Character shapes are either circular or curved, forming a loop. The filled area within the closed loop is to be calculated. The junction, which is the interface between two regions of a character, helps discover the slant angle.

The universe of discourse is selected for feature extraction, based on the positions of different line segments in the character image. Each zone extracts features, and features from all the zones in the skeleton are extracted. The different features taken include the total number of lines (horizontal, vertical, right diagonals, and the

Fig. 1 Sample input image

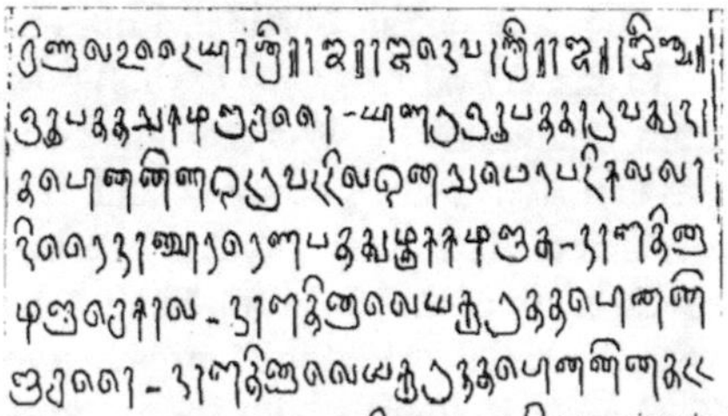

Fig. 2 Binarized image

norm_horizonta	norm_vertical	norm_rightslan	norm_leftslan	filled_area	eccentricity	extent	orientatior	character
0.124	0.4651	0.2229	0.1647	0.0112	0.25	0.1961	0.0088	1
0.1167	0.5417	0.17	0.1433	0.013	0.2525	0.1975	0.0087	1
0.36	0.455	0.175	0.0124	0.0608	0.1593	0.1511	0.0189	1
0.5116	0.3277	0.148	0.0142	0.1244	0.2326	0.1131	0.0185	1
0.1152	0.5176	0.1836	0.166	0.0111	0.25	0.1968	0.0093	1
0.113	0.2609	0.4058	0.2058	0.0107	0.2224	0.1981	0.0166	1
0.3811	0.1731	0.1469	0.0124	0.2519	0.197	0.0087	0	1
0.3811	0.1731	0.1469	0.0124	0.2519	0.197	0.0087	1	1
0.122	0.247	0.3963	0.2195	0.0113	0.1772	0.1414	0.0201	1
0.1312	0.5539	0.1326	0.1077	0.0157	0.1746	0.1191	0.0133	1
0.6366	0.1765	0.0901	0.0592	0.0328	0.1052	0.0954	0.0316	1
0.3321	0.3553	0.1226	0.1296	0.0202	0.2551	0.1473	0.0074	1
0.1111	0.4012	0.2531	0.2222	0.0112	0.1766	0.1593	0.0179	1
0.1031	0.5402	0.229	0.1066	0.015	0.1715	0.2108	0.018	1
0.3462	0.3085	0.1719	0.1358	0.0162	0.25	0.1493	0.0071	1
0.3462	0.3085	0.1719	0.1358	0.0162	0.25	0.1493	0.0071	1
0.3619	0.2739	0.1455	0.1339	0.0164	0.2638	0.1594	0.0067	1
0.3193	0.2062	0.1881	0.2142	0.019	0.2608	0.1501	0.0068	1
0.1324	0.4679	0.2198	0.1465	0.0153	0.2663	0.159	0.0066	1
0.1686	0.5288	0.1381	0.1242	0.0183	0.3445	0.1189	0.0084	1
0.2069	0.4797	0.1485	0.099	0.048	0.1648	0.1839	0.0318	1
0.2411	0.4649	0.155	0.1007	0.0192	0.2271	0.1495	0.009	1
0.3217	0.278	0.2282	0.1441	0.0163	0.23	0.1439	0.0086	1

Fig. 3 Stored feature vector into the table

left diagonals), the length of lines (horizontal, vertical, the right diagonals, and left diagonals), and the area of the skeleton. Figure 3 represents the feature vector of an inscription characters image. These feature vectors were imported into the network as an input.

5 Classification by Artificial Neural Network (ANN)

The neural network reduces manual training and recognizes the output easily. The basic requirement of a neural network is that it needs the data necessary for accurate training. It is used to classify inscription scripts. In this section, the ANN is trained using the backpropagation algorithm. After converting the Tamil inscriptions, the matrices are fed to the ANN as input. After the feedforward algorithm lays the groundwork for the working of the neural network, the backpropagation (BP) algorithm undertakes training, calculates errors, and modifies weights. The BP algorithm starts with computing the output layer, which is the only one where the desired output is available. In this research, the result of the BP-ANN is a matrix of 1×64. The output obtained from the BP-ANN can be used to obtain characters.

In the neural network, the leftmost nodes assign initial weights randomly and train continuously. Weights are added to and stored in the centermost nodes. Several nodes are generated and fall between 0 and 1, based on the threshold generated. The final weights are compared with the input, and the process is reversed to minimize errors. Finally, based on the weights, the vectors are compared with the test data and the corresponding character displayed if the weight matches. The ANN is trained with the inscriptions feature vector and gives good classification accuracy. The flow of the classification module is shown in Fig. 4.

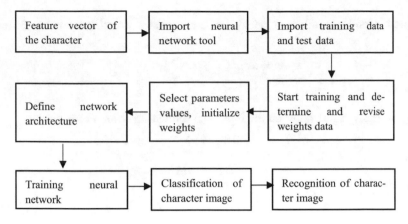

Fig. 4 Flow of classification module

Pseudocode

Step 1: Weights are initialized (randomly).
Step 2: Propagation entries are presented to the input layer:

$$XXi = E; \tag{1}$$

The spread to the hidden layer is made using the following formula:

$$Y = f\left(\sum_{i=1}^{7} X_i^* W_{ji} + X_0\right) \tag{2}$$

Then, we adopt below equation from the hidden layer to the output layer:

$$Z_k = f\left(\sum_{i=1}^{3} Y_i^* W_{kj} + Y_0\right) \tag{3}$$

where X_0 and Y_0 are scalars, and (x) is the activation function:

$$(x) = 1/1 + e^{-}; \tag{4}$$

Step 3: Back propagation of error at the output layer, the error between the desired output S_k and Z_k output is calculated by

$$E_k = Z_k(Z_k - 1)(S_k - Z_k) \tag{5}$$

In the hidden layer, the error calculation is propagated by using the below formula:

$$Y_j = f_j(1 - Y_j) \sum_{k=1}^{7} W_{kj} E_k \qquad (6)$$

Step 4: The weights between the input and hidden layer are corrected by

$$DW_{ij} = nX_i \Gamma_j \qquad (7)$$

$$DY_0 = nF_j \qquad (8)$$

The connections between the input layer and the output layer are changed as follows:

$$DW_{kj} = nY_j E_k \qquad (9)$$

N is a parameter to be determined empirically
Step 5: Loop to step 2 until a stop criterion (error threshold, the number of iterations).

6 Performance Metrics

This proposed system evaluated the performance of classification model by calculating the common performance measure including precision, recall, and f-measure. Here, the results of 24 characters classified are shown in Table 1. Using ANN, the overall accuracy rate obtained is 80%.

$$\text{Precision} = \frac{\text{True positive}}{\text{True positive} + \text{False positive}} \qquad (10)$$

$$\text{Recall} = \frac{\text{True positive}}{\text{True positive} + \text{False negative}} \qquad (11)$$

$$\text{f - measure} = \frac{2 * \text{precision} * \text{Recall}}{\text{precision} + \text{Recall}} \qquad (12)$$

Table 1 Performance metrics of character recognition using ANN

Ancient Tamil characters from stone	Modern Tamil characters	ANN Classification results			Ancient Tamil characters from stone	Modern Tamil characters	ANN Classification results		
		Precision	Recall	Accuracy			Precision	Recall	Accuracy
Vowels					**Consonants**				
[glyph]	ஷ	89.19	82.50	78.01	[glyph]	ல	68.75	84.62	65.00
[glyph]	ஜீ	75.36	81.25	75.36	[glyph]	ண	92.86	75.12	76.47
[glyph]	இ	91.55	85.53	81.52	[glyph]	ழ	81.34	93.43	79.37
[glyph]	ல	80.65	89.29	77.50	[glyph]	ற	80.37	87.71	75.16
[glyph]	ர	81.25	92.86	80.30	[glyph]	ப	80.29	85.71	73.68
[glyph]	ஐ	80.21	93.24	82.14	[glyph]	ட	79.87	85.36	74.50
[glyph]	ஓ	72.50	74.36	87.88	[glyph]	ப	84.38	73.67	71.00
Consonants					[glyph]	ந	79.87	85.36	74.50
[glyph]	க	92.86	76.47	75.12	[glyph]	வ	72.22	86.67	72.00
[glyph]	நி	70.50	86.54	69.61	[glyph]	ஞ	92.31	79.00	82.14
[glyph]	க	72.22	86.67	72.00	[glyph]	ஜ	85.32	80.13	79.21
[glyph]	ஞ	87.50	81.16	76.40	[glyph]	ஐ	76.92	96.77	77.78
					[glyph]	ன	71.83	85.02	77.80

7 Conclusion

This paper presents the implementation of ANN, since there is no standard methods or algorithms developed for recognizing character from Tamil stone inscriptions. The Features are extracted from preprocessed character by considering the local as well as global feature and that is given as an input to the neural networks, which predicts the character better, since more features are extracted from each character. Totally, 24 characters each with 100 numbers were trained, in that 24 characters, 21 characters are correctly predicted, and achieved recognition rate accuracy is 80%. In future work, the training characters have to be increased to achieve furthermore higher accuracy rate.

References

1. Durga KD, Maheswari PU, "Insight on character recognition for calligraphy digitization. In: International conference on technological innovations in ICT for agriculture and rural development (TIAR), pp- 78–83. IEEE, Chennai (2017).
2. Beulah PM, Sahana R (2015) Efficient modelling technique for classification and transliteration of ancient stone inscription. ARPN J En Appl Sci 10(7):2897–2902
3. Elakkiya V, Muthumani I, Jegajothi M (2017) Tamil text recognition using KNN classifier. Adv Nat Appl Sci 11(7):41–46
4. Mari SS, Raju G (2015) Modified view-based approaches for handwritten Tamil character recognition. ICTACT J Image Video Process 6
5. Manigandan T, Vidhya V, Dhanalakshmi V, Nirmala B (2017) Tamil character recognition from ancient epigraphical inscription using OCR and NLP. In: 2017 International conference on energy, communication, data analytics and soft computing, IEEE, pp 1008–1011
6. Shanthi N, Duraiswamy K (2010) A novel SVM-based handwritten Tamil character recognition system. Pattern Anal Appl 13(2):173–180
7. Mahalakshmi M, Sharavanan M (2013) Ancient Tamil script and recognition and translation using LabVIEW. In: International conference on communication and signal processing, IEEE, pp 1021–1026
8. Vellingiriraj EK, Balamurugan M, Balasubramanie P (2016) Text analysis and information retrieval of historical Tamil Ancient documents using machine translation in image zoning. Int J Lang Lit Linguist 2(4):164–168
9. Seethalakshmi R, Sreeranjani TR, Balachandar T, Singh A, Singh M, Ratan R, Kumar S (2005) Optical character recognition for printed Tamil text using Unicode. J Zhejiang Univ Sci 6(11):1297–1305
10. Bharath A, Sriganesh M (2012) HMM-based lexicondriven and lexicon-free word recognition for online handwritten Indic scripts. IEEE transactions on pattern analysis and machine intelligence 34.4, pp 670–682
11. Karunarathne KGND, Liyanage KV, Ruwanmini DAS, Dias GKA, Nandasara ST (2017) Recognizing ancient Sinhala inscription characters using neural network technologies. Int J Sci Eng Appl Sci 3
12. Ruwanmini DAS, Liyanage KV, Karunarathne KGND, Dias GKA, Nandasara ST (2016) An architecture for an inscription recognition system for Sinhala epigraphy. Int J Res Granthaalayah 4(12):48–64

13. Rajithkumar BK, Mohana HS (2014) Template matching method for recognition of stone inscripted Kannada characters of different time frames based on correlation analysis. Int J Electr Comput Eng 4(5):794
14. Pal U, Roy PP, Tripathy N, Lladós J (2010) Multi-oriented Bangla and Devnagari text recognition. Pattern Recogn 43(12):4124–4136
15. Durga KD, Maheswari PU (2018) Digital acquisition and character extraction from stone inscription images using modified fuzzy entropy-based adaptive thresholding. Soft Comput 23 (8):2611–2626

An Enhanced Computer Vision Algorithm for Apple Fruit Yield Estimation in an Orchard

R. Thendral and D. Stalin David

1 Introduction

India is the main producer of vegetables and fruits, which has put up the nation's development by raising the export quantity of an agricultural commodity. Early fruit-yield forecasting plays a vital role in productive and sustainable management of orchards. It is useful for growers and packers to plan for adequate labor and storage space, to obtain the appropriate number of bins for harvest, and to develop an orderly marketing plan. It is therefore a prerequisite for all partners in the food chain: orchard owners, shippers, retailers, and trade who all need information on fruit quantity [1]. Current yield estimation is a labor-intensive, time-consuming, inaccurate, and expensive operation since it is manually performed by counting the number of individual fruits.

In this proposed work, accurate yield estimation system was developed with the use of computer vision system. If we count the average number of apples per tree, then we can easily calculate the total number of apples per acre by multiplying apple fruit per tree by trees per acre. This research was conducted to develop an intelligent computer vision-based algorithm to automatically recognize and count the number of ripe apples on a tree in the natural daylight illumination condition. Images used to recognize fruits on trees can be differentiated into two groups based on the capturing method. First group of authors [2, 3] is based on additional lighting condition while capturing the images. The lighting conditions are selected with care to suppress surrounding light scatter and promote regions of interest. Some researchers also use nighttime because in this condition, light pollution can be avoided. The second group of authors [4–6] used CCD digital camera for taken an image without any artificial light sources. A review of fruit detection techniques can be found in [7]. Many researchers have developed a lot of the fruit segmentation

R. Thendral (✉) · D. S. David
Department of Computer Science and Engineering, IFET College of Engineering,
Gangarampalaiyam, Tamil Nadu 605108, India

© The Author(s), under exclusive license to Springer Nature Singapore Pte Ltd. 2022   263
R. R. Raje et al. (eds.), *Artificial Intelligence and Technologies*,
Lecture Notes in Electrical Engineering 806,
https://doi.org/10.1007/978-981-16-6448-9_27

algorithms, and a few are listed here. The combination of two features such as color and texture were used by the researchers in [8] for fruit recognition. A computer vision system capable of detecting mature fruits is presented in [9, 10]. In their work, the fusion of Intensity and color-based methods are used for fruit recognition. The author in [11, 12] used shape information to recognize fruits on tree. Combined shape and texture features are used by the researchers in [13] to recognize fruits. However, their approaches provide higher false positives because of other parts of the plant

Many researchers have focused on developing automation systems for specific crops. For apple fruit, researches have addressed some problems such as accurate detection [14], diameter estimation [15], apple picking [16], or automated pruning [17]. This work focuses on establishing fruit detection and counting for yield estimation. The overall objective of this proposed method is to develop an effective algorithm for detecting the number of ripe fruits from a tree in the natural daylight condition by using DWT filter and color-based segmentation methods. The details of the materials and methods are presented in Sect. 2. The details of the experimental results and discussion are presented in Sect. 3. Finally, conclusions of the proposed ripe fruits detection method are presented in Sect. 4.

2   Materials and Methods

2.1   Image Denoising Methods

This paper compares six different image denoising methods and picks the best filter dependent upon the preprocessing statistical measurement.

Mean Filter. It is a window filter of linear class, which smoothed image data. This filter function as low-pass one. The essential thought is for any element in the image takes an average across its neighborhood. A linear filter of size $(2N + 1) \times (2N + 1)$, with specified weights $w(s, t)$, for $s, t = -N,\ldots, N$, then the resultant equation is given by

$$f(x, y) = \sum_{s=-N}^{N} \sum_{t=-N}^{N} w(s, t) im(x + s, y + t) \tag{1}$$

The two-dimensional digital mean filter can be expressed as

$$w(s, t) = \begin{cases} \frac{1}{(2N+1)^2}, & -N \leq s, t \leq -N \\ 0, & \text{otherwise} \end{cases} \tag{2}$$

If the value of N is 1, then the 3×3 window in which averaging is carried out the mean filter output equation is expressed as

$$f(x,y) = \frac{1}{9} \sum_{s=-1}^{1} \sum_{t=-1}^{1} im(x+s, y+t) \tag{3}$$

Gaussian Filter. The Gaussian smoothing operator performs a weighted average of surrounding pixels based on the Gaussian distribution. It preserves boundaries and edges better than other filters. The two-dimensional digital Gaussian filter can be expressed as

$$w(s,t) = \frac{1}{2\pi\sigma^2} \exp\left(-\frac{s^2 + t^2}{2\sigma^2}\right) \tag{4}$$

If the value of σ is 0.5, then the window size is 3×3 and the resultant equation is expressed as

$$f(x,y) = \sum_{s} \sum_{t} w(s,t) im(x+s, y+t) \tag{5}$$

Median Filter. This filter protects edges while removing noise. For a filter of size $(2N + 1) \times (2N + 1)$, the output is defined for $x, y = (N + 1),\ldots, (n - N)$.

$$f(x,y) = \text{median}\{im(x+s, y+t) : s, t = -N, \ldots, N\} \tag{6}$$

Bilateral Filter. In this filter, edges are preserved well while noise is averaged out. At a pixel area x, this filter output is expressed as follows

$$J(x) = \frac{1}{c} \sum_{y \in N(x)} e^{\frac{-\|y-x\|^2}{2\sigma_d^2}} e^{\frac{-|I(y)-I(x)|^2}{2\sigma_r^2}} I(y) \tag{7}$$

Here the resulting pixel intensity is $J(x)$ and spatial neighborhood of pixel is $N(x)$, controlling the fall-off of weights in spatial and intensity domains are represented by σ_d and σ_r, respectively; $I(x)$ and C are the normalization constant.

$$c = \sum_{y \in N(x)} e^{\frac{-\|y-x\|^2}{2\sigma_d^2}} e^{\frac{-|I(y)-I(x)|^2}{2\sigma_r^2}} \tag{8}$$

Adaptive Filter. This filter minimizes a mean square error (MSE) value between an original image and the restored image. This filter can be expressed as

$$w(u,v) = \frac{H * (u,v)}{|H(u,v)|^2 + K} \tag{9}$$

where $k = \frac{P_{n(u,v)}}{P_{s(u,v)}}$, the degradation function and complex conjugate of degradation function is denoted by $H(u, v)$ and $H^*(u, v)$, respectively. Power spectral density of noise and un-degraded image is denoted by $P_n(u, v)$ and $P_s(u, v)$.

Discrete Wavelet Transform. It is an orthogonal function which can be applied to suppress noise without losing relevant image features. It catches both frequency and location information from the input image. The first step of wavelet-based procedure for denoising the image is choosing a wavelet filter (e.g., Haar, symlet, etc…) and number of levels for the decomposition. Here, we use Daubechies wavelet filter and single-level decomposition. Then compute the 2D-DWT of the noisy image. The next step is a threshold the non-LL sub-band. Here, we use soft thresholding method to the sub-band coefficients. Then perform the 2D-IDWT (inverse discrete wavelet transform) on the non-modified approximation LL sub-band image and the modified detail sub-band image to get the denoised image (Fig. 1).

$$\varphi(n_1, n_2) = \varphi(n_1)\varphi(n_2) \tag{10}$$

$$\Psi^{(H)}(n_1, n_2) = \Psi(n_1)(n_2) \tag{11}$$

$$\Psi^{(V)}(n_1, n_2) = \varphi(n_1)\Psi(n_2) \tag{12}$$

$$\Psi^{(D)}(n_1, n_2) = \Psi(n_1)\Psi(n_2) \tag{13}$$

In the above equations, $\Phi(n_1, n_2)$ represents approximation image, $\psi^{(H)}(n_1, n_2)$ represents horizontal detail image, $\psi^{(V)}(n_1, n_2)$ represents vertical detail image, and $\psi^{(D)}(n_1, n_2)$ represents diagonal detail image. For an DWT image function, $S(n_1 \times n_2)$ of size $N_1 \times N_2$ can be expressed as

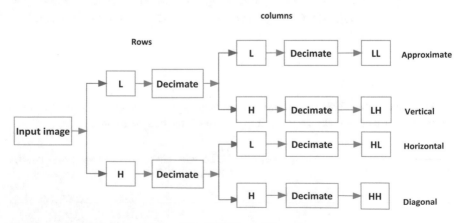

Fig. 1 Analysis filtering operation of the DWT in single level

$$W_\varphi(j_0, k_1, k_2) = \frac{1}{\sqrt{N_1 N_2}} \sum_{n1=0}^{N_1-1} \sum_{n2=0}^{N_2-1} S(n_1, n_2) \varphi_{j_0, k_1, k_2}(n_1, n_2) \qquad (14)$$

$$W_\Psi^i(j_0, k_1, k_2) = \frac{1}{\sqrt{N_1 N_2}} \sum_{n1=0}^{N_1-1} \sum_{n2=0}^{N_2-1} S(n_1, n_2) \Psi_{j_0, k_1, k_2}^i(n_1, n_2) \qquad (15)$$

Here, j_0 represents any starting scale, which may be treated as $j_0 = 0$. For a 2D IDWT image function $S(n_1 \times n_2)$ can be expressed as

$$S(n_1, n_2) = \frac{1}{\sqrt{N_1 N_2}} \sum_{k_1} \sum_{k_2} W_\varphi(j_0, k_1, k_2) \varphi_{j_0, k_1, k_2}(n_1, n_2) + \frac{1}{\sqrt{N_1 N_2}} \sum_{i=H,V,D} \sum_{j=0}^{\infty} \sum_{k_1} \sum_{k_2} W_\Psi^i(j, k_1, k_2) \Psi_{j, k_1, k_2}^i(n_1, n_2)$$

$$(16)$$

Let $L_0(z)$ is the low pass analysis filter, $L_1(z)$ is the low-pass synthesis filter, $H_0(z)$ is the high-pass analysis filter, and $H_1(z)$ is the high-pass synthesis filter, respectively. These filters have to fulfill the following two conditions

$$L_0(-z)L_1(z) + H_0(-z)H_1(z) = 0 \qquad (17)$$

$$L_0(z)L_1(z) + H_0(z)H_1(z) = 2z^{-d} \qquad (18)$$

The proposed wavelet-based method is tested for various threshold criteria and uses soft threshold to provide smoothness and better edge conservation, avoiding the discontinuous character of the hard threshold methods. Soft threshold replaces the discontinuous function of hard threshold by a continuous function ST, such as

$$ST(x) = \begin{cases} x, & \text{if } |x| \geq T \\ 2x - T, & \text{if } T/2 \leq x < T \\ T + 2x, & \text{if } -T < x \leq -T/2 \\ 0, & \text{if } |x| < T/2 \end{cases} \qquad (19)$$

This soft threshold function ST does not overstate the gap between significant and insignificant transform values.

2.2 Evaluation Measure

In this paper, we compare the different image denoising techniques by using their quality parameters such as root mean square error (RMSE) and peak signal-to-noise ratio (PSNR) value. The best denoising condition has the low RMSE value and higher PSNR value.

The sample input image of apple is shown in Fig. 2. The various filters are applied into this input image. The statistical measurement of these filters is

Fig. 2 Sample input image

presented in Table 1. This experimental statistical evaluation shows that 2D-DWT decomposition and reconstruction removes noise more efficiently than the other filtering methods. In the next step of image segmentation, this filtering output serves as input because it retains the required features without any loss. The 2D-DWT image filtering method is used for the better removal of noise and retains their features without any loss in the given input image. In the next step of image segmentation, this filtering output serves as input. Figure 3 shows the complete block diagram of the proposed technique.

3 Results and Discussion

Segmentation is used to detect the fruit region from the other background regions. The contrast between the fruit and background region is not good in this RGB color image. So, this RGB color image was transformed into YIQ color model [18, 19]. The following equation converts an RGB color space to YIQ

$$Y = 0.3R + 0.59G + 0.11B \tag{20}$$

$$I = 0.6R - 0.28G - 0.32B \tag{21}$$

$$Q = 0.21R - 0.52G + 0.31B \tag{22}$$

Table 1 RMSE, PSNR value between original and denoise input image

Methods	RMSE	PSNR (dB)
Mean filter	0.02	26.27
Gaussian filter	0.01	34.83
Median filter	0.02	27.18
Bilateral filter	0.03	34.78
Weiner filter	0.03	26.28
DWT filter	0.01	36.66

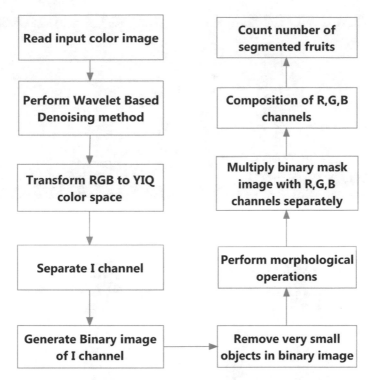

Fig. 3 Block diagram for algorithm to detect ripe fruits from the tree in the natural environment of a fruit orchard

Figure 4 show the histogram plot of *YIQ* components independently. In this diagram, the '*I*' component has the pixels of fruit region and small amount of background region. The other color components have large amount of background regions. For this reason, '*I*' component image (Fig. 5a) is used for this image segmentation procedure. This separated '*I*' component monochrome image is converted into binary (Fig. 5b) where the detected pixels are denoted by the value of '1' and the remaining pixels were denoted by the value of '0'.

Morphological operations are applied in this binary image to enhance the result. Morphological erosion followed by morphological fill operation is performed in order to get the complete fruit region. After all this operation, the ripe fruit regions are present in the image but still some regions are available those are not required. These unwanted regions are eliminated by removing the very small objects present in the binary image. The resultant image (Fig. 5c) contained the recognized fruit regions only. The resultant image (Fig. 5d) shows the total number of fruits in the given input sectional tree image.

This proposed fruit detection algorithm is validated by twenty images of apple fruits on tree collected randomly from internet, and each image contains more fruits. In this research, a close correlation was found between the manual count and using

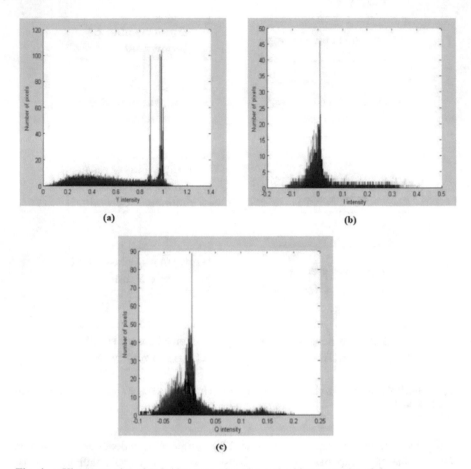

Fig. 4 **a** Histogram plot of Y, **b** histogram plot of I, and **c** histogram plot of Q

this proposed algorithm (Table 2). The results show that this fruit detection algorithm has the accuracy of 93% with an overall error rate of 1.5%. Figure 6 shows the result of our proposed algorithm.

4 Conclusions

In this paper, a computer vision algorithm was developed in order to detect the apple fruits on tree. This yield measurement algorithm consists of 2D-DWT filter for improving the image quality and segmentation methods for recognizing the fruits. Result shows that this algorithm has an accuracy of 93% and an error rate of 1.5%. Future work would integrate this computer vision algorithm to robot harvesting manipulator.

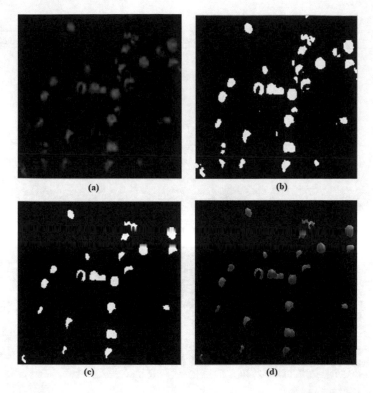

Fig. 5 a '*I*' color component image. **b** '*I*' component converted into a binary image. **c** Remove small objects and perform morphological operations. **d** Extracted fruit region shows the number of apples

Table 2 Results of computer vision detection versus manual count of different fruits from the tree

Tree	Manual count	Algorithm count	Difference	False detected
1	34	32	2	0
2	15	15	0	0
3	46	48	2	2
4	26	23	3	0
5	31	30	1	0
6	54	53	1	1
7	18	18	0	0
8	20	19	1	1
9	42	41	1	0
10	35	33	2	0
11	24	23	1	0
12	39	38	1	0
13	21	21	0	0

(continued)

Table 2 (continued)

Tree	Manual count	Algorithm count	Difference	False detected
14	48	46	2	0
15	41	40	1	1
16	32	33	1	2
17	52	50	2	3
18	16	15	1	0
19	50	44	6	0
20	35	35	0	0

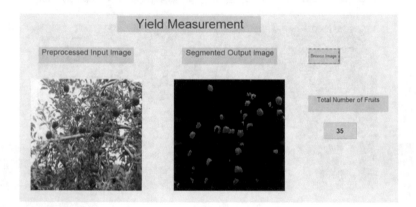

Fig. 6 Result of proposed algorithm

References

1. Gong A, Yu J, He Y, Qiu Z (2013) Citrus yield estimation based on images processed by an Android mobile phone. Biosyst Eng 115(2):162–170
2. Robot harvesting of apples (1984) D'ESNON, G. Proc AgrMation I:210–214
3. Sites P, Delwiche M (1988) Computer vision to locate fruit on. A tree. Trans ASAE 31: 257–263
4. Zaho J, Tow J, Katupitiya J (2005) On-tree fruit recognition using texture properties and color data. In: Conference on intelligent robots and systems, Edmonton, Alberta, Canada, pp 263–268
5. Stajnko D, Lakota M, Coevar MH (2004) Estimation of number and diameter of apple fruits in an orchard during the growing season by thermal imaging. Comput Electron Agric 42:31–42
6. Juste F, Sevilla F (1991) Citrus: a European project to study the robotic harvesting of oranges. In: 3rd international symposium on fruit, nut and vegetable harvesting mechanization, pp 331–338
7. Jimenez AR, Ceres R, Pons JL (2000) A survey of computer vision methods for locating fruit on trees. ASAE 43(6):1911–1920
8. Arivazhagan S, Newlin SR, Nidhyanandhan SS, Ganesan L (2010) Fruit recognition using color and texture features. J Emerg Trends Comput Inf Sci 1(2):90–94

9. Slaughter D, Harrel RC (1987) Color vision in robotic fruit harvesting. Trans ASAE 30(4): 1144–1148
10. Whitaker D, Miles GE, Mitchell OR (1987) Fruit location in a partially occluded image. Trans ASAE 30(3):591–597
11. Benady M, Miles GE (1992) Locating melons for robotic harvesting using structured light. ASAE Paper No.:92-7021
12. Qiu W, Shearer SA (1991) Maturity assessment of broccoli using the discrete Fourier transform. ASAE Paper No. 91-7005, St. Joseph, MI
13. Cardenas-Weber M, Hetzroni A, Miles GE (1991) Machine vision to locate melons and guide robotic harvesting. ASAE Paper No. 91-7006
14. Nicolai H, Pravakar R, Volkan I (2019) A comparative study of fruit detection and counting methods for yield mapping in apple orchards. J Field Robot. https://doi.org/10.1002/rob.21902
15. Apolo-Apolo OE, Pérez-Ruiz M, Martínez-Guanter J, Valente J (2020) A cloud-based environment for generating yield estimation maps from apple orchards using UAV imagery and a deep learning technique. Front Plant Sci 11:1086
16. Sarabu H, Ahlin K, Hu AP (2019) Leveraging deep learning and rgb-d cameras for cooperative apple-picking robot arms, 2019 ASABE Annual International Meeting. American Society of Agricultural and Biological Engineers
17. He L, Schupp J (2018) Sensing and automation in pruning of apple trees: a review. Agronomy 8(10)
18. Thendral R, Suhasini A (2017) Automated skin defect identification system for orange fruit grading based on genetic algorithm. Curr Sci 112(8):1704–1711
19. Thendral R, Suhasini A (2015) A comparative analysis of different color spaces for recognizing orange fruits on tree. ARPN J Eng Appl Sci 10(5):2258–2262

PlantBuddy: An Android-Based Mobile Application for Plant Disease Detection Using Deep Convolutional Neural Network

Saiful Islam Rimon⊚, Md. Rakibul Islam⊚, Ashim Dey⊚, and Annesha Das⊚

1 Introduction

Bangladesh, being an agricultural country, depends heavily on the production of crops to feed its people and maintain the economy. Around 70% population of Bangladesh lives in rural zones. More than 50% of Bangladesh's workforce is directly employed by agriculture. A greater part of the rural population depends on agriculture for most of their earnings [1]. The country needs a huge supply of crops every year. Plant diseases hamper the production of crops often. As a result, the price of food gets higher and poor people have to stay half-fed or unfed. Hence, plant diseases have become a great threat to crops. Moreover, due to the illiteracy of farmers, most of the time they cannot understand what is wrong with the crops. This causes them great sorrow. With the increase in the human populace and the decrease in croplands, the production rate of crops is in danger. Preventing plant diseases is one of many ways to keep the crop production rate above the necessary margin.

Recent advancements in technology have created remarkable opportunities in developing countries like Bangladesh. Android mobile phones are now cheap and affordable for lower-earning people. An automated system can greatly help the farmers to diagnose crop diseases easily and take actions accordingly to avoid the

S. I. Rimon (✉) · Md. R. Islam · A. Dey (✉) · A. Das
Chittagong University of Engineering and Technology, Chittagong 4349, Bangladesh
e-mail: u1604061@student.cuet.ac.bd

A. Dey
e-mail: ashim@cuet.ac.bd

Md. R. Islam
e-mail: u1604060@student.cuet.ac.bd

A. Das
e-mail: annesha@cuet.ac.bd

© The Author(s), under exclusive license to Springer Nature Singapore Pte Ltd. 2022   275
R. R. Raje et al. (eds.), *Artificial Intelligence and Technologies*,
Lecture Notes in Electrical Engineering 806,
https://doi.org/10.1007/978-981-16-6448-9_28

waste of crops. Hence, we got motivated to create an android application that can detect the diseases of plants from captured images of leaves.

Researchers around the world have taken the plant disease problem seriously and are searching for ways to prevent plant diseases and detect them at an early stage. Many researchers have approached various techniques to identify plant diseases. Image processing, machine learning, deep learning are some of the techniques being used to try to solve the problem [2]. By using TensorFlow, the models trained using deep convolutional neural network (CNN) can be used in wide varieties of mobile devices, even low-end devices. TensorFlow lite models use very little hardware resources which is why it is supported in most mobile devices [3].

This study aims to develop a system for the detection of plant diseases through deep CNN and image processing. Furthermore, the study aims to use the developed model to create a user-friendly android application named "PlantBuddy" to detect plant disease from images captured with the camera of the phone and give solutions to cure the disease. Plants show a range of symptoms such as colored spots or streaks occurring on the leaves when they are diseased. As the disease progresses, the visual symptoms change their color, shape, and size [4]. With the help of deep CNN, we can train these patterns and create a model to recognize them. By using the application, farmers can capture the images of plant leaves using any mobile camera of average quality. The image is then processed and cross-matched with the integrated model to identify the disease and provide solutions based on the detected disease. Thus, the farmers can save their time and money as well as their crops.

The rest of the paper is summarized as follows. Related literature and studies are presented in Sect. 2. The proposed methodology is described in Sect. 3. In Sect. 4, the results are presented. And finally, the conclusion to the paper is described in Sect. 5.

2 Related Literature and Studies

The literature survey is conducted to compare different methodologies previously proposed for identifying plant diseases using deep learning and image processing. Many studies have proposed different solutions to detect diseases.

In [4], an IoT-based system was presented that can automatically detect and identify plant leaf diseases. The authors used sensor devices to collect images of plants and plant leaves. Image processing, k-means clustering algorithm, and artificial neural networks are used. The invented device classifies diseases based on monitoring the temperature, humidity, and moisture of the plant with an accuracy over 90%.

In [5], the authors proposed a system that uses an edge detection technique to detect the diseased zones of the plant or fruit. Images of the fruit are captured first. Then, image segmentation is done using a segmentation technique. Afterward, the edges of the diseased zones are calculated in pixels. Depending on the number of pixels, the rate of contamination in the plant or fruit is provided. The control and treatment methods are provided hinged on the affected disease of the fruit.

In [6], deep learning neural network algorithm was explored. TensorFlow is used to process the data to be usable for training. Through the use of deep learning and neural network algorithm and based on the F1 score, the model for detecting plant diseases is built. An application was implemented and evaluated by specialists in this area of study. The accuracy of their developed model is mentioned to be 80%.

In [7], different kinds of deep learning model architectures were implemented, which are based on CNN architectures, to identify plant diseases with leaf images of healthy or diseased plants. VGG CNN architecture performed the best. It accomplished the accuracy of approximately 99% in classifying 17,548 plant leaf images.

In [8], the authors focused on recognizing a paddy plant disease using image processing. The system takes the paddy leaf image as an input and converts the RGB image to grayscale. Then, the morphological opening operation is applied to reduce noise and finally image segmentation. After these processes, they find the infected region of the paddy leaf.

In [9], a mobile application was developed for plant disease recognition using image processing which analyzes the color patterns of the diseased marks in plant leaves and bodies. The dataset images were captured under laboratory conditions with the help of a digital camera. The distance matrix is employed to calculate the distance between each pair of species. Image segmentation is used to partition the image of a plant into distinct regions containing each pixel with similar attributes. Mainly, the k-means clustering algorithm is implemented to identify the diseases. The authors claimed to have achieved 90% accuracy for their model using a small training set.

In [10], the authors showed effective and correct plant disease detection and identification techniques through the use of image processing in MATLAB. K-means clustering algorithm and multi-support vector machine (SVM) methods are used which are organized for both plant and fruit disease identification. Image segmentation and feature extraction are used to prepare the images for training.

In [11], the authors presented a study on various disease identification methods that are utilized in detecting plant diseases. They also described a method for image segmentation which is used to detect and identify plant diseases. The accuracy of the presented system is described to be approximately 95%. But, the number of sample images used in the system is only 60 for four different species.

In [12], a model is trained on images of plant leaves using the deep CNN with the goal to classify both crop species and the identity of the diseases. The proposed model can classify very quickly which is ideal for implementing into an application. However, the accuracy gained on the images which are not from the dataset is mentioned to be just above 31%.

In [13], different procedures to segment the diseased area of the plants were explained. The authors studied various feature extraction and identification methods that are used for extracting the features of the diseased leaf, additionally identifying the plant diseases. The utilization of artificial neural network methods for the detection of plant diseases like back-propagation algorithm and SVM is discussed.

3 Methodology

3.1 System Overview

The overview of the proposed system is shown in Fig. 1. The user of the application captures the image of a leaf of the subject plant using the phone camera. The captured image is then processed using the trained model integrated into the "PlantBuddy" application. The application then gives results based on the accuracy from the image whether the plant is healthy or diseased. The result will show the specific disease name as well as the solutions to cure the disease. There is also a feature to call the nearest agriculture department if the situation is too worse to handle by the user alone.

3.2 System Development Method

The development method of the system is depicted in Fig. 2. It shows the gradual steps of the system from training the model to implementing it in the application.
    The detailed system development steps are described as following:

Dataset Collection and Load. A public dataset of 54,305 images of plant leaves is utilized for training the model. The dataset is created by PlantVillage and can be found in Kaggle. The resolution of the images is 256 × 256 pixels. The images are saved in jpeg format. The train–validation split is 80–20%, which makes 43,456 images for training the model and 10,849 images for validation. The dataset contains images of 14 different species of crops and 26 different diseases. Figure 3 shows some samples of the dataset images, and Table 1 shows the description of the dataset.

Fig. 1 System overview

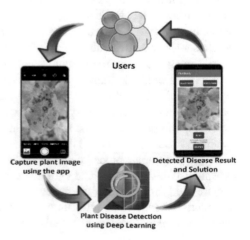

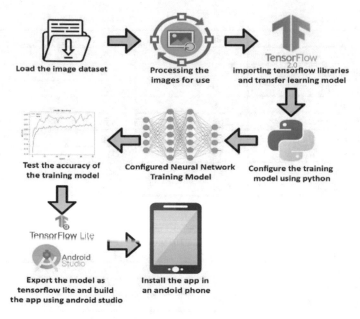

Fig. 2 System development method

a) Apple scab   b) Grape black rot   c) Tomato early blight   d) Potato late blight

Fig. 3 Sample images of the dataset

Image Preprocessing. Image preprocessing is employed to process digital images using algorithms to be usable for the system. It cleans the noise and distortion in images. Data generators are set up to read images from the source folder. Data augmentation is used to process the images in Keras. This technique enhances the condition of dataset images, so that deep CNN models can be built utilizing them. It applies a series of random transformations such as resizing, rotation, RGB to grayscale, and mixing images [14]. The images are then normalized in the range [0, 1] and resized to the required size for the network.

Choosing and Configuring the Architecture Model. An architecture model is required to train the model. TensorFlow Hub is used to import the preferred architecture to our model. TensorFlow, developed by Google, is a software library

Table 1 Dataset description

Class	Training images	Validation images	Total
Apple scab	504	126	630
Apple black rot	497	124	621
Cedar apple rust	220	55	275
Apple healthy	1316	329	1645
Blueberry healthy	1202	300	1502
Cherry healthy	684	170	854
Cherry powdery mildew	842	210	1052
Corn gray leaf spot	411	102	513
Corn common rust	954	238	1192
Corn healthy	930	232	1162
Corn northern leaf blight	788	197	985
Grape black rot	944	236	1180
Grape black measles	1107	276	1383
Grape healthy	339	84	423
Grape leaf blight	861	215	1076
Orange citrus greening	4405	1101	5506
Peach bacterial spot	1839	459	2298
Peach healthy	289	72	361
Pepper bell bacterial spot	797	199	996
Pepper bell healthy	1182	295	1477
Potato early blight	801	200	1001
Potato healthy	122	30	152
Potato late blight	800	200	1000
Raspberry healthy	297	74	371
Soybean healthy	4072	1018	5090
Squash powdery mildew	1468	367	1835
Strawberry healthy	365	91	456
Strawberry leaf scorch	888	221	1109
Tomato bacterial spot	1702	425	2127
Tomato early blight	800	200	1000
Tomato healthy	1273	318	1591
Tomato late blight	1529	382	1911
Tomato leaf mold	761	190	951
Tomato septoria leaf spot	1416	354	1770
Tomato spider mites	1341	335	1676
Tomato target spot	1124	280	1404
Tomato mosaic virus	299	74	373
Tomato yellow leaf curl	4286	1071	5357
Total	43,456	10,849	54,305

that is free and open-source. It is used for dataflow and differentiable programming [15]. We evaluate the applicability of deep CNN in our study. Particularly, this work focuses on two architectures, namely MobileNet V2 and Inception V3.

MobileNet V2 is a family of neural network architectures for efficient on-device image classification and related tasks. MobileNets come in various sizes controlled by a multiplier for the depth in the convolutional layers. They can also be trained for various sizes of input images to control inference speed [16]. On the other hand, Inception V3 is another neural network architecture for image classification. It is a CNN which has 48 layers. The softmax layer computes loss [17]. The selected model is then configured to be trained using deep learning and neural network algorithm. Using TensorFlow Hub to import the model reduces the training time for large datasets as it supports transfer learning with Keras API. Several parameters such as learning rate and epoch are set in this step.

We have to select a model that is accurate and lightweight. Lightweight model means that the exported model is low in size which is favorable for building a mobile application. Most deep learning models can take from 14 Megabytes to around 500 Megabytes in general [18]. Large model size is not preferable for mobile applications.

Checking the Accuracy of the Model. The configured model is trained for several epochs to check accuracy. It is important to check over-fitting and under-fitting in this step, which is observed by plotting a graph of accuracy and loss of the model, shown later. To achieve the highest possible accuracy, the model is trained several times tweaking a few parameters.

Exporting the Model and Building the PlantBuddy Application. In the last step, the trained model is exported to tflite model, and the android application, along with the features, is developed using android studio. The application is then ready to be installed in an android smartphone to use.

4 Results and Performance Analysis

4.1 Comparison of the Model Architectures

Both the CNN models are trained multiple times by varying several parameters such as batch size, learning rate, and epoch. Batch size is the number of examples used in one step of training. The learning rate decides the step size at each step while moving toward a minimum loss. Epoch refers to the total number of times the entire dataset which was trained through the deep learning algorithm. After enough experimentation, the models have shown the best results for batch size of 64, learning rate of 0.001, etc. The models are trained for 10 epochs. The performance of the models is given in Table 2.

Table 2 Architecture performance

Architecture	Training accuracy (%)	Training loss (%)	Validation accuracy (%)	Validation loss (%)	Model size (MB)
MobileNet V2	97.23	22.93	95.75	28.65	11
Inception V3	90.69	51.59	91.81	48.02	90

Presented metrics include:

Training Accuracy. It is the percentage of successful classification in the training dataset. This means the amount of correctly identified images from the dataset.

Training Loss. The average mistakes of the model in the training dataset. This means the average loss per batch, out of batches of the training dataset.

Validation Accuracy. It is the percentage of successful classification of classes in the validation dataset.

Validation Loss. This indicates the average loss per batch, out of batches of the validation dataset.

Model Size. It indicates the size of the exported trained model in megabytes (MB).

The performance of the MobileNet V2 architecture is shown in Fig. 4a. On the other hand, the performance of the Inception V3 architecture is shown in Fig. 4b. It can be seen that MobileNet V2 performs better than Inception V3 architecture for the same parameters. So, MobileNet V2 is chosen to build the final model.

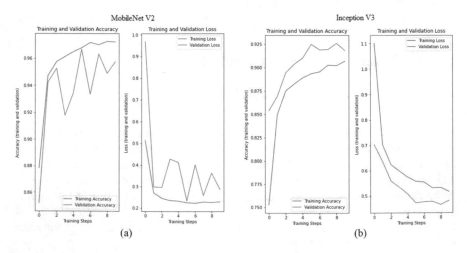

Fig. 4 Accuracy and loss graph of architectures **a** MobileNet V2 and **b** Inception V3

4.2 Accuracy of the Chosen Model

The accuracy of the chosen model on four randomly chosen images from the validation dataset is shown in Fig. 5. It can be seen that in three cases, the accuracy of the detected class is above the average training accuracy which is approximately 97%. In one case, the accuracy is below the average training accuracy. This determines that the performance of the selected model is acceptable and is ready to be used to build the mobile application.

4.3 PlantBuddy Application User Interface and Features

The user interface and features of the mobile application are shown in Fig. 6. The application has multiple pages and features. They are as follows.

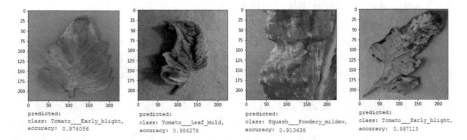

predicted:
class: Tomato___Early_blight,
accuracy: 0.974056

predicted:
class: Tomato___Leaf_Mold,
accuracy: 0.986278

predicted:
class: Squash___Powdery_mildew,
accuracy: 0.913438

predicted:
class: Tomato___Early_blight,
accuracy: 0.997113

Fig. 5 Checking the accuracy of the model on 4 random images

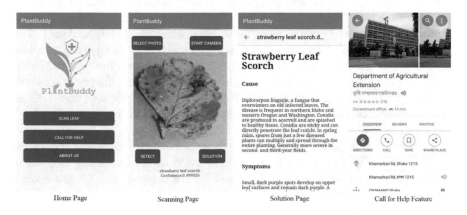

Home Page   Scanning Page   Solution Page   Call for Help Feature

Fig. 6 Mobile application interface and features

Home Page. The application shows the home page when first opened. There are three buttons here—Scan Leaf, Call for Help, and About Us. The "About Us" button shows the basic information about the application.

Scan Leaf. This button takes the user to the image scanning page. It is the main module of the application. There are two ways to load an image, either capturing a new image with the camera or load a previously captured image from the gallery. After that, pressing the "Detect" button shows the detected disease of the plant and also the accuracy of the detection.

Solution. After detecting the image, pressing the "Solution" button takes the user to the solution page where the treatment of the disease is described.

Call for Help. The "Call for Help" button takes the user to Google Maps and shows the nearest agricultural office based on GPS tracking. This feature can be used when the user cannot deal with the disease alone.

5   Conclusion and Future Work

In this study, an android-based mobile application has been developed to detect and identify plant diseases using deep CNN. The model has been trained using a large public dataset of plant images and has attained an accuracy of around 97% which indicates the acceptability of the model. The trained model takes an image of a plant leaf as input, processes it, and uses deep CNN algorithm to detect and identify plant disease. The model is exported to the android application. The application also provides necessary steps to prevent and cure the diseases. Due to the availability and low price of android mobile devices in recent times, even the lower-earning people like farmers can easily afford android devices. The proposed application of this study runs and works well in low-end devices as well. Moreover, the application interface is very easy to understand. So, it is quite practical for the farmers to be able to use the application. Farmers in rural areas can use it to detect plant diseases accurately and take action accordingly. It will help to avoid disaster in food production, thus increasing gross food production. The dataset images used in the study are captured in indoor lighting and environment. Hence, the performance of the proposed model decreases around 10–20% when images are captured outdoor. This limitation can be overcome by using a dataset of images that are captured in a real environment.

This work can be further explored to add new features for making the application more user-friendly. There is also a scope to build the system using a different dataset or images of other species of plants. Multiple language support can make it usable in different countries.

References

1. Bangladesh: Growing the economy through advances in agriculture. https://www.worldbank.org /en/results/2016/10/07/bangladesh-growing-economy-through-advances-in-agriculture, last accessed 2020/10/09
2. Patel A, Joshi M (2017) A survey on the plant leaf disease detection techniques. IJARCCE 6:229–231. https://doi.org/10.17148/ijarcce.2017.6143
3. Yunbin D (2019) Deep learning on mobile devices—a review. SPIE Defense + Commercial Sensing. https://doi.org/10.13140/RG.2.2.15012.12167
4. Thorat N, Nikam S (2017) Early disease detection and monitoring largefield of crop by using IoT. (IJCSIS) Int J Comput Sci Inf Secur 15:236–248
5. Sharath DM, Kumar SA, Rohan MG, Prathap C (2019) Image based plant disease detection in pomegranate plant for bacterial blight. In: 2019 international conference on communication and signal processing (ICCSP), pp 0645–0649. IEEE, Chennai, India. https://doi.org/10.1109/iccsp.2019.8698007
6. Valdoria J, Caballeo A, Fernandez B, Condino J (2019) iDahon: an android based terrestrial plant disease detection mobile application through digital image processing using deep learning neural network algorithm. In: 4th international conference on information technology (InCIT), pp 94–98. IEEE, Bangkok, Thailand. https://doi.org/10.1109/incit.2019.8912053
7. Ferentinos K (2018) Deep learning models for plant disease detection and diagnosis. Comput Electron Agric 145:311–318. https://doi.org/10.1016/j.compag.2018.01.009
8. Mahalakshmi J, Shanthakumari G (2017) Automated crop inspection and pest control using image processing. Int J Eng Res Dev 13:25–35
9. Thakre G, More A, Gajakosh K, Yewale M, Shamkuwar D (2017) A study on real time plant disease diagnosis system. Int J Adv Res Ideas Innov Technol 3:1118–1124
10. Raut S, Fulsunge A (2017) Plant disease detection in image processing using MATLAB. Int J Innov Res Sci Eng Technol 6:10373–10381. https://doi.org/10.15680/IJIRSET.2017.0606034
11. Singh V, Misra A (2017) Detection of plant leaf diseases using image segmentation and soft computing techniques. Inf Process Agric 4:41–49. https://doi.org/10.1016/j.inpa.2016.10.005
12. Mohanty S, Hughes D, Salathé M (2016) Using deep learning for image-based plant disease detection. Frontiers Plant Sci 7. https://doi.org/10.3389/fpls.2016.01419
13. Khirade S, Patil A (2015) Plant disease detection using image processing. In: 2015 international conference on computing communication control and automation, pp 768–771. IEEE, Pune, India. https://doi.org/10.1109/iccubea.2015.153
14. Shorten C, Khoshgoftaar T (2019) A survey on image data augmentation for deep learning. J Big Data 6. https://doi.org/10.1186/s40537-019-0197-0
15. Abadi M et al (2016) 12th USENIX symposium on operating systems design and implementation (OSDI '16), pp 265–283. USENIX Association, TensorFlow: a system for large-scale machine learning
16. Howard A et al (2017) MobileNets: efficient convolutional neural networks for mobile vision applications. arXiv: 1704.04861
17. Szegedy C, Vanhoucke V, Ioffe S, Shlens J, Wojna Z (2016) Rethinking the inception architecture for computer vision. In: 2016 IEEE conference on computer vision and pattern recognition (CVPR), pp 2818–2826. IEEE, Las Vegas, NV, USA. https://doi.org/10.1109/CVPR.2016.308
18. Team K (2020) Keras documentation: Keras applications. https://keras.io/api/applications, last accessed 2020/11/23

Developing a Cyber-Physical Laboratory Using Internet of Things

Ankush Handa and S. Sofana Reka

1 Introduction

Laboratory experiments are very important in the learning process of a student. Indeed, in the case of courses in electronics, the students need to gain hands-on exposure to hardware components—their usability, performance and limitations. Traditionally, laboratory experiments entail the process of carrying out a given procedure in a physical laboratory space within a given duration, e.g. verifying truth tables of digital gate ICs and frequency response of CE amplifier. This often restricts the learning process as the experiments are time bound. Also, in several places of learning, the laboratories are closed outside the college working hours further restricting the students. The proposed cyber-physical laboratory will comprise the necessary test hardware/equipment in a portable form so that the student can perform the experiments anywhere. The author in [1] tells us about the different roots in CIRP community which are related to collaborating computational entities which are in intense connection with the surrounding physical world. From [2], the author explains the interconnection and interoperability among different devices and also big data analysis. Reference [3] gives us a clear picture about the background and some technologies related to IoT with their concepts and objectives. The potentials regarding the application of IIoT are discussed in [4] which state the developments of it along with digital factory and cyber-physical systems. Established science and technology framework is explained in [5], in which the five elements have the common properties with the restructured cyber-physical science in IoT. In paper [6], the security concerns of using an IoT system to develop more defensible and survivable systems are addressed. Building a proper framework provides an organized presentation of an analysis methodology as [7]. Author from [8] presents his views about the current state of the industrial application of agent technology in CPSs and visions the way to face the challenges faced by emerging

A. Handa · S. S. Reka (✉)
School of Electronics Engineering, Vellore Institute of Technology, Chennai, India

© The Author(s), under exclusive license to Springer Nature Singapore Pte Ltd. 2022
R. R. Raje et al. (eds.), *Artificial Intelligence and Technologies*,
Lecture Notes in Electrical Engineering 806,
https://doi.org/10.1007/978-981-16-6448-9_29

CPS challenges. A new version of IEC 61131 is used for PLC targets that support OOPS; Java is used for embedded boards in [9]. The author [10] believes that apart from the challenges faced while using the IoT devices such as safety and security, the processing of data that for further knowledge is a key problem as well. Author in [11] cites that for IoT embedded systems to work better, it is the communication medium that has to be the most efficient. In paper [12] gives an insight about how cyber-physical systems and external world can be connected using a laboratory set-up. Authors in [13] show us the advantage of linking cyber-physical system with any other system to refine its uses. Whenever we design a system to satisfy our needs or to automate a process, one matter that always bothers us is the energy efficiency of that system which is addressed in [3–9]. The components are interfaced to the Internet through an embedded processing platform, wherein a user (in this case, the instructor) will be able to provide the necessary input signals to the hardware set-up and read the output back for verification. There are indeed limitations to this architecture with regard to the feasibility and the variety of experiments possible.

1.1 The Need of Virtualization

In order to plan efficient resource utilization, the underlying hardware and architecture should be exposed to its best possibility. For example, bringing in 1000 computers and doing minimal task of writing documents on them would be sheer wastage of money.

However, in a more generic approach, we could use virtual machines or let one machine be assigned for one major task. In the world of conventional learning, hardware has become cheaper and more powerful than any wild imagination in the past couple of years, but still the requirement of infrastructure to settle them is inevitable. Here comes the need of virtualizing.

It is not a new concept brought into the market but has stayed here for a long time now. The world of computation has warmly welcomed virtualization in order to make the best use of the underlying hardware. The challenge has now shifted towards making the laboratories, which make a significant part of the learning a virtual one. With the advent of virtualization, the students are not time bounded for performing the experiments in the scheduled or the assigned hours. In order to meet the timeline unknowingly shifts its concentration from learning-based approach to a result-based approach. This is an effort targeting on restoring the primary purpose of laboratory-based learning but on a secondary note also allows institutes to escape the heavy investment costs for harnessing a good output from these physical laboratories.

2 Methodology

To demonstrate a proof of concept, we focus on laboratory experiments in the digital electronics undergraduate laboratory course, wherein digital gate ICs and MSI ICs (like adders, counters, shift registers) are used to design, implement and verify binary functions, both sequential and combinational, on breadboards. We propose to use a Raspberry Pi 3 as the hardware interface between the circuit built on the breadboard (called hardware under experimentation or HUE henceforth) and the Internet. The reason for choosing an RPi3 is due to the availability of a large number of GPIO pins, availability of both 3.3 and 5 V supply voltage, on-board Wi-Fi and Ethernet, and an easy programming interface.

The GPIO pins of the RPi3 are connected to the I/O pins of the entire gate ICs. If the nodes in the HUE are more than the GPIO pins available to the user, digital multiplexers are used. The signal levels of the various components should be considered in order to make the above set-up functional. The RPi3 can also be used to provide power to the digital ICs. A test bench has to be written in Python to activate the necessary input signals to the HUE (configured as output in RPi3). The resulting output of the HUE is fed to the RPi3 for verifying the digital logic. This test bench depends on the HUE, which will change for each experiment. Once the student has hooked up a HUE to the RPi3, the instructor should be able to access the RPi3 over a webserver and run the necessary test bench.

The code takes in the input from both of these files, and as per the generated input, every gate is matched with an expected output and if the result differs, it generates an error stating that particular pin might be stuck at '0' or '1'. This tells the user about the location of the error. Thus, the user can either verify the connections or reinstall that new pin. In this work, the process downloaded the OAuth file which is used for the verification for the client to provide necessary keys in order to establish the connection. It actually provides the Internet accessible PI to download these files from the Google Drive. The reason we opt for the Google Drive is that it helps to elevate the capex for the project. Post-authorization, we download the config.py and test vector files which contain the user-fed definition of the input and the output pins and the test vectors on which the circuit has to be analysed. These test benches also contain the truth tables for the circuit. The outputs of the circuit are uploaded to the Google Spreadsheet which tells the teacher of the experiment performed.

3 Design Approach

In this work, the proposed design is developed in the following algorithm steps.

The PI is connected to the cloud, i.e. it has an Internet connection. The student makes the necessary connections in order to implement the circuit. The performer of the experiment, i.e. student, sends or uploads the netlist and the input and the

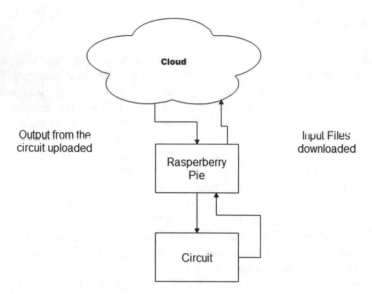

Fig. 1 Design approach of the model with the proposed

output pins and hence allowing the PI to interpret the designated pin. The PI is then authenticated with the OAuth2.0 protocol which requires client ID password and token number. The Pi then downloads the necessary authentication tokens and test vector files and then run them on the pi in order to get the output from the circuit. The output is then uploaded on the cloud with the Google Spreadsheets, and the output is verified (Fig. 1).

3.1 Authentication Algorithm for the Remote Devices

The request to the cloud first needs to be verified using the OATH verification. The OATH of verification requires a client ID and client secret key which needs to be mad at the server end in order for a successful session formation. However, in our case, the server or the Google Drive or client ID is already present as the client. json in the Raspberry Pi.

For an initial set-up, Google Drive client ID and secret key have to be present at the students' Raspberry Pi so that a secure connection can be established between the Google server and the remote device that is Raspberry Pi 3. In our prototype, user authentication 2.0 primarily revolves around the authentication and authorization of the user and granting access to come for establishing that particular session. If the access token is valid, then the resource server in our case Google Drive will let the remote device that is the Raspberry Pi to download the scripts. As per the above-mentioned steps for required authentication the script will not be available for testing the particular experiment.

3.2   Post-authentication

This Raspberry Pi 3 then connects and downloads the scripts which are essential for running the test bench over the decided on predefined input and output pins by the user that is the student and uploaded on the cloud. We also have a file named config.py which consists of the initial information that needs to be provided to the Raspberry Pi which consists of how many test vectors how many number of inputs how many number of output and weapons which are connected as inputs and pins which are connected as output The point to be noted here that the input and output lists of the pins as described above are with respect to the user. Hence, the input with respect to the user will be considered as output for Raspberry Pi, and output with respect to the administrator or the user of the Raspberry Pi be configured as input for the PI.

3.2.1   Test Vector File

The first three columns are the inputs, the fourth column denotes the sum, and the fifth column denotes the carry column. The circuit has been designed for a full adder circuit so the test factor is stored in a file named test vector.txt; in this file, there are four columns. The first three columns are the inputs, the fourth column denotes the sum, and the fifth column denotes the carry column. As per the result the input pins are been obtained by the values described in the first three columns respectively and the outputs are taken from the prescribed output pins and are related with the expected output given in IV and V column of the file. We also define the variable which counts the total number of errors which occurred or are different from the expected output (Fig. 2).

Fig. 2 Test vector file

```
0 0 0 0 0
0 0 1 1 0
0 1 0 1 0
0 1 1 0 1
1 0 0 1 0
1 0 1 0 1
1 1 0 0 1
1 1 1 1 1
```

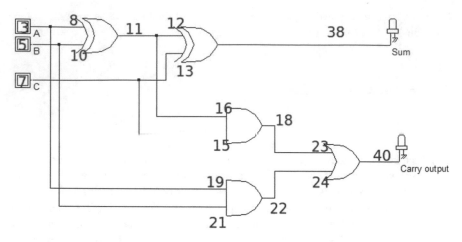

Fig. 3 Netlist mapping

4   Results and Discussion

The entire process is done by involving the troubleshooting of the circuit for CPS model. If the variable used for counting the differences in the actual and the current output has a value greater than 1, then the teacher can tell the particular student or provide an offline script which can be run on the Raspberry Pi and help the student to get first-level diagnosis of where the system or the circuit would have gone wrong. In order to model this diagnosis operation, we use a concept known as fault tree analysis and implementation (Fig. 3).

Suppose for a full adder circuit, every gate is input, i.e. all the inputs and outputs are fed as inputs to the Pi and are analysed on per-day basis for example, and gate will have two inputs and one output. All the elements results out irrespective of input or output is fed as input to the Raspberry Pi. As of now, the student needs to model the circuit and give every pin a number which is associated with the circuit and maps to the pins on the Raspberry Pi. Once the given test vectors are analysed on the predefined circuit, the output vectors are uploaded on Google Spreadsheet. In order to upload to the Google Spreadsheet, we need the faculties or the receivers' email ID and we also need the authentication token in order to establish a secured session (Fig. 4).

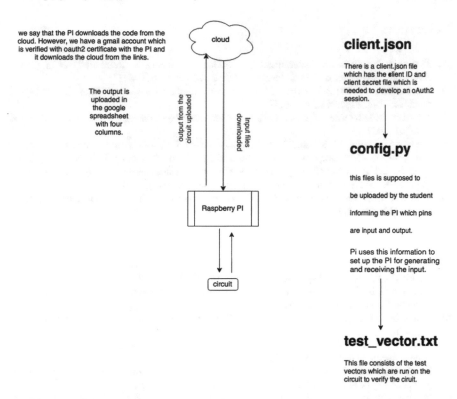

Fig. 4 Model upload of the virtual set-up

5 Conclusion

In this work, a cyber-physical approach of the circuit design approach with IoT approach was proposed which verifies the need of student and the faculty with virtual laboratory set-up in an easier approach. The model approaches the primitive level of understanding and final set-up of troubleshooting in the more dynamical way. The entire model has been tested and elaborated with various cases with IoT approach which is the needed approach for a virtual set-up in a long process.

References

1. Monostori L, Kádár B, Bauernhansl T, Kondoh S, Kumara S, Reinhart G et al (2016) Cyber-physical systems in manufacturing. CIRP Ann 65(2):621–641
2. Liu C, Jiang P (2016) A cyber-physical system architecture in shop floor for intelligent manufacturing. Procedia Cirp 56:372–377
3. Ma HD (2011) Internet of things: objectives and scientific challenges. J Comput Sci Technol 26(6):919–924
4. Jeschke S, Brecher C, Meisen T, Özdemir D, Eschert T (2017) Industrial internet of things and cyber manufacturing systems. In: Industrial Internet of Things, pp 3–19. Springer, Cham
5. Ning H, Liu H (2015) Cyber-physical-social-thinking space based science and technology framework for the Internet of Things. Sci China Inf Sci 58(3):1–19
6. Ross RS, McEvilley M, Oren JC (2016) Systems security engineering: considerations for a multidisciplinary approach in the engineering of trustworthy secure systems (No. Special Publication (NIST SP)-800-160)
7. Griffor ER, Greer C, Wollman DA, Burns MJ (2017) Framework for cyber-physical systems: volume 1, overview (No. Special Publication (NIST SP)-1500-201)
8. Leitao P, Karnouskos S, Ribeiro L, Lee J, Strasser T, Colombo AW (2016) Smart agents in industrial cyber–physical systems. Proc IEEE 104(5):1086–1101
9. Thramboulidis K (2015) A cyber-physical system-based approach for industrial automation systems. Comput Ind 72:92–102
10. Ochoa SF, Fortino G, Di Fatta G (2017) Cyber-physical systems, internet of things and big data
11. Akyildiz IF, Kak A (2019) The internet of space things/cubesats: a ubiquitous cyber-physical system for the connected world. Comput Netw 150:134–149
12. Rojas RA, Rauch E, Vidoni R, Matt DT (2017) Enabling connectivity of cyber-physical production systems: a conceptual framework. Procedia Manuf 11:822–829
13. Wang W, Hong T, Li N, Wang RQ, Chen J (2019) Linking energy-cyber-physical systems with occupancy prediction and interpretation through WiFi probe-based ensemble classification. Appl Energy 236:55–69

Hierarchical Attention-Based Video Captioning Using Key Frames

Munusamy Hemalatha and P. Karthik

1 Introduction

Video captioning is the process of generating a description for the video for better understanding of the video. Video captioning is most important to make better communication between the peoples. Video captioning used in social media like YouTube and Facebook plays a major role for better understanding of the video even if the audio is not clear in a video. The automatic description of a video using proper natural language sentences is a very challenging task. But the video captioning is useful in solving a wide range of applications like generating automatic subtitles of the videos, helping the visually challenged people, video retrieval, video tagging, etc. The image or video captioning is based on the deep learning method, which uses the long short-term memory (LSTM) as decoder. The long short-term memory is used to generate the caption for the video. Any video has continuous frames which usually contain redundant information. In this work, we identify only the key frames and use them for generating the captions. The rest of the paper is organized as follows—Section 2 discusses the literature review of the survey. Section 3 discusses the proposed framework. Section 4 describes the implementation details. Section 5 discusses the evaluation of the region-driven approach. Section 6 presents the conclusion.

2 Literature Survey

Many algorithms have been proposed for image captioning using LSTM as encoder and decoder [8, 9]. Long-term recurrent convolutional network proposed in [3] focuses on the model based on long-term recurrent convolutional networks

M. Hemalatha (✉) · P. Karthik
Anna University, Chennai, Tamil Nadu, India

© The Author(s), under exclusive license to Springer Nature Singapore Pte Ltd. 2022   295
R. R. Raje et al. (eds.), *Artificial Intelligence and Technologies*,
Lecture Notes in Electrical Engineering 806,
https://doi.org/10.1007/978-981-16-6448-9_30

(LRCNs). The model is used for visual recognition and description. It uses both the recursion and convolution layers for generating the descriptions. They propose a visual ConvNet to encode the video features of the input video. These features are provided as input to the LSTM for decoding the features and generate the descriptions. The encoded features are a deep state vector, and it represents the input video. Here motion features are extracted from video using C3D [7].

The hierarchical recurrent neural networks [13] are used to generate one or multiple sentences to describe a realistic video. The hierarchical framework contains two generators: a sentence generator which generates single-line descriptions, a paragraph generator which generates multiple sentences describing the video. They have developed a soft attention model, which applies attention on the different parts of the input video. The single-line description generated is captured by paragraph generator. The output of the sentence generator and paragraph generator is combined to develop a new initial state for the new sentence generator.

Video captioning approach in [5] uses attention LSTM and semantics concepts to generate descriptions. Attention is applied on the output of LSTM model to generate English sentence for a video. The LSTM model uses the multimodal features obtained from the video to generate descriptions that include the semantic keywords. The visual and semantic features are combined and projected into a common space for generating a generic representation which will be used for generating the descriptions of the video.

Describing video with attention-based bidirectional LSTM was proposed [1] focuses on bidirectional long short-term memory for video captioning. The model will preserve the global temporal and visual information. To focus on the particular object by applying a soft attention mechanism using adjusts the weights in dataset. The temporal attention contains two stages they are fusion to forward and backward passes and generate a caption to form a sentence word by word. There have been various approaches also used to develop the model for image and video captioning like [2, 4, 6, 10–12, 14].

3 Proposed Framework

The proposed approach analyses the video frames to find the object of interest and then extract the parameters of the object. The proposed method used here is key frame and hierarchical attention for key frame feature, object feature, and semantic feature. Key frame is a method in which an informative frame is selected from a video clip. The key frames are selected based on structural similarity index (SSIM). Figure 1 is the architecture of hierarchical attention-based video captioning using on the key frames.

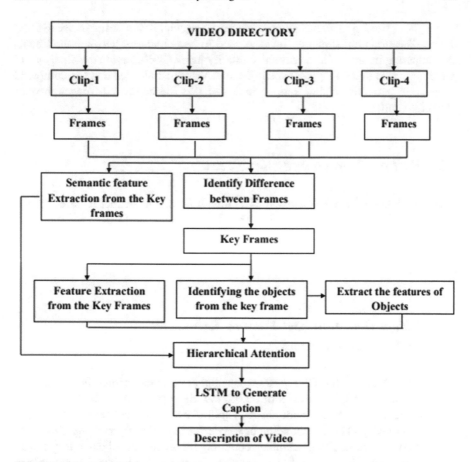

Fig. 1 Architecture of the proposed system

3.1 Identifying Key Frames

The video file is given as input for identifying the initially all videos are converted into frames and apply adaptive image thresholding to identify lighting conditions of the image. The threshold value is applied for each pixel in image, and it is classified into three different arguments for image thresholding. The first one for source of image and second argument contains information about the threshold to classify the pixel value and third argument for maximum value applied to pixel value.

$$\text{SSIM}(x, y) = \frac{\left(2\mu_x\mu_y C_1\right) + \left(2\sigma_{xy} + C_2\right)}{\left(\mu_x^2 + \mu_y^2 + C_1\right) + \left(\sigma_x^2 + \sigma_y^2 + C_2\right)}$$

The converted frames are given as input to structural similarity index (SSIM) model. The structural similarity index is used to find similarity between the frames by comparing in order. The structural similarity index (SSIM) compares images and generates value in the range between −1 and 1 where 1 indicates that the compared images are perfectly similar and −1 indicates that the compared images are not perfectly similar.

3.2   Feature Extraction of Key Frames

The convolutional neural network (CNN) model is used to extract features for the informative frames by using ResNet. The ResNet model is loaded; for the keras deep learning library, it uses the pre-trained weights from ImageNet for feature extraction of the informative frames. The frames and folder id are given as input to the keras.

3.3   Object Detection and Feature Extraction from the Key Frames

You only look once (YOLO) algorithm uses the key frames to identify object. You only look once (YOLO) is series of end-to-end deep learning model used for the object detection. The given input image is split into various grid cells by the CNN. From each input grid cell, a bounding box is derived and the object is classified into one of its type. The resulting candidate bounding boxes are consolidated to generate a final prediction. These bounding boxes are later processed to identify the objects present in image or the frames of a video.

The convolutional neural network (CNN) model is used to extract the feature from the objects based on the informative frames using the concept of ResNet. The ResNet is a model that can be loaded; for the keras deep learning library, it uses the pre-trained weights from ImageNet for feature extraction of the informative frames.

3.4   Language Model

We apply hierarchical attention on the key frame features, object features, and semantic features. The hierarchical attention determines the importance of each of the features for generating captions. Long short-term memory (LSTM) model is used as decoder to generate descriptions. The output of the hierarchical attention model is provided as input to the LSTM layer. The LSTM model generates one word at each time step. We also apply the Bahdanau attention [15] on the output of the LSTM while generating descriptions. The process is continued till, and end of sequence token is generated by the LSTM. We use categorical cross-entropy loss while training the model.

4 Experiments

4.1 Dataset

The MSVD dataset is developed by Microsoft, here MSVD stands for Microsoft Research video description Corpus. It totally consists of around 2000 video clips each of 5–10 s duration. It consists of around 40 thousand sentences describing the videos. The videos belong to various categories capturing a wide range of activities. We follow the standard data split for training with 60% videos for training, 20% videos for validation, and 20% videos for testing.

4.2 Implementation Details

The environment used to implement the proposed approach is Keras in anaconda with TensorFlow as backend. For running resource, anaconda notebook is used. The dataset is randomly divided into 80% for training, 10% for validation, and 10% for test for captioning on MSVD dataset. The extracted feature of informative frame is stored in a pickle file, object feature, semantic feature, and train id and description file which contains the name of the video file to map with feature and description file which contain the possible descriptions of the visual content. The process uses the keras and long short-term memory models to load the text document and train the model.

5 Results and Discussion

5.1 Experimental Results

The extracted features of the informative frame, object feature, and semantic features are given as an input to the LSTM model. The model is trained along with train id and description file. Figure 2 is the graph for model accuracy and loss function of this model, and Table 1 represents the training and testing results.

To evaluate the generated captions based on handcrafted representations image, three handcrafted representations are considered including S2VT, BiLSTM, and

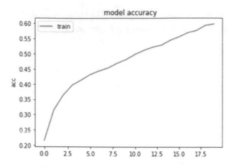 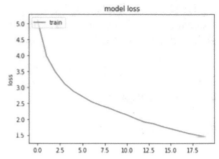

Fig. 2 Graph for model accuracy and loss

Table 1 Training and testing results

Heading level	Training scores				Testing scores			
	BLEU-1	BLEU-2	BLEU-3	BLEU-4	BLEU-1	BLEU-2	BLEU-3	BLEU-4
Without hierarchical attention on key frames	76.8	71.4	67.8	57.2	74.6	64.0	59.7	48.7
Hierarchical attention on key frames	85.6	78.0	73.6	64.6	83.1	73.2	67.0	56.3

ReBiLSTM. The results of BLEU score obtained in methods involving handcrafted methods are given in Table 2.

The trained model is evaluated using the LSTM architecture. The trained model is used to check the accuracy value for key frame identifying for video captioning. The description file, extracted feature of informative frames, and trained model are given as input to the LSTM to generate caption for the video. The BLEU score values are used to compare the original description with generated description. The generated BLEU score values are BLEU1 [86.1], BLEU2 [79.2], BLEU3 [75.0], and BLEU4 [65.5], which is higher than existing system.

Figure 3 shows objects of key frames. You only look once (YOLO) algorithm is used to identify the object from the key frames. Figure 4 shows the captions generated for the videos.

Table 2 Comparison results of other models

Methods	BLEU-1	BLEU-2	BLEU-3	BLEU-4
S2VT	79.9	60.5	48.3	36.5
BiLSTM	79.4	60.5	48.6	35.6
ReBiLSTM	79.0	60.5	48.4	37.3

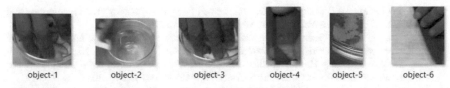

| object-1 | object-2 | object-3 | object-4 | object-5 | object-6 |

Fig. 3 Objects of informative frame

```
video-1 startseq a man is playing with a shapes of a sheet endseq
video-2 startseq a man is washing violin endseq
video-3 startseq a woman is drinking endseq
video-4 startseq a man playing playing endseq
video-5 startseq a little is washing vegtable board endseq
video-6 startseq a kitten is washing asleep endseq
video-7 startseq a man is wallaby board endseq
video-8 startseq a girl is talking endseq
video-9 startseq a woman is slicing a carrot endseq
video-10 startseq a woman is slicing a carrot endseq
```

Fig. 4 Generated caption for video

6   Conclusion

The proposed model uses the natural language for generating the description for video clips. In other video captioning model, they use each and every frame in a video to generate a caption. Instead of using each and every frame in video, a key frame is used to generate the caption. Structural similarity index (SSIM) is used to identify the key frames. SSIM approach is used to identify the difference between the frames, by comparing each frame. A threshold value is applied to get the key frames. The features are extracted from the key frame. Identifying the objects from the key frames then apply feature extraction for the objects in key frames and also extract the semantic feature. The key feature, object feature and semantic feature, is trained with description file and trained id using long short-term memory (LSTM) model. The results show that the proposed model provides better captions than the other video captioning algorithms.

References

1. Bin Y, Yang Y, Shen F, Xie N, Shen HT, Li X (2019) Describing video with attention-based bidirectional LSTM. IEEE Trans Cybern 49(7):2631–2641
2. Dahiya D, Issac A, Dutta MK, Ríha K, Kriz P (2018) Computer vision technique for scene captioning to provide assistance to visually impaired. In: 41st international conference on telecommunications and signal processing (TSP), Athens, pp 1–4

3. Donahue J et al (2017) Long-term recurrent convolutional networks for visual recognition and description. IEEE Trans Pattern Anal Mach Intell 39(4):677–691
4. Gan Z et al (2017) Semantic compositional networks for visual captioning. In: IEEE conference on computer vision and pattern recognition (CVPR), Honolulu, HI, pp 1141–1150
5. Gao L, Guo Z, Zhang H, Xu X, Shen HT (2015) Video captioning with attention-based LSTM and semantic consistency. IEEE Trans Multimedia 19(9):2045–2055
6. Karpathy A, Fei-Fei L (2015) Deep visual-semantic alignments for generating image descriptions. In: Proceedings of the IEEE conference on computer vision and pattern recognition, pp 3128–3137
7. Tran D, Bourdev L, Fergus R, Torresani L, Paluri M (2015) Learning spatiotemporal features with 3D convolutional networks (2015). In: IEEE international conference on computer vision (ICCV), Santiago, pp 4489–4497. https://doi.org/10.1109/ICCV.2015.510
8. Venugopalan S, Rohrbach M, Donahue J, Mooney R, Darrell T, Saenko K (2015) Sequence to sequence—video to text. In: IEEE international conference on computer vision (ICCV), Santiago, pp 4534–4542
9. Vinyals O, Toshev A, Bengio S, Erhan D (2015) Show and tell: a neural image caption generator. In: IEEE conference on computer vision and pattern recognition (CVPR), Boston, MA, pp 3156–3164
10. Xu Y, Han Y, Hong R, Tian Q (2018) Sequential video VLAD: training the aggregation locally and temporally. IEEE Trans Image Process 27(10):4933–4944. https://doi.org/10.1109/TIP.2018.2846664
11. Wu Q, Shen C, Wang P, Dick A, Hengel AVD (2018) Image captioning and visual question answering based on attributes and external knowledge. IEEE Trans Pattern Anal Mach Intell 40(6):1367–1381. https://doi.org/10.1109/TPAMI.2017.2708709
12. Yang Y et al (2018) Video captioning by adversarial LSTM. IEEE Trans Image Process 27(11):5600–5611
13. Yu H, Wang J, Huang Z, Yang Y, Xu W (2016) Video paragraph captioning using hierarchical recurrent neural networks. In: IEEE conference on computer vision and pattern recognition (CVPR), Las Vegas, NV, pp 4584–4593
14. Yu N, Hu X, Song B, Yang J, Zhang J (2018) Topic-oriented image captioning based on order-embedding. IEEE Trans Image Process 28(6):2743–2754. https://doi.org/10.1109/TIP.2018.2889922
15. Bahdanau D, Cho K, Bengio Y (2016) Neural machine translation by jointly learning to align and translate. arXiv-1409.0473

Skin Cancer Prediction Using Machine Learning Algorithms

Arun Raj Lakshminarayanan, R. Bhuvaneshwari, S. Bhuvaneshwari,
Saravanan Parthasarathy, Selvaprabu Jeganathan,
and K. Martin Sagayam

1 Introduction

Skin cancer is most prevalent cancer, with 3.5 million cases every year, worldwide. In every 57 s, an individual succumb to skin cancer. Skin cancer turns up when mutation takes place in the DNA of certain cells. People with excessive exposure to sunlight, radiation, having lesions, inherited cancer genes are more likely to develop a skin cancer. Basal cell carcinoma, Squamous cell carcinoma and melanoma are the three types of skin cancer. Basal cell cancer is the widely recognized one, which is generally curable. Squamous cell carcinoma is aggressive and spread rapidly to other parts of the body if untreated. Melanoma is the most serious type of cancer which causes bleeding or oozing and may lead to death. The early diagnosing is very much needed in determining the disease and getting appropriate treatment, increasing the survival rate of 94%.

Dermatologist screens the disease by examining with naked eyes followed by biopsy. But, even highly skilled dermatologist has less than 80% chance in diagnosing the cancer cells accurately. Hence, the computer aided image analysis algorithms are employed to detect the disease at the earliest. Here, we develop an automated system for the prediction of dermatological diseases. Our model is designed in such a way that takes image as an input, preprocessing the images and extracting required features, training the model and use machine learning algo-

A. R. Lakshminarayanan (✉) · R. Bhuvaneshwari · S. Bhuvaneshwari · S. Parthasarathy · S. Jeganathan
B.S. Abdur Rahman Crescent Institute of Science and Technology, Chennai, India
e-mail: arunraj@crescent.education

S. Parthasarathy
e-mail: saravanan_cse_2019@crescent.education

K. M. Sagayam
Karunya Institute of Technology and Sciences, Coimbatore, India
e-mail: martinsagayam@karunya.edu

© The Author(s), under exclusive license to Springer Nature Singapore Pte Ltd. 2022   303
R. R. Raje et al. (eds.), *Artificial Intelligence and Technologies*,
Lecture Notes in Electrical Engineering 806,
https://doi.org/10.1007/978-981-16-6448-9_31

rithms to predict the output. Here, we have compared multiple algorithms like AdaBoosting algorithm, Gradient Boosting algorithm and Decision tree with the Convolutional Neural Network. Computers are playing an essential role in the identification of issues in the human body.

The cancer-influenced area in the image is termed as "lesion". The objective of this study is to predict the presence of skin cancer in the human skin using various machine learning algorithms. The dataset is classified into two parts one for training and the other for testing. The output determines the presence or the absence of cancer, which is melanoma or non-melanoma.

2   Literature Survey

Mirbeik-Sabzevari and Negar Tavassolian [1] utilized the strategy of wave skin malignancy imaging in their paper to foresee the skin disease tissue apparitions in which the skin ghosts of millimetre waves are made to interface with the human skin and tumours so that, they are utilized to distinguish the millimetre-wave skin malignant growth. The ordinary and harmful skin tissues are tried utilizing the blend which contains de-ionized water, oil, gelatin and so forth. The permittivity of the ghosts is estimated utilizing the thin structure open-finished coaxial test with millimetre-wave vector organize analyser. These outcomes show that the skin is new or not over the whole recurrence go. They have additionally determined the infiltration profundities of millimetre waves for both typical and harmful skin are determined. This paper at last confirms that the waves can profoundly infiltrate through the human skin by influencing the epidermis of the skin.

Bumrungkun et al. [2] proposed a method to detect the skin cancer using SVM and snake model. Since the skin cancer is one of the causes for increase in deaths among all cancer, this paper tries to detect cancer by applying all the techniques in the cancer image which has asymmetry, colour variation, etc. To use these features easily a method called image segmentation is introduced here for automatic skin cancer detection. Support Vector Machine (SVM)-based segmentation of image is useful for finding the required parameters for snake algorithm.

Alquran et al. [3] used classification and detection of melanoma skin cancer using SVM. The melanoma skin cancer easily spreads to other parts of the body in which the image processing is very important for every disease. The non-invasive computer vision automatically analyses the image and evaluates it. It involves steps such as preprocessing, segmentation, asymmetry, border, colour, etc. using PCA the total dermoscopy score is calculated. They have classified using SVM and have shown the accuracy result of about 92.1%.

Yaroslavsky et al. [4] proposed the consolidated optical and terahertz imaging for intraoperative outline of non-melanoma skin malignancy strategy. Here, they have told the significance of the reason for non-melanoma skin diseases. Precise intraoperative outline of skin malignant growth was established by joining the terahertz and optical imaging in new toughness extractions. The tissue picture is contrasted and their particular histopathology. This paper shows that the blend of light optical imaging is more efficient than the cross-enraptured nonstop wave terahertz imaging when contrasted and the devices utilized for controlling the careful extraction.

The authors Mansutti et al. [5] detected the skin cancer using a method called millimetre-wave substrate integrated waveguide probe. To obtain suitable performance, the device is relied on conventional waveguides, which makes the fabrication process very costly. Here, they have given the design of a Substrate Integrated Waveguide (SIW), which avoids the fabrication problems. Based on the dielectric property of the skin the probe is excited using a micro-strip. It can detect the tumours in thin and thick skin at the early stage, which is as small as 0.2 mm diameter.

The authors Divya et al. [6] presented a soft computing approach-based segmentation and analysis of skin cancer in which they have explained the importance of skin melanoma analysis. An algorithm, Firefly-based Tsallis function and ACS convert the region of cancer into an image. The authors have considered the DERMQUEST database for segmentation and analysis process. This paper provides better results on the precision, sensitivity and accuracy, by image quality measures that are well known.

Neha and Kaur [7] have detected skin cancer using a device called wearable antenna. This method uses a high gain producing wearable antenna for detection. It merges with the electric property of the human skin and it operates in X-band. It allows to sense differences in the properties of skin tissues through a band of frequency. This antenna is designed using EM wave solver. This method improves the gain and achieves size reduction using metamaterials.

The authors Mane and Shinde [8] came with an approach called dermoscopy images method for detecting the melanoma skin cancer. Biopsy method detects the skin cancer, which was very painful. To overcome this challenge, the authors decided to detect the melanoma using the computer-based diagnosis called the dermoscopy. The captured image is undergone the preprocessing, segmentation and feature extraction which is used to get the segmented lesion followed by classification by SVM. This method shows that SVM with linear kernel provides us with the correct accuracy.

Barata and Marques [9] diagnosed the skin cancer using the deep learning techniques with the hierarchical architectures. A deep learning architecture performs diagnosis. This paper shows the benefits of using dermoscopy images in a ordered way also provided that the evaluation criteria are taken into account for the diagnosis in deep learning method.

Osipovs et al. [10] introduced a system called the cloud infrastructure for detecting the scalable skin cancer. In order to increase the diagnostic quality response, they have used image acquisition under special multispectral illumination for obtaining skin images. These images are analysed by the cloud. The cancer research team processes the image using mat lab algorithms. Using this method does not need to apply every algorithm to the respective architecture of the device. This architecture gives fast scaling for real-time requirements, it also uses central load balancing server, which requests and sends image to be processed to less loaded mat lab station. If load is high, the server launches additional processing mat lab station.

3 Proposed System

The focus of this work is to predict the skin cancer with improved accuracy and classify it into melanoma or non-melanoma. The proposed model-driven architecture uses the deep Convolution Neural Network (CNN). The images were preprocessed and divided into training and testing datasets. In the training phase, the CNN equipped model learnt the patterns from the dataset. The count of convolutional layer would be decided based on the error rate. The model would predict the melanoma by appraising the testing dataset.

Firstly, the image of skin had been given as an input along with other parameters like age and gender. After uploading the image, it is under process of prediction. Every uploaded image will have fit with frames. Each frame is noted as '1' or '0'. '0' indicates the particular part of unaffected skin; '1' indicates the particular part of affected skin. Then, the areas, which were noted as '1' had been separated from the areas that are noted as '0'. The areas marked as '1' were only subjected for further prediction. Based on the comparison of resulted image with training set, the prediction would be given as melanoma or non-melanoma. This CNN model would be more helpful to identify the skin cancer disease at the earlier stage.

4 System Architecture

See Fig. 1.

5 Results and Evaluation

Figure 1 represents the functional flow of this study; Figs. 2 and 3 indicate the correlation between attributes in the dataset. In Fig. 2, the image in left is the original and the glare removal is yet to be done. The outcome of glare removal process is in the right side. Figure 3 illustrates the before and after segmentation of images. The black colour backgrounded indicates the absence of skin cancer. Whereas, the white coloured centre part shows the presence of skin cancer. The area in which the melanoma is present is noted as '1'. Similarly, the area in which the melanoma is not present is noted as '0'. Figure 4 is the original image which is used for the prediction. Figure 5 encapsulates result of averaging which is done for enhancing the image. Figures 6 and 7 exemplify the blurring and median blurring

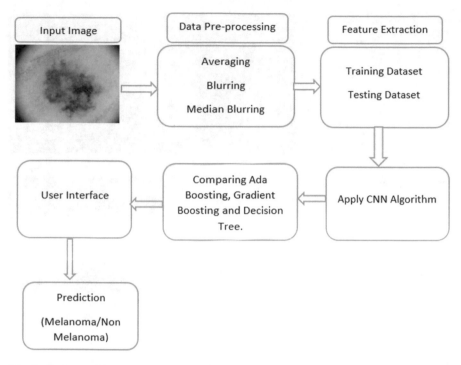

Fig. 1 System architecture

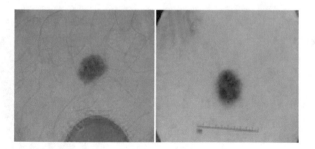

Fig. 2 Before and after Glare removal

operations. These processes remove noises from the image and its edges. Figure 8 shows the final outcome of the preprocessing phase. The focus of this research work is to implement classification algorithms and analogizes the denouements. The dataset was fractionated into training and testing portions in the ratio of 80/20. Convolutional Neural Network, AdaBoosting, Gradient Boosting and Decision Tree were used to predict the presence of melanoma.

Fig. 3 Before and after
segmentation

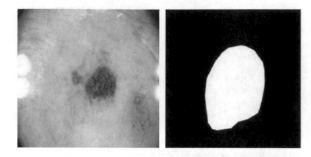

Fig. 4 Original image

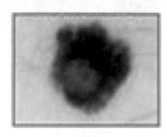

5.1 Preprocessing Steps

See Figs. 5, 6, 7 and 8.

Fig. 5 Averaging

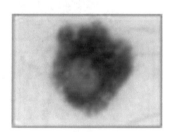

Fig. 6 Blurring

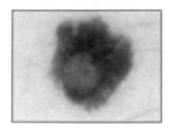

Fig. 7 Median blurring

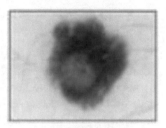

Fig. 8 Resulting image

5.2 Prediction Accuracy

The accuracy connotes the appropriately predicted values. Figure 9 delineates the accuracy of each algorithms tested. The gradient boosting algorithm excelled others by producing 85% of accuracy. The decision tree and Convolutional Neural Network delivered 83% and Adaboosting made it with 70%, respectively.

Fig. 9 Accuracy of various learning techniques

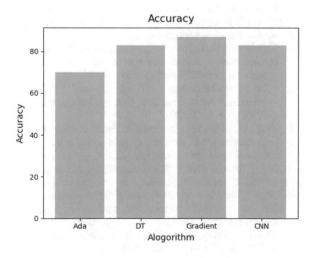

6 Conclusion

The proposed system is having the Convolutional Neural Network classifiers to break down and anticipate whether the given information picture is been influenced by malignant growth tumours or not. This approach improves the general efficiency and furthermore decreases the computational time. Consequently, there is a requirement for increasingly solid and exact frameworks that can be advantageous to both experienced and beginner doctors.

In the current clinical setting, dermatologists assess the situation and control the unexpectedness. This procedure is emotional, non-quantitative and blunder inclined. Other than the issues, clinical pictures are difficult to track down while ensuring the secrecy of the patients. The advantage of CNN is that it detects the importance features automatically without any human supervision. It is an effective method, which could be used in any type of prediction problem involving image data as an input. In future, we plan to work on increasing the accuracy of CNN and finding skin cancer as early as possible.

References

1. Mirbeik-Sabzevari A, Tavassolian N (2019) Ultrawideband, stable normal and cancer skin tissue phantoms for millimetre-wave skin cancer imaging. IEEE Trans Biomed Eng 66 (1):176–186
2. Bumrungkun P, Chamnongthai K, Patchoo W (2018) Detection skin cancer using SVM and snake model. In: 2018 international workshop on advanced image technology (IWAIT), Chiang Mai, pp 1–4
3. Alquran H, Qasmieh IA, Alqudah AM, Alhammouri B, Alawneh E, Abughazaleh A, Hasayen F (2017) Skin cancer detection and classification using SVM. In: IEEE Jordan conference on applied electrical engineering and computing technologies (AEECT), pp 1–5, Oct 2017
4. Yaroslavsky AN, Joseph C, Patel R, Fan B, Musikansky A, Neel VA, Giles R (2016) Intraoperative delineation of nonmelanoma skin cancer. In: International conference laser optics (LO), vol 7(5), pp 295–303
5. Mansutti G, Mobashsher AT, Abbosh AM (2017) Millimeter-wave substrate integrated waveguide probe for near-field skin cancer detection. In: Australian microwave symposium (AMS), pp 1–5, Oct 2017
6. Divya G, Uniya D, Sivakumar R, Sundaravadivu K (2017) Soft computing approach based segmentation and analysis of skin cance'. In: International conference on computer communication and informatics (ICCCI), pp 1–5, Jan 2017
7. Neha, Kaur A (2016) Wearable antenna for skin cancer detection. In: 2nd international conference on next generation computing technologies (NGCT), pp 197–201, Oct 2016
8. Mane S, Shinde S (2018) A method for Melanoma skin cancer detection using dermoscopy images. In: Fourth international conference on computing communication control and automation (ICCUBEA), pp 1–6, Aug 2018
9. Barata C, Marques JS (2019) Deep learning for skin cancer diagnosis with hierarchical architectures. In: IEEE 16th international symposium on biomedical imaging (ISBI), pp 841–845, Apr 2019
10. Osipovs P, Bliznuks D, Kuzmina I (2018) Cloud infrastructure for skin cancer scalable detection system. Adv Wirel Opt Commun (RTUWO) 50–54

Review on Technological Advancement and Textual Data Management Algorithms in NLP and CBIR Systems

M. Diviya and A. Karmel

1 Algorithms for Handling Textual Features

The technique of synthesizing a text data or retrieving an image in response to the text data is implemented in various fields of machine learning, such as text mining, text synthesizing, image captioning, image processing, animation, and graphics. Different strategies have been used and yet are continued to be found in the developmental stage. The current situation is overloaded with a wide pool of data scattered across the globe. While many algorithms and methods are emerging, it is yet to be researched and explored in the field of science. The explanations for this are learning about image similarities, user experience with a program, database requirement, semantic difference, and spatial reference distance with image and its attributes, and image resolution retrieved from database [1, 2]. A CBIR system [3] must acquire the content-oriented functionality of its image database in order to scan for images based on their content; this method is known as feature extraction. Photos are created automatically from the term produced that is shown in various applications such as art creation, creation of cartoon images, generation of computer-aided images, and generating realistic photographs. Improvements to technological facts may be used in a variety of applications, including speech synthesis, image and video generation, semantic image processing, style transfer, and so on. Image synthesis has attracted much attention since the introduction of GANs [2, 4] and served as a prominent component of the expert systems. Even though, GANs are known to be difficult to train when implemented in an attempt to generate high-resolution images. This paper provides a comprehensive overview of state-of-the-art techniques and algorithms for handling text-based inputs and how GANs-based approaches could be embedded in real-time image processing,

M. Diviya (✉) · A. Karmel
SCOPE, VIT Chennai Campus, Chennai, Tamilnadu, India
e-mail: diviya.m2019@vitstudent.ac.in

© The Author(s), under exclusive license to Springer Nature Singapore Pte Ltd. 2022   311
R. R. Raje et al. (eds.), *Artificial Intelligence and Technologies*,
Lecture Notes in Electrical Engineering 806,
https://doi.org/10.1007/978-981-16-6448-9_32

image-to-image translation, face-to-face ageing, and 3D image synthesis applications. Experimental findings illustrate state-of-the-art output by textual data, CBIR, and GANs.

1.1 Methodologies to Handle Textual Input in Natural Language Processing

Each researcher who works with text data takes up with segmentation and tagging which are the basic components of text processing. Figure 1.depicts the various algorithms involved in handling textual data. In [5], the authors showed up with two representations to analyze POS tagging of languages that lack corpus for training and research performance. The languages which are being studied are Picard, Occitan, and Alsatian. The three languages have close proximity to the French language, so it was initially taken for study with the available corpus based on the source language and based on the study of the target regional languages corpus. Figure 2 shows various POS tagging algorithms employed. The tagger and HMM of BiLSTM CRF (conditional random fields) POS were used, and tagging of POS was done. The results have been improved when transposition is applied with the tagger and when the isometric alignment is used to project the source language vectors to the target language. The classification of the provided input for spam detection [6] is a challenging activity when considered in all languages. Researchers aimed at detecting spam images based on Arabic text embedded in social networks. A deep convolution recurrent neural network [7] followed by NLP preprocessing of text using NLTK toolkit and classification algorithm. The analysis was performed using the following datasets such as Twitter image, ICDAR 2017, Twitter cropped image, a synthetic word, and Arabic synthetic dataset. The results are as follows: for ICDAR 2017 and Twitter images, the best accuracy was achieved with the highest F1 score of 73%. To derive the word's POS tag, the model uses the posterior probability of the morphological component and stem component. The remaining corpus parameters are used to optimize the probability. The results are evaluated using the measures recall and precision, and the F-measure. The program reached 95.52% accuracy. Even many taggers and stemmers have been developed to rely on input language and corpus for each of their performances. The authors began a survey of [8, 9] Indian language POS taggers. The taggers are developed using the hidden Markov model (HMM), the support vector model (SVM), rule-based approaches, and maximum entropy (ME), and conditional random field (CRF). The accuracy of those algorithms are optimistic. The algorithm operates by removing affixes since it is simpler and requires no dictionary. The method overcomes the disadvantages of iterative rule-based affix stripping. The search for information revolved around the data without taking into account the semantic values. The key consideration [10] is to enhance conceptual-based quest by involving structural ontological knowledge which combines concepts and relationships. The primary aim of semantic-oriented knowledge search is to extract information based on the synonym of the query terminologies or the contextual dependence of the terms. The proposed method considers Kohler et al. concepts by including sibling information to the index. The method's accuracy and recall deduce to 0.56 and 0.42. Domain-specific issues are the big limitation to this approach; another significant application of NLP in recent times is the summarization of text [11]. The algorithm operates by using stochastic neural network to solve the drawbacks of conventional approaches already in use. The algorithms functions by considering input factors such as sentence position related to paragraph, number of named entities, frequency of term and inverse document frequency, and number of numerals involved in the sentence construction. The thesis concentrates [12] on resolving the uncertainty in the classification of symbols that generated confusion in Tamil language based on online handwritten words. The researchers have discussed the issue of idiosyncrasies in Indic scripts. The suggested technique works by segmenting terms into symbols by identifying words in turn by SVM. Bigram language model and a classifier which outperforms the regional model boost the performance. The main idea of the work is to present the post-processing techniques [13] essential for overcoming the partial disambiguation with confused symbols in the language models. Ultimately, the ambiguity that occurs with identical series of symbols can be discovered by observing dynamic time wrapping scheme [14, 15] using minor variations of the characters. The dataset was collected with a size of 15,000 words from high school and college students. Several preprocessing steps to arrive at feature vectors for the classifier were carried out proceeding to classification. Support vector machine (SVM) acts as the primary classifier followed by the bigram language model. The model functions by measuring the probability of next symbol appearing in comparison to the preceding symbol by following Markovian dependence. The primary objective [16] is to build a model that fits well for heterogeneous modalities that suit text and picture. The author points out the need for cross-model retrieval and outlined the specific applications that rely on these algorithms. The overall architecture is composed of two networks designed to integrate text and image. The initial layers had RESNET as a full-connected network for image embedding. Gated recurrent device was used as one layer for text embedding, and three fully connected layers were used. To achieve the precision in prediction, the authors considered triplet loss and adversarial error. The Flicker30k dataset and MS-COCO dataset were in testing where the findings are positive for the model suggested by the researchers. Lemmatization [17] is another essential process in the processing of text which helps to recognize the root word in the given word. Lemmatization plays a critical function in languages that are sensitive to context. The investigators used BiLSTM. The system suggested is experimented in four languages, such as Hindi, Marathi, Spanish, and French. The model is composed of two LSTMs for edit trees classification [13] and word embedding.

Edit trees have been developed that are generalized according to the suffix length and the substring prefix which is the longest sequence. The organization of edit trees employs distribution of conditional probability, and each lemma depends on the specific context of the word [18]. For each character level, the words in the

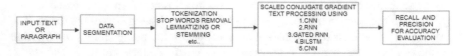

Fig. 1 Overview of handling textual input by NLP algorithms

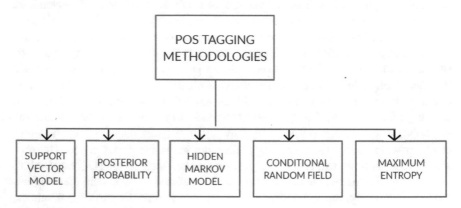

Fig. 2 POS Tagging algorithms

specified sample adopt one hot encoding. The main character from word-level embedding serves as an input to the CNN for enduring positional embedding. The provided model outperforms the previous models with an accuracy of over 80% in three out of four languages [19]. Scaled conjugate gradient algorithm was used by researchers to arrive at a quicker training period during which the network's accuracy was improved [20]. The architecture was developed to support broad handwritten text recognition datasets. The approach begins with the preparation of the segmentation image [21] followed by the extraction of the feature. In this process, different critical features such as transformation feature, sliding window feature, contour feature, and, finally, shadow feature are extracted. A minimum of 100 features is compiled and fed in to the next stage of neural network training using SCG as it increases processing speed in a better way. The accuracy of the algorithms is brought in Fig. 3, which says about the performance level of the algorithms. For the process Powell-Beale, Fleture-Reeves, and scaled conjugate gradient algorithm, the efficiency is higher. Another important observation was the Hessian matrix error value was shown to be accurate and positive in all iterations. The words are accepted with 98.91% accuracy, and 99.02% accuracy is reached in case of character level. There is even further research that brings out the information on the algorithms for processing NLP.

CBIR Techniques for Text Recognition: Each CBIR work indulges in text or image or scenic view, and so on. Figure 4 outlines the CBIR query processing. In [21], the researchers suggested a new approach for overcoming the current binarization algorithm in scene picture and video frame text recognition. Based on the

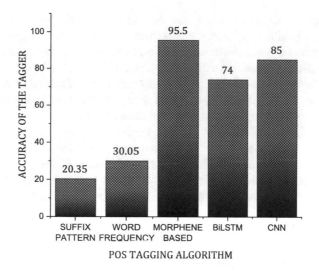

Fig. 3 Accuracy of the POS taggers

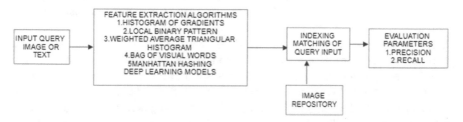

Fig. 4 Query processing by CBIR system

discrete wavelet transform and SVM classifier, the appropriate color channel is selected. Gabor filters perform the texture analysis of the image. HMM uses the pyramidal histogram of oriented gradient features for feature extraction and word recognition. The output is validated using ICDAR 2003, ICDAR 2015 dataset, which is powerful for the identification of scene text images. The performance is highest when using HMM and PHOG, resulting in 78.44 and 75.41% efficiency, respectively. Many researchers are working on [22] image retrieval, the algorithm based on the weighted average triangular histogram [23]. The image attributes are generally based on the location of the objects and the area comprising the image. Using the visual word model pocket, the picture is described as a less histogram collection of order [24, 25] which improves the efficiency and effectiveness of the retrieval process. Nevertheless, the shadow aspect is that it lacks the given image's spatial attributes. Weighted average triangular histograms of visual terms are included in the suggested solution, which incorporates the spatial quality of image [3] that prevents overfitting while the dictionary scale becomes high and overcomes semantic gap issues. The BOVW is represented [24, 26, 27] by using the available quantization algorithm to render the feature vectors of the image, here, using k-means to create a visual vocabulary. The best map obtained in retrieval is 87.85 in the Coral-A picture dataset for a dictionary scale of 600 with 10 semantic level. The best map value of 84.38 for the dictionary size of 800 is obtained with the Coral-B image dataset. Scene 15's most dynamic data collection with dictionary size of 1000 visual words yields the best figure of about 81.94.

Reference [24] CATIRI: An efficient method for context- and image-based retrieval work has been performed to improve retrieval efficiency through indexing, and CATIRI follows a three-phase solution framework with a new indexing structure called MHIM tree (Manhattan Hashing Inverted Index and M tree). A top-k query algorithm was developed. Hamming distance has been used in previous methods to calculate the similarity points in a hash space that breaks the neighborhood structure in the original feature space, but such a restriction is overuse of Manhattan hashing. The dimensionality is further reduced by iterative quantization followed by quantization of Manhattan [28] the hash code is generated as a result. The defined M tree [29] is a balanced B tree used to index multidimensional, generic metric space; here, the algorithm generates a bounding sphere by grouping neighboring objects. A novel approach to image retrieval [26] based on content. The suggested work addresses the issue when retrieval happens in a complex environment. Reference [4] for regional descriptors, the feature vector is extracted using HSV color space which is a well-known technique. The visual content of the image is represented by the shape and color. The image's form function is segmented using the clustering algorithm fuzzy C-means also known as soft clustering or soft-k-mean. A combination of the individual image applied with an area mask and the number of regions collected by segmentation [30] produce the first attribute vector. The application image feature vector and the database image feature vector are compared, and the discrepancy and threshold are determined. Figure 5 screens the performance of feature extraction algorithms.

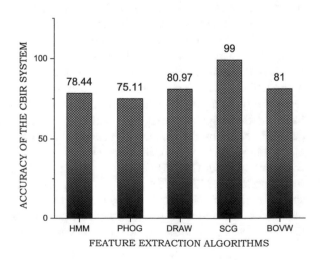

Fig. 5 Performance of feature extraction algorithms

Deep learning for content-based image retrieval [25] the authors expose the semantic difference that occurs from image resolutions obtained and human high-end resolution. So, they came out with the interest in methodologies of deep learning to bridge the semantic gap. In their comprehensive study [23], the researchers addressed distance metric learning methodologies under global supervised approaches and local supervised approaches. These methods of DML learning often employ batch learning methods. But the efficiency goes high when it is taken in the case of using deep learning techniques for retrieval [31]. For instance, the authors quoted from the work carried out on the ImageNet dataset using deep learning techniques that gained the top 5 test error rate of 15.3% by considering over 1000 image classes. The error rate has reduced to 13.34% when teaching is performed concurrently [3]. The author throws a lime light to reduce the semantic gap that exists during the retrieval of images, which in turn enhances the retrieval algorithms' performance. The methodologies suggested using an efficient fusion technique by combining the function descriptor speed up robust features (SURF) and gradient histogram (HOG). The HOG descriptors are applied to extract global features consisting of semantic information. The extracted features are concatenated vertically to create a fusion of visual terms. Using SVM, the image detection is achieved by applying the Helliger Kernel Function. The Euclidean distance is used to measure the similarity scores. Based on mean average accuracy, the effects are calculated. The increased size of vocabulary improves the efficiency. The study is conducted for Coral 1000 dataset, Coral 1500 dataset, and Coral 5k dataset.

GANs A New Era of Image Synthesizing Applications Based on Textual Features. In recent years, several researchers who were attracted by the deep neural nets implemented GANs [27]. Figure 6 shows working of GAN network. When GANs are used in the retrieval of images, it outperforms but still lacks the high-quality image generation. GANs are used in the proposed two-stage system to generate a high quality image. Stage I GAN [30] interprets the shape and color of the scene based on the description given in the text. The Stage II GAN further interprets the effects of Stage I GAN and produces a high-resolution image. A tree-like structure is generated using several generators and discriminator. The biggest stumbling block can occur when we attempt to produce high-resolution pictures.

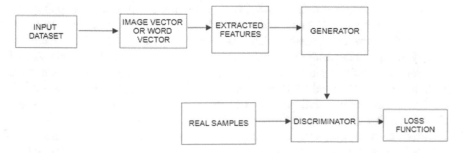

Fig. 6 Generative adversarial network

In Stage I, two-stage GANs are used to generate images from text descriptors through sketch refinement followed by Stage II GAN to learn and capture the text information lost in Stage I GAN. Many applications began using (GANs) [31] for tasks related to information recovery. In recent times, GANs have been used in deep learning in image synthesis and retrieval. While information retrieval [2] takes place in a hierarchical data delivery situation, GANs are equipped in such a way as to tackle the retrieval efficiency in discrete data as well. The authors concentrated in the proposed work on developing the GAN framework and its data generation. They have also developed an open-source platform such as IRGAN and texygen to support research using GANs. The study outlines [31] the system work which improves the training of generative adversarial networks (GANs). It consists of constructing a block of images [32] resulting from the textual input provided in the various GANs versions. The class output is divided into random negative, easy negative, hard negative, semi-hard negative, semi-soft negative, and easy-to-hard negative. The researchers have proposed the generative model for text generation [33] and categorization using adaptive learning and reinforcement learning [23]. The datasets used are Amazon review dataset, Yelp review dataset, Stanford sentiment tree dataset, NEWS dataset which crawled from NY Times, Reuters and USA Today, and Emotion dataset. A sub-dataset was developed for theoretical research based on these datasets. The phrases from the dataset are used to pretrain the generator [34]. The discriminator is labeled with the generated phrases and the actual phrases. Eventually, the classifier is programmed to identify the phrases. The system is trained using RL and GAN and devoid of RL [1]. The AttnGAN generative network may synthesize fine-grained image details based on user-specified textual information. The system consists of a text encoder and image encoder connected through attentional generative network. The text is given as an input [35] the word characteristics are collected and transformed into a word vector matrix. The sentence vector is transformed to the conditioning vector. The paradigm of focus includes two inputs from image features and one from word features. The R-precision is used as the parameter of measurement. The current model implements the state-of-the-art GAN models. The inception score is 14.4% higher than other ones.

The region-word similarity score of each word overcomes the constraint of prior study [23]. Based on cosine similarity, the image region and the sentence are matched. For Flicker30k dataset, MS-COCO dataset, the approach was applied and the findings are encouraging. The approach [35] relies on zero-shot learning, meaning that attribute learning takes place on the go and the system learns to identify new inputs accordingly without any training. The methodology consists of two classes, a trained class label and another class that contains unseen label attributes. The model used CUB birds and NAB dataset to study and reached 43.74 and 35.58% accuracy, respectively. The authors came up [34] with the neural network architecture—deep recurrent attentive writer (DRAW) to produce images. The given network works by combining the human eye mechanism by considering a sequential variational self-encoding framework for building complex image iteration. The algorithm builds on the state of the art for MNIST generative models,

Fig. 7 Accuracy of GAN network

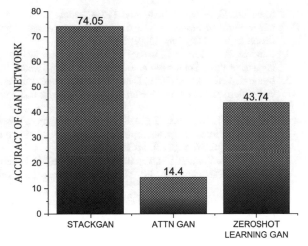

PERFORMANCE OF GAN NETWORK

and by working on the datasets of street-view house numbers. The image quality of output images, which cannot be separated with the naked eye from real data. Figure 7 depicts the performance accuracy of the various GAN Network.

2 Conclusion

The analytical results given by the author give the researchers limelight in deciding on the methodologies that work best for handling textual inputs based on the textual features. Furthermore, each method has its own turbulence to deploy with. Another significant fact is that algorithm selection depends on the dataset under analysis, the time, and consistency of the preparation. The survey provides a snapshot of methodologies that produce optimal outcomes and also quirks about the new retrieval era of GANs that supports the next level of image processing and text-based inputs. The inference provides the users with details on text handling and the different algorithms involved in its speed from the table above. The efficiency of the methods reviewed gives recommendations on algorithms to effectively handle textual inputs.

References

1. Xu T, Zhang P, Huang Q, Zhang H, Gan Z, Huang X, He X (2017) Attngan: Fine-grained text to image generation with attentional generative adversarial networks, In: Proceedings of the IEEE conference on computer vision and pattern recognition, pp 1316–1324
2. Karras T, Aila T, Laine S, Lehtinen J (2018) Progressive growing of gans for improved quality, stability, and variation. In: 6th international conference on learning representations:1710.10196

3. Mehmood Z, Abbas F, Mahmood T, Javid MA, Rehman A, Nawaz T (2018) Content-based image retrieval based on visual words fusion versus features fusion of local and global features. Arab J Sci Eng 43(12):7265–7284
4. Magistry P, Ligozat A-L, Rosset S (2019) Exploiting languages proximity for part-of-speech tagging of three French regional languages. Lang Resour Eval 53(4):865–888
5. Imam N, Vassilakis V (2019) Detecting spam images with embedded arabic text in twitter. In: 2019 international conference on document analysis and recognition workshops (ICDARW), pp 1–6
6. Cai J, Li J, Li W, Wang J (2018) Deep learning model used in text classification. In: 2018 15th international computer conference on wavelet active media technology and information processing (ICCWAMTIP), pp 123–126
7. Ganesh J, Parthasarathi R, Geetha T, Balaji J (2014) Pattern based bootstrapping technique for Tamil POS tagging. In: Mining intelligence and knowledge exploration. Springer, pp 256–267
8. Thenmalar S, Geetha T (2014) Enhanced ontology-based indexing and searching. Aslib J Inf Manag
9. Priyadharshan T, Sumathipala S (2018) Text summarization for Tamil online sports news using NLP. In: 2018 3rd international conference on information technology research (ICITR), pp 1–5
10. Karthieswari R, Sreethivya M, Kumar DR, Soman K (2014) Tamil characters classification using random kitchen sink algorithm (RKS). In: Proceedings of the 2014 international conference on interdisciplinary advances in applied computing, pp 1–5
11. Liu R, Zhao Y, Wei S, Zheng L, Yang Y (2019) Modality-invariant image-text embedding for image-sentence matching. ACM Trans Multimedia Comput Commun Appl (TOMM) 15(1): 1–19
12. Thomas M, Ryan C, Alexander F, Hinrich S (2015) Joint lemmatization and morphological tagging with LEMMING. In: Proceedings of the 2015 conference on empirical methods in natural language processing, pp 2268–2274
13. Niels R, Vuurpijl L (2005) Dynamic time warping applied to Tamil character recognition. In: Eighth international conference on document analysis and recognition (ICDAR'05), pp 730–734
14. Sundaram S, Ramakrishnan AG (2009) An improved online Tamil character recognition engine using post-processing methods. In: 2009 10th international conference on document analysis and recognition, pp 1216–1220
15. Chakrabarty A, Chaturvedi A, Garain U (2019) CNN-based context sensitive lemmatization. In: Proceedings of the ACM India joint international conference on data science and management of data, pp 334–337
16. Chel H, Majumder A, Nandi D (2013) Faster training algorithms in neural network based approach for handwritten text recognition. Int J Image Process (IJIP) 7(4):358
17. Chel H, Majumder A, Nandi D (2011) Scaled conjugate gradient algorithm in neural network based approach for handwritten text recognition. In: International conference on computational science, engineering and information technology, pp 196–210
18. Xu B, Huang K, Liu C-L (2010) Similar handwritten Chinese characters recognition by critical region selection based on average symmetric uncertainty. In: 2010 12th international conference on frontiers in handwriting recognition, pp 527–532
19. Blumenstein M, Verma B (1999) Neural-based solutions for the segmentation and recognition of difficult handwritten words from a benchmark database. In: Proceedings of the fifth international conference on document analysis and recognition. ICDAR'99 (Cat. No. PR00318), pp 281–284
20. Bhunia AK, Kumar G, Roy P, Balasubramanian R, Pal U (2018) Text recognition in scene image and video frame using Color Channel selection. Multimedia Tools Appl 77(7): 8551–8578

21. Mehmood Z, Mahmood T, Javid MA (2018) Content-based image retrieval and semantic automatic image annotation based on the weighted average of triangular histograms using support vector machine. Appl Intell 48(1):166–181
22. Reed S, Akata Z, Yan X, Logeswaran L, Schiele B, Lee H (2016) Generative adversarial text to image synthesis. In: Proceedings of the 33rd international conference on machine learning, PMLR 481605.05396
23. Zeng M, Yao B, Wang Z-J, Shen Y, Li F, Zhang J, Lin H, Guo M (2019) CATIRI: an efficient method for content-and-text based image retrieval. J Comput Sci Technol 34(2):287–304
24. Singh N, Singh K, Sinha AK (2012) A novel approach for content based image retrieval. Procedia Technol 4:245–250
25. Wang X-Y, Yu Y-J, Yang H-Y (2011) An effective image retrieval scheme using color, texture and shape features. Comput Stand Interfaces 33(1):59–68
26. Zhang H, Xu T, Li H, Zhang S, Wang X, Huang X, Metaxas DN (2018) Stackgan++: Realistic image synthesis with stacked generative adversarial networks. IEEE Trans Pattern Anal Mach Intell 41(8):1947–1962
27. Cao X, Chen L, Cong G, Jensen CS, Qu Q, Skovsgaard A, Wu D, Yiu ML (2013) Spatial keyword querying. In: International conference on conceptual modeling, pp 16–29
28. Mirza M, Osindero S (2014) Conditional generative adversarial nets. arXiv preprint arXiv:1411.1784
29. Kulis B, Grauman K (2009) Kernelized locality-sensitive hashing for scalable image search. In: 2009 IEEE 12th international conference on computer vision, pp 2130–2137
30. Goodfellow I, Pouget-Abadie J, Mirza M, Xu B, Warde-Farley D, Ozair S, Courville A, Bengio Y (2014) Generative adversarial nets. In: Advances in neural information processing systems, pp 2672–2680
31. Vendrov I, Kiros R, Fidler S, Urtasun R (2015) Order-embeddings of images and language. ICLR,1511.06361
32. Li Y, Pan Q, Wang S, Yang T, Cambria E (2018) A generative model for category text generation. Inf Sci 450:301–315
33. Gregor K, Danihelka I, Graves A, Rezende DJ, Wierstra D (2015) Draw: a recurrent neural network for image generation. In: Proceedings of the 32nd international conference on machine learning, PMLR 371502.04623
34. Ioffe S, Szegedy C (2015) Batch normalization: accelerating deep network training by reducing internal covariate shift. In: Proceedings of the 32 nd international conference on machine learning, Lille, France. JMLR: W&CP volume 37 1502.03167
35. Zhu Y, Elhoseiny M, Liu B, Peng X, Elgammal A (2017) A generative adversarial approach for zero-shot learning from noisy texts. In: Proceedings of the IEEE conference on computer vision and pattern recognition, pp 1004–1013

Arduino Board-Based Wireless Controlled Seed Sowing Robot

M. Sugadev, T. Ravi, Anugula Venkatesh Kumar, and T. Ilayaraja

1 Introduction

Robotics plays an important role in modern world, and they assist humans in all aspects of life. Application of robotics in industries substantially increases the production and improves quality of manufactured goods [1]. The advancement in artificial neural network field further improves the precision at which robots executes their tasks and its decision making capability [2]. The application of robotics in agricultural sector along with human work force will lead to precision farming that could save enormous electrical power and optimum utilization of water and other resources [3]. Robots are widely used in ploughing, seed sowing, watering of plants, spraying, power management, pesticides and fertilizers spraying, weed removing, etc. In agricultural sector, many farmers go by traditional sowing method and face some difficulties. Due to shortage in agricultural workers and increase in large-scale farming, robotics has been widely used. Some recent work done on agricultural robotics are briefly described here.

Nithish Kumar et al. (2020) presented a GSM-based automated seed sowing device using ultrasonic sensors for obstacle detection, but no provision is provided for remote adjustment of seed spacing and depth [4]. Sangole et al. (2020) proposed a semi-automatic Arduino-based agricultural bot for seed sowing [5]. Madhu and Patil Raj Kumar (2017) has proposed a seed sowing machine, operated manually. This machine can sow any type of seeds and maintain equal space between the seeding points. The cost of this method is less and useful for small-scale farmers. It can be used for cultivating crops like cauliflower, potato, radish, etc. It also incorporates a seed-metering device to know the amount of seeds in the tank. At a set proportion, the seed will fall on the ground according to the rotation of the wheel. There is a disc, which is used to allow only one seed for one rotation [6].

M. Sugadev · T. Ravi · A. V. Kumar (✉) · T. Ilayaraja
Department of ECE, Sathyabama Institute of Science and Technology, Jeppiaar Nagar, Sholinganallur, Chennai 600119, India

© The Author(s), under exclusive license to Springer Nature Singapore Pte Ltd. 2022   323
R. R. Raje et al. (eds.), *Artificial Intelligence and Technologies*,
Lecture Notes in Electrical Engineering 806,
https://doi.org/10.1007/978-981-16-6448-9_33

Nivash and Kavin (2018) presented a method to reduce wastage of the seeds at the time of sowing. It is a hand-operated machine that can sow the seed and spray fertilizer. It has a shaft and fork pipe (handle) which is connected to the wheel. The seed box is fixed to the shaft of the wheel [7]. Rahul and Ganesh (2017) developed a methodology to uniformly distribute the seeds and at desired rate. The principle of machine is to push forward by using handle, and then, first wheel starts rotating and gear is mounted on axel of wheel; it is to rotate, and its rotation is then transferred to pinion through the chain rule. The rotary motion of the pinion moves upwards and downwards which then reciprocate the piston single acting mounted on top of storage tank [8].

Bhushan and Durgesh Varma (2018) worked on adjustable speed rate and eliminating human involvement in the functioning of the machine. The size of the machine is moderate and consists of auto seed feeding system. It also includes obstacle detection unit and change its course direction upon finding any obstacles in the field like stones, trees, etc. A chain is connected to the seed meter for rotatory motion, and when the farmer puts the seeds in hopper, seeds drops into the seed meter, and then, seed meter rotates in clockwise or anti-clockwise direction and seeds then drops into the seed pipe which is connected to the furrow opener for planting a seed. A furrow metal is used to cover the dropped seed with soil. A DC motor is connected to the spur and worm gear for steering mechanism. A microcontroller unit controls the entire system and control instructions are sent to the machine through a mobile phone [9].

Song Yuqiu and Xin Mingjin (2016) have discussed about seed sowing machine design for cultivation of paddy. Rice is the staple food for the Chinese people. Sprouted rice seed broadcasting is a unique but simple agricultural technology based on direct sowing and potted seedling broadcasting. This minimizes the amount of seed required, reduces sowing time, less labour force, cost less and increased production efficiency than manual transplanting of rice seedling.

The drawbacks of manual sprouted rice seeds broadcasting are also analysed in this paper. The aerodynamic property of sprouted rice seed was analysed and evaluated to obtain parameters for developing sprouted seeds broadcasting machine and to obtain uniform distribution of pneumatically broadcasted sprouted seeds. The results proved that the floating speed is related to the length of sprout and kernel weight of the sprouted seeds. The regression equation of the floating coefficient vary with the ratio of sprout length (SL) to kernel weight (KW), and it can be used to determine the floating coefficient and floating speed of sprouted seeds within certain limit. The distribution uniformity test of pneumatic broadcasting with a machine for seedling broadcasting was conducted, and the results showed that the uniformity can meet the requirement of agronomy specification and it had no mechanical damage to the sprout [10].

From the study of recent works on seed sowing machines, it is observed that the existing models can sow seeds, but the depth at which the seeds are sowed and distance between the seeding position are not easily adjustable. Moreover, the designs are larger and requires heavy electrical power for its working.

2 Proposed Seed Sowing Robot Components

The proposed model of the seed sowing robot is shown in Fig. 1. It consists of drilling unit, seed dropping unit and movement control unit. This machine can be used for small-scale planting of seeding for crops such as ground nuts, onion and radish. The functions of each block of the system are described below.

2.1 ATMEGA Microcontroller

The entire actions of the robot are controlled using ATMega-328 RISC architecture based 8-bit microcontroller. It consists of 23 GPIO pins, 6-channel 10-bit analog digital converter, 2 kB SRAM, 1 kB EEPROM, PWM modules, I2C interface, SPI interface, programmable serial USART, etc. and operates with 1.8–5.5 V DC supply. Atmega-328 usually runs with 16 MHz external crystal connected to it. It can also be programmed to make use of 8 MHz internal RC clock generator to avoid the external crystal oscillator. It also provides direct PWM output signals at its output ports to facilitate easy speed control of DC motors. Figure 2 shows the pin out of ATMega328 controller and its mapping with Arduino Uno I/O pinout [2, 11, 12].

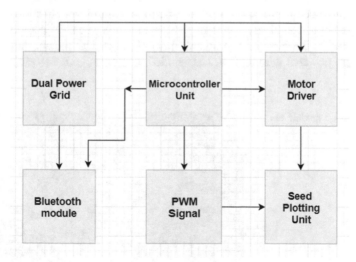

Fig. 1 Block diagram of seed sowing robot

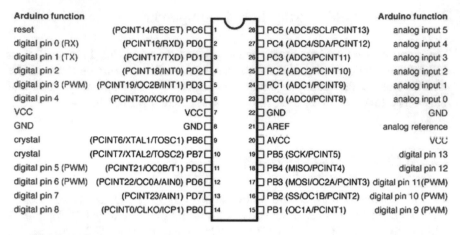

Fig. 2 ATMega328 controller and Arduino Uno I/O mapping

2.2 DC Motors and H-Bridge Module

Four-geared DC motors with 300 rpm and 21 kg cm torque are used for navigating the robot in the desired direction. Two 7.5 A H-bridge modules (shown in Fig. 3) control the four DC motors fixed at the bottom metal frame, which holds the entire structure of the robot. IC LM298 used in the H-bridge module provides interfacing between the microcontroller and DC motors. It protects the microcontroller from back emf produced in the motor coil and helps to control the direction of motors simply by altering the binary inputs to H-bridge module. LM298 also does current amplification of the binary drive signals [13, 14]. The DC motors and H-bridge modules are powered with a 12 V/7AH sealed maintenance free (SMF) battery.

Fig. 3 H-bridge module pin out

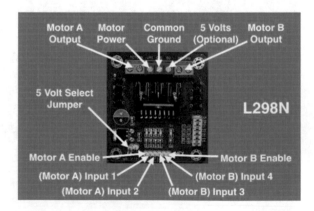

2.3 Dual Power Grid (12 V, 6 V)

The power supply unit provides regulated 12 V for DC motors and 6 V for microcontroller circuit and sensor modules. The dual power grid uses 12/7AH SMF battery. SMF batteries are also called as value regulated lead acid (VRLA) batteries, and they do not any emit any fumes or gases on a continuous basis. Figure 4 shows a typical 12 V/7.5AH SMF battery [15, 16].

2.4 Pulse Width Modulation (PWM)

The ATMEGA-328 microcontroller-based Arduino Uno board has six I/O ports that can provide PWM output signals. By varying the width of the PWM pulse signal, the speed of the DC motors and the angular position of the servomotor shaft are conveniently controlled [1, 12]. Figure 5 shows PWM outputs of ATMega328 controller with different duty cycle.

2.5 Bluetooth Module (HC05)

HC-05 is 2.4 GHz adaptive frequency hopping-based Bluetooth transceiver with − 80 dBm sensitivity and 4 dBm RF output power covering a maximum distance of 10 m. The HC05 module connected to the microcontroller unit facilitates control of various operations of the robot using any Bluetooth-enabled smart phone.

Fig. 4 SMF battery

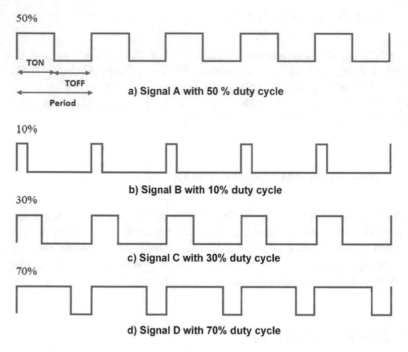

Fig. 5 PWM outputs for different duty cycle

2.6 Servomotor

A 3 kg cm servomotor is used for opening and closing the valve of the seed dropping unit, and a 12 kg cm servomotor is used for position control of the drilling unit for lifting up or pushing down of the drilling bit against the soil surface to make pits to drop seeds. The angular position of the servomotor shaft can be set between 0° and 180° by setting the duty cycle of the PWM input control signal. The servomotor plays a key role in the functioning of seed sowing assembly. This feature helps to control the rate of seed dropping. Figure 6 shows the servomotor used to control the degree of opening of the lid fixed at the outlet of seed feed line pipe.

3 Seed Sowing Mechanism

The proposed system uses a drilling bit to make holes on the soil. The drill bit attached to a servomotor controls the movement of the drill bit in the vertical axis. The drill bit penetrates the ground to make a small pit or hole to the required depth and moves out. The seeds are filled in the seed tank container that consists of an

Fig. 6 Servomotor controlled
lid and drill bit

opening port controlled by servomotor. The amount of time and degree of opening
of the lid fitted at the end of seed feed line determines the number seeds dropped in
the hole or pit. After dropping the seeds, the robot moves forward to the next seed
sowing location. Meanwhile, a metal plate attached at the rear side bottom of the
robot covers sowed seed with soil as the robot moves forward. Figure 7 below
shows the seed sowing assembly used in the proposed robot.

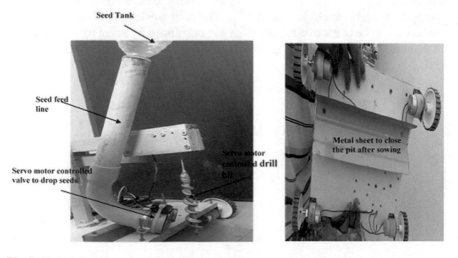

Fig. 7 Proposed seed sowing mechanism

3.1 Operations of the Seed Sowing Robot

A unique and customized GUI based on Android is developed to control various functions of the robot. By installing this application on any smart Android mobile, all the functions of the robot is remotely controlled from a maximum distance of 10 m. Some of the buttons created in the GUI and their functions are given below:

"F"—move robot in forward direction.

"B"—move robot in backward direction.

"L"—move robot to left side.

"R"—move robot to right side.

"CW"—drill the soil in clockwise direction (move downwards).

"AW"—drill the soil in anti-clockwise direction (move upwards).

The flowchart in Fig. 8 illustrates the complete sequence of actions performed by the proposed seed sowing robot. The code is written in C++ and compiles using Arduino IDE open-source software.

The completed prototype model of the proposed seed sowing robot is shown in Fig. 9.

4 Results and Discussions

The fabricated prototype model of the seed sower shown in Fig. 9 is tested for all its functionalities with an Android mobile software interface. The operations of the seed sowing robot were easily controlled using a mobile phone through a custom designed Android interface. The robot navigated smoothly on the field in automatic mode by avoiding obstacles by itself. However, seed sowing function is not automated, and it has to be carried out by issuing commands from the mobile device. Its overall weight is around 5 kg. The seed sowing robot moves over flat soil bed. It can be made to work on rough soil conditions by using rocker bogie like mechanism. The seed feeding pipe diameter may be varied according to the type of seed.

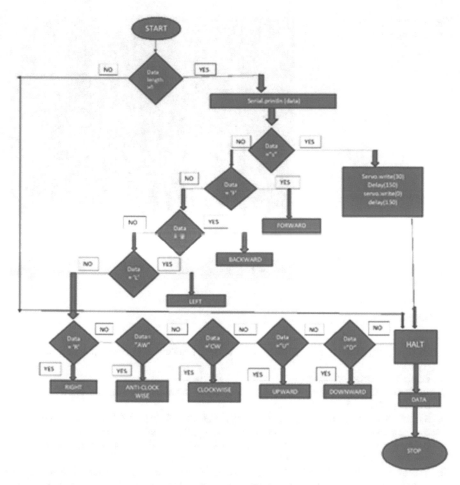

Fig. 8 Sequence of operations of the seed sowing robot

5 Conclusion

In this paper, the design methodology and working principle of a prototype seed sowing robot is discussed. With this machine, we could sow seeds at desired rate and control the depth and distance between seeding locations. The use of drilling system enables the robot to sow seeds on different soil conditions. The operational speed of the proposed prototype model can be improved by using higher ampere rating power supply. The work can be further enhanced by making it fully autonomous by incorporating computer vision and path planning algorithms.

Fig. 9 Prototype model of
seed sowing robot

References

1. Pundkar MR, Mahelle AK (2015) A seed-sowing machine: a review. Int J Eng Soc Sci 3 (03):68–74
2. https://dronebotworkshop.com
3. Joshua R, Vasu V, Vincent P (2010) Solar sprayer—an agriculture implement. Int J Sustain Agric 2(01):16–19
4. Sangole MK, Patil DP, Dhamane KA, Jathar RK, Kardile RS (2020) Semi-automatic seed sowing robot. Int J Innov Technol Exploring Eng (IJITEE) 9(06):1185–1187
5. Nitish kumar K, Balaji R, Raviteja S, Sakshi K, Akhila K (2020) Automated seed sowing robot. Int J Adv Res Ideas Innov Technol 6(2):460–464
6. Kasturi ML, Thorat Swapnil V, Patil Girish V, Patil Rajkumar N (2017) Design and fabrication of seed sowing machine. Int Res J Eng Technol (IRJET) 4(09):704–707
7. Nivash and Kavin (2018) Design and modification of automatic seed sowing machine. Int J Sci Res Dev 6(02):617–618
8. Gune GP, Adalinge NB, Lavate GB, Mane RR (2017) Design and manufacturing of seed sowing machine. Int J Adv Res Ideas Innov Technol 3(07):705–708
9. Deshmukh B, Varma D (2018) Fabrication and implementation of automatic seed sowing machine. Int J Eng Sci Res Technol 03:274–281
10. https://www.quora.com
11. Zanwar SR, Kokate RD (2012) Advanced agriculture system. Int J Robot Autom (IJRA) 1 (02):107–112
12. Zhang B, Xin M, Song Y, Wu L, Ren W (2017) Design of a manually operated rice seed rope planter. Int Agric Eng J 26(01):104–111
13. Marode RV, Tayade GP, Agrawal SK (2013) Design and implementation of multi seed sowing machine. Int J Mech Eng Robot Res 2(04):422–429

14. Kyada AR, Patel DB (2014) Design and development of manually operated seed planter machine. In: 5th international all India manufacturing technology, design and research conference (AIMTDR 2014), IIT Guwahati, India, pp 591–596
15. Sridhar HS (2013) Development of single wheel multi use mutually operated weed remover. Int J Mod Eng Res 3(06):3836–3840
16. Ramesh D, Girish kumar HP (2014) Development of agriculture seeding equipment. Int J Informative Futuristic Res 1(10):133–138

Information Retrieval Using n-grams

Mansi Sood, Harmeet Kaur, and Jaya Gera

1 Introduction

Information retrieval can be defined as finding data relevant to an enquiry commonly called as query in today's digital world. The concept exists since ages evolving from the past eras of finding information manually to creating Boolean indexes to more informative inverted indexes, storage-intensive positional indexes, vector space model, support vector machines etc.

Information can be categorized as structured and unstructured data. Structured data will be having a proper format like SQL databases, for example, and unstructured data is like a random text excerpt from some document. Unstructured data, however, is an ambiguous umbrella term as no meaningful data can strictly be called unstructured. Any random document found on internet can be considered semi-structured if we can identify a title, heading, header, footer and paragraphs in it. Information retrieval majorly deals with digging this semi structured, structured and unstructured data to answer the queries posed by any user.

Information retrieval systems need an effective mechanism to store and represent the plethora of documents commonly called as corpus so as to process and answer a query in real time. Answering a query in real time poses a direct question on the representation of documents in the corpus. Inverted index is the most widely used representation. It consists of a dictionary of terms, also known as vocabulary, prepared after preprocessing documents in the corpus. Corresponding to each term in vocabulary, there exists a list of documents such that term appears one or more than one time in those documents. This list is known as postings list. Inverted

M. Sood · J. Gera
Shyama Prasad Mukherji College, University of Delhi, Delhi, India
e-mail: jayagera@spm.du.ac.in

H. Kaur (✉)
Hansraj College, University of Delhi, Delhi, India
e-mail: hkaur@hrc.du.ac.in

© The Author(s), under exclusive license to Springer Nature Singapore Pte Ltd. 2022   335
R. R. Raje et al. (eds.), *Artificial Intelligence and Technologies*,
Lecture Notes in Electrical Engineering 806,
https://doi.org/10.1007/978-981-16-6448-9_34

indexes can be enhanced into positional indexes to record the positions of occurrence of terms in the documents.

Queries, often Boolean in nature, seek the presence of particular terms in the document or could be complex, enquiring the existence of more than one terms in proximity or continuity to each other.

n-grams is a contiguous occurrence of n terms in a text. For example, "save water save life" is a four-gram phrase. A query searching for such a phrase should be answered keeping in mind that all these four terms may occur in a document in different positions. However, only those documents that contain all these four terms in continuity should be reported as relevant to the query. Such queries are also termed as phrase queries. Proximity queries enquire about occurrence of terms in proximity to each other. For example, finding documents where terms water and life occur at a distance of four to five words from each other. "Water is precious for life" is a match in this case.

Searching n-grams or terms in close proximity to each other in a set of documents can have a lot of applications in today's data-intensive era. For example, plagiarism detection, paraphrase detection [1], text compression [2], text classification [3], sentiment classification [4], feedback analysis etc., to name a few.

This paper proposes an algorithm to find the occurrence of n-gram in the corpus of documents indexed using a positional index. It suggests two optimizations as well on the proposed algorithm which can significantly reduce the number of positions to be searched for existence of n-gram query terms.

This paper is organized as follows. Section 2 covers the related work in this area and applications of text processing as well as n-gram searching. Section 3 describes the proposed algorithm for n-gram searching and related methodology. Section 4 gives the conclusion and scope of future work in this area. The proposed algorithm shall be implemented and analyzed for its time and memory requirements in the future and shall also be compared with other methods of n-gram searching as well to establish its efficiency.

2   Related Works

Information retrieval has been a major focus area for researchers. Many techniques are widespread in this field like Boolean incidence matrix, inverted index, positional index, vector space models, probabilistic models, support vector machines and other machine learning-based techniques like clustering, classification, etc. [5–11]. Inverted index is a simple yet powerful technique to search contents relevant to a query. Reference [12] presents a method based on inverted index to improve teleconsultation system SOFIA. This system answers health-based questions raised by any solicitant by first checking if the question is duplicate, i.e., has been answered in past, retrieving the existing answer in this case or should be answered by forwarding to a professional capable of answering the question.

Text processing has become a fundamental to overcome the information overload problem in the information retrieval domain. Reference [13] presents a neural network-based text processing system that works on raw sentences to build a language independent sub-word tokenizer. Reference [14] uses deep neural network to classify short texts often found on social media networks and microblogging sites. Classification can help to understand the context of short texts and identify the similarity in them. For example, "upcoming iphone model" and "new apple mobile" are related to each other [15]. Reference [16] presents a graph model for text processing to differentiate meaningless text from meaningful ones.

Applications based on searching n-grams and identifying words in proximity to each other demands access to a tremendous amount of data. Not just the documents, web pages and other e-contents but also to a lot of positional data inside them. Handling this much magnitude of data requires an effective data representation and data access methods. Reference [17] uses B-trees over GPU architecture to maximize parallelism and reduce the memory requirements for n-gram searching. Reference [18] has developed an n-gram search tool on entire Wikipedia corpus. This tool can accept wildcard characters, PoS tags like noun phrase, verb phrase etc., in the search query and returns matched documents reporting frequency of occurrence of n-grams and its context in just a fraction of a second.

There are many applications that make use of n-gram searching in background to accomplish their objectives. Chatbot is a well-known application used by uncountable websites to answer user queries and give a personalized response to the users. Reference [19] proposes the use of n-gram matching and other machine learning-based techniques to understand a user's intention behind a message and also to correct typographical errors. References [20, 21] presents a system based on support vector machine classifier and n-gram matching to differentiate the tweets written by a human user and those written by bots. Reference [21] also attempts to identify the gender of the user. Reference [22] proposes classification of URLs into malicious and benign URLs by applying n-gram matching technique. This helps in preventing cyber-attacks where users are tricked by attackers to click on a malicious URL.

3 Proposed Work and Methodology

3.1 Inverted and Positional Index

Building inverted index for a collection of documents in the corpus requires tokenization of text into lexemes called tokens. These tokens are further preprocessed into individual terms using the following preprocessing steps [5]:

Remove stop words—Stop words are the terms that occur very frequently in the text. For example, to, is, are, for, etc. As these words are very common, their presence does not give any meaningful insight into the data. However, they can increase the index size significantly. Hence, they can be removed safely.

Normalization—Normalization means transforming superficially different tokens into standard form. For example, "India" and "india" are treated as same tokens. Similarly, U.N.E.S.C.O. and UNESCO should be treated as same.

Stemming—Reducing words to their root form. For example, "faster" should be reduced to "fast," and "planning" should be reduced to "plan."

Terms obtained after preprocessing are then sorted lexicographically. Every document in the corpus is allocated a unique document identifier (*DId*) to create a postings list in sorted order of *DId*. Every term in the dictionary points to a combination of two things—first, the document frequency specifying the number of documents this term appears in the corpus and second, the sorted list of *DId* where it appears, i.e., the postings list.

Positional index holds an extra piece of information in the postings list. Along with the *DId*s, it also records the number of occurrences of term in that document (commonly known as term frequency) and a sorted list of occurrence positions of this term in the document. Figure 1 illustrates a positional index built on three sample documents. *df* refers to document frequency. A node in the figure captures the following three information: *DId*, number of occurrences of the term in this document, i.e., term frequency and positions of the occurrences. For example, term "life" has a document frequency of three as it occurs in all three documents, occurs once at position one in document one and so the term frequency in document one is one.

3.2 Answering Phrase Queries Using n-grams

Many a times we would want to search for existence of specific n-grams in the corpus instead of individual terms. For example, "back to square one," "late better than never," "US presidential elections," "Dropping air quality index in Delhi" etc. Clearly, searching individual terms would not suffice. For a document to be relevant, terms in the phrase need to occur in continuity. Positional indexes play a key role in answering such queries.

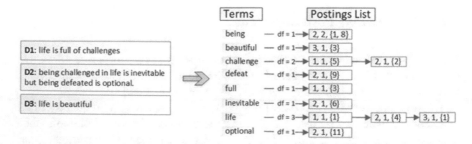

Fig. 1 Positional Index built on a sample of three documents. A node in postings list captures document identifier, term frequency and positions of occurrences

This paper presents an algorithm to search the positional index to find the occurrence of an n-gram in the corpus. The algorithm consists of two procedures **Find_Match_Docs** and **Find_n-gram**.

Find_Match_Docs—Compute a set S of documents each containing all the n terms of the n-gram query (these n terms may or may not occur in continuity).

1. Assuming terms are stored in lexicographical order in the positional index, search for n terms q_1, q_2... q_n of n-gram query and store the first elements of their respective posting lists in P_1, P_2...P_n. P_i will point to the first document (in sorted order of DId) containing query term q_i.
2. Identify the query term T with minimum document frequency (D) while retrieving posting lists P_1, P_2...P_n in step 1. This can be done by simply keeping track of minimum document frequency seen so far and a pointer *minpost* to point to the corresponding posting list with minimum document frequency.
3. Starting with documents in the *minpost* list (i.e., the postings list of a query term out of n-grams having minimum document frequency), compare the documents in this *minpost* list and all other postings lists P_1, P_2...P_n (except the one pointed by *minpost*) and create an intersection set S of documents that contain all the n-gram terms q_1, q_2... q_n.

Find_n-gram—In the document set S (received from Find_Match_Docs), find the set M of documents each having all the n terms of n-gram query in continuity.

1. In the postings list P_1, jump to a document number d such that it is the first document in P_1 that is also a member of S. This implies document d contains all the query terms q_1, q_2... q_n. We need to check if d has all these terms in continuity or not. Let *pos* be the first occurrence position of q_1 in this document d.
2. To match the n-gram, document d should have term q_2 at position *pos* + 1. To verify this, search for document d in the posting list P_2 and then search for position *pos* + 1 in its list of occurrences. If it is there, look for document d in P_3 and position *pos* + 2 in the occurrence position list and so on for *pos* + n − 1 position in the occurrence position list for d in P_n.
3. If at any time searched position is not found in occurrence position list corresponding to document d in any of the posting lists, restart at the next position of q_1 in the document d. However, if we were at the last position of q_1 in the document d, then Find_n-gram moves to next document d' in postings list P_1 that is also a member of set S.

Figure 2 presents an example run of the proposed algorithm for n-gram query "being challenged" on the positional index built in Fig. 1 and pseudocode for the above two procedures **Find_Match_Docs** and **Find_n-gram**.

Optimizations in Find_n-gram—If at any time position to be searched in postings list P_i is greater than the last occurrence position of q_i in document d, then terminate search for this document and move to next document d' in postings list P_1 that is also a member of set S.

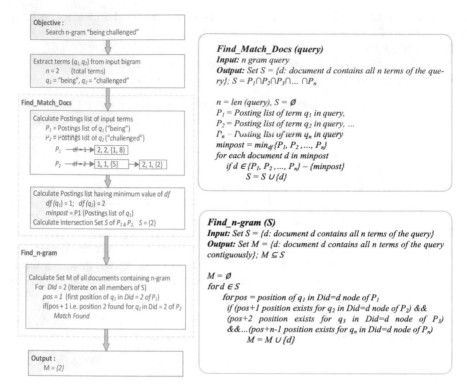

Fig. 2 Search n-gram query "being challenged" in positional index shown in Fig. 1, Pseudocode for Find_Match_Docs and Find_n-gram

Another optimization is instead of working on the occurrence positions in document d of postings list P_1, Find_n-grams should look for postings list P_i corresponding to query term q_i with minimum number of occurrences in document d (term frequency). This will minimize the overall number of positions to be verified for the contiguous existence of n-grams in the document.

These two optimizations can improve the running time of the proposed algorithm as the number of possible positions searched for existence of n-gram can be significantly reduced after applying these optimizations.

4 Conclusion and Future Work

This paper proposed an algorithm to search the existence of n-grams in a collection of documents, commonly known as corpus, using a positional index. Positional index maintains a list of all the preprocessed terms occurring at some positions in any of the documents in corpus. Along with these terms, it stores a postings list for each term listing the number of documents in which it appears and the corresponding document identifiers. For every document, it further lists the position of occurrences of the term in that document. This paper suggested two optimizations as well on the proposed algorithm which can significantly reduce the number of positions to be searched for existence of n-gram query terms.

In future, comparison of the proposed algorithm with the existing algorithms with respect to the memory and time requirements shall be undertaken. Based on the results, its application for paraphrase detection, building dictionaries for sentiment analysis, classification based on existence of certain specific phrases etc., shall be explored.

References

1. Hunt E, Janamsetty R, Kinares C, Koh C, Sanchez A, Zhan F, Ozdemir M, Waseem S, Yolcu O, Dahal B, Zhan J, Gewali L, Oh P (2019) Machine learning models for paraphrase identification and its applications on plagiarism detection. In: Proceedings—10th IEEE international conference on big knowledge, ICBK 2019, pp 97–104
2. Nguyen VH, Nguyen HT, Duong HN, Snasel V (2016) n-gram-based text compression. Comput Intell Neurosci 1–11. https://doi.org/10.1155/2016/9483646
3. Burdisso SG, Errecalde M, Montes-y-Gómez M (2020) τ-SS3: a text classifier with dynamic n-grams for early risk detection over text streams. Pattern Recognit Lett 138:130–137. https://doi.org/10.1016/j.patrec.2020.07.001
4. Tripathy A, Agrawal A, Rath SK (2016) Classification of sentiment reviews using n-gram machine learning approach. Expert Syst Appl 57:117–126. https://doi.org/10.1016/j.eswa.2016.03.028
5. Manning CD, Raghavan P, Schütze H (2008) Introduction to information retrieval. Cambridge University Press, USA
6. Croft B, Metzler D, Strohman T (2009) Search engines: information retrieval in practice. Addison-Wesley Publishing Company, USA
7. Berger A, Lafferty J (1999) Information retrieval as statistical translation. In: Proceedings of the 22nd annual international ACM SIGIR conference on research and development in information retrieval, pp 222–229. Association for Computing Machinery, New York, NY, USA
8. Furnas GW, Deerwester S, Dumais ST, Landauer TK, Harshman RA, Streeter LA, Lochbaum KE (1988) Information retrieval using a singular value decomposition model of latent semantic structure. In: Proceedings of the 11th annual international ACM SIGIR conference on research and development in information retrieval, pp 465–480. Association for Computing Machinery, New York, NY, USA
9. Banawan K, Ulukus S (2018) The capacity of private information retrieval from coded databases. IEEE Trans Inf Theory 64:1945–1956. https://doi.org/10.1109/tit.2018.2791994
10. Nanda R, John AK, Di Caro L, Boella G, Robaldo L (2017) Legal information retrieval using topic clustering and neural networks. In COLIEE@ ICAIL (pp 68–78)
11. Kamakshaiah K, Seshadri R (2017) Prototype survey analysis of different information retrieval classification and grouping approaches for categorical information. In: 2017 international conference on intelligent computing and control (I2C2), pp 1–7
12. Oliveira Serra H, Bezerra Maia L, Salomon A, da Silva Lima N, de Sousa Silva R, Bezerra Maia A, Rocha Araújo A, Maia Sousa T, Viegas Dias L, Alves Montes M (2019) Information retrieval based on a search engine with inverted index for Sofia (online system for interactive improvement of primary care). In: INTED2019 proceedings, pp 9279–9283. IATED

13. Kudo T, Richardson J (2018) SentencePiece: a simple and language independent subword tokenizer and detokenizer for Neural Text Processing. CoRR. abs/1808.0
14. Zhan J, Dahal B (2017) Using deep learning for short text understanding. J Big Data 4:34. https://doi.org/10.1186/s40537-017-0095-2
15. Gasser R, Rossetto L, Schuldt H (2019) Towards an all-purpose content-based multimedia information retrieval system. CoRR. abs/1902.03878
16. Grigoryeva EG, Kochetova LA, Pomelnikov YV, Popov VV, Shtelmakh T (2019) V: towards a graph model application for automatic text processing in data management. IOP Conf Ser Mater Sci Eng 483:12077. https://doi.org/10.1088/1757-899x/483/1/012077
17. Bogoychev N, Lopez A (2016) N-gram language models for massively parallel devices. In: ACL (1)
18. Sekine S, Dalwani K (2010) Ngram search engine with patterns combining token, POS, Chunk and NE Information. In: LREC
19. Tedjopranoto ML, Wijaya A, Santoso LH, Suhartono D (2019) Correcting typographical error and understanding user intention in Chatbot by combining N-gram and machine learning using schema matching technique. Int J Mach Learn Comput 9:471–476
20. Pizarro J (2019) Using N-grams to detect Bots on Twitter. In: CLEF
21. Vogel I, Jiang P (2019) Bot and gender identification in twitter using word and character N-grams notebook for PAN at CLEF 2019
22. Hai T, Hwang S (2018) Detection of malicious URLs based on word vector representation and ngram. J Intell Fuzzy Syst 35:1–13. https://doi.org/10.3233/JIFS-169831

An Automated Decision Support Systems Miner for Intuitionistic Trapezoidal Fuzzy Multiple Attribute Group Decision-Making Modeling with Constraint Matrix Games

P. John Robinson, Deng-Feng Li, and S. Samuel Nirmalsingh

1 Introduction

Decision making is a process of gathering information, valuing situation, generating and choosing alternatives, and implementing the result. Many researchers have contributed to the field of intuitionistic fuzzy sets and problems dealing with MAGDM [1, 8, 10–17]. Decision Support Systems (DSS) are used to reduce errors and increase the performance of the decision and hence have immense importance in the field of artificial intelligence. DSS is an information system that supports decision making by accessing large volumes of information generated from various related information systems involved in organizational business processes. A DSS is intended to support the decision-maker but won't replace the role of decision-maker in solving problems. DSS is built upon summary information, exceptions, patterns, and trends using the analytical models. DSS can be either fully computerized or human-powered, or combination of both. Linear programming approaches for constraint matrix games and various other forms of game theory [2–7, 9] are found to be helpful in solving unknown information in MAGDM problems. The automated DSS-Miner algorithm proposed in this paper uses Apriori algorithm which will remove the insignificant decision variables from the system.

P. J. Robinson · S. S. Nirmalsingh (✉)
PG and Research Department of Mathematics, Bishop Heber College,
Tiruchirappalli, India

D.-F. Li
School of Management and Economics, University of Electronic Science
and Technology of China, Chengdu, China

2 Intuitionistic Trapezoidal Fuzzy Power Generalized Aggregation (*ITFPGA*) Operator

Let $\tilde{a}_j = \left(\left[a_j^1, a_j^2, a_j^3, a_j^4 \right]; \mu_j, v_j \right)$ be a group of Intuitionistic Trapezoidal Fuzzy Numbers (*ITFNs*), and its weight vector be $\omega = (\omega_1, \omega_2, \ldots, \omega_n)^{\mathrm{T}}$ where $\omega_j \geq 0, j = 1, 2, \ldots, n$ and $\sum_{j=1}^{n} \omega_j = 1$. Then *ITFPGWA* operator is defined as follows [8]:

$$ITFPGWA(\tilde{a}_1, \tilde{a}_2, \ldots, \tilde{a}_n) = \left(\sum_{j=1}^{n} \omega_j \left(1 + T(\tilde{a}_{ij}) \right) \tilde{a}_j^{\lambda} / \sum_{j=1}^{n} \omega_j \left(1 + T(\tilde{a}_{ij}) \right) \right)^{1/\lambda},$$

$$\text{where} \quad \lambda \in (0, +\infty).$$

$$(1)$$

3 The Automated DSS-Miner Algorithm for Trapezoidal IFS

Let $A = \{A_1, A_2, \ldots, A_m\}$ be a discrete set of alternatives, and $G = \{G_1, G_2, \ldots, G_n\}$ be the set of attributes, $\gamma = (\gamma_1, \gamma_2, \ldots, \gamma_n)$ is the weighting vector of the attributes, where $\gamma_j \in [0, 1], \sum_{j=1}^{n} \gamma_j = 1$. Let $D = \{D_1, D_2, \ldots, D_t\}$ be the set of decision-makers, with weight vector $\omega = (\omega_1, \omega_2, \ldots, \omega_n)$. The algorithm is as follows:

Step 1: Calculate the supports

$$\text{Sup}\left(\tilde{r}_{ij}^k, \tilde{r}_{ij}^l \right) = 1 - d\left(\tilde{r}_{ij}^k, \tilde{r}_{ij}^l \right), \quad k, l = 1, 2, \ldots, p,$$

$$(2)$$

where $d\left(\tilde{r}_{ij}^k, \tilde{r}_{ij}^l \right)$ is the Hamming distance, and find

$$T\left(\tilde{r}_{ij}^k \right) = \sum_{\substack{l = 1 \\ l \neq k}}^{p} \text{Sup}\left(\tilde{r}_{ij}^k, \tilde{r}_{ij}^l \right),$$

$$(3)$$

and the weights can be calculated as:

$$w_{ij}^k = \gamma_k \left(1 + T\left(\tilde{r}_{ij}^k \right) \right) / \sum_{k=1}^{p} \gamma_k \left(1 + T\left(\tilde{r}_{ij}^k \right) \right),$$

$$(4)$$

Step 2: Aggregate the information using ITFPGWA operator, find $T(\tilde{r}_{ij})$ using (3) and the weights are given by:

$$\varpi_{ij} = \omega_j\left(1 + T(\tilde{r}_{ij})\right)/\sum_{j=1}^{n} \omega_j\left(1 + T(\tilde{r}_{ij})\right), \tag{5}$$

Step 3: Calculate the comprehensive evaluation of each alternative using

$$\tilde{z}_i = ITFPGWA(\tilde{r}_{i1}, \tilde{r}_{i2}, \ldots, \tilde{r}_{in}). \tag{6}$$

Step 4: Utilize the Apriori algorithm and the DSS-Mining rule to identify the closely related itemsets, to remove the insignificant items from the domain.

DSS-Miner Algorithm:

C_k: fuzzy itemsets of size k

L_k: frequent fuzzy itemset of size k

  Input { *ITF Decision Matrices*}

    Calculate {*ITFPGWA*}

      Produce {*Individual Preference Decision Matrix*}

      Produce {*Collective Overall Preference Decision Matrix*}

L_1 = {frequent fuzzy items};

For (k = 1;L_k! = $\varnothing$; k + +)**do begin.**

    While $f$support($item - set$) $\geq$ *Threshold* **do**

C_{k+1} = fuzzy items produced from L_k; //that is Cartesian product $L_{k-1} \times L_{k-1}$ hence eliminating any k-1 itemsets which are not frequent //

    Calculate {*Correlation coefficient between itemsets $K(F_A, F_B)$*}

      Calculate {*Median of all $K_{MED}(F_A, F_B)$*}

    While $K(F_A, F_B) > K_{MED}(F_A, F_B)$ **do**

For each operation in domain **do** add the count of all.

Itemsets in C_{k+1} that are contained in t.

L_{k+1} = itemsets   in C_{k+1} with $f$supp($item - set$) $\geq$ *Threshold*   and $K(F_A, F_B) > K_{MED}(F_A, F_B)$.

    End

    Return $\cup_k L_k$;

Step 5: Using the correlation coefficient Formula [11] find the relationship between the remaining decision variables and the positive ideal solutions for intuitionistic trapezoidal sets, $\tilde{r}^+ = ([1, 1, 1, 1]; 1, 0)$:

$$K_{TzIFS}^+(\tilde{r}_i, \tilde{r}^+) = \frac{C_{TzIFS}(\tilde{r}_i, \tilde{r}^+)}{\sqrt{E_{TzIFS}(\tilde{r}_i).E_{TzIFS}(\tilde{r}^+)}}, \tag{7}$$

Step 6: Alternatives are ranked A_i (i = 1, 2, ..., m) according to the highest correlation coefficient obtained and select the best one.

4 Interval-Valued Constraint Matrix Games and Auxiliary Linear Programming Models

Let us consider a type of constraint matrix game with interval-valued payoffs with pure strategy sets S_1 and S_2 and mixed strategy sets Y and Z for player I and player II. The weighting vector information are given by the decision-makers in the form of a constraint matrix games for the MAGDM model involved. A payoff matrix of player I is concisely expressed as $\tilde{I} = (\tilde{I}_{ij})_{m \times n} = \left(\left[\underline{a}_{ij}, \overline{a}_{ij} \right] \right)_{m \times n}$ where $\tilde{I}_{ij} = [\underline{a}_{ij}, \overline{a}_{ij}](i = 1, 2, \ldots, m; j = 1, 2, \ldots, n)$ are closed intervals, $\underline{a}_{ij}$ and $\overline{a}_{ij}$ are the lower bound and the upper bound of the interval $\tilde{I}_{ij}$, respectively. Let $\delta_1, \delta_2, \delta_3$, and δ_4 be the four strategies of both player I and player II. The mixed strategies of player I and player II must satisfy the constraint conditions $40y_1 + 20y_2 + 60y_3 + 30y_4 \leq 70$ and $30z_1 + 40z_2 + 50z_3 + 10z_4 \leq 60$, respectively. The set of constrained strategies are expressed as follows:

$$Y = \{y | 40y_1 + 20y_2 + 60y_3 + 30y_4 \leq 70, y_1 + y_2 + y_3 + y_4 \leq 1,$$
$$-y_1 - y_2 - y_3 - y_4 \geq -1, y_1, y_2, y_3, y_4 \geq 0\} \tag{8}$$

and

$$Z = \{z | -30z_1 - 40z_2 - 50z_3 - 10z_4 \geq -60, z_1 + z_2 + z_3 + z_4 \leq 1,$$
$$-z_1 - z_2 - z_3 - z_4 \geq -1, z_1, z_2, z_3, z_4 \geq 0\} \tag{9}$$

The interval-valued payoff matrix $\tilde{I}$ of player I is given as follows:

$$\tilde{I} = \begin{pmatrix} [15, 23] & [-17, -10] & [8, 15] & [-6, -2] \\ [-11, -6] & [12, 18] & [-9, -3] & [18, 25] \\ [20, 28] & [15, 19] & [26, 32] & [-40, -30] \\ [-14, -9] & [-40, -36] & [-8, -4] & [30, 35] \end{pmatrix} \tag{10}$$

where the element $a_{11} = [15, 23]$ indicates the profit of player I which is between 15 and 23. Coefficient matrices and vectors of the constraint sets for player I and player II are obtained as follows:

$$B = \begin{pmatrix} 40 & 1 & -1 \\ 20 & 1 & -1 \\ 60 & 1 & -1 \\ 30 & 1 & -1 \end{pmatrix}, E^T = \begin{pmatrix} -30 & 1 & -1 \\ -40 & 1 & -1 \\ -50 & 1 & -1 \\ -10 & 1 & -1 \end{pmatrix} \tag{11}$$
$$\text{and } c = (70, 1, -1)^T, d = (-60, 1, -1)$$

The linear programming models using the upper and lower bounds can be constructed as follows:

$$\max\{-60\bar{x}_1 + \bar{x}_2 - \bar{x}_3\}$$

$$\text{s.t}\begin{cases} -30\bar{x}_1 + \bar{x}_2 - \bar{x}_3 - 23\bar{y}_1 + 6\bar{y}_2 - 28\bar{y}_3 + 9\bar{y}_4 \leq 0; \ -40\bar{x}_1 + \bar{x}_2 - \bar{x}_3 + 10\bar{y}_1 - 18\bar{y}_2 - 19\bar{y}_3 + 36\bar{y}_4 \leq 0 \\ -50\bar{x}_1 + \bar{x}_2 - \bar{x}_3 - 15\bar{y}_1 + 3\bar{y}_2 - 32\bar{y}_3 + 4\bar{y}_4 \leq 0; \ -10\bar{x}_1 + \bar{x}_2 - \bar{x}_3 + 2\bar{y}_1 - 25\bar{y}_2 + 30\bar{y}_3 - 35\bar{y}_4 \leq 0 \\ 40\bar{y}_1 + 20\bar{y}_2 + 60\bar{y}_3 + 30\bar{y}_4 \leq 70; \ \bar{y}_1 + \bar{y}_2 + \bar{y}_3 + \bar{y}_4 \leq 1; \ -\bar{y}_1 - \bar{y}_2 - \bar{y}_3 - \bar{y}_4 \leq -1; \ \bar{x}_1, \bar{x}_2, \bar{x}_3, \bar{y}_1, \bar{y}_2, \bar{y}_3, \bar{y}_4 \geq 0 \end{cases}$$

Solving in linear programming, we get:

$$\bar{y}^* = (0.294, 0.5047, 0.1681, 0.0332)^{\mathrm{T}} \tag{12}$$

$$\max\{-60\underline{x}_1 + \underline{x}_2 - \underline{x}_3\}$$

$$\text{s.t}\begin{cases} -30\underline{x}_1 + \underline{x}_2 - \underline{x}_3 - 15\underline{y}_1 + 11\underline{y}_2 - 20\underline{y}_3 + 14\underline{y}_4 \leq 0; \ -40\underline{x}_1 + \underline{x}_2 - \underline{x}_3 + 17\underline{y}_1 - 12\underline{y}_2 - 15\underline{y}_3 + 40\underline{y}_4 \leq 0 \\ -50\underline{x}_1 + \underline{x}_2 - \underline{x}_3 - 8\underline{y}_1 + 9\underline{y}_2 - 26\underline{y}_3 + 8\underline{y}_4 \leq 0; \ -10\underline{x}_1 + \underline{x}_2 - \underline{x}_3 + 6\underline{y}_1 - 18\underline{y}_2 + 40\underline{y}_3 - 30\underline{y}_4 \leq 0 \\ 40\underline{y}_1 + 20\underline{y}_2 + 60\underline{y}_3 + 30\underline{y}_4 \leq 70; \ \underline{y}_1 + \underline{y}_2 + \underline{y}_3 + \underline{y}_4 \leq 1; \ -\underline{y}_1 - \underline{y}_2 - \underline{y}_3 - \underline{y}_4 \leq -1 \\ \underline{x}_1, \underline{x}_2, \underline{x}_3, \underline{y}_1, \underline{y}_2, \underline{y}_3, \underline{y}_4 \geq 0 \end{cases}$$

Solving in linear programming, we get:

$$\underline{y}^* = (0.3076, 0.5003, 0.1587, 0.0334)^{\mathrm{T}} \tag{13}$$

Similarly, another set of the linear programming models using the upper and lower bounds can be constructed as follows:

$$\min\{70\bar{s}_1 + \bar{s}_2 - \bar{s}_3\}$$

$$\text{s.t}\begin{cases} 40\bar{s}_1 + \bar{s}_2 - \bar{s}_3 - 23\bar{z}_1 + 10\bar{z}_2 - 15\bar{z}_3 + 2\bar{z}_4 \geq 0; \ 20\bar{s}_1 + \bar{s}_2 - \bar{s}_3 + 6\bar{z}_1 - 18\bar{z}_2 + 3\bar{z}_3 - 25\bar{z}_4 \geq 0 \\ 60\bar{s}_1 + \bar{s}_2 - \bar{s}_3 - 28\bar{z}_1 - 19\bar{z}_2 - 32\bar{z}_3 + 30\bar{z}_4 \geq 0; \ 30\bar{s}_1 + \bar{s}_2 - \bar{s}_3 + 9\bar{z}_1 + 36\bar{z}_2 + 4\bar{z}_3 - 35\bar{z}_4 \geq 0 \\ -30\bar{z}_1 - 40\bar{z}_2 - 50\bar{z}_3 - 10\bar{z}_4 \geq -60; \ \bar{z}_1 + \bar{z}_2 + \bar{z}_3 + \bar{z}_4 \geq 1; \ -\bar{z}_1 - \bar{z}_2 - \bar{z}_3 - \bar{z}_4 \geq -1 \\ \bar{s}_1, \bar{s}_2, \bar{s}_3, \bar{z}_1, \bar{z}_2, \bar{z}_3, \bar{z}_4 \geq 0 \end{cases}$$

Solving in linear programming, we get:

$$\bar{z}^* = (0.0925, 0.0539, 0.4861, 0.3675)^{\mathrm{T}} \tag{14}$$

$$\min\{70\underline{s}_1 + \underline{s}_2 - \underline{s}_3\}$$

$$\text{s.t}\begin{cases} 40\underline{s}_1 + \underline{s}_2 - \underline{s}_3 - 15\underline{z}_1 + 17\underline{z}_2 - 8\underline{z}_3 + 6\underline{z}_4 \geq 0; \ 20\underline{s}_1 + \underline{s}_2 - \underline{s}_3 + 11\underline{z}_1 - 12\underline{z}_2 + 9\underline{z}_3 - 18\underline{z}_4 \geq 0 \\ 60\underline{s}_1 + \underline{s}_2 - \underline{s}_3 - 20\underline{z}_1 - 15\underline{z}_2 - 26\underline{z}_3 + 40\underline{z}_4 \geq 0; \ 30\underline{s}_1 + \underline{s}_2 - \underline{s}_3 + 14\underline{z}_1 + 40\underline{z}_2 + 8\underline{z}_3 - 30\underline{z}_4 \geq 0 \\ -30\underline{z}_1 - 40\underline{z}_2 - 50\underline{z}_3 - 10\underline{z}_4 \geq -60; \ \underline{z}_1 + \underline{z}_2 + \underline{z}_3 + \underline{z}_4 \geq 1; \ -\underline{z}_1 - \underline{z}_2 - \underline{z}_3 - \underline{z}_4 \geq -1 \\ \underline{s}_1, \underline{s}_2, \underline{s}_3, \underline{z}_1, \underline{z}_2, \underline{z}_3, \underline{z}_4 \geq 0 \end{cases}$$

Solving in linear programming, we get:

$$\underline{z}^* = (0.0992, 0.0827, 0.4745, 0.3436)^{\mathrm{T}} \tag{15}$$

5 MAGDM with Weights Derived from Solving Interval-Valued Constraint Matrix Games

A collaborator wants to choose one of the best institutions in the country among the set of four, $\{A_1, A_2, A_3, A_4\}$, and the attributes to consider are C1—Campus Location, C2—Career Opportunities, C3—Academic Quality, C4—Campus Environment. Based on these four attributes, the four experts $\{e_1, e_2, e_3, e_4\}$ evaluated the efficiency of the four institutions. Suppose that $\gamma = (0.294, 0.5047, 0.1681, 0.0332)$ (12) is the expert weight vector, and $\omega = (0.0925, 0.0539, 0.4861, 0.3675)$ (14) is the attribute weight vector, which we have obtained by solving the interval-valued constraint matrix game (upper bound) given by the decision-maker. The Decision matrix given by the experts with trapezoidal intuitionistic fuzzy data are given in Table 1.

By applying the automated DSS-Miner algorithm for trapezoidal IFS, all other values are computed and the matrix is as follows:

$$R = \begin{bmatrix} ([0.0934, 0.2144, 0.3243, 0.4604]; 0.6394, 0.2523) & ([0.1957, 0.2599, 0.4644, 0.5323]; 0.6304, 0.2579) \\ ([0.2746, 0.5191, 0.6527, 0.9407]; 0.5331, 0.2675) & ([0.3336, 0.4529, 0.7103, 0.8625]; 0.5805, 0.2129) \\ ([0.3156, 0.4162, 0.5191, 0.6407]; 0.6487, 0.1426) & ([0.0999, 0.3756, 0.7367, 0.8976]; 0.6063, 0.2589) \\ ([0.0000, 0.1101, 0.3095, 0.4650]; 0.7836, 0.1481) & ([0.1377, 0.3160, 0.4884, 0.6438]; 0.7483, 0.1304) \end{bmatrix}$$

$$\begin{bmatrix} ([0.4386, 0.5733, 0.7503, 0.9362]; 0.6048, 0.2660) & ([0.3543, 0.5808, 0.6684, 0.8821]; 0.7102, 0.0661) \\ ([0.4690, 0.6089, 0.6779, 0.8088]; 0.7466, 0.0928) & ([0.3390, 0.5527, 0.8375, 0.8989]; 0.6864, 0.1956) \\ ([0.2424, 0.4549, 0.5535, 0.8331]; 0.5191, 0.3036) & ([0.2515, 0.4021, 0.4960, 0.6209]; 0.6145, 0.2073) \\ ([0.0118, 0.2164, 0.3647, 0.5630]; 0.6990, 0.1034) & ([0.0174, 0.1958, 0.2290, 0.3763]; 0.6307, 0.0933) \end{bmatrix}$$

The computed correlation coefficient $K(\tilde{z}_i, \tilde{z}^+)(i = 1, 2, 3, 4)$ values are:

$$K(\tilde{z}_1, \tilde{z}^+) = 0.9343, K(\tilde{z}_2, \tilde{z}^+) = 0.9576, K(\tilde{z}_3, \tilde{z}^+) = 0.8848, K(\tilde{z}_4, \tilde{z}^+)$$
$$= 0.9477,$$

The ranking according to highest correlation coefficient $A_2 > A_4 > A_1$.
The best alternative is A_2.

Suppose that $\gamma = (0.3076, 0.5003, 0.1587, 0.0334)$ (13) is the expert weight vector, and $\omega = (0.0992, 0.0827, 0.4745, 0.3436)$ (15) is the attribute weight vector, which we have obtained by solving the interval-valued constraint matrix game (lower bound) given by the decision-maker. Then the ranking is $A_2 > A_4 > A_1$, And the best alternative is A_2. It can be observed that the decision variable A_3 is less important and has been automatically removed by the proposed DSS-Miner algorithm. The ranking obtained from MAGDM with weights derived from solving interval-valued constraint matrix games using upper bound and MAGDM with weights derived from solving interval-valued constraint matrix games using lower bound are compared in Table 2.

The ranking of the decision variables solved using *ITFPGWA* operator using both upper bound and lower bound weights are found to be consistent.

Table 1 Decision matrix

R¹	([0.12,0.15,0.23,0.26];0.65,0.24)	([0.35,0.39,0.46,0.48];0.64,0.25)	([0.34,0.46,0.54,0.62];0.64,0.28)	([0.48,0.54,0.58,0.62];0.61,0.18)
	([0.32,0.45,0.56,0.67];0.55,0.23)	([0.54,0.62,0.71,0.82];0.72,0.16)	([0.46,0.51,0.53,0.68];0.74,0.15)	([0.27,0.35,0.48,0.52];0.68,0.24)
	([0.24,0.26,0.27,0.28];0.52,0.24)	([0.35,0.51,0.64,0.73];0.51,0.28)	([0.21,0.35,0.47,0.52];0.42,0.46)	([0.39,0.46,0.51,0.63];0.74,0.18)
	([0.11,0.14,0.24,0.38];0.54,0.16)	([0.28,0.34,0.41,0.47];0.76,0.17)	([0.14,0.28,0.43,0.57];0.59,0.35)	([0.19,0.24,0.27,0.37];0.51,0.28)
R²	([0.17,0.24,0.27,0.36];0.64,0.25)	([0.25,0.27,0.36,0.37];0.64,0.27)	([0.36,0.42,0.51,0.65];0.52,0.29)	([0.32,0.47,0.52,0.68];0.74,0.03)
	([0.26,0.37,0.41,0.58];0.51,0.29)	([0.25,0.31,0.48,0.52];0.43,0.45)	([0.35,0.48,0.49,0.54];0.76,0.05)	([0.42,0.54,0.73,0.75];0.72,0.16)
	([0.34,0.38,0.42,0.45];0.72,0.12)	([0.14,0.28,0.49,0.54];0.59,0.31)	([0.25,0.38,0.41,0.69];0.54,0.27)	([0.31,0.35,0.37,0.38];0.54,0.21)
	([0.14,0.17,0.25,0.28];0.87,0.15)	([0.27,0.39,0.48,0.58];0.72,0.29)	([0.04,0.18,0.24,0.37];0.72,0.06)	([0.25,0.36,0.37,0.42];0.63,0.05)
R³	([0.24,0.28,0.34,0.37];0.64,0.25)	([0.32,0.37,0.54,0.68];0.57,0.24)	([0.37,0.43,0.67,0.78];0.72,0.24)	([0.45,0.56,0.58,0.67];0.78,0.09)
	([0.18,0.27,0.35,0.42];0.54,0.28)	([0.41,0.42,0.45,0.57];0.63,0.04)	([0.33,0.35,0.52,0.55];0.74,0.21)	([0.52,0.67,0.69,0.74];0.58,0.24)
	([0.17,0.26,0.37,0.54];0.57,0.12)	([0.43,0.51,0.63,0.79];0.75,0.15)	([0.23,0.34,0.37,0.48];0.56,0.23)	([0.34,0.56,0.67,0.78];0.56,0.23)
	([0.15,0.27,0.39,0.49];0.74,0.13)	([0.25,0.28,0.37,0.39];0.79,0.01)	([0.12,0.17,0.24,0.31];0.78,0.07)	([0.12,0.23,0.24,0.34];0.78,0.08)
R⁴	([0.25,0.35,0.47,0.52];0.51,0.48)	([0.34,0.41,0.58,0.61];0.68,0.24)	([0.52,0.68,0.72,0.84];0.73,0.07)	([0.25,0.38,0.49,0.52];0.52,0.43)
	([0.23,0.25,0.34,0.39];0.67,0.23)	([0.45,0.52,0.61,0.76];0.71,0.18)	([0.14,0.28,0.43,0.51];0.57,0.31)	([0.37,0.39,0.42,0.45];0.67,0.23)
	([0.14,0.23,0.28,0.37];0.72,0.06)	([0.31,0.35,0.54,0.64];0.72,0.12)	([0.41,0.57,0.63,0.75];0.72,0.19)	([0.12,0.23,0.45,0.56];0.56,0.34)
	([0.04,0.17,0.23,0.34];0.56,0.12)	([0.18,0.24,0.37,0.49];0.76,0.15)	([0.42,0.54,0.57,0.61];0.74,0.05)	([0.56,0.67,0.68,0.78];0.56,0.23)

Decision matrix of Expert 1 (R¹), Decision matrix of Expert 2 (R²), Decision matrix of Expert 3 (R³), Decision matrix of Expert 4 (R⁴)

Table 2 Comparison of the ranking

Weights used in MAGDM	Final ranking
Probabilities of the strategies used to attain the upper bound $\bar{v}$ of player I & $\bar{w}$ of player II	$A_2 > A_4 > A_1$
Probabilities of the strategies used to attain the lower bound $\underline{v}$ of player I & $\underline{w}$ of player II	$A_2 > A_4 > A_1$

6   Conclusion

In this work, the ITFPGWA operator is used in aggregating the information given by the experts in a decision-making analysis. The proposed DSS-Miner removes the less important variable from the decision system automatically. The alternatives are ranked according to the highest correlation coefficient of its corresponding intuitionistic trapezoidal fuzzy number which is obtained by the aggregation of the decision matrices, and its respective weights are computed from the interval-valued constraint matrix game using simplex method. In the future, various linear programming techniques can be incorporated for solving matrix games like fuzzy interval-valued matrix game, fuzzy matrix games using ranking techniques, matrix games with fuzzy payoffs to derive the decision-making weights for MAGDM problems.

References

1. Atanassov KT, Gargov G (1989) Interval-valued intuitionistic fuzzy sets. Fuzzy Sets Syst 3:343–349
2. Larbani M (2009) Non-cooperative fuzzy games in normal form: a survey. Fuzzy Sets Syst 160:3184–3210
3. Li DF (2008) Lexicographic method for matrix games with payoffs of triangular fuzzy numbers. Int J Uncertain Fuzziness Knowl Based Syst 16:371–389
4. Li DF (2010) Mathematical-programming approach to matrix games with payoffs represented by Atanassov's interval-valued intuitionistic fuzzy sets. IEEE Trans Fuzzy Syst 18:1112–1128
5. Li DF, Cheng CT (2002) Fuzzy multi objective programming methods for fuzzy constrained matrix games with fuzzy numbers. Int J Uncertain Fuzziness Knowl Based Syst 10:385–400
6. Li DF, Nan JX (2009) A nonlinear programming approach to matrix games with payoffs of Atanassov's intuitionistic fuzzy sets. Int J Uncertain Fuzziness Knowl Based Syst 17:585–607
7. Li DF, Nan JX (2014) Linear programming Technique for solving interval-valued constraint matrix games. J Ind Manag Optim 10:1059–1070
8. Liu PD, Liu Y (2014) An approach to multiple attribute group decision making based on intuitionistic trapezoidal fuzzy power generalized aggregation operator. Int J Comput Intell Syst 7(2):291–304
9. Liu ST, Kao C (2009) Matrix games with interval data. Comput Ind Eng 56:1697–1700
10. Robinson JP, Amirtharaj ECH (2012) A search for the correlation coefficient of triangular and trapezoidal intuitionistic fuzzy sets for multiple attribute group decision making. Communications in computer and Information Sciences-283, Springer, pp 333–342

11. Robinson JP, Amirtharaj ECH (2014) MAGDM-miner: a new algorithm for mining trapezoidal intuitionistic fuzzy correlation rules. Int J Decis Support Syst Technol 6(1):34–59
12. Robinson JP, Jeeva S (2019) Intuitionistic trapezoidal fuzzy MAGDM problems with Sumudu transform in numerical methods. Int J Fuzzy Syst Appl 8(3):1–46
13. Wan JQ, Zhang ZH (2008) Programming method of multi-criteria decision-making based on intuitionistic fuzzy number with incomplete certain information. Control Decis 23:1145–1148
14. Wei GW (2010) Some arithmetic aggregation operators with intuitionistic trapezoidal fuzzy numbers and their application to group decision making. J Comput 5:345–351
15. Xu ZS (2011) Approaches to multiple attribute group decision making based on intuitionistic fuzzy power aggregation operators. Knowl-Based Syst 24:749–760
16. Xu ZS, Yager RR (2010) Power-geometric operators and their use in group decision making. IEEE Trans Fuzzy Syst 18:94–105
17. Yager RR (2001) The power average operator. IEEE Trans Fuzzy Syst Man Cybern-Part A: Syst Humans 31:724–731

Disaster Mitigation Using a Peer-to-Peer Near Sound Data Transfer System

R. Padma Priya⊙, Ritumbhara Bhatnagar⊙, and Shaaran Lakshminarayanan⊙

1 Introduction

When we consider managing and saving crowds in the times of crisis, for instance indoor fire hazards or natural calamities (floods, earthquakes, etc.), the lack of access to crucial information and failure to access help at a concerned area normally happens due to compromised network towers or servers. Additionally, in view of the ongoing COVID-19 pandemic and pre-empting any other pandemic in the near future, we have taken into consideration the concept of social distancing to manage crowds in places of hazards effectively. We aim to solve this problem with the help of largely, near sound data transfer (NSDT) technology along with peer-to-peer (P2P) networking.

Sound is a great medium to transfer data when it comes to proximity, taking in view the high speed and no need of an external setup. This is the reason many industries are using NSDT to transfer crucial data within their own premises.

NSDT concerns itself with transmission of data over near sound frequencies, for instance over ultraviolet rays. This technology has gained popularity over the past three to four years, and such efforts to increase data transfer rate using NSDT have been promising, through one such technology "ChirpCast: Data Transmission via Audio" [1]. Chirp Software Development Kit (SDKs) have made it very easy for developers to send data over audio. Data is provided to the SDKs in the form of an array of bytes [1]. Transmission of the data can be via audible or inaudible ultrasonic audio depending on the configuration of the Chirp SDK [2]. Another breakthrough, LISNR is an advanced, near ultrasonic, ultra-low power data transmission technology that enables fast, reliable, and secure communication between devices via a speaker and/or microphone [3]. A device enabled with LISNR soft-

R. Padma Priya (✉) · R. Bhatnagar · S. Lakshminarayanan
School of Computer Science and Engineering, Vellore Institute of Technology, Vellore, Tamil Nadu, India
e-mail: padmapriya.r@vit.ac.in

ware can detect, transmit, and receive data from LISNR audio technology within audio range. Once a LISNR signal is detected, the SDK demodulates the data sent over the audio and performs the specified function on the device based on the data [3]. LISNR's ultrasonic platform consists of 3 core product optimizations—close range data transmissions called "Point," mid-range transmissions called "Zone" and long range transmissions called "Radius" [3]. NSDT has also disrupted the financial industry, in the form of Google's Tez (Google Pay in India) transferring payment data from one device to another by the use of its speakers and microphone [4] This information is also highlighted in Table 1.

Apple and Google together have developed a contact tracing technology. The two tech giants have used Bluetooth technology to enable the same, using a strong security driven backend [5]. Singapore uses Trace Together to manage contact tracing to evade COVID-19 spread [6]. Whereas, India used Aarogya Setu App for COVID-19 contact tracing and mitigation [7]. India has also effectively proposed a 4Ts approach tracing, tracking, testing, and treating to deal with COVID-19.

We have proposed an effective model for contact tracing and mitigation, which is more accessible to the user, by making use of an amalgamation of NSDT and peer-to-peer networking, so that internet and Bluetooth connectivity never pose a plausible necessity in terms of safety from COVID-19.

This paper is organized as follows. Section 2 portrays the literature review. In Sect. 3, we explain about the methodology used. Two scenarios are proposed, and their application has been discussed in Sect. 4.

2   Literature Review

Over the past few years, milestones have been achieved in both NSDT and P2P networks. Let us discuss a few of these that we have come across in order to receive aid for our own proposition.

Authors in [1] investigated a few balance methods for sending and getting information utilizing sound waves through item speakers and underlying PC mouthpieces. Requiring just that PC clients run a little application, the framework effectively gives powerful room-specific broadcasting at information paces of 200 bit/s.

Table 1 Details of various currently prevailing companies that make use of NSDT technology

Company	NSDT use case	Description
Chirp	Chirp Audio SDK	A SDK which allows people to transfer textual information through sound [2]
LISNR	LISNR Sonic SDK	A SDK which allows people to transfer files, textual information, and security keys through sound [3]
Google	Google Pay (Tez)	Transfers payment data from one device to another by the use of its speakers and microphone [4]

Data in the form of music is proposed in paper [8]. In this work, the authors have proposed a sound information concealing framework that shrouds data into signals not known previously. The framework can shroud information into unrecorded music, or encompassing sounds when all is said in done and can be utilized to convey data acoustically starting with one gadget then onto the next.

Diving deeper into Near Field Communication (NFC) we come across "Dhwani" in paper [9]. They address the test of empowering NFC-like ability on the current base of cell phones. To this end, Dhwani, a novel, acoustics-based NFC framework that utilizes the receiver and speakers on cell phones, accordingly killing the requirement for any specific NFC equipment [9]. Another breakthrough has been achieved through "Acoustic communication system using mobile terminal microphones," where in DoCoMo has built up an information transmission innovation called "Acoustic OFD" that inserts data in discourse or music and communicates those sound waves from an amplifier to a receiver [10].

Use of the ultrasonic spectrum of rays for data transmission as suggested by the authors of paper [11], implements a close ultrasonic correspondence convention in the 18.5–20 kHz band, which is unintelligible to most people, utilizing ware cell phone speakers and mouthpieces to communicate and get signals [11].

When it comes to data security in the light of NSDT, in paper work [12], authors have designed and proposed "EnGarde" (a small system) that can be stuck on the rear of a telephone to give the capacity to stop malevolent communication [12]. EnGarde is altogether detached and collects power through a similar NFC source that it monitors, which makes the equipment plan moderate, and encourages inevitable incorporation with a mobile [12]. L. Deshotels, author of paper [13] actualized an ultrasonic modem for android and found that it could impart signs up to 100 feet away [14]. Moreover, likewise confirmed that this assault is viable with the transmitter within a pocket. Android gadgets with vibrators can deliver short vibrations which make detached sound [14].

If we focus on mobile devices, authors of paper [15] proposed android-based middleware that goes about as the interface point between versatile hubs and higher application layers for versatile pervasive computing [15]. The middleware basically targets supporting and upgrading the conventions for direct P2P correspondence among clients in the troupe versatile climate [15]. Additionally, the paper presents a conversation of the accessible P2P middleware for the versatile climate alongside its applications [15]. Moreover, "Enabling Mobile Peer-to-Peer Networking" presents a P2P file distributing network with an upgraded system for mobile networks [16]. It examines the pertinence of current P2P methods for asset access and intercession with regards to 2.5G/3G portable organizations. They explore a versatile P2P engineering that can accommodate the decentralized activity of P2P document offering to the interests of organization administrators, e.g., control and execution and depends on edonkey convention [16].

In view of usage of P2P networks for disaster management, paper [17] investigated the Geo Collaborative applications utilized for disaster recovery. As catastrophe the executives naturally happens in exceptionally unique conditions, these applications experience the ill effects of lack of a reasonable connection with the server system. The server in this case is the only carrier of information [17]. They propose to effectively use P2P networking in order to interconnect various nodes. Subsequently, one dynamic association between each node and the server room in turn is adequate to sustain the fiasco that has taken place [17]. Authors in [6] have proposed image stitching, google API-based street view retrieval, and distant matrix APIs-based solutions aiding the first responders to save the lives of the victims in the disaster areas. Also in paper [18], authors have explored the Google voice over protocols-based IoT framework for detecting emergency situations in smart office environments.

A huge part of the proposition from the authors of paper work [13] deals with a highly accurate, efficient, and reliable location system. It presents a novel ultrasonic area framework which uses broadband transducers [13]. They depict the transmitter and beneficiary equipment and portray the ultrasonic channel data transmission [13].

Our proposition seeks to aid during pandemic times thus aids in following social distancing norms, for which inspiration is drawn from paper [19], which intends to give an idea to develop networking related to proximity and distance, dependent on high-recurrence sound waves transmitted and caught by the speaker and receiver found on phones [19].

Considering accessibility and ease of usage, wearable devices assist further in passing on crucial information while not making users absent from the real world. Authors of [20] center on the plan and advancement of an ETSI M2M Gateway (GW) on a cell phone, started up in a smartphone [20]. G.E. Santagati and T. Melodia in "U-wear" presented in paper [14], recommend a wearable gadget UWear which comprises of a lot of programming characterized multi-layer functionalities that can be executed on broadly useful preparing units, e.g., chip, microcontrollers or FPGAs, among others, to empower arranged activities between wearable gadgets outfitted with ultrasonic networks [21].

Finally, in paper [22], the authors have examined the essentials of ultrasonic engendering in tissues and investigate significant tradeoffs, including the decision of a transmission recurrence, transmission force, and transducer size [22].

The papers we reviewed, proposed an imminent solution without having a proper directive to the implementation, in addition to that usage of the NSDT and peer-to-peer technology in combination wasn't discussed appropriately. Both of these technologies have a great combined impact in various fields of engineering, especially in disaster mitigation systems. In our paper, we have discussed these technologies and their implementation in detail.

3   Proposed Mitigation Scenarios Using Peer-to-Peer Near Sound Data Transfer System

In this section of the paper, we present the basic implementation of a peer-to-peer system in a mobile device for two scenarios. Exit path retrieval and indoor mitigation is considered as our first scenario. An outdoor environment with COVID-19 victims presence detection is considered as our second scenario. Unlike traditional P2P methods, we have used the "Data Over Sound" technique to transmit data between the individuals simultaneously. The data to be transferred is first transformed into byte sized arrays and then converted into an audio signal, which can be transmitted by any device with a speaker and consequently received by any device with a microphone. As the data over sound technique employs a near ultrasonic sound wave, it is thus possible to transfer data reliably over distances of several meters and even in noisy everyday environments. All the transmissions take place entirely through audio signals (near ultrasonic sound waves) and so, no internet connection or cellular connectivity is required to transmit the data.

In our current scenario, we're exploiting the same technique of Near Sound Data Transfer (NSDT) for disaster management through an app. When a user installs the app for the first time, the user will be prompted with a form where he/she will be required to enter their details. This information is then compressed and stored in the user's device. During a calamity, the app when prompted starts streaming the data to nearby devices forming a mesh network. The data comprises key information such as name, last known location, and phone number to help the rescue team take reliable actions in a much lesser time.

3.1   Exit Path Retrieval and Indoor Mitigation: (Scenario 1)

When a disaster occurs in a closed highly dynamic environment, despite a well laid disaster management plan there's always widespread panic among the people which leads to chaos and stamping. Our proposed solution can greatly aid this problem by providing reliable information towards identifying an optimal exit path in indoor situations to the victims and the volunteers to take necessary actions at a faster pace.

1. The person's GPS coordinates are logged in real time by the app.
2. Anyone in the affected area reaching the exit can share their route to other victims through the app.
3. The app uses a NSDT-based peer-to-peer network to transmit the information to all the victims without any cellular connectivity.
4. In order to reduce chaos and jamming, the app checks for newer routes and transmits the information to the people for better crowd awareness and management.
5. When a person reaches an area with cellular connectivity, the location data along with the data of the victims in the calamity is transferred to the cloud.
6. The victims and the volunteers can thus take faster and reliable actions with the acquired data. Readers are requested to look Figs. 1, 2 and 3 to acquire detailed information on the above mentioned steps visualized through relevant flowchart, system diagram and demo scene captured from the implemented app respectively.

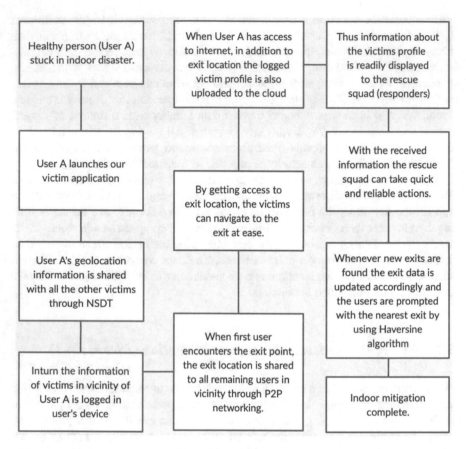

Fig. 1 Flowchart depicting indoor disaster mitigation during a pandemic using the aid of a mobile application. Implemented using NSDT and P2P networking

3.2 Enhancing Biosecurity by Exploiting Near Sound Data Transfer: (Scenario 2)

The 2019 coronavirus sickness brought about by serious intense respiratory condition has risen exponentially influencing all districts of the world. More than 14.3 million cases have been accounted for equivalent to eighteenth July 2020. By every logical means, the world is going through a COVID-19 pandemic. Without any drug mediation, the best way to secure oneself and the others against COVID-19 is to diminish the blending of conceivable irresistible individuals through early ascertainment or decrease of contact with other sound people.

We have proposed a solution that exploits the technique Near Sound Data Transfer (NSDT) to alert a healthy individual whenever a possible vector is in vicinity. The risk analysis for each individual was outsourced from Aarogya

Fig. 2 System diagram depicting disaster mitigation during a pandemic using the aid of a mobile application. Implemented using NSDT and P2P networking

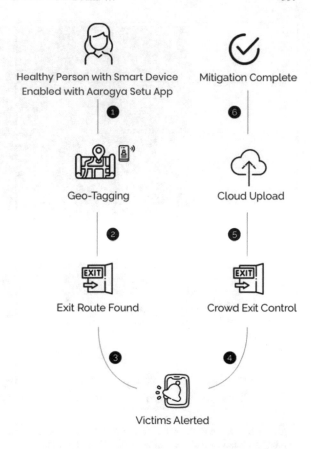

Healthy Person with Smart Device Enabled with Aarogya Setu App

Mitigation Complete

Geo-Tagging

Cloud Upload

Exit Route Found

Crowd Exit Control

Victims Alerted

Setu [23]. Aarogya Setu is a COVID-19 contact tracing app developed, approved and used in India.[1]

4 Conclusion

In this paper, we have come up with an efficient solution to replace cellular communications during a disaster. Near Sound Data Transfer (NSDT) has unique capabilities that provide the potential for it to become a vital device-to-device (D2D) connectivity solution, particularly in times of disaster and calamities. Moreover, the concept of opportunistic networks on top of the NSDT technology holds great potential in building complex offline decentralized systems of the future that would create a greater impact in times of distress.

---

[1] Also readers are requested to look Figs. 4 and 5 to acquire detailed information towards the implementation of our proposed methodology using NSDT.

Fig. 3 Live Demo of the First Responders Application developed, healthy and unhealthy vectors can be identified in real time through the same. Red dots depict "Rescue Squads," blue dots depict victims, green dots are for "Paramedic," whereas orange dots are for "Fire Department" in the place of calamity

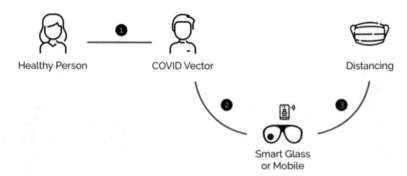

Fig. 4 (i) Healthy person comes in the vicinity of a possible vector. (ii) The mobile app recognizes the vector and sends an alert to the device. (iii) The Healthy Person is alerted and takes the necessary measures

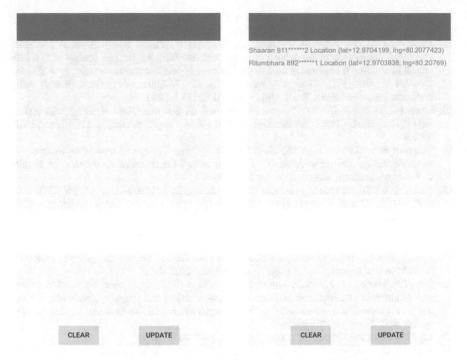

Fig. 5 Application developed on NSDT displaying identities of humans in vicinity of the carrier of the smartphone along with location coordinates. Note: The readers are reminded about the disabled internet connection in the smartphone

References

1. Iannacci F, Huang Y (2015) ChirpCast: data transmission via audio
2. G Elert. 44, 1999 (2003) Frequency range of human hearing. C D'Ambrose. The Physics Factbook, 43
3. Androutsellis Theotokis S, Spinellis D (2004) A survey of peer-to-peer content distribution technologies. ACM Comput Surv 36(4):335–371
4. BBC Article Page, https://www.bbc.com/news/technology-53146360. Last Accessed 2020/09/25
5. Dailt Mail Article, Google Tez uses ultrasonic sound for data transmission, https://www.dailymail.co.uk/sciencetech/article-4896748/Google-Tez-payment-app-uses-ultrasonic-sound.html#:~:text=Google%20has%20unveiled%20a%20new,heard%20by%20t
6. Padmapriya R, Nikhil AS (2016) A novel mobile cloud augmented emergency path finder algorithm using image stitching and google API to guide disaster victims to safety, IJPBS
7. Hazas M, Ward A (2002) September. A novel broadband ultrasonic location system. In: International conference on ubiquitous computing, pp 264–280. Springer, Berlin, Heidelberg
8. Novak E, Tang Z, Li Q (2018) Ultrasound proximity networking on smart mobile devices for IoT applications. IEEE Internet Things J 6(1):399–409
9. Nandakumar R, Chintalapudi KK, Padmanabhan V, Venkatesan R (2013) Dhwani: secure peer-to-peer acoustic NFC. ACM SIGCOMM Comput Commun Rev 43(4):63–74

10. Matsuoka H, Nakashima Y, Yoshimura T (2006) Acoustic communication system using mobile terminal microphones. NTT DoCoMo Tech J 8(2):2–12
11. Gummeson JJ, Priyantha B, Ganesan D, Thrasher D, Zhang P (2013 June) EnGarde: Protecting the mobile phone from malicious NFC interactions. In: Proceeding of the 11th annual international conference on Mobile systems, applications, and services, pp 445–458
12. Getreuer P, Gnegy C, Lyon RF, Saurous RA (2017) Ultrasonic communication using consumer hardware. IEEE Trans Multimedia 20(6):1277–1290
13. Lazic N, Aarabi P (2006) Communication over an acoustic channel using data hiding techniques. IEEE Trans Multimedia 8(5):918–924. https://doi.org/10.1109/TMM.2006. 879880
14. Santagati GE, Melodia T (2015, May) U-wear: software-defined ultrasonic networking for wearable devices. In: Proceedings of the 13th annual international conference on mobile systems, applications, and services, pp 241–256
15. Deshotels L (2014) Inaudible sound as a covert channel in mobile devices. In: 8th {USENIX} workshop on offensive technologies ({WOOT} 14). url: https://www.usenix.org/conference/woot14/workshop-program/presentation/deshotels
16. Chirp Wikipedia HomePage. https://en.wikipedia.org/wiki/Chirp_(company. Last Accessed 2020/09/10
17. Apple News, https://www.apple.com/in/newsroom/2020/04/apple-and-google-partner-on-covid-19-contact-tracing-technology/, Last Accessed 2020/09/28
18. Priya RP, Marietta J, Rekha D, Mohan BC, Amolik A (2019) IoT-based smart office system architecture using smartphones and smart wears with MQTT and Razberry. In: Proceedings of the 2nd international conference on data engineering and communication technology, pp 623–632. Springer, Singapore
19. LISNR HomePage. https://lisnr.com/. Last Accessed 2020/09/10
20. Santagati GE, Melodia T, Galluccio L, Palazzo S (2013) Ultrasonic networking for e-health applications. IEEE Wirel Commun 20(4):74–81
21. Oberender JO, Andersen FU, de Meer H, Dedinski I, Hoßfeld T, Kappler C, Mäder A, Tutschku K (2004 June) Enabling mobile peer-to-peer networking. In: International workshop of the EuroNGI network of excellence, pp 219–234. Springer, Berlin, Heidelberg
22. Kellerer W, Schollmeier R, Wehrle K (2005) Peer-to-Peer in mobile environments. In: Steinmetz R, Wehrle K (eds) Peer-to-Peer systems and applications, pp 401–417
23. Livemint Article, https://www.livemint.com/ai/artificial-intelligence/how-aarogya-setu-app-works-and-how-it-helps-fight-covid-11594512597402.html, Last Accessed 2020/09/28

Rainfall-Based Crop Selection Model Using MapReduce-Based Hybrid Holt Winters Algorithm

V. Kaleeswaran, S. Dhamodharavadhani, and R. Rathipriya

1 Introduction

Water is essential for all life processes, cannot be an alternative, and is used for transportation; it is a source of energy and serves many useful purposes like agriculture and industry-related activities. Rain is the major source of irrigation in the India, and it is directly connected with the crop production. Yields of crops, especially in rain-fed region, depend on rainfall patterns. Therefore, it is very important to predict rainfall quantity from historical rainfall data using statistical methods [1]. Another important factor is soil moisture that purely depends on the rainfall. It also play vital role in the crop production. Therefore, rainfall prediction is the most important area and needy tool. It has very good impact in determining the food production. The most popular and frequently used data mining algorithms are K-means, C4.5, SVM, EM, Apriori, AdaBoost, PageRank, Naive Bayes, KNN, and CART. The authors of [2] provided detailed explanations of each of these methods and discussed their impact on current and future research in these methods. The rainfall forecasting for boyolai, central java, Indonesia is made by using the Z-Score model and Winters triple exponential smoothing. The authors predicted the weather based on climate classification using the Oldman method [3]. The weather-based crop yield prediction model was proposed in [4] to forecast the yield of 12 major California crops. But very few crop yields were predicted with high accuracy, suggesting that crop yield variability should be captured for more accurate forecasting. The time series analysis using HW algorithm is used to identify level, trend, and seasonality in the rainfall dataset and forecast the districtwise rainfall in Tamilnadu, India, based on these information [5]. In this paper, forecasting the districtwise rainfall using hybridization of KNN with MapReduce-based HW. MapReduce framework is used here for the effective initialization process of HW algorithm iteratively. The experimental study is designed with different initializa-

V. Kaleeswaran · S. Dhamodharavadhani · R. Rathipriya (✉)
Department of Computer Science, Periyar University, Salem, India

© The Author(s), under exclusive license to Springer Nature Singapore Pte Ltd. 2022
R. R. Raje et al. (eds.), *Artificial Intelligence and Technologies*,
Lecture Notes in Electrical Engineering 806,
https://doi.org/10.1007/978-981-16-6448-9_37

tion parameters like alpha, beta, and gamma of MapReduce-based HHW. The results showed that these parameters have a greater impact on the forecasting quality [6].

2 Methods and Materials

2.1 KNN

K-nearest neighbors (KNN) algorithm uses object or data similarity to predict the new data, i.e., a value is assigned to the predicted data based on how closely it matches the data in the training set [7]. The steps of the algorithm are given as following Table 1.

2.2 Holt Winter Algorithm (HW)

The Holt Winters algorithm smooth's a time series data for forecasting the values [8, 9]. The work flow of the HW is shown in Fig. 1

Where α, β, and γ are smoothing constants.

t is the time period, Y_t is the actual observed values, S is the length of seasonality.

L is the level component, M_t is the trend component, S_t is the seasonal component.

F_{t+r} is the forecast for t periods ahead.

Table 1 KNN algorithm steps	
	Step: 1 Load the rainfall database. Divide the database into training and test datasets
	Step: 2 Initialize the K value
	Step: 3 For each data in the test data do
	(a) Calculate the distance between the test data and each record of the training data using distance measurements such as Euclidean distance
	(b) Arrange the distance value in the ascending order
	(c) Select the top K value from the sorted vector
	(d) Finally, assign a class to the test data
	Step: 4 Stop the process

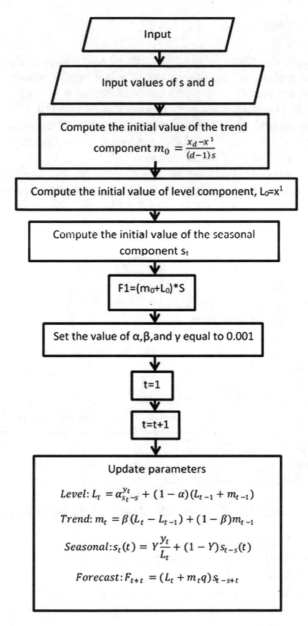

Fig. 1 Holt Winter method

2.3   MapReduce

Google introduced MapReduce paradigm, a distributed data processing framework. It is useful for processing large amounts of data in a parallel, reliable, and efficient way in cluster environments. It works by breaking down every process into two stages. It consists of two phases, namely Map and Reduce [10, 11].

Map() function reads the data block from the given files. It processed data block and generates the key-value pairs as intermediate outputs. The output of a Mapper is given as the input to the Reducer function [10, 11].

In Reducer phase, the Reducer gets input as the key-value pair from multiple map tasks [10, 11]. Then, the Reducer aggregates those intermediate data tuples into a smaller set of key-value pairs which is the final output. Holt winter Map Reduce pseudocode as shown in Fig. 2.

```
Function MapReduce_Holt_Winter (data_files)
Key: District
Value: Predict Values
Rain_data = datastore (rainfall dataset)
Map_output = MapReduce (Rain_data)
return Map_output
-----------------------------------------------------------------------------------
Function Map (data)
Key: District
Value: Monthly Rainfall for each district
add (mkey, mval)            // Generate Key Value Pairs (one for each record)
-----------------------------------------------------------------------------------
Function Reduce ({mkey, mval})
Key: District
Value: Predicted values of the district

Combine the values {mval1, mval2...} of each unique key (mkey) together

If data has No Trend and No Season
        Apply the Simple Exponential Smoothing
Else if data has No Season
        Apply the Holt's Linear Trend Exponential Smoothing
Else
Then
        Apply the Holt's Winter Seasonal Exponential Smoothing
        Predicted values of all months in each district
add (district, {predict_Y and measures})
// Generate Key Value Pairs (one for each district)
-----------------------------------------------------------------------------------
```

Fig. 2 Holtwinter MapReduce pseudocode

2.4 Performance Measures

Table 2 shows the Performances measures and its descriptions.

3 Proposed Methodology: Rainfall-Based Crop Selection Model

The proposed model is developed for the crop section based on the weather data like temperature and rainfall. The dataset taken for sowing is unsupervised data, which is converted into supervised data by applying the thumb rule of different crop and KNN. The thumb rule for crop selection is mentioned in Fig. 2.

The steps of the proposed MapReduce-based hybrid Holt Winter algorithm with KNN are shown in Fig. 3. The implementation of MapReduce-based HHW algorithm for rainfall dataset is done with the support of the MATLAB 2019 toolbox. It is used to predict the rainfall from time series districtwise rainfall dataset of Tamilnadu, India. In MATLAB, MapReduce() function has input arguments, namely Datastore, Map() function, and Reduce() function.

- A datastore is a storage variable for storing the rainfall dataset in the MATLAB toolbox.
- A Map () reads the rainfall dataset from the given input files and stored in the data store for further processing. It processed these data into key-value pairs where the district is the key; monthly rainfall is the value stored against each district. This <key-value> pair is called intermediate output. This intermediate output is provided as input to the next phase (i.e., Reducer() function). Every chunk of the data store called Map() function once and worked independently and parallelly.
- A Reducer() function calls the HW algorithm to predict the monthly rainfall by district from the intermediate <district, monthly rainfall> pairs of the Map (); outputs the final <district, monthly rainfall> pairs by aggregating the monthly rainfall against the unique key value. Typically, the Reducer() function is called once for each district in the unique key list and outputs the predicted monthly rainfall in <Key, Value> pair format (Fig. 3 and Table 3).

Crop prediction techniques like logistic regression and Holt Winter method are applied for preprocessed data to identify suitable crop planting or sowing for forecasting.

Table 2 Performances measures and its descriptions

Performance measures	Descriptions	Formula
RMSE	Root mean squared rrror (RMSE) is a polynomial scoring rule which also measures the error average size. This is the square root of the square differences measured between prediction and actual observation [12–15]	$RMSE = \sqrt{\frac{1}{n}\sum_{i=1}^{n}(P_i - A_i)^2}$ It represents as n = number of samples, P = predicted values, A = actual values
MAE	Mean squared error (MAE) means average of all absolute errors [12–15]	$MAE = \frac{1}{n}\sum_{i=1}^{n}|x_i - x|$ It represents as n = number of errors, $|x_i - x|$ = absolute error
MAPE	The mean absolute percentage error (MAPE) is the most common measure used to forecast error and works best if there are no extremes to the data (and no zeros) [12–15]	$MAPE = \frac{1}{n}\sum_{i=1}^{n}\left|\frac{(P_i - A_i)}{P_i}\right|$ It represents as n = number of samples, P = predicted values, A = actual values, t = time period
MAD	The mean absolute deviation is the mean difference between the each data point. It provides us with an idea of the variability in a dataset [12–15]	$MAD = \frac{\sum |x_i - \bar{x}|}{n}$ It represents as n = number of errors, $|x_i - x|$ = absolute error

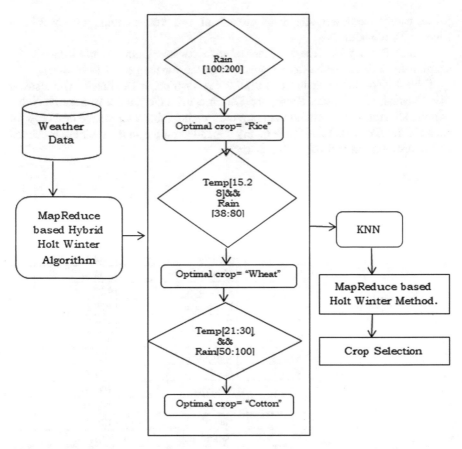

Fig. 3 MapReduce-based hybrid Holt Winter algorithm

Table 3 Hybrid Holt-Winter algorithm	Input: Rainfall, Temperature
	Output: Districtwise crop selection
	Step1: Generate thumb rule for crop selection using rainfall and temperature values
	Step2: Predicted rainfall = MapReduce (rrainfall dataset)
	Step3: Crop selection = KNN (predicted rainfall, temperature)
	Step4: Out the crop selection for each district

3.1 Results and Discussion

The predicted rain value is calculated using MapReduce-based Holt Winter algorithm. It is used for selecting the best crop for higher yield like rice, wheat, maize, jowar, barely, sunflower, sugarcane groundnut, tea, coffee, cotton, jute for 29 districts in Tamilnadu.

Dataset Description: The 11 years of districtwise monthly rainfall dataset for Tamilnadu, India, is taken from www.data.gov.in for the period 1992–2002.

Table 4 displays the predicted rainfall for 29 districts in Tamilnadu. The districts like Chennai, Coimbatore, Karur, and Dharmapuri have rainfall in negative value which indicating lower rainfall than previous years. Table 5 shows the best crop for each district. The crop selection is purely depended on the amount rainfall predicted and temperature in that respective district.

Table 4 Crop predictions using Holt Winters method

S. No.	District	Predicted values
1	Ariyalur	147.4
2	Chennai	−52.7
3	Coimbtore	−36
4	Cuddalore	42.3
5	Dharmapuri	−29.8
6	Dindigul	179
7	Erode	49.3
8	Kanchipuram	88.7
9	Karur	−74.8
10	Madurai	106.86
11	Nagapatinam	293.91
12	Namakkal	121.86
13	Perambalur	59.11
14	Pudukottai	15.07
15	Ramanathapuram	112
16	Salem	108.05
17	Sivagangai	173.5
18	Thanjavur	177.9
19	Nilgiris	261.3
20	Theni	11.83
21	Thiruvallur	90.65
22	Thoothukkudi	169.15
23	Tiruchirapalli	204.44
24	Tirunelveli	10.98
25	Tiruvannamalai	152.59
26	Vellore	201.12
27	Villuppurm	14.32
28	Virudhunagar	82.05
29	Thiruvarur	169.15

Table 5 List of districtwise best crop selection

S. No.	District	Crop selection
1	Ariyalur	Rice, Sunflower, Jute
2	Cuddalore	Sunflower, Wheat, Jowar
3	Dindigul	Rice, Tea, Coffee
4	Erode	Rice, Sunflower, Wheat, Jowar
5	Kanchipuram	Sunflower, Jowar, Maize, Groundnut, Cotton, Sugarcane
6	Madurai	Rice, Sunflower, Sugar
7	Nagapatinam	
8	Namakkal	Rice, Sunflower, Jute, Sugar
9	Perambalur	Sunflower, Jowar, Wheat, Groundnut, Cotton
10	Ramanathapuram	Rice, Sunflower, Sugarcane
11	Salem	Rice, Sunflower, Sugarcane
12	Sivagangai	Rice
13	Thanjavur	Rice
14	Nilgiris	Tea, Coffee
15	Thiruvallur	Sunflower, Jowar
16	Thoothukkudi	Rice
17	Tiruchirapalli	Rice, Sunflower, Sugarcane
18	Tirunelveli	Rice
19	Tiruvannamalai	Rice
20	Vellore	Rice, Sunflower, Sugarcane
21	Villuppurm	Sunflower, Jowar

$$\text{Level} : L_t = \alpha \frac{y_t}{s_{t-s}} + (1 - \alpha)(L_{t-1} + m_{t-1})$$

$$\text{Trend} : m_t = \beta(L_t - L_{t-1}) + (1 - \beta)m_{t-1}$$

$$\text{Seasonal} : s_t(t) = \gamma \frac{y_t}{L_t} + (1 - \gamma)s_{t-s}(t)$$

$$\text{Forecast} : F_{t+r} = (L_t + m_t q)s_{t-s+r}$$

3.2 Performance Analysis

The quality measurements like mean squared error (MSE), root square error (RMSE), mean absolute deviation (MAD), and mean absolute percentage error (MAPE) of Holt Winters method for prediction of rainfall is given in Table 2. Usually, RMSE is treated as the key performance indicator for prediction algorithm. The RMSE value of the MapReduce-based Holt Winter algorithm is 4.7 which is

Table 6 Performance
measurements

Error measurements	Value
Mean squared error (MSE)	22.3067
Root square error (RMSE)	4.7230
Mean absolute deviation (MAD)	0.0012
Mean absolute percentage error (MAPE)	0.0012

very low. Therefore, it is concluded that the proposed crop selection model is performing well in identifying the right crops for each district. It automatically increases the yield and profit (Table 6).

4   Conclusion

The rainfall-based crop prediction model using MapReduce-based hybrid Holt Winter algorithm has been presented in this paper. The proposed model developed in the MATLAB toolbox. It offered promising forecasting results for districtwise crop selection based on rainfall and temperature factors. In future, this work is extended with different optimized machine learning techniques to provide highly accuracy result.

Acknowledgements The authors acknowledges the UGC—Special Assistance Programme (SAP) for the financial support to her research under the UGC-SAP at the level of DRS-II (Ref.No. F.5- 6/2018/DRS-II (SAP-II)), July 26, 2018, in the Department of Computer Science, Periyar University.

References

1. Gallus WA (2010) Application of object-based verification techniques to ensemble precipitation forecasts. Weather Forecast 25(1):144–158
2. Ramakrishnan N (2009) C4.5. The top ten algorithms in data mining chapman & hall/CRC data mining and knowledge discovery series, pp 1–19
3. Hartomo KD, Prasetyo SYJ, Anwar MT, Purnomo HD (2019) Rainfall prediction model using exponential smoothing seasonal planting index (ESSPI) for determination of crop planting pattern. In: Computational intelligence in the internet of things advances in computational intelligence and robotics, pp 234–255
4. Lobell DB, Cahill KN, Field CB (2006) Weather-based yield forecasts developed for 12 California crops. Calif Agric 60(4):211–215
5. Sellam V, Poovammal E (2016) Prediction of crop yield using regression analysis. Indian J Sci Technol 9(38)
6. Prathyusha A (2018) A survey on prediction of suitable crop selection for agriculture development using data mining classification techniques. Int J Eng Technol 7(3.3):107
7. Al-Hafid DMS, Al-Maamary GH (2012) Short term electrical load forecasting using holt-winters method. AL-Rafdain Eng J (AREJ) 20(6):15–22

8. Drašković D, Radojičić V, Dobrodolac M (2019) Short-term forecasting of express service volume using Holt-Winters model. Tehnika 74(5):697–703
9. Baker RJ (2020) The application of selection indices in crop improvement. In: Selection indices in plant breeding, pp 13–22
10. Dhamodharavadhani S, Rathipriya R (2018) Region-wise rainfall prediction using mapreduce-based exponential smoothing techniques. In: Advances in intelligent systems and computing advances in big data and cloud computing, pp 229–239
11. Malewicz G (2011) Beyond MapReduce. In: Proceedings of the second international workshop on MapReduce and its applications—MapReduce '11
12. Dhamodharavadhani S, Rathipriya R (2021) Novel COVID-19 mortality rate prediction (MRP) model for India using regression model with optimized hyperparameter. J Cases Inf Technol (JCIT) 23(4):1–12. http://doi.org/10.4018/JCIT.20211001.oa1
13. Dhamodharavadhani S, Rathipriya R (2016) A pilot study on climate data analysis tools and software. In: 2016 Online international conference on green engineering and technologies (IC-GET), pp 1–5. http://doi.org/10.1109/GET.2016.7916863
14. Devipriya R, Dhamodharavadhani S, Selvi S (2021) SEIR model FOR COVID-19 Epidemic using DELAY differential equation. J Phy Conf Scr 1767(1):012005
15. Dhamodharavadhani S, Rathipriya R (2020) Forecasting dengue incidence rate in Tamil Nadu using ARIMA time series model. In: Machine learning for healthcare, pp 187–202
16. Sivabalan S, Dhamodharavadhani S, Rathipriya R (2020) Arbitrary walk with minimum length based route identification scheme in graph structure for opportunistic wireless sensor network. In: Swarm intelligence for resource management in internet of things, pp 47–63

IoT-Based Sheep Guarding System in Indian Scenario

Sudheer Kumar Nagothu and Chilaka Jayaram

1 Introduction

Sheep farming is a common practice in rural India. Each sheep costs around 10,000–15,000 INR, which is almost equal to 1/10th of Indian per capita income. When one of this animal is stolen or when it is lost, it will cause huge financial burden to the farmer. Monitoring these animals is an important task, which is fulfilled here using GPS, GPRS, and RFID.

The circuit attached to the leader of the mob is shown in Fig. 1. As shown in Fig. 2, the circuit consists of GPS, WiFi, and RFID reader. The power required to operate the circuit can be supplied with the help of a battery. All the animals in the mob except the leader consist of a passive RFID. The RFID reader provided on the leader's neck is used to read the RFID information about all the remaining animals.

2 Literature Survey

The GPS and GPRS integrated circuit can be used in various applications [1–3] such as taking attendance of students or monitoring people and guidance to cyclist. Certain investigation has also included weather data for watering the crop using GPRS [4]. To improve the accuracy of GPS, Kalman filter algorithm can be used [5, 6]. GPS and GPRS system can be used in applications such as toll collection, as an aid to partially vision impaired people, in smart watering system and as driving aid using Doppler sensor [7–10]. GPS and GPRS system is used as an anti-theft alerting system for animals and to monitor them [11].

S. K. Nagothu (✉) · C. Jayaram
RVR & JC College of Engineering, Chowdavram, Guntur, Andhra Pradesh 522019, India
e-mail: nsudheerkumar@rvrjc.ac.in

© The Author(s), under exclusive license to Springer Nature Singapore Pte Ltd. 2022   375
R. R. Raje et al. (eds.), *Artificial Intelligence and Technologies*,
Lecture Notes in Electrical Engineering 806,
https://doi.org/10.1007/978-981-16-6448-9_38

Fig. 1 System circuit

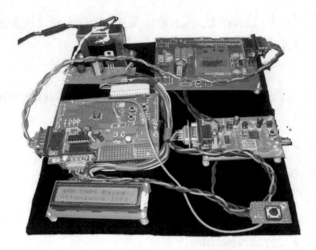

Fig. 2 Block diagram

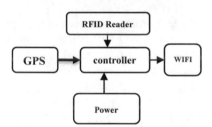

In the current research, an idea of monitoring and guarding the sheep using IoT is presented. The system can also be implemented using GPRS. A subscriber identifier module can be used to identify the animal in mob. Instead of using SIM for each animal in this research, we are using a RFID for each animal. The RFID is more economical and robust compared to subscriber identifier module. With evolvement of internet of things, it has been used in various applications [12–16]. Animals monitoring is one among them. With the help of IOT systems being used in animal guarding system, the burden of farmers can be reduced.

3 Working

A situational working model is shown in Fig. 3. The model shows animals in the mob are fitted with a passive RFID, and the leader is fitted with a GPS, WIFI, and RFID reader card which is shown in Fig. 1. The leader consists of an RFID reader, which will collect the information of all other animals through their passive RFID. When some animal misses from the mob, an alert will be sent to the owner about missing animals.

Fig. 3 Situational working models

A passive RFID is shown in Fig. 4. It will be stapled to the ear of sheep as shown in Fig. 5. RFID is a 16 digit number, and generally, they are unique. The passive RFID generally gets the power from RFID reader antenna.

Initially, danger areas such as roads, ponds, or agricultural field information can be entered into the server through their latitude and longitude. Always, the position of the mob (leader position) is compared with the stored information; when the mob enters these dangerous areas, an alert will be sent to the owner or shepherd for precautionary measures.

The animal details with their RFID information are stored in the database. When the leader tells about a missing animal with their RFID, it is possible to know the complete details as shown in Table 1. It is also possible to add a photo of the animal and its iris information. With the help of RFID, all the animal information in the mob will be sent to the server through WIFI [17]. This data is cross checked with pre-stored information. When information about a particular animal is not received in the past 30 min, an alert will be sent to the shepherd.

Fig. 4 RFID ear tag

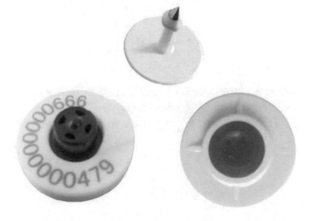

Fig. 5 RFID ear tag on sheep
ear

Table 1 Animal details

Sl. No	Name	Value
1	Animal RF ID	9,502,611,438,939,166
2	AGE	1.5 years
3	Date of birth	11/4/2015
4	Bread	Nellore
5	Male/Female	Female

Figure 6 gives a glimpse of working of model. The reader antenna generates the RFID signal which will hit the passive RFID on animals' ear. The passive RFID reflects the signal with a strength relatively less than reader antenna. The system working is shown in Fig. 7 in terms of flow diagram. A database of dangerous areas such as ponds, lakes, and highways will be created. The sheep mob details are collected from the leader, and if there is any missing animal, the shepherd will be alerted. Haversine formula (shown below) is used to compute the distance and angle between the current position and stored information, and when the distance is below 50 m, an alert will be sent to the shepherd.

$$a = \sin^2\left(\frac{\varphi_2 - \varphi_1}{2}\right) + \cos(\varphi_2) * \cos(\varphi_1)\sin^2\left(\frac{\lambda_2 - \lambda_1}{2}\right)$$

$$d = 2 * R_E * a\tan 2\left(\frac{\sqrt{a}}{\sqrt{(1-a)}}\right)$$

where λ_1, φ_1 is longitude and latitude of leader animal position, λ_2, φ_2 is longitude and latitude of dangerous areas position and R_E is radius of the earth (around 6,371,000 m), and all angles are in radians. Distance in meters is given by d.

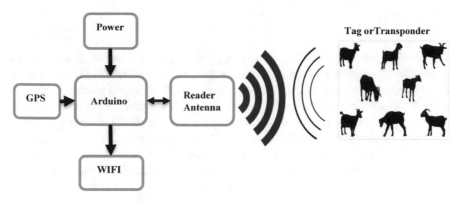

Fig. 6 RFID reader antenna getting information from Mob

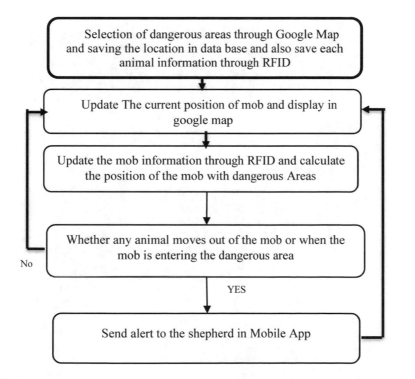

Fig. 7 Flow chart

4 Results

The position information of the mob in terms of latitude and longitude can be used to visualize the position of the mob using goggle earth as shown in Fig. 8. The app is implanted in Blynk platform [18]. To access the Blynk app, a generated token id is used and with help of the token the current position of mob in terms of latitude and longitude will be send to app. The program will check the current position of the mob with dangers areas which are already entered in to database. If the distance is less than 50 m, an alert will be send in terms of red LED in app as shown in Fig. 8. The system will update the app at every 10 min. From Fig. 8, it can be observed that there is a field for missing animals which will be filled with number of animals missed. If there are no missing animals, this field will be blank.

Fig. 8 App for sheep monitoring

5 Conclusion

Here, an idea is proposed to safeguard the sheep from missing or entering into a dangerous area with the help of GPS, WIFI, and RFID. By continuously monitoring the mob with the help of Google Earth maps, the burden of shepherd is reduced. It is also possible to take precautionary measures by the shepherd with the help of alert messages being received. Using Blynk mobile application, the farmer can continuously monitor the animals. A low cost and simple setup is designed to help the shepherd to monitor the animals. The system has certain limitations such as, animals with passive RFID should be in the vicinity of active RFID. If they are hided under bushes or something similar, it will be problematic to find the presence of animals.

References

1. Nagothu SK, Kumar OP, Anitha G (2014) Autonomous monitoring and attendance system using inertial navigation system and GPRS in predefined locations. In: 2014 3rd International conference on eco-friendly computing and communication systems, Mangalore, pp 261–265. https://doi.org/10.1109/Eco-friendly.2014.60
2. Nagothu SK, Anitha G, Annapantula S (2014) Navigation aid for people (joggers and runners) in the unfamiliar urban environment using inertial navigation. In: 2014 sixth international conference on advanced computing (ICoAC), Chennai, pp 216–219. https://doi.org/10.1109/ICoAC.2014.7229713
3. Nagothu, SK, Kumar OP, Anitha G (2016) GPS aided autonomous monitoring and attendance system. Procedia Comput Sci 87:99–104, ISSN 1877-0509, https://doi.org/10.1016/j.procs.2016.05.133
4. Nagothu SK (2016) Weather based smart watering system using soil sensor and GSM. In: 2016 World conference on futuristic trends in research and innovation for social welfare (Startup Conclave), Coimbatore, pp 1–3. https://doi.org/10.1109/STARTUP.2016.7583991
5. Rengarajan M, Anitha G (2013) Algorithm development and testing of low cost way point navigation system. Eng Sci Technol: An Int J 3(2):411–414
6. Rajaduraimanickam K, Shanmugam J, Anitha G (2005) ADDR-GPS data fusion using Kalman filter algorithm. In: 24th Digital avionics systems conference, vol 2, p 7. https://doi.org/10.1109/DASC.2005.1563447
7. Nagothu SK (2016) Automated toll collection system using GPS and GPRS. In: 2016 International conference on communication and signal processing (ICCSP), Melmaruvathur, Tamilnadu, India, pp0651–0653. https://doi.org/10.1109/ICCSP.2016.7754222
8. Nagothu SK, Anitha G (2016) INS—GPS integrated aid to partially vision impaired people using Doppler sensor. In: 2016 3rd international conference on advanced computing and communication systems (ICACCS), Coimbatore, pp 1–4. https://doi.org/10.1109/ICACCS.2016.7586386
9. Nagothu SK, Anitha G (2017) Low-Cost smart watering system in multi-soil and multi-crop environment using GPS and GPRS. In: Proceedings of the first international conference on computational intelligence and informatics, vol 507 of the series. Advances in intelligent systems and computing, pp 637–643. https://doi.org/10.1007/978-981-10-2471-9_61
10. Nagothu SK, Anitha G (2015) INS-GPS enabled driving aid using Doppler sensor. In: 2015 International conference on smart sensors and systems (IC-SSS), Bangalore, pp 1–4. https://doi.org/10.1109/SMARTSENS.2015.7873619

11. Nagothu SK (2018) Anti-theft alerting and monitoring of animals using integrated GPS and GPRS in Indian scenario. Pak J Biotechnol 15(Special Issue I):56–58
12. Nóbrega L, Tavares A, Cardoso A, Gonçalves P (2018) Animal monitoring based on IoT technologies. In: Proceedings of the IoT vertical and topical summit on agriculture, Tuscany, Italy, 8–9 May 2018
13. Nagothu SK (2021) GPS aided hassle-free cycling using sensors. In: Arunachalam V, Sivasankaran K (eds) Microelectronic devices, circuits and systems. ICMDCS 2021. Communications in computer and information science, vol 1392. Springer, Singapore. https://doi.org/10.1007/978-981-16-5048-2_35
14. Nagothu SK (2022) Smart student assessment system for online classes participation. In: Iyer B, Ghosh D, Balas VE (eds) Applied information processing systems. Advances in intelligent systems and computing, vol 1354. Springer, Singapore. https://doi.org/10.1007/978-981-16-2008-9_51
15. Nagothu SK (2021) Intelligent control of aerator and water pump in aquaculture using fuzzy logic. In: Arunachalam V, Sivasankaran K (eds) Microelectronic devices, circuits and systems. ICMDCS 2021. Communications in computer and information science, vol 1392. Springer, Singapore. https://doi.org/10.1007/978-981-16-5048-2_13
16. Nagothu SK (2021) ANFIS based smart wound monitoring system. In: Arunachalam V, Sivasankaran K (eds) Microelectronic devices, circuits and systems. ICMDCS 2021. Communications in computer and information science, vol 1392. Springer, Singapore. https://doi.org/10.1007/978-981-16-5048-2_15
17. https://www.exportersindia.com/beijing-raybaca-technology-co-ltd/electronic-animal-button-ear-tag-beijing-china-1841729.html
18. https://www.blynk.io/en

Comparative Analysis of Wireless Communication Technologies for IoT Applications

B. Shilpa⊚, R. Radha, and Pavani Movva

1 Introduction

In 1999, Kevin Ashton coined the word "Internet of things" when he was connecting various devices in supply chain management using RFID tags [1]. But the concept of connected devices is way back from 1832 as when the first electromagnetic telegraph was designed. IoT history was started after the invention of major component Internet in 1960s; thereafter, the growth of IoT gained rapid action over the next decades. Presently, we are living in the era of Internet of things, where each and every activity of living and non-living things are tracked and performed with necessary action. Internet of things is characterized in several ways by various researchers. Hear, two of simple and common definitions are: Vermesan et al. [2] described the Internet of things as an interface between the physical and the digital worlds. The digital world interacts with the physical world via a plethora of sensors and actuators. Another definition by Pena-Lopez et al. [3] describes the Internet of things as a concept in which computing and networking abilities are incorporated in any entity imaginable. With the use of internet of things, we are able to perform the complex task by simple objects which are combined with intelligence. Sensors, actuators, pre-processors, and transceivers are used to provide connectivity and intelligence in IoT devices [4].

B. Shilpa (✉) · P. Movva
Department of Electronics and Communication, IcfaiTech, Shankarpally, Hyderabad, India
e-mail: b.shilpa@ifheindia.org

P. Movva
e-mail: pavanimovva@ifheindia.org

R. Radha
School of Computer Science and Engineering, Vellore Institute of Technology, Chennai, India
e-mail: rradha@vit.ac.in

© The Author(s), under exclusive license to Springer Nature Singapore Pte Ltd. 2022   383
R. R. Raje et al. (eds.), *Artificial Intelligence and Technologies*,
Lecture Notes in Electrical Engineering 806,
https://doi.org/10.1007/978-981-16-6448-9_39

IoT is not a single innovation, it is a compound of different technologies and distributed among various industries. But all the IoT interventions share the same goal regardless of the application. Those unique goals are minimizing the power requirement, decreasing the complexity of design, minimizing the cost and area, effective utilization of resources and increasing overall efficiency of the system [5]. To enable IoT devices with all the mentioned requirements, various wireless technologies are emerged. As per the standard definitions given by IEEE, Wireless Personal Area Networks (WPAN), Low Rate Wireless Personal Area Networks (LRWPAN), Wireless Body Area Networks (WBAN), and Wireless Local Area Networks (WLAN) are classified as short distance protocols because of their most range of one kilo meter. While Wide Area Networks (WAN) and Low-Power Wide Area Networks (LPWAN) comes under long-range technologies as they support ranges of more than 1 km [6].

Short-range technologies are employed by critical IoT applications which have the requirements of low latency and high availability, where battery life is not considered as major parameter. In this paper under short-range technologies, Radio Frequency IDentification (RFID), Near Field Communication (NFC), Bluetooth Low Energy (BLE), Wireless HART, Zigbee (IEEE 802.15.4), Z-wave, IEEE 802.11ah (WiFiHalow), and ANT are investigated. Long-range technologies are ideal for massive IoT applications require broad coverage and long battery life. Under long-range technologies, LTE MTC Cat M1, EC-GSM-IoT, NB-IoT, 5G, LoRa, Sigfox, Weightless, Ingenu, and DASH7 are investigated.

2 Short-Range Wireless Technologies

Short-range technologies are used for applications which require connectivity within small area of coverage. These technologies are favourable to local area networks as these are easy to deploy, configure and to maintain. The features given by short-range technologies are high data rates, high availability, and occasional latency.

As per IEEE standards [7], we have various categories of short-range technologies like:

Wireless Personal Area Network (WPAN):

This kind of network is preferable for sending information over short distances which includes private, familiar class of devices. It establishes the network with devices surrounding the human body. WPANs are used in personal or wearable devices for implementing easy connectivity and power efficient solutions.

The IEEE 802.15.1 standard is defined for WPAN. One of the maximum used WPAN technology is Bluetooth. Its architecture and operation of physical, MAC layers are specified by this standard. Bluetooth has several versions from V1.0 to V5.0. In 2010, Bluetooth Low Energy (BLE) was launched with greater and more suitable capacities for devices with low power [8].

Low Rate Wireless Personal Area Network (LRWPAN):

The IEEE 802.15.4 standard specified for LRWPAN. LRWPAN represents low cost, very low-power, short-range wireless communications. It insists on easy installation, data reliability, extreme low cost, and decent battery life while being versatile protocol. By using this standard as base, many short-range technologies like Zigbee, Wireless HART, and 6LoWPAN are designed to operate within range of 10 m with 250 kbps of data rate.

Wireless Local Area Network (WLAN):

WLAN is a network of two or more devices connected using wireless communication to construct a Local Area Network (LAN). LAN referred to areas under particular limits such as a house, office, school, or any other building. The users in LAN are free to move in that limited area and stay connected to network by using WLAN technologies. WLANs can extend the area of coverage by deploying the gateways.

The IEEE 802.11 standard specifies the physical and MAC layer protocols for implementing WLAN. This is the utmost popular widely used standard for accessing Internet wirelessly. The primary standard was launched in 1997 and later supported with amendments and gained popularity as WiFi standard. Among the family of 802.11 standards, IEEE 802.11ah is currently most popular, as it defines the WLAN system working at sub-1 GHz license free bands. This standard is being used in large-scale sensor networks because the transmission range is improved comparatively with traditional standard [9].

2.1 Comparison of Short-Range Wireless Technologies

The available wireless technologies for short-range communication are investigated, and analysed. Every technology is having different features compare to others [10]. These technologies are compared by considering the major parameters like operating frequency, data rate, range, and power consumption in order to elect the best performed technology for the requirements of application. The comparisons are presented in Table 1.

3 Long-Range Wireless Technologies

In this section, we elaborate an outline of the wireless long-range technologies that currently exist [11, 12]. It is not fair to categorize those based on their exclusive or inclusive features because some may be member of both aspects, as LoRa does. Rather, we suggest collecting them focusing on which frequency spectrum they are using either the authorized cellular frequencies or the unlicensed ISM bands [13].

Table 1 Comparison of short-range wireless technologies

Name of the parameter	RFID	NFC	BLE	Zigbee	WiFiHaLow	WirelessHart	Z-Wave	ANT
Standardization	ISO/EPC	ISO/IEC 14,443	IEEE 802.15.1	IEEE 802.15.4	IEEE 802.11ah	IEEE 802.15.4 PHY HART MAC	Proprietary	Proprietary
Frequency	125–135 kHz, 13.56 MHz, 868–930 MHz, 2.45 and 5.8 GHz	13.56 MHz	2.4 GHz	2.4 GHz, 868 and 915 MHz	900 MHz	2.4 GHz	900 MHz	2.4 GHz
Range (meters)	0.1–5	0.1	50	10–100	Upto 1000	10–600	30	<30
Data rate	500 kb/s@ payload of 16–32 bits	106 kb/s or 212 kb/s or 424 kb/s 848 kb/s	1 Mb/s	20, 40, 250 Kb/s	150–400, 650–780 Kb/s	250 kb/s	9.6, 40, 100 Kb/s	1 Mb/s
Power consumption, mA	Varies with frequency	<5	<12.5	<40	low	<50µA	<23	<16
TX output power, dBm	1.5 mw	20 or 23	<19	–3 to 10	10 to 30	10	<0	–20 to 0
Multiplexing	TDMA	Antenna Multiplexing	FHSS	DSSS	OFDM	TDMA	FHSS	TDMA
Modulation	Proximity	ASK, BPSK	GFSK	OQPSK, BPSK	BPSK, QPSK, QAM	OQPSK, BSPK, ASK	FSK, GFSK	GFSK
Topology	P2P, Point to multipoint	Peer to peer	P2P, star	star, tree, mesh	One-hop	star, cluster mesh	star, mesh	P2P, star, tree, mesh
Packet length	16–64 kbps	segments	8–47 bytes	100 bytes	100 bytes	250 bits	255 bits	32 bit

Through using the licensed mobile frequencies, it certifies that the band is used lucidly through approved devices. However, the use of such frequencies involves subscription fees, and they could already be occupied due to the extensive usage of mobile devices, particularly in densely populated urban areas. On the alternative, the unlicensed ISM bands does not guarantee that the frequencies are used rationally, as everyone is approved for transmitting and several devices may intervene. The limitations of the duty cycle must be understood depending on the range of frequencies used, power and the physical positioning of the device. Around the same time, using unlicensed bands often means, we can easily extend the network with new base stations, build private networks, and use certain technologies for peer-to-peer communication. For two key purposes, one may wish to install one's own infrastructure: either because the region to be tracked is not served by any other technology or to retain control over infrastructure maintenance. LPWAN is currently powered to meet the needs of nascent IoT implementations to deliver a range of features like wide area communications and wide-scale networking for minimum power, small-cost and low-data tolerance devices. Typically speaking, LPWAN can be classified into two groups—unlicensed and licensed LPWAN. Unlicensed LPWAN technologies [14, 15] state that they use unlicensed spectrum frequencies over the ISM band. Acknowledgments for the use of the unlicensed band do not necessarily reimburse the unlicensed LPWAN suppliers for spectrum licenses. As a result, it reduces the cost of implementation. LPWAN also offers solutions using approved bands for long-range applications as per the 3rd Generation Partnership Project (3GPP). The solutions for the licensed LPWAN [16] are Extended Coverage GSM for Internet of things (EC-GSM-IoT), Long-Term Evolution Machine-Type Communications Category M1 (LTE MTC Cat M1), Narrowband IoT (NB-IoT) [17] and 5G [18].

3.1 Comparison of Long-Range Wireless Technologies

All the long-range technologies are studied and compared with specific parameters like frequency, data rate, range, power consumption, and battery life and presented in Table 2 [13], which will be feasible to understand about the performance of technology based on the parameters and can select the suitable technology based on the requirements.

4 IoT Applications

The IoT transforming traditional objects into smart objects by enabling physical things to sense and perform with the use of intelligence and technologies. By using the IoT technology, lifestyle of people become smart and secure. To improve overall

Table 2 Comparison of long-range wireless technologies

Name of the prameter	LoRa	Sigfox	Weightless	Dash7	LTE-M	EC-GSM	NB-IoT	5G
Standardization	LoRa WAN	Sigfox	Weightless SIG	ISO/IEC 18,000–7	3GPP	3GPP	3GPP	3GPP
Frequency	868 MHz 433/ 915 MHz 430 MHz	868 MHz 902 MHz	433/470/ 868 MHz, 915 GHz, 430 MHz	433/868/ 915 MHz	Licensed	Licensed	Licensed	24–100 GHz
Data rate	250 bps–50 kbps	100 bps (UL) 600bps (DL)	200 b/s–100 kb/s	167 kb/s	UL/DL: 1 Mb/s	70 kb/s	DL– 234.7 kb/s UL– 204.8 kb/s	20 gb/s
Range (km)	5(urban)15(rural)	10(urban) 50(rural)	2	5	11	15	15	1000 feet
TX output power, dBm	EU: 13 US: 20	EU:14 US: 21.7	17	433 MHz: +10 868/ 915 MHz: + 27	23	20 or 23	23	49
Band width per channel	0.3 MHz: 863– 870 MHz 2.16 MHz: 902– 928 MHz	100 Hz (600 Hz USA)	12.5 kHz	1.75 MHz	1.4– 20 MHz	200 kHz	18C kHz	1000/area
Modulation	Proprietary CSS	DBPSK (UL) GFSK (DL)	GMSK offset QPSK FMDA + TDMA	GFSK	OFDMA SC-FDMA	GMSK, 8PSK	GFSK BPSK	OFDM, CP-OFDM
Transmission technique	FHSS (Aloha)	UNB	TDMA	BLAST	FDD/TDD	FDD	FDD	FDD

(continued)

Table 2 (continued)

Name of the prameter	LoRa	Sigfox	Weightless	Dash7	LTE-M	EC-GSM	NB-IoT	5G
Battery operation	Many years	Many years	10 Years	Upto 10 years	+10 Years	+10 Years	+10 Years	+10 Years
Security	AES CCM 128	Key generation	AES-128	AES-128	AES 256	AES 256	NSA AES 256	EAP–AKA
Mobility	Yes	No	Yes	Yes	Yes (limited)	Yes	No	Yes

quality of life, different IoT applications are emerged almost all in every domain related to everyday life of a human being [19, 20].

There is no such domain where IoT cannot be implemented; it can be in every field; various IoT devices are developed to satisfy the customers and still developing as per customer specific requirements. Here in this paper, the numerous, diversified IoT applications are broadly categorized as follows:

4.1 Consumer Applications

Consumer application refers to the applications designed for the use of individual or a group of people for the ease of performing tasks. Many IoT applications are designed for consumer use, which includes smart home, wearables, general health care, asset tracking, remote access of workplace and entertainment appliances. Requirements of consumer applications include short-range, peer-to-peer communication, and small amount of data.

4.2 Commercial Applications

Commercial IoT is applicable for the environments where gathering of people on regular basis happens like offices, schools, hotels, hospitals, shopping center's etc. For satisfying the commercial requirements, many applications are available in present market such as smart office, location services, monitoring environment, connected lighting, asset tracking, medical health care and many more.

4.3 Industrial Applications

Industrial Internet of things (IIoT) is buzz word among different industries and IoT providers. IIoT exactly means to be industries with decades of expertise start developing or upgrading their hardware equipment's and software methods to integrate with IoT solutions [21]. IIoT devices gather the information from connected equipment's and analyzes the data to regulate and monitor the system. Gateways are most essential to provide IoT solutions to industries. Manufacturing units of any industry might be installed a decade back so to analyze the system and to provide scalability, we need to transfer the data to cloud for processing. Gateways act as a bridge between sensors or controllers and cloud. Gateways also used for data logging and processing solutions. The revolutions in present industrial fields are driving toward the smart industry which will be mention as fourth generation as Industry4.0 where every industry will be connected with the internet.

4.4 Infrastructure Applications

IoT takes a big part in developing, monitoring and controlling of any type of infrastructure. Infrastructure development and upgradation is essential to sustain for the technological changes. Infrastructure IoT can be applied for monitoring the structural health of buildings, bridges, and railway tracks, and it also helps to minimize the time and cost of infrastructure development [22].

5 Matching Application Requirements with Available Technologies

IoT is a collection of various technologies and endless applications. Billions of devices are connected to the internet to provide the smart services. The reason for the emergence of huge number of applications in IoT is that the requirements of applications vary from one domain to another. So there is a driving force toward development of new technologies to meet the application specific requirements. The existing traditional technologies also participating in this drive by updating like Bluetooth smart from classic Bluetooth, and Zigbee-IP from Zigbee to address the IoT requirements. New standard of technology such as IEEE 802.11ah is released for satisfying the key requirements of IoT like low-power consumption, reduced operational cost, low computations, and a wider coverage range.

Electing the perfect technology is the vital step of designing process of the application. To achieve this, some considerations must be in account like reliability, security, speed, power consumption, cost, etc. In previous sections of this paper, the various available technologies are described, and compared with metrics, later IoT applications are explained with their requirements. In this section, we are trying to match the application requirements with the technologies which help the customers to choose the best fit for their application.

However, all the technologies will have their own advantages and disadvantages. There will be always trade-off between power consumption, range, and data rate. As mentioned in the last section applications of IoT falls into different categories. Few require only peer -to-peer communication, other may operate in personal area network, and high end applications may involve heterogeneous devices over a wide area. So the technology suitable for one application may not be suitable for other applications. IoT develops an ecosystem to integrate the different technologies and devices to be communicated from anywhere and anytime.

For low-power applications, IEEE 802.11.ah is preferable over BLE and Zigbee as it able to cover the area of 1 km. For long range with low-power applications, LoRa is preferable with the range of 15 km. If application consists of large number of nodes, Zigbee may be the best choice due to its mesh networking and even provides high data rate. In long-range technologies, cellular technologies like NB-IoT have large network size but as it uses licensed frequencies which increases the cost involved. Whereas LoRa can also accommodate larger number of devices and comparatively with lesser cost but the QoS matters. In view of energy consumption, IEEE 802.11.ah is specifically designed for low-power consumption but Zigbee considerably showing less power requirements in congested networks. In long-range technologies, the low-power requirement will be satisfied by LoRa, contrarily NB-IoT is having high-power requirements. As per the applications and specific requirements, suggested technologies are tabulated in Table 3.

Table 3 Application specific requirements and suitable technologies

Domain	Application	Application specific requirements	Suitable technology
Smart home	Smart appliances	Peer-to-peer communications, short range, low data	BLE, Zigbee, IEEE 802.11.ah
	Intrusion detection	Large amount of data	WiFi
Smart city	Structural health monitoring	Long battery life, low amount of data	LoRa
	surveillance	High speed, QoS	Wifi, 5G
Smart Energy	Smart grid	Reliability, security	NB-IoT
	Renewable energy management	Reliability, security	NB-IoT
Retail	Smart Payments	Nearby communication	NFC
	Inventory Management	Identification and monitoring	RFID
Logistics	Shipment Monitoring	Long range, connectivity, Long battery life	LoRa
	Remote vehicle diagnosis	Frequent transmission, mobility	LoRa, NB-IoT
Agriculture	Pest control	Long range, connectivity	LoRa
	Cattle monitoring	Long range, connectivity	LoRa
Industry	Machine diagnosis	Short range	Zigbee, Wireless HART
	Indoor air quality	Low data, short range	Zigbee
Healthcare	Wearable's	Low-power consumption, less data	BLE
	Medical equipment monitoring	Identification	RFID
Environment	Weather monitoring	Reliability, security	NB-IoT
	Forest fire detection	Long range, connectivity, Long battery life	LoRa

6 Conclusion

The IoT is capable of becoming a transformative power, having a positive effect on the lives of millions around the world. It aims to bring about a big shift in the efficiency of individuals quality of life and of companies via a massive, regionally sensible network of smart devices. IoT has the capacity to allow extensions and advancements to simple offerings in delivery, logistics, defense, utilities, education, healthcare, and other areas. It also offering brand new surroundings for the development of applications. To participate in this ecosystem, we tried ourselves to study about the available wireless communication technologies for the IoT and made an attempt to indicate the excellent applicable technology for the utility primarily based on precise requirements of specific application. For this, we provided records of well-known wireless technologies within the market with the aid of classifying them depending on the variety and compared with their parameters; later, IoT application areas and their necessities have been discussed. And eventually, we provided excellent matched technology based on the utility particular necessities.

References

1. Liu T, Lu D (2012) The application and development of IoT. In: 2012 International symposium on information technologies in medicine and education, vol 2, pp 991–994, IEEE
2. Vermesan O, Friess P, Guillemin P et al (2011) Internet of things strategic research roadmap. Internet Things: Global Technol Societal Trends 1:9–52
3. Peña-López I (2005) Itu internet report 2005: the internet of things
4. Sethi P, Sarangi SR (2017) Internet of things: architectures, protocols, and applications. Hindawi J Electr Comput Eng 1:1–25
5. Ding J, Nemati M, Ranaweera C, Choi J (2020) IoT connectivity technologies and applications: a survey. IEEE Access 8:67646–67673
6. Oliveira L, Rodrigues JJPC, Kozlov SA, Rabêlo RAL,Victor Hugo C (2019) MAC layer protocols for internet of things: a survey. Molecular diversity preservation international
7. IEEE Computer Society (2003) IEEE standards IEEE 802.15.4-Part 15.4—wireless MAC and PHY specifications for low-rate wireless personal area networks—LR-WPANs, 2003th ed.; The Institute of Electrical and Electronics Engineers, Inc.: Piscataway, NJ, USA
8. Darroudi S, Gomez C (2017) Bluetooth low energy mesh networks: a survey. Sensors 17:06
9. Adame T, Bel A, Bellalta B, Barcelo J, Oliver M (2014) IEEE 802.11 AH: The WiFi approach for M2M communications. IEEE Wirel Commun 21:144–152
10. Elkhodr M, Shahrestani S, Cheung H (2016) Emerging wireless technologies in the internet of things: a comparative study. Int J Wirel Mobile Netw (IJWMN) 8(5)
11. Raza U, Kulkarni P, Sooriyabandara M (2017) Low power wide area networks: an overview. IEEE Commun Survey Tutorials 19:855–873
12. Bembe M, Abu-Mahfouz A, Masonta M, Ngqondi T (2019) A survey on low-power wide area networks for IoT applications. Telecommun Syst 71:249–274
13. Foubert B, Mitton N (2020) Long-Range wireless radio technologies: a survey. Future internet 12:13
14. Semtech LoRa Technology Overview. Available online: https://www.semtech.com/lora

15. SigFox Radio Access Network Technology. Available online: http://www.sigfox.com/en/sigfox-iot-radio-technology
16. Global mobile Suppliers Association (GSA) (2018) Evolution from LTE to 5G: global market status; technical report; GSA: Washington, DC, USA
17. Ratasuk R, Mangalvedhe N, Zhang Y, Robert M, Koskinen JP (2016) Overview of narrowband IoT in LTE Rel-13. In: Proceedings of the IEEE conference on standards for communications and networking (CSCN),Berlin, Germany, 31 October–2 November 2016, p. 1–7
18. Chettri L, Bera R (2020) A comprehensive survey on internet of things (IoT) towards 5G wireless systems. IEEE Internet Things J 7(1):16–32
19. Sharma V, Tiwari R (2016) A review paper on "IOT and it's smart applications. Int J Sci Eng Technol Res (IJSETR) 5(2)
20. Da Xu L, He W, Li S (2014) Internet of things in industries: a survey. IEEE Trans Ind Informat 10(4):2233–2243
21. Shilpa B, Mahamood MR (2017) Design and implementation of framework for smartcity using LoRa technology. Sreyas Int J Scientists Technocrats 1(11):36–43
22. Shilpa B, Malipatil S, Reddy J (2019) Lora technology based potholes and humps detection for smart city transportation.Int J Eng Adv Technol (IJEAT) 8(6):702–705

MADLI: Mixture of Various Automated Deep Learning Classification for Paddy Crop Images

A. Srilakshmi, K. Madhumitha, and K. Geetha

1 Introduction

This model arrives at the solution of resolving plant disease recognition by constructing a convolution neural network (CNN) and its pre-trained architectures which gave higher efficiency of 99%. Though there are many models such as SVM, Res-Net has been used to classify the same dataset of paddy leaf diseases, CNN gave the better accuracy. This model helps farmers to classify a large sample of data with accurate classification there by making it cost efficient and saves the loss that could possibly occur when the disease is left unrecognized. Res-Net is easy to train because it is faster and has a smaller number of parameters compared to VGG-16. The dataset has three main diseases, namely bacterial leaf blight, brown spot, and blast. The data's are equally divided among different classes by performing augmentation. Since there is equal number of augmented images, enough features will be extracted from the dataset. The cause of the disease may be because of any source such as meteorological factor as well as pests but detecting at earlier stages by classifying correctly would avoid the destruction of crops. The more the images for every class more accurate the result will be.

A. Srilakshmi (✉) · K. Madhumitha · K. Geetha
Sastra Deemed To Be University, Thanjavur, India
e-mail: srilakshmi@cse.sastra.edu

K. Geetha
e-mail: geetha@cse.sastra.edu

© The Author(s), under exclusive license to Springer Nature Singapore Pte Ltd. 2022   395
R. R. Raje et al. (eds.), *Artificial Intelligence and Technologies*,
Lecture Notes in Electrical Engineering 806,
https://doi.org/10.1007/978-981-16-6448-9_40

2 Background

Training a model using SVM as a classifier and different layers in CNN for classification of blast disease recognition and accuracy can be analyzed by confusion matrix [1]. A simple CNN model along with ResNet50 architecture is utilized for classification of multi-crop plant disease classification in an efficient manner [2]. Training a CNN model with VGG-Net and transfer learning lead to the required result with better accuracy and performance measures [3]. When there is a need for efficient classification rather than focusing on entire image in a dataset to analyze the type of the disease, S-CNN can be used which focus on the diseased part resulting a greater accuracy [4]. A two stage CNN architecture has been proposed which when compared to existing architecture such as squeeze net and VGG and other architectures lead a good result [5]. By preprocessing the images in a dataset such as compressing and rescaling, one can make a simple CNN architecture an efficient classifier [6]. A faster R-CNN architecture along with regional proposal neural network with CNN model is also one such useful model for disease recognition [7]. VGG-Net has fully connected layers with large number of parameters in order to increase efficiency it has to be removed. Xception-Net made up of inception network, NAS-Net can play a role of increasing accuracy rate. Dense-Net has less parameter and could solve the problem of vanishing gradient when neural network gets deeper [8]. SVM when used with VGG16, tuning parameters, back and forward propagation, Mobile-Net, and InceptionV3 layers can increase efficiency [9]. When a CNN model is pre-trained by extracting feature and transfer learning methodology, then if it is applied gives a more efficient model for classification [10]. Residual learning-based CNN can be appreciated for disease identification [11]. Even on using dataset of smart phone images in SSCNN and classified using Mobile-Net in CNN results in accurate model [12]. The efficient net proposes its own architecture which uses a new activation function called swish is applied which shows better accuracy than any other state-of-art models. The efficient net is also a replacement of complex models such as Mobile-Net, etc. [13]. CAAE (Conditional adversial auto-encoders) has been utilized to learn features so that with a less layers, i.e., zero or in few shots the system can be transferred into supervised learning algorithm which is helpful for further classification [14].

3 Materials and Methodology

The dataset has been collected from UCI repository with three different diseases. Basic preprocessing like flipping, resizing, and translation has been done to increase the size of dataset. The more the images the system will be trained well and predicts correctly. Depending upon the size of the image, various models can be used. Some of the CNN architectures like Google-Net and VGG and others have more layers which does not suits well small dataset. So the classification of images can be done with basic machine learning algorithms, and in case if the size of the image is increased after performing some preprocessing, then CNN fits well. The images have been reshaped and taken as RGB images without converting to grayscale or pencil sketch. Resizing has been done with dimensions of 300 × 300. The dataset has been divided into training, testing, and validation folders. Each in turn consists of three classes with 551 diseased leaf images. The original images have been taken as flipped, rotated, and cropped in order to increase the efficiency of our model.

The above images Fig. 1 are bacterial blight which is the major disease which occurs in almost all the crops. 1 to 5 are the preprocessed images. Other diseases such a blast and brown spot are also considered as most common diseases which affects the production of plant. Figure 2 shows the flipped, and the rotated (Fig. 3), translated images as (Fig. 4) and last preprocessing applied is background Subtraction (Fig. 5). It is also difficult to control those diseases and predict proper fertilizers to avoid these diseases. The above images are received after performing

Fig. 1 Original

Fig. 2 Flipping

Fig. 3 Rotation

Fig. 4 Translation

Fig. 5 Background
subtraction

flipping of images. Usually preprocessing helps in increasing the image size and to help the model to learn well. The images are stored separately as 70% for training and 30% for testing the CNN model.

4 Implementation

The dataset has been first trained with SVM which is a supervised learning algorithm. In order to transform the input to a particular form, different kernel can be chosen. It is the most commonly used supervised machine learning algorithm either for classification or regression. It is highly used in bio technology applications for gene prediction and in leaf disease classification problems. In this approach, we used radial basis kernel function (RBF) which suits well for classification. The RBF function suits well almost for all classification problems and helps in finding the hyper plane more accurately. The format of the RBF kernel depends upon the Euclidean distance as shown in the below formula:

$$K(m, m') = \text{exponential} \left[\frac{-\|m - m'\|^2}{2\sigma} \right] \tag{1}$$

The SVM runs by finding the feature vectors and finding proper boundary between the class labels. In this, RBF function mentioned in Eq. (1) is the parameter refers to Euclidean function and r is the radius. After running the model with the dataset, the accuracy is 83%, and if the size of the images is increased with other preprocessing techniques, then the performance degrades. Below table shows the accuracy and other performance measures. The output for nearly four iterations is made to run and finally reached 83% accuracy. Convolution neural network (CNN) is the first used to train the model. CNN is an ANN similar to the human nervous system which consists of different layers of interconnected neurons. CNN uses deep learning algorithms and is useful to classify images by assigning weights. It is used in many fields because of its efficiency to identify images, face recognition, digit analysis, etc. The preprocessed image is loaded to the model and basic layers are added. The basic layers include max-pooling, convolution layers, and dense layers. As the layers get fully connected, it leads to the disappearance of gradients. The extra VGG16 layer has been added to this model. It is able to find more features and, hence, performs well. The CNN with VGG-16 is compared with residual networks which makes difference in generating the performance measures. The residual networks by skipping some layers in between and adding additional weights in between. Figure 6 shows the CNN architectural sample model used to implement this classification.

Each layer corresponds to extracting different features to classify the diseased leaf. The model uses 2D convolution with 32 filters at the initial stage and implemented using keras. The parameters such as Adam optimizer, learning rate with 0.001 with stride equal to 2 is used. At the dense layer, the number of units used is 4 and for this multiclass classification softmax activation issued.

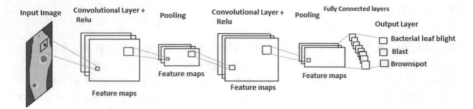

Fig. 6 CNN model

5 Results and Discussion

The results obtained from the models are shown below. The analogy of more the images more the results are accurate. Correct procedure has to be applied for performing preprocessing algorithms. In the below results shown for Res-Net is after preprocessing it. This model uses image data generator which performs random preprocessing by its own. The most significant characteristics or features are captured by the VGG is shown below as images in Fig. 7. The model itself performs preprocessing and model training with its own defined hyper parameters.

The above results are generated from CNN-VGG which gives accuracy in 5 epochs as 98%. Figure 7 shows confusion matrix of which all features are captured properly by the model and gives good accuracy. The Fig. 8 shows the confusion matrix of the VGG and Figs. 9 and 10 shows model accuracy and model loss.

SVM (Grid Search)

Table 1 shows the SVM performance of which yields its 84% accuracy. Though it takes time during training, it fills well to this dataset yielding better results. This when trained with more images after preprocessing the accuracy degrades.

Residual Net

Table 2 shows the performance of residual network which yields 99% accuracy. The accuracy of residual network is better than another model which yields 99.21%.

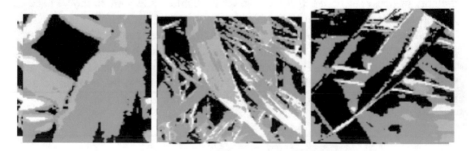

Fig. 7 VGG-16 processed image

Train for 56 steps, validate for 56 steps Epoch 1/5

56/56 - 12s - loss: 11.0469 - accuracy: 0.4937 - val_loss: 0.9354 - val_accuracy: 0.6032

.
.
.
.

Epoch 5/5

56/56 - 10s - loss: 0.2191 - accuracy: 0.919   - val_loss:
                                                              2

0.0771   val_accuracy: 0.9820

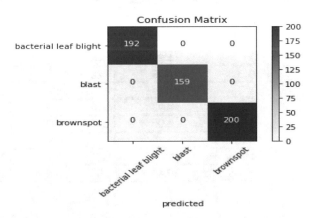

Fig. 8 Confusion matrix for VGG

Fig. 9 Model Accuracy

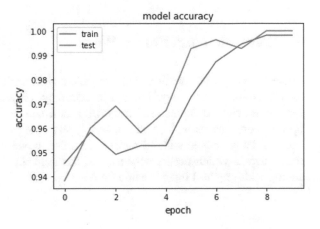

Fig. 10 Model loss

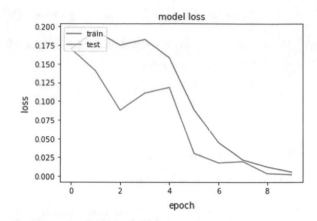

Table 1 SVM performance

	Precision	Recall	f1-score	Support
0	0.73	0.88	0.80	65
1	0.79	0.74	0.76	57
2	0.93	0.80	0.86	65
3	1.00	1.00	1.00	24
Accuracy			0.83	211
Macro avg	0.86	0.85	0.86	211
Weighted avg	0.84	0.83	0.83	211

Here vgg-16 also produces almost the same result but fluctuating when there is change in some image preprocessing.

6 Conclusion and Future Work

There are many CNN architectures available among which certain models suits well for almost many datasets yielding good accuracy. Here residual network suits well for any number of inputs, and it shows a constant learning with good result. In future, more images will be added, and appropriate preprocessing would be done and will be compared with all models. In order to add more classes, there may be some disease which occurs very least in the field, in that case optimal class balancing with required images could be done.

Table 2 Residual net performance

```
Train for 11 steps

Epoch 1/5
11/11 [==============================] - 159s14s/step -
loss: 1.2691 - accuracy:0.6548
Epoch 2/5

11/11 [==============================]    - 122s  11s/step -
loss: 0.3464 - accuracy: 0.8869

Epoch 3/5

11/11 [==============================]    - 107s  10s/step -
loss: 0.2173 - accuracy: 0.9345

Epoch 4/5

11/11 [==============================]    - 107s  10s/step -
loss: 0.1121 - accuracy: 0.9702

Epoch 5/5

11/11 [==============================]    - 108s  10s/step -
loss: 0.0611 - accuracy: 0.9921
```

References

1. Radhakrishnan S (2020) An improved machine learning algorithm for predicting blastdisease in paddy crop. Mater Today Proc 682–686
2. Picon A, Seitz M, Alvarez-Gila A, Mohnke P, Ortiz-Barredo A, Echazarra J (2019) Crop conditional convolutional neural networks for massive multi-crop plant disease classification over cell phone acquired images taken on real field conditions. Comput Electron Agric
3. Chen J, Zhang D, Sun Y, Nanehkaran YA (2020) Using deep transfer learning for image-based plant disease identification. Comput Electron Agric 173:2020
4. Sharma P, Paul Singh Berwal Y, Ghai W (2019) Performance analysis of deep learning CNN models for disease detection in plants using image segmentation. Inf Process Agric
5. Rahman CR, Arko PS, Ali ME, Iqbal Khan MA, Apon SH, Nowrin F, Wasif A (2020) Identification and recognition of rice diseases and pests using convolutional neural networks. Biosyst Eng 194:112–120
6. Lu Y, Yi S, Zeng N, Liu Y, Zhang Y (2017) Identification of rice diseases using deep convolutional neural networks. Neurocomputing 267:378–384
7. Ozguven MM, Adem K (2019) Automatic detection and classification of leaf spot disease in sugar beet using deep learning algorithms. Physica A: Statist Mech Appl
8. Waheed A, Goyal M, Gupta D, Khanna A, Hassanien AE, Pandey HM (2020) An optimized dense convolutional neural network model for disease recognition and classification in corn leaf. Comput Electron Agric 175

9. Agarwal M, Gupta SK, Biswas KK (2020) Development of efficient CNN model for tomato crop disease identification. Sustain Comput: Inf Syst
10. Sethy PK, Barpanda NK, Rath AK, Behera SK (2020) Deep feature based rice leaf disease identification using support vector machine. Comput Electron Agric 175
11. Karthik R, Hariharan M, Anand S, Mathikshara P, Johnson A, Menaka R (2019) Attention embedded residual CNN for disease detection in tomato leaves. Appl Soft Comput
12. Barman U, Choudhury RD, Sahu D, Barman GG (2020) Comparison of convolutional neural networks for smartphone image based real time classification of leaf disease 177
13. Ümit A, Uçar M, Akyol K, Uçar E (2020) Plant leaf disease classification using EfficientNet deep learning model. Ecol Inf 101182
14. Zhong F, Chen Z, Zhang Y, Xia F (2020) Zero-and few-shot learning for diseases recognition of *Citrus aurantium L.* using conditional adversarial autoencoders. Comput Electron Agric 179:105828
15. Zhong Y, Zhao M (2020) Research on deep learning in apple leaf disease recognition. Comput Electron Agric

Diabetic Retinopathy Diagnosis with InceptionResNetV2, Xception, and EfficientNetB3

Mukkesh Ganesh, Sanjana Dulam, and Pattabiraman Venkatasubbu

1 Introduction

Diabetic retinopathy (DR) is a diabetic condition that affects the eyes. Worldwide one-third of the estimated diabetic population show signs of DR. Elevated sugar levels in the blood can lead to the blockage of blood vessels in the retina. This condition is termed as non-proliferative DR (NPDR) which could worsen to proliferative DR (PDR). If left untreated, scar tissues stimulated by the growth of new blood vessels may cause the retina to detach from the back of your eye which can cause complete blindness. Hence, early diagnosis is critical to the mitigation of this medical condition. Over the past decade, deep learning (DL)-assisted diagnostic systems have risen in number and have outperformed the traditional image processing-based systems. From detecting cancerous tumors in lungs and breast scans to the diagnosis of COVID-19 from CT scans, this technology has gained wide acceptance within the medical field. Rapid innovation in the deep learning-based computer vision has given rise to powerful neural network architectures which have greatly enhanced the performance of these models. In much more recent years, researchers have started to use the power of transfer learning to make models converge faster and better for tasks that previously had limited training resources.

In this paper, we will be utilizing transfer learning to explore and compare the performance of state-of-the-art neural network architectures for diagnosing the severity of DR from retinal fundus images. For training and validating these

M. Ganesh (✉) · S. Dulam · P. Venkatasubbu
Vellore Institute of Technology, Chennai, Tamil Nadu, India
e-mail: g.mukkesh2017@vitstudent.ac.in

S. Dulam
e-mail: sanjana.dulam2017@vitstudent.ac.in

P. Venkatasubbu
e-mail: pattabiraman.v@vit.ac.in

© The Author(s), under exclusive license to Springer Nature Singapore Pte Ltd. 2022   405
R. R. Raje et al. (eds.), *Artificial Intelligence and Technologies*,
Lecture Notes in Electrical Engineering 806,
https://doi.org/10.1007/978-981-16-6448-9_41

models, we make use of the Kaggle dataset and test the performance of the same on the Messidor-2 dataset and Indian diabetic retinopathy image dataset (IDRiD).

2 Related Work

In [1], the authors used convolutional neural networks (CNNs) and achieved a validation sensitivity of 95% on the Messidor-2 dataset. They also explored the use of transfer learning with GoogLeNet and AlexNet and were able to achieve accuracies of 74.5% on 2-ary, 68.8% on 3-ary, and 57.2% on 4-ary classifications. In [2], Inception V3 architecture was pre-trained in various ways with a subset of the CHCF dataset, neural networks were compared, and a few were able to achieve an accuracy of 88% test set. Using the Kaggle dataset, [3] were able to achieve an accuracy of 75% and sensitivity of 95% on 5000 validation images. The study in [4] compared the performance of a novel DL algorithm for the detection of DR with previous results on the benchmark Iowa Detection Program (IDP) dataset. By replacing IDP's feature-based lesion detectors with CNN-based lesion detectors, the hybrid model achieved a sensitivity of 96.8% on the Messidor-2 dataset. The retrospective study of [5] analyzed posterior pole photographs of patients with diabetes. A randomly initialized GoogLeNet was trained on 95% of the photographs using manual modified Davis grading of three additional adjacent ones achieving an accuracy of 81%. AlexNet has been applied in [6] to enable an optimal DR computer-aided diagnosis (CAD) solution. The proposed system on the standard Kaggle fundus dataset, with LDA feature selection, reaches a classification accuracy of 97.93% and 95.26% with PCA. The Inception V4 model architecture in [7] was trained on a large dataset of more than 1.6 million retinal fundus images and was then tested on 2000 images; it showed a 5-class accuracy of 88.4%. An accuracy of 72.5% with VGGNet was achieved in [8] and a weighted fuzzy C-means algorithm was used to diagnose the severity of the disease. In [9], residual U-net architecture consistently outperformed the traditional non-residual U-net models in segmentation tasks. Carrera et al. [10] extracts features such as blood vessels and hard exudates that were used to train a support vector machine (SVM), which achieved a maximum sensitivity of 94.6% on the Messidor-2 dataset.

3 Dataset Description

The standard Kaggle dataset published with the consent of EyePACs in 2015 consists of 35,126 images which we made use of for training the models. The dataset of high-resolution retina images taken under varying conditions is labelled based on the severity of DR in each image on a scale ranging from 0 to 4, where 0 —no DR, 1—mild, 2—moderate, 3—severe, and 4—proliferative DR, being the most severe. The Messidor-2 dataset and IDRiD [11], which we used for evaluating the performance of our trained models, are significantly different from the Kaggle dataset. These variations help give us a realistic estimation of the models' performance since more often than not the images do not belong to the same distribution as the data on which the models were trained.

The Messidor-2 dataset is an aggregation of several DR eye evaluations, consisting of 1748 macula-centered eye fundus images. The 512 fundus images in Indian diabetic retinopathy image dataset (IDRiD) were captured by a retinal ophthalmologist in India. The Kaggle training data were split into 90% training and 10% validation, while the two remaining datasets were used to evaluate the models (Fig. 1).

4 Model Architecture

CAD systems have started to increasingly utilize deep learning methodologies over traditional image processing workflows. The recent architectures developed by computer vision and deep learning researchers offer better performance while using lesser computing resources.

In this study, we focus on three major architectures, namely Xception, InceptionResnetV2, and EfficientNetB3, and compare their performance on the task of DR diagnosis. In trained models, the initial convolution kernels have been shown to detect basic patterns such as edges. By initializing the weights of these network layers to values learnt from other tasks such as ImageNet and CoCo image classification challenges, the models can converge faster or sometimes even perform better with having limited data at hand. This process known as transfer learning is leveraged in this paper in an attempt to achieve better results while training these architectures for this specific task.

4.1 Xception

Xception (extreme inception) is an architecture proposed by Google as an improvement over its Inception V3 architecture. The original Inception architecture

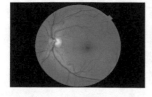 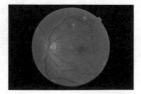 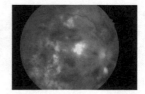

(a) No DR - Kaggle   (b) Moderate DR - Messidor-2   (c) Proliferative DR - IDRiD

Fig. 1 Sample images with varying severity of DR from different datasets

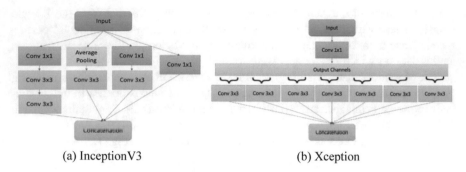

(a) InceptionV3   (b) Xception

Fig. 2 Differences in Inception and Xception architectures

used depth-wise convolution followed by a 1×1 convolution to modify output dimension. Depth-wise convolution involved channel-wise $N \times N$ spatial convolution. Depth-wise separable convolution on the other hand made use of 1×1 convolution before performing the $N \times N$ channel-wise spatial convolutions. This was shown to yield better results than the vanilla Inception model (Fig. 2).

4.2   InceptionResNetV2

Deeper neural networks often take longer to train and may fail to converge due to vanishing gradients. To mitigate these issues, Microsoft introduced the residual neural network architecture, which had skip connections between convolution blocks, which solves the vanishing gradient problem caused by the deeper architectures while also greatly speeding up the training process. InceptionResnetV2 takes this concept of skip/residual connections from ResNets and applies it to the Inception architecture, thereby enhancing the performance of the model (Fig. 3).

Fig. 3 InceptionResNetV2 architecture

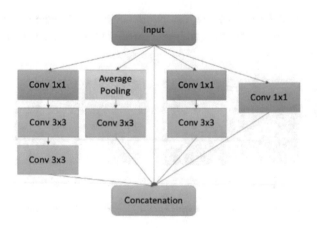

4.3 EfficientNetB3

The traditional practice for model scaling is to randomly increase the model depth or width or to use greater resolutions of input images for training and evaluation. This results in tedious manual fine-tuning and longer training times. Introduced by Google, EfficientNets are scaled in a more principled manner. The resulting architecture makes use of mobile inverted bottleneck convolution. Each stem of the eight EfficientNets contains seven blocks, each further consisting of an increasing number of subblocks from EfficientNet B0 to B7 (Fig. 4).

5 Implementation

Our study was conducted using the Kaggle environment. The Tensorflow framework was chosen to develop these deep learning models. The preprocessing stage consisted of resizing the images to 299×299. Tensorflow's image data generator class was used to augment our dataset by using a few transformations such as horizontal flip, zoom, shear, and brightness modifiers. For these experiments, we used ImageNet pre-trained models as our baseline architecture. The output layers were removed, and a two-dimensional global average pooling layer was added to reduce the dimensions to the number of channels. We added another layer of 256 neurons, after which a dropout layer with a dropping probability of 0.4 was used for better generalization. The output layer was decided based on the classification problem at hand and used softmax activation. Nesterov implemented Adam was chosen as the optimizer, while categorical cross-entropy loss was chosen as the loss function for our models.

Previous studies have performed both binary and 5-ary classification on this dataset. For the task of binary classification, classes 0–2 (no DR, mild DR, and

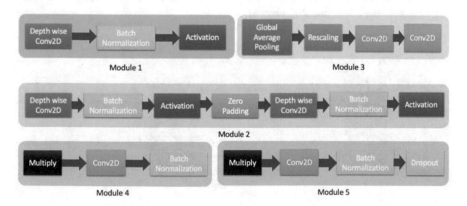

Fig. 4 Modules in EfficientNets

Moderate DR) and classes 3 and 4 (severe DR and proliferative DR) were combined respectively. Hence, two versions of these models were developed for both binary and 5-label classification.

6 Results and Discussion

The variations that arise from different scanning techniques and imaging procedures invariantly hamper the real-world application of these DL systems as these models which are trained to detect patterns in a particular distribution of images might not be able to replicate similar results under varying conditions. 3511 images from the Kaggle dataset were used for validation, while the real-world performance of the model was tested on images from both the Messidor-2 and IDRiD. 2-ary and 5-ary classifications were conducted, and the accuracies are compared in Fig. 5.

The inherent difficulty of 5-ary classification is amplified by the complexity of IDRiD, which was shown in the IDRiD grand challenge [12] held in 2018. A baseline accuracy of less than 60% was achieved by most of the participating teams. The accuracies achieved by the models discussed in this paper are comparable to those of the top teams from the grand challenge. Our best 5-ary classification model, InceptionResNetV2, was able to achieve a higher accuracy of 64% on this dataset while providing similar performance on class-wise sensitivity and specificity (Table 1).

The Xception model was able to reach the highest accuracy of 69% on 5-ary classification of the Messidor-2 dataset and is tied with EfficientNetB3 on 2-ary classification with an accuracy of 97%. It also achieves the highest accuracy of 84% on the 2-ary classification of IDRiD, while also offering significantly higher sensitivity and specificity (Table 2).

Apart from achieving the highest accuracy of 64% on the 5-ary classification on IDRiD, the InceptionResNetV2 model also offers high sensitivities on most of the classes. In the task of 2-ary classification, it achieves an exceptionally high sensitivity of 99% on Messidor-2 (Table 3).

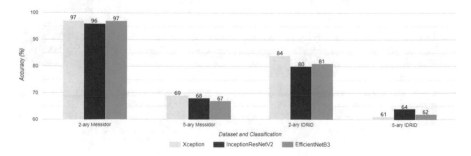

Fig. 5 Comparison of model accuracies

Table 1 Xception model performance

Dataset	5-ary				2-ary			
	Class	Spec	Sens	Acc	Class	Spec	Sens	Acc
Messidor-2	0	74	90	69	0	97	99	97
	1	0	0					
	2	54	66		1	81	59	
	3	78	41					
	4	53	46					
IDRiD	0	74	89	61	0	94	83	84
	1	0	0					
	2	62	54		1	68	86	
	3	49	29					
	4	42	72					

Table 2 InceptionResNetV2 model performance

Dataset	5-ary				2-ary			
	Class	Spec	Sens	Acc	Class	Spec	Sens	Acc
Messidor-2	0	72	89	68	0	97	99	96
	1	0	0					
	2	57	62		1	75	58	
	3	65	77					
	4	58	31					
IDRiD	0	76	90	64	0	92	79	80
	1	0	0					
	2	66	57		1	62	83	
	3	46	62					
	4	52	41					

Table 3 EfficientNetB3 model performance

Dataset	5-ary				2-ary			
	Class	Spec	Sens	Acc	Class	Spec	Sens	Acc
Messidor -2	0	69	98	67	0	97	99	97
	1	0	0					
	2	63	46		1	83	56	
	3	77	13					
	4	50	49					
IDRiD	0	75	88	62	0	92	81	81
	1	0	0					
	2	59	57		1	64	72	
	3	64	15					
	4	37	85					

EfficientNetB3, while having only 12 million parameters when compared to the 22 million and 55.8 million parameters of Xception and InceptionResNetV2, respectively, can achieve comparable results with faster inferences, training, and convergence rates making it a viable solution for automated diagnostic systems. It achieved an accuracy of 62% and 81% on the 5-ary and 2-ary classification of IDRiD. The general trend of achieving higher sensitivity (85% on class 4 IDRiD) at the cost of a lower specificity (37% on class 4 IDRiD) is observed here too.

7 Conclusion and Future Enhancements

Diagnosis of DR at its earlier stages greatly increases the probability of successful recovery. Given this, the need for efficient diagnosis systems is highly essential for which various state-of-the-art deep learning techniques are increasingly being adopted. The time taken to train, validate, and test the performance of these models is critical due to its direct impact on the research throughput in this field. Our study shows that EfficientNets can be used as an alternative to the previous state-of-the-art model architectures as it converges faster, takes less time to train, and produces comparable results while utilizing lesser computing resources. Transfer learning with this model also greatly speeds up the convergence rate.

Although binary classification produces better results, 5-ary classification should be given more attention since diagnosing DR at early stages would help greatly reduce the risk of it developing into more severe stages. This multi-label classification suffers from a larger loss in accuracy due to the lack of sufficient images in the mild to moderate categories of DR. In the future, we plan to extend our work by evaluating the performance of these models with augmented datasets and also make use of undersampling to see if we can mitigate this particular drawback. We also plan to explore image processing techniques such as denoising to make these models more robust.

References

1. Lam C et al (2018) Automated detection of diabetic retinopathy using deep learning. In: AMIA summits on translational science proceedings 2018, p 147
2. Kanungo YS, Srinivasan B, Choudhary S (2017) Detecting diabetic retinopathy using deep learning. In: 2017 2nd IEEE international conference on recent trends in electronics, information & communication technology (RTEICT). IEEE
3. Pratt H et al (2016) Convolutional neural networks for diabetic retinopathy. Procedia Comput Sci 90:200–205
4. Abràmoff MD et al (2016) Improved automated detection of diabetic retinopathy on a publicly available dataset through integration of deep learning. Invest Ophthalmol Vis Sci 57(13): 5200–5206
5. Takahashi H et al (2017) Applying artificial intelligence to disease staging: deep learning for improved staging of diabetic retinopathy. PloS one 12(6):e0179790

6. Mansour RF (2018) Deep-learning-based automatic computer-aided diagnosis system for diabetic retinopathy. Biomed Eng Lett 8(1):41–57
7. Sayres R et al (2019) Using a deep learning algorithm and integrated gradients explanation to assist grading for diabetic retinopathy. Ophthalmology 126(4):552–564
8. Dutta S et al (2018) Classification of diabetic retinopathy images by using deep learning models. Int J Grid Distrib Comput 11(1):89–106
9. Alom MZ et al (2018) Recurrent residual convolutional neural network based on u-net (r2u-net) for medical image segmentation. arXiv:1802.06955
10. Carrera EV, Andrés G, Ricardo C (2017) Automated detection of diabetic retinopathy using SVM. In: 2017 IEEE XXIV international conference on electronics, electrical engineering and computing (INTERCON). IEEE
11. Porwal P et al (2018) Indian diabetic retinopathy image dataset (IDRiD): a database for diabetic retinopathy screening research. Data 3(3):25
12. Porwal P et al (2020) IDRiD: diabetic retinopathy–segmentation and grading challenge. Med Image Anal 59:101561

Human Emotion Detection Using Convolutional Neural Networks with Hyperparameter Tuning

Aparna Chaparala[ID]

1 Introduction

Human emotion recognition has been an active research area for the past few years, due to the increasing demand for applications in perceptual and cognitive sciences and affective computing. It has become an essential component for fields such as computer animations, sociable robots, and neuromarketing. Human emotions can be recognized by using facial expressions and vocal tones. According to Kaulard et al. [1], nonverbal components convey two-thirds of human communication, while verbal components convey only one-third. Various kinds of data including physiological signals, such as electromyograph (EMG), electrocardiogram (ECG), and electroencephalograph (EEG), can also be considered as input for the emotion recognition process. Among these, the facial image is the promising input type as it is noninvasive and provides an ample amount of information for expression recognition. Emotions can be categorized into three types: basic emotions (BEs), compound emotions (CEs), and micro-expressions (MEs). Basic emotions cover neutral, anger, disgust, fear, surprise, sadness, and happiness.

Two categories of approaches for facial expression recognition (FER) are in use: conventional approaches and deep learning-based approaches. When compared to deep learning-based techniques, conventional techniques are advantageous as they require less computational power. Hence, no additional infrastructure is needed. Input images having illumination changes, occlusion, and deflection of the head may influence the face detection task performance and reduce the accuracy of FER. Conventional techniques are not suitable for noisy input data. Deep learning-based techniques address these issues. Of late, convolutional neural networks (CNNs) were proven effective for face detection [2]. As CNNs contain deep layers and use

A. Chaparala (✉)
RVR&JC College of Engineering (A), Chowdavaram, Guntur, AP, India
e-mail: aparna@rvrjc.ac.in

© The Author(s), under exclusive license to Springer Nature Singapore Pte Ltd. 2022 415
R. R. Raje et al. (eds.), *Artificial Intelligence and Technologies*,
Lecture Notes in Electrical Engineering 806,
https://doi.org/10.1007/978-981-16-6448-9_42

elaborate designs, they can ably handle noisy data automatically [3]. CNNs proved to exhibit better performance than conventional methods for the FER task [4, 5]. The performance of CNN highly depends on the choice of its hyperparameters. It is possible to enhance the CNN's performance by optimizing hyperparameters such as the number of hidden layers, units in each layer, filters, size of the filter, batch size, and learning rate. The present work considers the optimization of hyperparameters that describe the CNN structure. Grid search and random search techniques are commonly used for this purpose [6]. Each of these techniques has its limitations, and both need more time and domain expertise for identifying ideal hyperparameter values. Metaheuristic-based approaches can address these shortcomings as they are stochastic approximation methods. The present work employed the differential evolution (DE) algorithm for tuning the selected hyperparameters.

2 Related Work

Kim et al. [7] proposed to train multiple CNNs. They have shown an improvement in training by changing the network topology and random weight initialization. An interesting method for selecting the CNN structure was presented by Gao et al. [8]. They proposed gradient priority particle swarm optimization (GPSO) with gradient penalties for tuning CNN architecture. Experimental results have shown that the proposed method has gained competitive prediction performance for the emotion recognition task. Bergstra and Bengio [6] proposed to employ a grid or random search for tuning hyperparameters. Since the number of hyperparameters is large, testing is computationally expensive. Snoek et al. [9] have addressed the limitations of trial and error-based techniques for hyperparameter optimization. They have proposed a Bayesian optimization framework. Bochinski et al. [10] have shown that evolutionary algorithms can outperform the existing hyperparameter optimization methods.

3 Methodology

Benchmark dataset for facial expression recognition is split into training set (TS) and testing set (TE). For ensuring that samples of all classes get selected, stratified sampling without replacement is used. Selected samples from TS generate tuning set (TUS). Tuning set is further divided into TUS1 and TUS2. TUS1 is used for hyperparameter optimization. TUS2 is used for validating the outcome of optimization. Differential evolution is performed until the termination condition is met. CNN is trained using the outcome of DE on TS. The holdout method is used for assessing the performance of the trained model. After training, the model's performance is assessed by using TE. Table 1 specifies the architecture of the convolutional neural network used in the present work (Fig. 1).

Table 1 Configuration of convolutional neural network

Convolutional layers	Six convolutional layers are used with a filter size of 64 for first two layers, 128 for next two layers, and 256 for the last two layers. The kernel size is set to 3 × 3 for all convolutional layers. ReLU activation function is used
Max pooling layer	Two max pooling layers are used. First layer after 2 convolutional layers and second layer after next four convolutional layers with a dropout rate of 20%. Each layer is two dimensional and uses a pool size of 2 × 2
Fully connected layers	Two fully connected layers are used. Flattened output of previous layers is given as input to first fully connected layer. A dropout rate of 40% is used. The second fully connected layer with a softmax activation function is the output layer

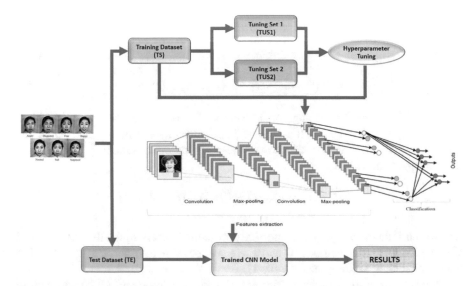

Fig. 1 Architecture of the proposed model

Hyperparameter Tuning—Metaheuristic optimization techniques proved to yield better results when the search space is large and complex [11]. Since the number of hyperparameters is large in CNN, tuning them is computationally expensive. Hence, the proposed model determines the optimal network topology by using the differential evolution (DE) algorithm. A simple, yet powerful, population-based stochastic search technique, differential evolution (DE) [12], has gained much attention and a wide range of successful applications [13, 14], due to its simplicity, ease in the implementation, and quick convergence.

The hyperparameters considered for tuning using DE include number of convolutional layers, filter size, stride, dropout rate, and batch size. A vector comprising the above-mentioned parameters is used as a chromosome for the DE algorithm. Precision and recall values for each of the six basic emotions are

Fig. 2 Differential evolution algorithm scheme

calculated by using the confusion matrix. Fitness function is defined as F = AvgPrec + AvgRec, where AvgPrec is the average of precision values computed for each basic emotion. Likewise, AvgRec is the average of recall values. DE aims to improve the existing solution using the techniques of mutation, recombination, and selection. The general paradigm of differential evolution is shown in Fig. 2.

Initialization—Creation of a population of individuals. The ith individual vector (chromosome) of the population at current generation t with d dimensions is as follows

$$Z_i(t) = \left[Z_{i,1}(t), Z_{i,2}(t), \ldots, Z_{i,d}(t)\right] \tag{1}$$

Mutation—A random change of the vector Z_i components. For each individual vector $Z_k(t)$ that belongs to the current population, a new individual, called the mutant individual, U is derived through the combination of randomly selected and pre-specified individuals.

$$U_{k,n}(t+1) = Z_{m,n}(t) + F * \left(Z_{i,n}(t) - Z_{j,n}(t)\right) \tag{2}$$

where the indices m, n, i, j are uniformly random integers mutually different and distinct from the current index 'k' and F is a real positive parameter, called mutation factor or scaling factor (usually ϵ [0, 1]).

Recombination (Crossover)—Merging the genetic information of two or more parent individuals for producing one or more descendants. Binomial crossover is used in the present work. The binomial or uniform crossover is performed on each component n (n = 1, 2, ..., d) of the mutant individual $U_{k,n}(t + 1)$. For each component, a random number 'r' in the interval [0, 1] is drawn and compared with the crossover rate (CR) or recombination factor (another DE control parameter), CR $\in$ [0, 1]. If $r <$ CR, then the nth component of the mutant individual $U_{k,n}(t)$ will be selected; otherwise, the nth component of the target vector $Z_{k,n}(t)$ becomes the nth component.

$$U_{k,n}(t+1) = \begin{cases} U_{k,n}(t+1), & \text{if } \text{rand}_n(0, 1) < \text{CR} \\ Z_{k,n}(t), & \text{otherwise} \end{cases} \tag{3}$$

Selection—Choice of the best individuals for the next cycle. If the new offspring yields a better value of the objective function, it replaces its parent in the next generation; otherwise, the parent is retained in the population, i.e.,

$$Z_k(t+1) = \begin{cases} U_k(t+1), & \text{if } f(U_k(t+1)) > f(Z_k(t)) \\ Z_k(t), & \text{if } f(U_k(t+1)) < f(Z_k(t)) \end{cases} \tag{4}$$

where f is the objective function to be minimized. It can be inferred that DE is a powerful population-based heuristic search technique that has empirically proven to be very robust for global optimization over continuous spaces. As the number of control parameters in DE is very few compared to other algorithms, DE is effective and efficient and thus can be treated as a widely applicable approach for solving real-world problems [13, 14].

4 Experimentation

For experimentation, two benchmark datasets, CK+ and Japanese Female Facial Expressions (JAFFE), are used.

CK+ Dataset: This dataset has 593 image sequences representing seven basic expressions (happiness, sadness, surprise, disgust, fear, anger, and neutral) of 123 models. Since the work is focused on recognition of six basic expressions, neutral expression images were ignored. Out of 593, 309 sequences have validated emotion labels that belong to one of the six previously mentioned emotions. They were selected by excluding other sequences. From each image sequence, last two frames were selected making a dataset of 618 images.

Japanese Female Facial Expressions (JAFFE): The JAFFE dataset has 213 images of ten female Japanese models. Each image represents one of the seven basic emotions (including neutral emotion). Images pertaining to neutral expression arc not uscd.

The proposed model is implemented using Keras with a TensorFlow back end in Python 3.6. Experiments are conducted on the selected datasets. Seventy percentage of the samples are used for training, and remaining 30% is used for testing. For validating hyperparameter tuning, 20% of the samples from training dataset are used. The samples are selected by using stratified sampling. Tables 2, 3, 4, and 5 show the confusion matrices of the two datasets used with and without hyperparameter tuning. Prediction accuracies for CK+ dataset and JAFFE dataset are depicted in Fig. 3. For both the datasets, optimization of hyperparameters has improved the accuracy of all basic emotions except fear. Fear has least impact of optimization. For JAFFE dataset, accuracy is decreased by 1%. Proposed model has improved the overall classification accuracy by 4.32% for CK+ dataset and 3.78% for JAFFE dataset.

Table 2 Confusion matrix for CK+ dataset without hyperparameter tuning

	Anger	Disgust	Fear	Happy	Sad	Surprise
Anger	0.7437	0.0834	0.0749	0.0428	0.0552	0
Disgust	0.0875	0.7535	0.0526	0.0187	0.0877	0
Fear	0.1476	0	0.7073	0	0.1019	0.0432
Happy	0.0425	0.0521	0.0472	0.7984	0.0582	0.0016
Sad	0.0637	0.0274	0.1126	0	0.7144	0.0819
Surprise	0.0553	0.0897	0	0.0404	0	0.8146

Table 3 Confusion matrix for CK+ dataset with hyperparameter tuning

	Anger	Disgust	Fear	Happy	Sad	Surprise
Anger	0.7943	0.1033	0.1024	0	0	0
Disgust	0.0948	0.7842	0.027	0	0.094	0
Fear	0.02	0	0.728	0	0.102	0.15
Happy	0.032	0.048	0.029	0.865	0	0.026
Sad	0	0.039	0.122	0	0.7451	0.0939
Surprise	0.0658	0.0597	0	0	0	0.8745

Table 4 Confusion matrix for JAFFE dataset without hyperparameter tuning

	Anger	Disgust	Fear	Happy	Sad	Surprise
Anger	0.6227	0.1026	0.0721	0	0.1789	0.0237
Disgust	0.1285	0.6815	0.0928	0	0.0427	0.0545
Fear	0.0098	0	0.6978	0	0.1352	0.1572
Happy	0.0236	0.0594	0.0587	0.7046	0.0216	0.1321
Sad	0.0252	0.1486	0.1734	0	0.6288	0.024
Surprise	0.0245	0.0438	0.0364	0.1058	0.0467	0.7428

Table 5 Confusion matrix for JAFFE dataset with hyperparameter tuning

	Anger	Disgust	Fear	Happy	Sad	Surprise
Anger	0.6543	0.1165	0.0687	0	0.1605	0
Disgust	0.1346	0.7374	0.0834	0	0	0.0446
Fear	0.0198	0	0.6878	0	0.1352	0.1572
Happy	0	0.0484	0.0297	0.8154	0	0.1065
Sad	0	0.1439	0.1839	0	0.6358	0.0364
Surprise	0	0.0361	0.0265	0.1278	0.0351	0.7745

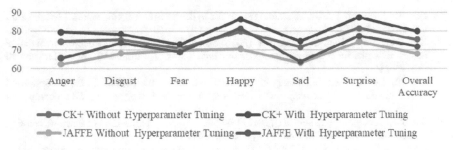

Fig. 3 Classification accuracies for CK+ and JAFFE datasets

5 Conclusion

The present study proposes to optimize the convolutional neural network hyperparameters for improving the human emotion recognition rate from facial expressions. Conventional techniques fail to offer good classification accuracy for noisy input data. As CNNs contain deep layers, they can handle noisy data and are proven suitable for facial expression recognition. However, CNNs demand high computation power making their applicability limited. The performance of CNN highly depends on the choice of its hyperparameters. To enhance the CNN performance for facial expression recognition, its hyperparameters are optimized using the DE algorithm. CK+ and JAFFE datasets are used for assessing the tuned model's performance. The results obtained have shown that hyperparameter tuning has improved the overall accuracy by 4%.

Acknowledgements The author wishes to thank the management of RVR&JC College of Engineering, for funding the present work.

References

1. Kaulard K, Cunningham D, Bülthoff H, Wallraven C (2012) The MPI facial expression database—a validated database of emotional and conversational facial expressions. PLoS ONE 7(3):e32321
2. Li H, Lin Z, Shen X, Brandt J, Hua G (2015) A convolutional neural network cascade for face detection. In: Proceedings of IEEE conference on computer visual and pattern recognition, pp 5325–5334
3. LeCun Y, Bengio Y, Hinton G (2015) Deep learning. Nature 521:436–444
4. Cirean DC, Meier U, Masci J, Gambardella LM, Schmidhuber J (2011) Flexible, high performance convolutional neural networks for image classification. In: Proceedings of the twenty-second international joint conference on artificial intelligence (IJCAI'11), vol 2, AAAI Press, Barcelona, Catalonia, Spain, pp 1237–1242
5. Ko BC (2018) A brief review of facial emotion recognition based on visual information. Sensors 18(2):401

6. Bergstra J, Bengio Y (2012) Random search for hyper-parameter optimization. J Mach Learn Res 13:281–305
7. Kim BK, Roh J, Dong S-Y, Lee S-Y (2016) Hierarchical committee of deep convolutional neural networks for robust facial expression recognition. J Multimodal User Interfaces 10 (2):173–189
8. Gao Z, Li Y, Yang Y, Wang X, Dong N, Chiang HD (2020) A GPSO-optimized convolutional neural networks for EEG-based emotion recognition. Neurocomputing 380:225–235
9. Snoek J, Larochelle H, Adams RP (2012) Practical bayesian optimization of machine learning algorithms. Adv Neural Inf Process Syst 2951–2959
10. Bochinski E, Senst T, Sikora T (2017) Hyper-parameter optimization for convolutional neural network committees based on evolutionary algorithms. In Proceedings of the 2017 IEEE international conference on image processing (ICIP), Beijing, China, pp 3924–3928
11. Radhika S, Chaparala A (2018) Optimization using evolutionary metaheuristic techniques: a brief review. Brazilian J Oper Prod Manage 15(1):44–53
12. Price K, Storn R (1995) Differential evolution—a simple and efficient adaptive scheme for global optimization over continuous spaces. International Computer Science Institute, Berkeley. Berkeley, CA
13. Sajja R, Rao CS (2014) A new multi-objective optimization of master production scheduling problems using differential evolution. Int J Appl Sci Eng 12(1):75–86
14. Radhika S, Rao CS, Pavan KK (2013) A differential evolution based optimization for Master production scheduling problems. Int J Hybrid Inf Technol 6(5):163–170

An Optimal Steering Vector Generation Using Chaotic Binary Crow Search Algorithm for MIMO System

P. Sekhar Babu, P. V. Naganjaneyulu, and K. Satya Prasad

1 Introduction

Beam forming is a radio frequency management technique in which an access point makes use of various antennas to transmit the same signal. Beam forming is considered a subset of smart antennas or advanced antenna systems. In general, beam forming uses multiple antennas to control the direction of a wave front by appropriately weighting the magnitude and phase of individual antenna signals in an array of multiple antennas. The applications of beam forming are found in numerous fields such as radar, seismology, sonar, and wireless communications [1, 2]. The MIMO radar performs multiple transmissions and has receiving antennas that has received enduring attention from researchers due to its capacity of providing diversity to enhance system performance [3]. One of the most important types of radar is the MIMO radar that has influenced many researchers to choose this topic for research in recent years [4, 5]. The collocated MIMO radar recognizes number of targets which is having a flexible number of transmit antennas for transmitting beam patterns in both uniform and directional beam patterns [6]. The omnidirectional beam pattern is transmitted for independent orthogonal waveforms, while the generated beam patterns are depending on the transmit wave feature [7, 8]. Consequently, the data rate of the signal decreases due to the side information of the receiver antenna, and the transmitted signals cannot be correctly reconstructed if the transmitted side information is corrupted [9]. Some of the well-known methods for channel detection in MIMO beam forming are estimated by using covariance

P. S. Babu (✉)
ECE Department, UCEK, JNTUK, Kakinada, Andhra Pradesh, India

P. V. Naganjaneyulu
Sri Mittapalli College of Engineering, Tummalapalem, N16, Guntur, Andhra Pradesh, India

K. S. Prasad
Rector of Vignan's Foundation for Science, Technology & Research, Guntur, Andhra Pradesh, India

© The Author(s), under exclusive license to Springer Nature Singapore Pte Ltd. 2022   423
R. R. Raje et al. (eds.), *Artificial Intelligence and Technologies*,
Lecture Notes in Electrical Engineering 806,
https://doi.org/10.1007/978-981-16-6448-9_43

matrix and steering vector, joint beam forming, space–time block codes (STBCs) with the pilot symbols tracked radar target. The MIMO system performance degradation is obtained because of the finite number of training iterations, and the signal steering vector was mismatched with the desired signal by corrupting the training data [10]. However, the beam forming technique in MIMO is a challenging process as it creates difficulty during non-convex optimization due to the non-convex constraints imposed by the phase shifters. In addition to this, the design of such beam formers for frequency selective channels is very difficult [11].

The present research overcomes time-variant distortions by developing an efficient algorithm called as CBCSA. In this work, artificial neural network (ANN) is tested and trained for the generation of steering vectors. The dimension of the virtual steering vector of the MIMO radar is relatively large, and thus, the autocorrelated matrix is estimated by using a low-complexity method which improved the computational efficiency of the adaptive beam forming algorithm [12]. The generated steering vectors are evaluated for determining the particular channels for beam forming. The proposed chaotic binary crow search algorithms (CBCSA) effectively predict the various channels for beam forming, and also it resolves the optimization problems [13, 14].

This research paper is organized as follows. In Sect. 2, numerous research papers on channel estimation in MIMO radars using beam forming techniques are discussed. The detailed explanation about the proposed system is given in Sect. 3. In addition, Sect. 4 illustrates the quantitative analysis and comparative analysis of the proposed system. The conclusion is made in Sect. 5.

2 Literature Review

The existing models that were involved in channel estimation using the beam forming in the MIMO radars are given as follows.

Qian [15] developed a robust adaptive beam former design for multiple-input multiple-output (MIMO) radar systems. The beam former system was designed with the help of beam forming weight vector. The desired signal was analyzed with the help of steering vectors, and the shrinkage estimator computed the interference due to plus noise covariance matrix. This model estimated the covariance matrix using rank-constrained minimization method for better performance. However, an average run time consumed by the developed method was more when compared to the existing methods.

Li [16] developed a joint optimization of hybrid beam forming in two stages for multiuser massive MIMO system in frequency division duplexing (FDD) mode. The weighted conditional average mean square error minimization (WAMMSE) algorithm was developed under the kronecker channel. The kronecker channel model showed the strongest eigenbeams of the receive correlation matrix. The optimal analog combiner maximized the intra-group interface to intergroup interface with addition to noise ratio. The advantage of the developed scheme was it showed the best performance even at the lower frequencies. However, the base station does not know the instantaneous information of the intergroup interference.

Zhou [17] developed a colocated MIMO radar waveform optimization with receive beam forming. An optimization process was performed for the receiver end of MIMO radar to analyze the beam forming operation. The receiver end of the system showed optimization criteria during beam forming operation. The models showed that correlation side lobes at the receiver end were suppressed for lower side lobe level. The Doppler problem was not examined for the system even when the generalization of the criterion case was not much difficult. However, the waveform optimization algorithm could be integrated with the hybrid algorithm for a better performance output.

3 Proposed Method

The aforementioned problems in the existing methods are overcome by proposing an efficient CBCSA optimization algorithm in this research. The wireless channels are predicted accurately on the basis of statistical fading past value and linear filter that predicts the channels for evaluation. In this section, the ANN is used for optimization with the chaotic binary crow search algorithm (CBCSA) to determine the accuracy of the channel and overcomes the optimization problem present in the process [18].

3.1 Feature Selection Using Chaotic Binary Crow Search Algorithm

Crow search algorithm is the most important existing optimization algorithm that suffers from low convergence rate and entrapment in the local optima. The research work considered a novel metaheuristic optimizer, namely CBCSA to overcome the optimization problem. The steering vectors from the downlink and uplink channels are adjusted to obtain maximum signal-to-noise ratio (SNR).

The output is achieved by minimizing the entire interference at the output array by eliminating the entire unwanted array and thereby it continually maintains a constant signal of interest. The flowchart of the proposed chaotic binary crow search algorithm is shown in Fig. 1. In Fig. 1, NP is known as the new position, MP is known as memory position, and AP is the awareness probability.

In this section, the randomly generated variables are used for updating the crow position and are substituted by chaotic variables. The optimal solution is influenced by variables substitution that updates the crow position. The convergence rate, chaotic sequence, and chaotic maps are used for generating optimal solution and a combination of such chaos is used for CBCSA. The research uses 10 different

Fig. 1 Flowchart of the
proposed chaotic binary CSA

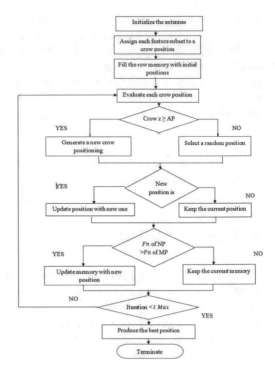

chaotic maps for performing optimization process such as Chebyshev, sinusoidal, circle, sine, tent, gauss, piecewise, singer, logistic, and iterative. The different chaotic maps used are improved significantly for the convergence rate, and performance of CSA is explained in the following section. The chaotic sequences are combined with the CSA approach as described in Eq. 1.

$$
y^{j,i+1} = \begin{cases} y^{j,t} + C_j \times fl^{j,t} \times (N^{z,t} - y^{j,t}), & C_z \geq AP^{j,t} \\ \text{Choose a rand position,} & \text{otherwise} \end{cases}
\tag{1}
$$

Here, C_j is the chaotic map value at jth iteration.
C_z is the chaotic map value obtained at zth iteration.
AP is the awareness probability.

The proposed CBCSA optimization algorithm is a feature selection algorithm that optimizes the generated steering vectors. In CBCSA, binary form is obtained as the solution pool where the results obtained are in the form of {0, 1}. The agents are transferred continuously to the binary space by using the equations as shown in (2) and (3)

$$y^{j,t+1} = \begin{cases} 1 & \text{if}(s(y^{j,t+1})) \geq \text{rand}() \\ 0 & \text{otherwise} \end{cases} \tag{2}$$

where

$$s(y^{j,t+1}) = \frac{1}{1 + e^{10(y^{j,t+1} - 0.5)}} \tag{3}$$

Here,

rand() is the randomly generated number from uniform distribution [0,1] and $y^{j,t+1}$ is the binary position updated at the t iteration for s steering vector. The proposed CBCS algorithm is designed based on wrapper method and implemented as a feature selection algorithm. The chaotic sequence is embedded in the iterations generated during binary searching. The dataset is described by using the optimal feature using CBCSA. The feature selection is used for improving the classification performance, reduction in the computation cost, and length of the feature subset. The detailed description is as follows.

3.2 Parameter Initialization

The proposed CBCSA is used to obtain optimal parameters which are given in Table 1. The CBCSA starts at the beginning and adjusts the parameters and initializes the crow positions randomly in the search space. Each position of the feature subset consists of different number of features having distinct length. The CBCSA is initialized by setting the parameter in Table 1.

Table 1 Parameter settings for CBCSA

Parameter	Value
Number of crows (M)	30
Awareness probability (AP)	0.1
Flight length (fl)	2
Lower bound	0
Upper bound	1
Maximum number of iterations (tMax)	50
Random number (R_j)	[0, 1]
Total number of dimension (D)	= Total number of features

Fitness function

The crow positions at each iteration are computed using the fitness function F_{n_t}. The data are randomly divided into two different parts such as testing and training. The proposed CBCSA sets M to 10 that ensures the stability of the obtained results. The proposed model aims for two main objectives, namely evaluating classification accuracy and the selecting number of features. The adopted fitness function combined two objectives into one weight factor as shown in Eq. 4.

The classification accuracy calculates the number of correctly classified sequences to the total number of instances. The KNN classifier is used for the proposed model, wherein $k = 3$ having mean absolute distance.

Where L_f is the selected feature subset length.
L_t—denotes the number of features used totally.
w_f—weighted factor having [0,1] value that controls the number of selected features and classification accuracy.

The main aim of the research is to increase the accuracy, so it generally sets the weight factor value to 1. But in the proposed method, the weight factor is set to 0.8. The best solution for the proposed method is obtained by maximizing the accuracy and by minimizing the number of selected features.

$$F_{n_t} = \text{maximize}\left(\text{Accuracy} + w_f \times \left(1 - \frac{L_f}{L_t}\right)\right) \qquad (4)$$

The crow position is updated using proposed CBCSA, and termination criteria are performed. The optimization process terminates when the iterations are reached to the maximum number which leads to the best solution.

3.3 Training and Testing the Data Using Artificial Neural Network

The ANN uses one continuous function for its description that is obtained as the small error using the training process. ANN values are updated for performing reduction in errors by obtaining possible values. The error result is obtained by analyzing the difference between the target attributes and the ANN output value.

The ANN has many distinct nodes set parallel for processing simple units that are structured and is used in network topology. The ANN has three different layers such as input, hidden, and output layers, where each layer consists of nodes linked with weights. The Neural Network weights are linked to the next-later nodes and likewise form stacked network for all the nodes of ANN. The bias nodes consist of single output node which is connected to that of remaining leftover nodes. The nodes are having connections with the bias nodes which are obtained from leftover nodes and are known as bias weights.

The generated weights are connected to each node x and are summed to indicate the activation function for input value of the node such as sigmoid function, sign function, and step function. The sigmoid function is most commonly used activation function which is shown in Eq. 5.

$$f(x) = \frac{1}{1+e^{-x}} \tag{5}$$

Here, the error is computed by using Eq. 2 which measures the differences for obtaining actual output (O) and desired output (T) and gives rise to a feedforward which is shown in Eq. 6.

$$\text{Error} = \frac{1}{2} \Sigma\Sigma(T - O)^2 \tag{6}$$

Figure 2 shows the block diagram of the beam forming. The outputs generated from the CBCSA feature selection algorithm are the steering vectors, and the generated array output is shown in Eq. 7. The average total output power P for ith array is expressed in Eq. 7.

$$P_i E\left[y_i(n)^2\right] = E\left[w_{i,k}^H(n)x(n)x^H(n)w_{i,k}(n)\right] \tag{7}$$

where the $w_{i,k}^H$ indicates the weights of steering vector for the ith user index and H is the Hermitian conjugate transpose.

$y_i(n)$ is an output array output signal and E is the mean operator defined in Eq. 8.
$$R = E\left[x(n)x(n)^H\right] \tag{8}$$

The autocorrelation matrix (R) which corresponds to an input signal receives an output power averagely for an array as shown in Eq. 9

$$P_i = w_{i,k}^H R_{w_{i,k}}(n) \tag{9}$$

The final steering weight of the vector maximizes SNR of the output array and minimizes the problem as per Eq. 10

$$w_{i,1}...w_{i,k}, P_{1,}...P_k, \left(\sum_{i=1}^{K} P_i\right) \quad \text{Subject to SNR} \geq \delta_i \tag{10}$$

where δ_i is the SNR value having lower dB, P_i is the transmission power for the ith user, and $w_{i,k,}$ is the optimum weight value of beam forming vector for ith user. This proposed research aims to obtain steering vectors with minimum power during entire transmission and maintains the SNR above the threshold. Thus, beam

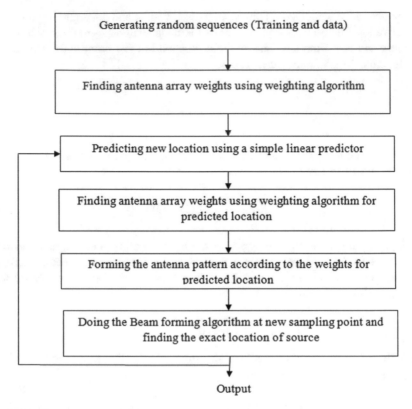

Fig. 2 Block diagram of the beam forming

forming is obtained from antenna arrays, and optimum weights are computed by using adaptive beam forming algorithms. The better performance and capacity are provided based on MIMO beam forming. From the setup of the proposed method, beam forming adjusts steering vectors for obtaining a maximum SNR.

4 Result and Discussion

This section detailed about the experimental result and discussion of the proposed algorithm. In this research, MATLAB (version 2019) was utilized for experimental simulations with 8 GB RAM, 3.0 GHz Intel i5 processor, and one TB hard disk. In this section, CBCSA algorithm is evaluated for various parameters such as bit error rate (BER), standard deviation (SD), symbol error rate (SER), and signal-to-noise ratio (SNR).

4.1 Performance Measures

- **Bit Error Rate:**

Bit error rate (BER) is a measure of the number of the bit errors occur in a given number of bit transmissions. The expression for BER is shown in Eq. 11.

$$BER = \frac{\text{Total number of error bits}}{\text{Total number of bits transmitted}} \tag{11}$$

- **Standard deviation:**

The standard deviation is a measure of the amount of variation or dispersion of a set of values. The expression for the standard deviation is expressed in Eq. 12

$$\sigma = \sqrt{\frac{\sum(x_i - \mu)^2}{N}} \tag{12}$$

σ is population standard deviation, N is the size of the population, x_i is each value from the population, and μ is the population mean.

- **Symbol Error Rate:**

Symbol error rate (SER) is defined as the number of changes in symbols, waveform, or signaling events across the transmission medium per time unit using a digitally modulated signal or a line code. The SER is expressed as shown in Eq. 13.

$$SER = \frac{\text{Number of symbols in error}}{\text{Total number of Transmitted symbols}} \tag{13}$$

- **Signal-to-Noise Ratio:**

Signal-to-noise ratio is defined as the ratio of the signal power to the noise power such as background noise or unwanted input. The SNR is expressed as shown in Eq. 14.

$$SNR = \frac{P_{signal}}{P_{noise}} \tag{14}$$

4.2 Quantitative Analysis

The performance measures SNR, SER, and BER are calculated for the proposed CBCSA.

The proposed CBCSA optimization algorithm performs the optimization process and thus generates the steering vectors for the determining the significant channel for beam forming. The symbol error rate across all channels is calculated across SNR, and bit error rate is also calculated across all the channels. Table 2 shows the performance measures BER and SNR obtained for the proposed CBCSA, and the performance measures SER and SNR obtained for the proposed CBCSA are tabulated in Table 1. Tables 2 and 3 show an improved SNR of 23.96 dB when the BER is 10^{-4} and the SNR of 34.95 when the SER is 10^{-10}.

The tabulation of amplitude values obtained for the proposed CBCSA method is shown in Table 4, and Fig. 3 shows the graph obtained for maximum errors for the proposed method.

4.3 Comparative Analysis

In this section, the proposed CBCSA and the existing particle swarm optimization and bacterial foraging optimization (PSO-BFO) algorithm are evaluated for various parameters. The steering vectors obtained from the proposed CBCSA algorithm showed better performance than BFO techniques obtained for same computational

Table 2 Performance measures BER and SNR obtained for the proposed CBCSA

SER	10^{-2}	10^{-3}	10^{-5}	10^{-8}	10^{-9}	10^{-10}
SNR (dB)	−10	−5	−1.09	9.89	20	34.95

Table 3 Performance measures SER and SNR obtained for the proposed CBCSA

BER	10^{0}	10^{-1}	10^{-2}	10^{-3}	10^{-4}
SNR (dB)	0	11.45	14.07	18.76	23.26

Table 4 Tabulation of amplitude values obtained for the proposed CBCSA method

Samples	5	10	15	20	25	30	35	40	45	50
Amplitude	1	2	6	1	1	0	1	7	0	4

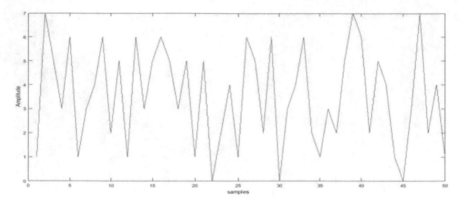

Fig. 3 Maximum errors obtained for the proposed CBCSA method

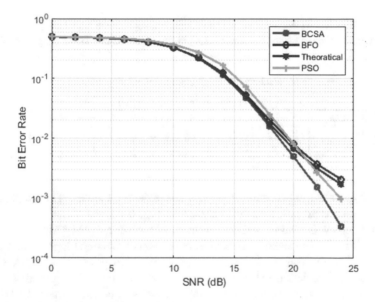

Fig. 4 Comparison graph for SNR against BER

load. The comparison graph for existing PSO-BFO and the proposed CBCSA against SER is shown in Fig. 4. From the results, it is observed that the SNR has received higher dB of noise and is plotted against BER.

Figure 5 shows the graph of the maximum errors obtained in the proposed method. The tabulation of maximum errors obtained for the proposed and existing method is shown in Table 5.

The amount of SNR obtained at the receiver end for the modulated signals shows the quality of the received output. The SNR performance obtained higher values at the receiver when the proposed model is developed. The proposed CBCSA showed

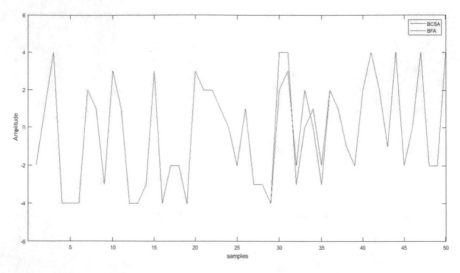

Fig. 5 Graph obtained for the maximum errors in the proposed and the existing methods

Table 5 Maximum errors obtained for the proposed and existing method

Samples	5	10	15	20	25	30	35	40	45	50
Amplitude (BFA)	−4	3	1.56	3.58	−3	2.48	−3	−2	−1.5	4
Amplitude (CBCSA)	−5	2	1.3	3.06	−4	4	−2	−1	−1.3	5

better SNR values when compared with the existing models. The graph for the existing and the proposed model is shown in Fig. 6

The standard deviation obtained from the existing and the proposed model is shown in Fig. 7. The standard deviation performance obtained higher values at the receiver when the proposed model is developed. The proposed CBCSA showed better standard deviation values when compared with the existing models.

The comparison graph for the existing PSO-BFO and the proposed CBCSA is plotted against SNR. From the results, it was observed that the actual SNR received is 2% higher when compared to SER. Figure 8 shows the comparison for SNR against SER. The comparison analysis for the SNR against SER for the proposed and the existing methods is shown in Table 6. The SNR values are plotted against the SER values ranging from 10^0 to 10^{-10}. As the value of SER decreases, the SNR will increase.

From the results obtained from the proposed system, it shows that the proposed system has better SNR when compared to the existing BFO-, PSO-based system. The importance of the proposed CBCSA was that it achieved better values of steering vectors and also these vectors were trained into ANN to achieve better SNR values. The proposed CBCSA research handles both time-variant distortion

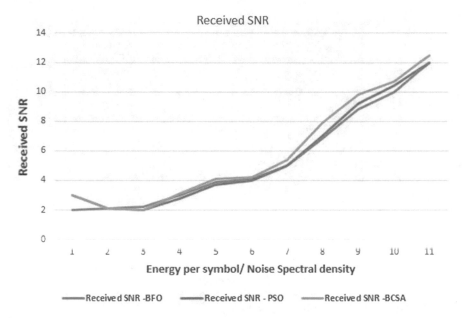

Fig. 6 Graph obtained for the received SNR performance measure for the proposed and the existing methods

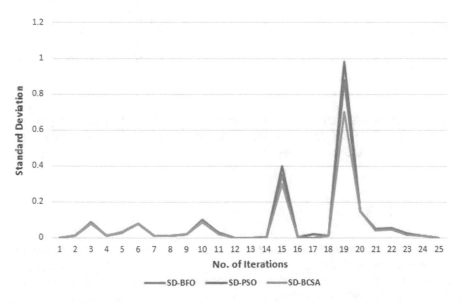

Fig. 7 Graph obtained for the standard deviation for the proposed and the existing methods

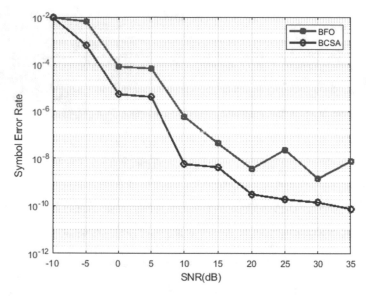

Fig. 8 Comparison graph for SNR against SER

Table 6 Comparison of SNR against SER for the proposed and the existing model

SER	10^0	10^{-2}	10^{-4}	10^{-5}	10^{-6}	10^{-8}	10^{-10}
Existing BFO (SNR dB) [5]	−4.8	0	5.6	10.54	23.75	35.46	36.53
Proposed CBCSA (SNR dB)	−14.75	−10	−4.56	−1.09	7.78	9.89	34.59

and also overcomes the optimization problems to reduce the complexity of the system. As the SER decreases, the SNR was increasing in dB which gives improvement in the convergence speed and decreases the time consumption.

5 Conclusion

The main objective of the beam forming is to obtain steering and weight vectors that maximize the SNR rate. The beam forming for the MIMO system provided better results for the determination of suitable channel during signal transmission. The proposed CBCSA algorithm is mainly focused on channel selection of the MIMO wireless transmission system, and it is important to train the CBCSA with ANN parameters to effectively producing channels for beam forming. The limitation of optimization problem during time-variant distortions in existing model BFO-PSO is overcome by CBCSA algorithm that obtained steering vectors as features and those trained into ANN for channel determination. The optimization is performed using a proposed CBCSA algorithm that improved the convergence speed, and the accuracy of the neural network was obtained by training the steering vectors during the feature selection process. The obtained results show that the proposed CBCSA achieved better SNR ratio when compared to the existing PSO and BFO algorithm. The future work can be extended by implementing hybrid PSO and CSA algorithm for overcoming optimization problem.

References

1. Li X, Wang B (2019) Hybrid MIMO phased array radar with center-spanned subarrays. IEEE Access 7:166883–166895
2. Liu F, Masouros C, Li A, Ratnarajah T (2017) Robust MIMO beam forming for cellular and radar coexistence. IEEE Wirel Commun Lett 6(3):374–377
3. Elbir AM, Mishra KV (2019) Joint antenna selection and hybrid beamformer design using unquantized and quantized deep learning networks. IEEE Trans Wirel Commun 19(3):1677–1688
4. Hung YC, Tsai SHL (2014) PAPR analysis and mitigation algorithms for beam forming MIMO OFDM systems. IEEE Trans Wirel Commun 13(5):2588–2600
5. Babu PS, Naganjaneyulu PV, Prasad KS (2019) Adaptive beam forming of MIMO system using optimal steering vector with modified neural network for channel selection. Int J Wavelets, Multi Resolut Inf Process 18(1):1941006
6. Payami S, Sellathurai M, Nikitopoulos K (2019) Low-complexity hybrid beam forming for massive MIMO systems in frequency-selective channels. IEEE Access 7:36195–36206
7. Deng H, Geng Z, Himed B (2016) MIMO radar waveform design for transmit beam forming and orthogonality. IEEE Trans Aerosp Electron Syst 52(3):1421–1433
8. Niesen U, Unnikrishnan J (2019) Joint beam forming and association design for MIMO radar. IEEE Trans Signal Process 67(14):3663–3675
9. Huang J, Su H, Yang Y (2019) Robust adaptive beamforming for MIMO radar in the presence of covariance matrix estimation error and desired signal steering vector mismatch. IET Radar Sonar Navig 14(1):118–126
10. Lan L, Liao G, Xu J, Zhang Y, Liao B (2020) Transceive beam forming with accurate nulling in FDA-MIMO radar for imaging. IEEE Trans Geosci Remote Sens 58(6):4145–4159
11. Raghavan V, Cezanne J, Subramanian S, Sampath A, Koymen O (2016) Beam forming tradeoffs for initial UE discovery in millimeter-wave MIMO systems. IEEE J Sel Top Sign Process 10(3):543–559
12. Huang J, Su H, Yang Y (2018) Low-complexity robust adaptive beam forming method for MIMO radar based on covariance matrix estimation and steering vector mismatch correction. IET Radar Sonar Navig 13(5):712–720
13. Wang X, Hassanien A, Amin MG (2018) Dual-function MIMO radar communications system design via sparse array optimization. IEEE Trans Aerosp Electron Syst 55(3):1213–1226
14. Zhang W, Vorobyov SA (2015) Joint robust transmit/receive adaptive beam forming for MIMO radar using probability-constrained optimization. IEEE Signal Process Lett 23(1):112–116
15. Qian J, He Z, Zhang W, Huang Y, Fu N, Chambers J (2018) Robust adaptive beam forming for multiple-input multiple-output radar with spatial filtering techniques. Signal Process 143:152–160

16. Li Z, Han S, Sangodoyin S, Wang R, Molisch AF (2018) Joint optimization of hybrid beam forming for multi-user massive MIMO downlink. IEEE Trans Wirel Commun 17(6):3600–3614
17. Zhou S, Lu J, Varshney PK, Wang J, Ma H, Liu H (2020) Colocated MIMO radar waveform optimization with receive beam forming. Digital Signal Process 98:102635
18. Sayed GI, Hassanien AE, Azar AT (2019) Feature selection via a novel chaotic crow search algorithm. Neural Comput Appl 31(1):171–188

IoT-Based Auto-Disinfectant Sprinkler System for Large Enclosed Space

K. S. Ackshaya Varshini, T. Aghil, G. Anuradha,
Y. Ashwin Ramanathan, G. Suganya⊙, and K. Karunamurthy⊙

1 Introduction

Sanitization is maintaining public health conditions by adopting stringent hygienic measures, and this activity is essential for the well-being of the society. This process of sanitization helps to maintain good health and increases the lifespan of humans on earth. As a result of inadequate sanitization, around 827,000 people die per year [1]. During this COVID-19 pandemic period, it is our responsibility to make sure that everything around us is cleansed and sanitized [2]. This pandemic situation taught the human community to be extra careful in all our movements, especially while dealing with people through physical means [3]. To ensure safety, the World Health Organization (WHO) insists on the sanitization of all places usable in real time. Hence, sanitization becomes mandatory and is becoming a regular activity [4]. There are many sanitization mechanisms and devices [5–7] available in the market which help to disinfect the surroundings in different ways. Pandya et al. [6] discussed a prototype that could help an individual to protect himself using an automatic sanitizer spray system. This system is equipped with a sanitizer sensing unit that works on solar power. Kodali et al. [7] designed a solar-powered sanitized toilet as part of a smart city project. The authors used ICT techniques with IoT to design a self-cleaning toilet that requires less water and is free from water clogging. Gnanasekar et al. [8] discussed a smart system to protect the hospital environment from overflowing bins.

The sanitization approach discussed in the literature limits to a localized area or a group of people. When the area to be sanitized is large, sanitizing the region or area puts a lot of hardship. But, with the smooth transition to adopt to the new-normal life through unlocking measures by the government such as opening malls, theaters,

K. S. A. Varshini · T. Aghil · G. Anuradha · Y. A. Ramanathan · G. Suganya ·
K. Karunamurthy (✉)
Vellore Institute of Technology Chennai, Chennai, Tamilnadu, India
e-mail: karunamurthy.k@vit.ac.in

© The Author(s), under exclusive license to Springer Nature Singapore Pte Ltd. 2022   439
R. R. Raje et al. (eds.), *Artificial Intelligence and Technologies*,
Lecture Notes in Electrical Engineering 806,
https://doi.org/10.1007/978-981-16-6448-9_44

and educational institutions, the need for sanitization in a proper way for a large area becomes inevitable. Internet of things (IoT) made things possible and has brought control of the entire world into a single point [6]. Taking advantage of this, and using information and communication technologies (ICT), in this research article, a user-friendly system of sanitizing without human intervention for making life simpler and safer [7–10] is discussed. There are many solutions suggested by researchers, but the methods available in the literature have limitations such as manual intervention, heavy lifting of sanitizing equipment, and time-consuming to sanitize nook and corner of the place [11].

2 Proposed Architecture

IoT-based autosprinkler system works by receiving inputs from even a remote place through a Web-based portal controlled by an administrator. This sprinkler system model is attached at the center of the ceiling of an enclosed space such as a seminar hall or auditorium. The size of the sprinkler depends on the shape and size of the hall to be sanitized. The sprinkler has 40 holes and can vary based on the diameter of the sprinkler and the amount of disinfectant. Spray nozzles atomize the sanitizer in the form of mist. The container for holding the disinfectant/sanitizer is connected to the nozzle through a pipeline. The sprinkler is also attached to a motor capable of rotating through 360°.

This setup is controlled by a user interface that could send and receive data from sprinklers through the cloud. The Wi-Fi module ESP8266 sends the data to the cloud by booting the module to serial mode which enables modifications using "AT" commands which are basic communication commands used (*Thingspeak* cloud, an open-source software is used to collect and view the data in real time through various forms like charts and graphs) [12]. Ultrasonic sensors are used to determine the distance that the mist spray will cover so that the entire area is covered. The If-This-Then-That (IFTTT) protocol is used to inform the administrator about the time since the last usage and also the amount of disinfectant available in the storage container [13]. Once the setup is done, the sprinkler will start working according to the time interval which is entered through a Web site by the administrator.

The disinfectant spray will automatically stop its operation after the specified time, and this is communicated to the administrator. The device operates on both battery and electric power sources to provide 24 × 7 availability.

2.1 Design and Assembly

The various components that are chosen for arrangement including base, nozzle, rotating axle, DC motors, ultrasonic sensors, ESP8266, and Arduino UNO are identified through literature study.

Figure 1 represents the flow of activities to assemble a sanitizing system. The components of the device are 3D printable and made up of recyclable PLA plastics, so the overall cost of the device is minimum, and this helps in mass production of the device in a lesser time duration. Figure 2 shows the design of the components of the disinfectant autosprinkler and the assembly of the device.

Figure 3 depicts the circuit design of autosprinkler system. A motor is placed for facilitating the spinning action that covers 360° ensuring that maximum distance is covered by the disinfectant mist droplet in the particular region [13]. A pump motor is connected to activate the sprinkler atomization. ESP8266 helps in getting the input from the user in order to program the device to sanitize in regular intervals [10, 13].

2.2 Distance Covered

The sprinkler is provided with four ultrasonic sensors at the edges to determine the distance the disinfectant will reach [7] and to calculate the area of coverage.

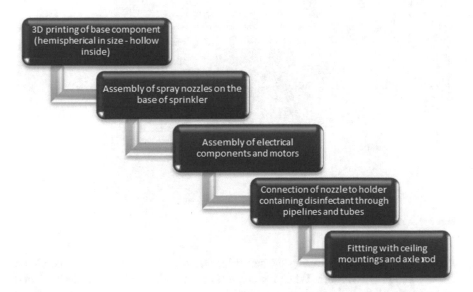

Fig. 1 Sprinkler—assembly steps

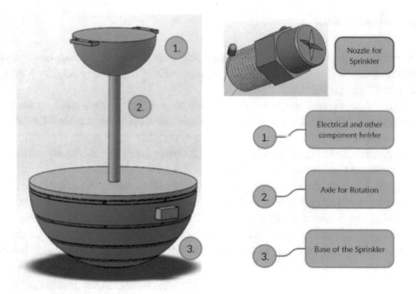

Fig. 2 3D view of sprinkler

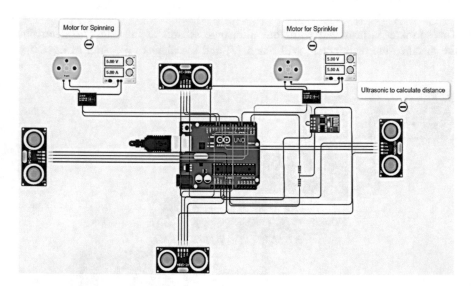

Fig. 3 Circuit diagram of autosprinkler

The power required for driving the sprinkler and the DC motor speed depends on the area of spray coverage. When the content of the sprinkler is below 20% of total capacity, the user will get an alert for refilling.

Figure 4 indicates the distance covered by the spray and the notations used in Eq. 1. The area of coverage and number of sprinklers required are calculated using Eq. 1.

$$A = (X + Y) * (P + Q) \tag{1}$$

where

A—Area of the region covered by the disinfectant,
X, Y—distance reached by the spray along Y-direction
P, Q—distance reached by the spray along X-direction.

Figure 5 depicts the graphs generated by four ultrasonic sensors placed on the disinfectant autosprinkler to find the maximum distance that will be covered in the sanitization operation for a closed square area hence resulting in similar graphs.

2.3 Threats to Validity

The device proposed is controlled by a remote Web site, and hence security is a major concern as the data sent and received are from the cloud. The chance for vulnerability attacks is more while using open-source software compared to licensed software. So, an open-source update plan is required, and an individual is assigned to watch for published vulnerabilities as-and-when it is necessary, and to test the integrity of the information and deploy an update to reduce the user's risk.

Fig. 4 Distance coverage by autosprinkler

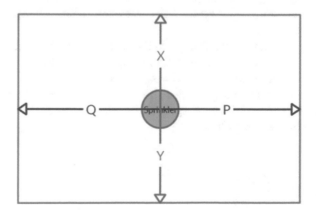

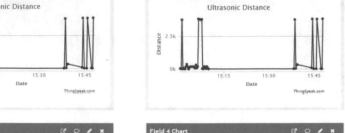

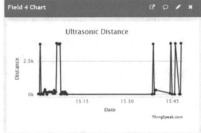

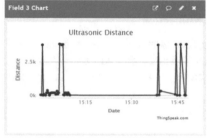

Fig. 5 Ultrasonic distance covered by sprinkler

3 Conclusion and Future Works

Complete sanitization of the space is possible with minimal human involvement and with a satisfactory level using this proposed system. This system functions on the data such as the start time, end time, and duration of the spray. These data are provided on a Web site, which also gives an alert when it has been too long (3 h approx.) since the last sanitization process occurred. The plastic used in manufacturing this device uses recycled PLA plastic, which makes the device eco-friendly.

In future work, the refilling of the disinfectant can be automated with mobile storage tanks, and artificial intelligence (AI) can be implemented when the system is made available for commercial use. The device can be made to receive crowd size as an input along with the size of the room to sprinkle the disinfectant. It can also be integrated with a fire sprinkler system and can act as both disinfectant sprinkler and fire rescue sprinkler.

References

1. World-Health-Organization, Sanitation. https://www.who.int/news-room/fact-sheets/detail/sanitation
2. WHO. Water, sanitation, hygiene and waste management for the COVID-19 virus
3. Singh T, Mahajan R, Bagai D (2016) Smart waste management using wireless sensor network. Int J Innovative Res Comput Commun Eng 4(6):10343–10347
4. WHO. http://www.who.int/gpsc/5may/automated-hand-hygiene-monitoring.pdf
5. Karimpour N, Karaduman B, Ural A, Challenger M, Dagdeviren O (2019) IoT based hand hygiene compliance monitoring. In: International symposium on networks, computers and communications, Istanbul, Turkey, pp 1–6
6. Pandya S, Sur A, Kotecha K (2020) Smart epidemic tunnel: IoT-based sensor-fusion assistive technology for COVID-19 disinfection. Int J Pervasive Comput Commun
7. Kodali RK, Ramakrishna PS (2017) Modern sanitation technologies for smart cities. In: IEEE region 10 humanitarian technology conference, Dhaka, pp 706–709
8. Gnanasekar Λ, Nivedheetha Λ (2018) IoT based hospital sanitation system. In: Int J Curr Eng Sci Res Comput Sci Eng
9. Kumar N, Vuayalakshmi B, Jenifer P, Shankar A (2016) IOT based smart garbage alert system using Arduino UNO
10. Nicole D (2020) Sustainable cities and the internet of things (IOT) technology, pp 39–47
11. Vermesan O, Friess P, Guillemin P, Gusmeroli S (2011) Internet of things strategic research agenda. In: Internet of things -global technological and societal trends, River Publishers
12. Stephen S, Geofrey KR, Margaret M (2012) Hygiene and sanitation in public eating places in one municipal health system of Uganda. Erudite J Med Med Sci Res 1(1):1–8
13. Lee I, Lee K (2015) The Internet of Things (IoT): applications, investments, and challenges for enterprises. Bus Horiz 58(4):431–440

Implementation of Pupil Dilation in AI-Based Emotion Recognition

K. S. Maanav Charan, Alenkar K. Aswin, K. S. Ackshaya Varshini, and S. Kirthica

1 Introduction

Emotion is associated with various neurophysiological outcomes that are abstract in nature which include one's feeling, internal thoughts, and expression of other general behavioral responses that include happiness, sadness, angriness, and many other states of feelings [1]. Even though emotions are something one knows well to express, they are yet something we cannot define evidently [2]. Emotions as a whole express one's neurological response to the situation in that surrounding environment and based on the events transpired or with which individuals as well [3]. Emotion recognition thus is the process of identifying an individual's emotion. It is the process of studying and identifying an individual's temporal emotional state based on various factors [1].

Emotion recognition is an upcoming matter of interest in the rapidly progressing world. Emotion recognition will be needed in various studies as it can help better detect the underlying notions of a human mind and state feeling without being verbally said [3, 4]. The emotion recognition studies can help understand the abstract concept of human emotion which till date nobody has been able to define. As a state of feeling, emotion can only be felt, but to make more advancements in bioscience, it is vital for us to in-depth understand this concept.

The first system of emotion recognition dates back to 1971, proposed by Ekman [5], which relies on expressions to differentiate the primary six emotions. Ever the majority of researches concentrated on a system that relies on facial expressions and speech [6] as it is easy to process [2, 7]. But all these features can be controlled by

K. S. M. Charan · A. K. Aswin · K. S. A. Varshini
School of Mechanical Engineering, VIT Chennai, Chennai, India
e-mail: maanavcharan.ks2018@vitstudent.ac.in

S. Kirthica (✉)
School of Computer Science and Engineering, VIT Chennai, Chennai, India
e-mail: s.kirthica@gmail.com

© The Author(s), under exclusive license to Springer Nature Singapore Pte Ltd. 2022   447
R. R. Raje et al. (eds.), *Artificial Intelligence and Technologies*,
Lecture Notes in Electrical Engineering 806,
https://doi.org/10.1007/978-981-16-6448-9_45

us, and the results could be tampered easily. Sometimes same facial expressions could mean a different emotion. Such variables may mislead the system and will not be sufficiently effective. By considering another variable that is the pupil dilation, this disadvantage can be overcome since it is controlled by the body's autonomic nervous system, and the chances of misleading are less [8].

Emotions can be exhibited by an individual anytime and every time. It is one psychological response to the surrounding environment. With a modernizing world, study of emotion can help solve human problems and help understand human psychology more precisely and also play a significant role when combined with AI. Thus, emotion stimuli alongside pupil dilation can help study the neuro pattern evolved when one responds to the surroundings and provide a basis for further in-depth analysis of human emotions.

2 Methodology

In this system, pupil dilation is considered as a variable in addition to the existing facial emotion recognition system to give more accurate results. Pupil dilation is the process in which the pupil changes its size due to factors such as emotions or change in light intensity. They can portray our emotions and feelings very accurately. Pupil dilation occurs when stimulation of the automatic nervous system's sympathetic branch takes place [9]. But the pupil response to any emotional state occurs at a very small scale with changes around one or half millimeters [10]. We could reach to a conclusion about the emotional state of the individual on the basis of pupil dilation [4]. But since the variation is too small, it cannot always be captured using a camera. Hence, a different approach is taken in order to calculate the dilation.

2.1 Pupil Size Calculation

For the calculation of pupil size, two variables, heart rate and galvanic skin response, are considered. Not many researches are based on these factors but one such research which was done recently in 2018 by Wang et al. [11] gives a co-relation between these three factors which can be used for our model for the pupil size calculation. Multiple regression analysis was done trial-by-trial basis, and the following equation was given by their model.

$$\text{Size of Pupil} = (A * \text{Heart Rate}) + (B * \text{Galvanic Skin Response}) + C \quad (1)$$

In Eq. 1, the values A, B, and C are the constant linear weights that the model generated, which are the external factors affecting the pupil dilation. Heart rate is defined as the number of times one's heart beats within a minute [12].

When conflicted to a specific emotion, one's heart beat also changes accordingly, and these continuous variations caused in electrical characteristics of the skin are referred to as electro dermal activity. Thus, galvanic skin response is a term included within electro dermal activity that refers to the variations in the sweat gland activity in response to the basis of emotional state of the individual. They are all linked and associated with our autonomic nerve system. Galvanic skin responses are associated with sympathetic and heart rate with parasympathetic actions. When any parasympathetic action takes place, heart rate and pupil size decrease [13]. Similarly, certain changes in the pupil size can be observed when a sympathetic action takes place.

2.2 Working of the Pupil Dilation Emotion Recognition System

Initially, the heart rate and the galvanic skin conductance of the person are measured. These values are then processed and used to calculate the pupil size with the help of the mentioned equation [9, 12]. The calculated value is taken over to the next step classed classifier selection which recognized the emotion with the given set of values. Various classifiers are present for this purpose of emotion recognition [14]. Out of all, the K-nearest neighbors (KNNs) algorithm will be most suitable, as this type of algorithm is mainly applicable for the classification and regression predictive purposes [15]. Finally, the emotion is recognized and given as output (Fig. 1).

In state of anger or fear, the heart beat tends to increase because of the stress induced. In state of amusement or happiness, the heart beat is significantly lower than in case of fear. The root means square successive differences, an index of heart rate variability is larger in the case of amused condition than in fear and angered conditions [12]. It could therefore be said that the selective activation of

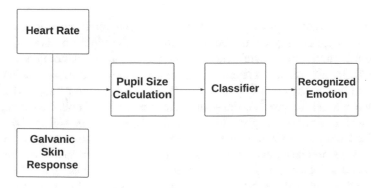

Fig. 1 Schematic flowchart of working of an emotion recognition system based on pupil dilation

parasympathetic nervous system is due to a decreased heart rate during amused state. When amused, the feeling of well-being stimulates the sympathetic inhibition which helps to reserve mental and physiological resources. Similarly, in the cases of anger and fear, a low parasympathetic activity coupled with uninhibited sympathetic activities induced by these emotions leads to an increase in the heart rate of an individual [7].

The emotional arousal of one changes on the basis of their surroundings. Also, emotions like happy, sad, angry, and scary lead to change in the sweat glands and provide the galvanic skin response factor for the existent state of emotion [3]. Human body sweat glands are regulated by automatic nervous system. In automatic nervous system, the sympathetic branch is responsible for this response when an emotional arousal occurs, thus leading to the sweat glands activating and causing the electrical variance. Galvanic skin response signal is detected by attaching electrodes to the second and third finger. The applied current variance across the two electrodes is used to measure galvanic skin response. Recent developments along the field allow the usage of bracelets and watches in the place of electrodes. All these variations and changes in the skin conductance and heart rate will cause a subsequent change in the pupil size [8]. And for different emotions, the particular range of pupil size will vary accordingly. Researches prove that an arousing bigger emotion or stimuli can increase the pupil dilation [16]. Negative emotions such as crying will end up with a larger pupil dilation than positive emotions [4]. Apart from emotion recognition, when performing effortful tasks, the pupil size increases [14]. Audio- and video-based emotion recognition is developing these days, and pupil dilation will make it more efficient [13]. For example, it is simpler and easier to recognize 'fear' through audio-based emotion recognition, but to understand the feeling of 'disgust' accurately, pupil dilation is used [17]. Hence, combining the pupil dilation input with existing video-based emotional recognition will result in improved accuracy.

3 Results and Discussions

In addition to face emotion recognition, when this pupil dilation method for emotion recognition is used, the output results can be highly accurate than compared to their individual results. The two systems are therefore combined to create an overall, highly efficient system. Fig. 2 explains how the pupil dilation works along with face recognition to predict emotion in an accurate manner [18].

In face recognition process, the spatial data captured and collected is converted into 4D feature which is the given to the classifier as input. The collected data is the normalized and is divided into frames [11, 18]. Each frame covers a part of the face like eyebrows, forehead, eye, and cheek. In each of these frames, the features are consolidated together to form a set of data vector. From these data, features are selected and extracted. A separate classifier is assigned for each frame [15].

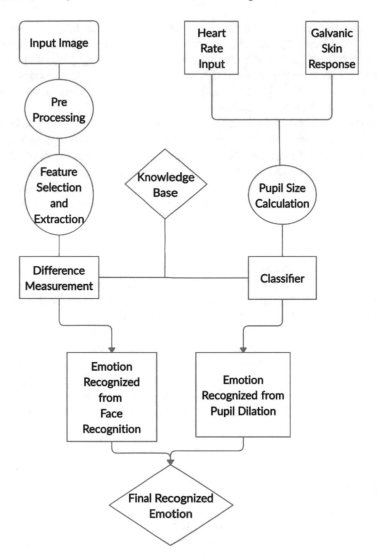

Fig. 2 Schematic flowchart of working of an emotion recognition system based on both facial recognition and pupil dilation

These classifiers utilize the features extracted and significantly determine the emotion from the frames [19]. The emotion that was extracted from the majority of the frame is then declared as the emotion evaluated from the facial recognition [20].

When it comes to pupil dilation, it is calculated using heart rate and galvanic skin response. Heart rate is captured and is multiplied with external factors that affect the rate, and the same is done with galvanic skin response. These values end up in the size of the pupil [10]. The size obtained by the calculation is then used by another classifier to decide the emotion. For example, when the pupil size is high compared to normal values, it is an extreme emotion. This emotion is then compared with the output emotion obtained from the face recognition, and in case of variation in results obtained, pupil dilation results are given more weightage as it gives accurate emotional state of a person. This is how pupil dilation is used efficiently in emotion recognition.

Individuals are known to be linked to cognitive processing during tasks that do not involve emotional stimuli when subjected to affective changes in pupil sizes [16]. But evidence found says that emotional processing recruits cortical regions normally associated with cognition, and hence, pupil responses can avert the underlay affective processing [13]. Studies reveal that emotional arousal can be stimulated by both visual and auditory factors, causing a greater increase in pupil size than emotionally neutral stimuli. Researches indicate that negatively balanced stimuli cause greater dilation of the pupil than positive stimuli [17]. On the other hand, using galvanic skin response on the basis of stimulation of sweat glands, we understand the emotional status of the individual [10].

If quantitative data is added to the study of emotional arousal by using values of skin conductance or the number of peaks of galvanic skin response, it allows us to understand human behavior and emotions precisely. Hence, galvanic skin response can help with its data to put a more precise and in detail view of the emotional recognition than pupil dilation all alone. Likewise, analyzing an individual's heart rate gives a sense of the individual's emotional state [2]. As said above in different state of emotions—heart rate is found to be different, and thus, alongside galvanic skin response, these together can help to put in perspective the emotion recognition in a more precise manner than solely relying on the outcome of the pupil dilation alone.

4   Conclusion

As this technologically advanced world progresses rapidly, we tend to cogitate and find more unique ways to make our lives more efficient and easier. In such situation for future betterments, the study of emotion recognition does play a vital role. Thus, it makes it more pertinent to invest in this concept of emotional recognition. Emotional recognition will play a major bearing in both technological and psychological fields. Technologically, this will have a major bearing in human interface system. In psychological aspect to help understand more precisely about human emotions and solutions to individuals who are struggling with the emotions.

It is said that an emotional state of an individual does have an influence on their heart rate and galvanic skin response, and thus, usage of those will play a major role in recognition of emotion which will prove to be more efficient, accurate, and reliable than the existing methodologies. Thus, emotion recognition is key to future progress, and the inclusion of parameter like GSR and heart beat alongside pupil dilation makes the system more efficient and reliable.

As a future work, this method of emotion recognition can be made more accurate and practical with advancement in technologies by combining face recognition, pupil dilation, and heart rate using artificial intelligence. Brain actions may also be detected using advanced methodologies to predict the emotion more precisely.

References

1. Andreassi JL (2013) Psychophysiology: human behavior and physiological response. Psychology Press
2. Singh D (2012) Human emotion recognition system. Int J Image, Graph Sign Process 4 (8):50–56
3. Salah AA, Kaya H, Gürpınar F (2019) Video-based emotion recognition in the wild. In: Multimodal behavior analysis in the wild, pp 369–386 Academic Press
4. Zhang B, Provost EM (2019) Automatic recognition of self-reported and perceived emotions. In: Multimodal behavior analysis, pp 443–470. Academic Press
5. Ekman P, Friesen WV (1971) Constants across cultures in the face and emotion. J Pers Soc Psychol 17(2):124
6. Nwe TL, Wei FS, De Silva LC (2001) Speech based emotion classification. In: Proceedings of IEEE region 10 international conference on electrical and electronic technology, pp 297–301
7. Lee CM, Narayanan SS, Pieraccini R (2002) Classifying emotions in human-machine spoken dialogs. In: Proceedings IEEE international conference on multimedia and expo, pp 737–740
8. Oliva M, Anikin A (2018) Pupil dilation reflects the time course of emotion recognition in human vocalizations. Scientific Reports 8:1–10
9. Why Do Pupils Dilate in Response to Emotional States? Online: https://www.scientificamerican.com/article/eye-opener-why-do-pupils-dialate/
10. Kret ME (2018) The role of pupil size in communication. Is there room for learning? Cogn Emot 32(5):1139–1145
11. Wang C-A, Baird T, Huang J, Coutinho D, Brien DC, Munoz DP (2018) Arousal effects on pupil size, heart rate, and skin conductance in an emotional face task. Front Neurol 1029
12. Akselrod S, Gordon D, Ubel FA, Shannon DC, Berger AC, Cohen RJ (1981) Power spectrum analysis of heart rate fluctuation: a quantitative probe of beat-to-beat cardiovascular control. Science 213(4505):220–222
13. Bradley MB, Miccoli LM, Escrig MA, Lang PJ (2008) The pupil as a measure of emotional arousal and automatic activation. Psychophysiology 45(4):602–607
14. Black MJ, Yacoob Y (1995) Tracking and recognizing rigid and non-rigid facial motions using local parametric model of image motion. In: Proceedings of the international conference on computer vision, pp 374–381
15. William R, Sherman AB, Craig (2018) Understanding virtual reality 2nd edn, pp 190–256. Morgan Kaufmann
16. Loewenfeld IE, Bremner FD, Smith SE (1999) The pupil: anatomy, physiology, and clinical applications, p 2278
17. de Gee JW, Knapen T, Donner TH (2014) Decision-related pupil dilation reflects upcoming choice and individual bias. Proc Nat Acad Sci 111(5):618–625
18. Busso C, Deng Z, Yildirim S, Bulut M, Lee CM (2004) Analysis of emotion recognition using facial expressions, speech and multimodal information. In: Proceedings of the 6th international conference on multimodal interfaces, pp 205–211. Association for Computing Machinery, State College, PA, USA

19. De Silva LC, Ng PC (2000) Bimodal emotion recognition. In: Proceedings fourth IEEE international conference on automatic face and gesture recognition, pp 332–335

20. Yoshitomi Y, Kim SI, Kawano T, Kilazoe T (2000) Effect of sensor fusion for recognition of emotional states using voice, face image and thermal image of face. Robot and human interactive communication. In: Proceedings of 9th IEEE international workshop

A Generalized Comprehensive Security Architecture Framework for IoT Applications Against Cyber-Attacks

M. Nakkeeran and Senthilkumar Mathi

1 Introduction

By 2025, the Gartner prediction of IoT connected devices will grow to 38.6 billion and by 2030 will hit about 50 billion [1]. The growing IoT deployments in various fields have made our lives easy with automation and sophistication.

The IoT age contributes to a massive explosion of devices, unparalleled data volume and time, and a multitude of forms of interconnectivity, and puts exceptional loads on the network. Vulnerabilities of IoTs/IIoTs identification, authentication, and data integrity require robust protection approaches, such as having a holistic perception of protection, detection of anomaly and abuse, adaptive policy updates, software testing, and maintaining the appropriate modifications which will be vital to ensuring security [2]. Besides, it would also require the screening and classification of packets for their level of security and to ensure that they are processed accordingly. Dynamic flow control would be required to detect/mitigate malicious traffic flows. Policy implementation has to be federated. Dynamic encryption should be provided, and integrity audits should also be performed. It should offer to mitigate vulnerabilities and counter attacks such as the injection of information attacks and the manipulation of services. Advancements in the evolving paradigm of machine learning (ML)/deep learning (DL) provide an extensive defensive solution against cyber-attacks in the IoT/cyber-physical environment rather than firewalls and signature-based intrusion detection system (IDS) [3–8].

Although IoT has a tremendous influence on every aspect of life, there are many security concerns and difficulties in IoT systems while implementation. Security

M. Nakkeeran (✉) · S. Mathi
Department of Computer Science and Engineering, Amrita School of Engineering,
Coimbatore, Amrita Vishwa Vidyapeetham, Ettimadai, India
e-mail: m_nakkeeran@cb.students.amrita.edu

S. Mathi
e-mail: m_senthil@cb.amrita.edu

© The Author(s), under exclusive license to Springer Nature Singapore Pte Ltd. 2022   455
R. R. Raje et al. (eds.), *Artificial Intelligence and Technologies*,
Lecture Notes in Electrical Engineering 806,
https://doi.org/10.1007/978-981-16-6448-9_46

causes prominent challenges in IoT systems, software, and platforms [9]. Addressing this critical IoT issue, this paper provides comprehensive end-to-end security solutions for several security issues and challenges. Full-fledged IoT security needs to consider the basic key principles of data confidentiality (C), integrity (I), availability (A), access control, identification capabilities, as well as different dimensions of six IoT security principles as explained in Sect. 5.3. After analyzing some of the most recent IoT studies, this paper presents a potential IoT security framework in several areas of technological, educational, and industrial sectors.

This paper is split into five sections as follows: In Sect. 2, the literature survey depicts the main issues and challenges related to IoT security considerations. Section 3 reveals security issues and challenges in IoT. Section 4 outlines the security attacks in IoT layers. Section 5 provides approaches to the solution in a comprehensive manner. In the end, Sect. 6 draws the conclusion and future work.

2 Literature Survey

The evolving IoT field and its potential impact on various sectors not only minimize the cost but also provide a transparent backdoor for cyber-attackers to easily intrude into IoT network. As per Gartner reports, around 5.5 million "events" are linked to everyday life. Hackers start targeting IoT devices with weak security in botnets and other loop holes in the network. Due to increased physical and digital threats, there is always a demand for full-fledged security solutions to safeguard IoT devices from cyber-attackers.

Based on the findings of the research, it is important to plan and enforce an effective framework for IoT solutions that can guarantee: CIA triads in heterogeneous environments. Security threats are a vital issue to be considered for the minimal ability of the systems being connected wirelessly while deploying IoT systems with wireless sensor connectivity, edge networks, and application accessibility which become a challenging task for monitoring and controlling [10]. Security concerns such as botnets, denial-of-service (DoS)/distributed denial-of-service (DDoS) attacks, man-in-the-middle (MITM) attack, identity and data theft, social engineering, advanced persistent threats (APT), Ransomware, and remote recording hinder IoT deployment in application usage and restrict access to the user information and application management, etc.

Researchers and academic people in and around the world have provided an IoT security solution to specific attacks to specific devices with specific connectivity protocols like Zigbee, Wi-Fi, Bluetooth LE (BLE), etc. There are some solutions to mitigate the attacks over communication protocols such as MQTT, CoAP, and Web Sockets, and machine learning/deep learning techniques as a solution for IDS against cyber-attacks. There is no such comprehensive and unified IoT security solution to enhance the IoT devices to be highly secure in an intelligent way with a cross-layer approach.

IoT security solutions need many improvements at different levels in the architecture. At present, only a few papers address multiple security issues with common IoT solutions. There is an immense need to draw a common security solution to cover most of the security issues in one picture. But there is a challenging task to implement security solutions for all types of attacks because of its computing capacity and battery power constraints. This paper discussed various security issues and attacks in IoT layers with the strategic comprehensive end-to-end security solution at four different levels with six security principles.

3 Security Issues and Challenges in IoT

Most IoT systems have fewer security measures against cyber threats. The IoT system's architecture makes these security flaws difficult to guarantee.

Security issues which can raze IoT applications include:

- Lack of established technology and market processes: Diverse expectations are increasing rapidly in number. This ambiguity will in effect help to implement bugs and create an opening for the attackers to penetrate the business.
- Weak guidelines on lifecycle maintenance and IoT system management.
- Concerns about physical stability. Lack of consensus on how to handle IoT edge applications for security and authorization.
- IoT-based security operation activities lack the best practices.
- Standards for auditing and logging of IoT components are not set.

IoT system stability needs to be ensured with the already said network security principles of CIA triads, authentication along with lightweight solutions, heterogeneity, key management, and policies since it is essential for all types of communications in the networks. Moreover, the IoT has several constraints and limitations in terms of component devices, computing power capabilities, and the ubiquitous design of IoT, which require broad studies in organizing security. These security principles need to be practiced for achieving a safe connectivity environment for the individuals, apps, processes, and other kinds of stuff in the IoT.

3.1 Smart City Threats

In the evolving age, IoT has become ubiquitous and the smart city strives to simplify and improve people's lives (Fig. 1). It consists of smart home, smart energy, smart grids, smart transportation, smart health, smart water management, etc. [11]. A smart city is divided into 4 layers as discussed in Sect. 5.

When constructing a smart city, obsolete technology may be required to be fused with IoT, and unknown threats need to be dealt with over security policies and procedures [12].

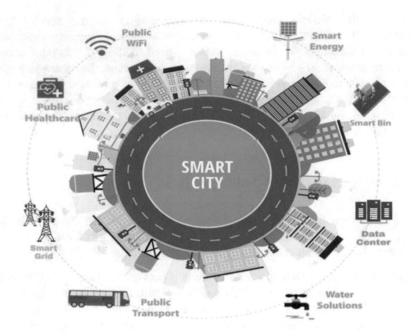

Fig. 1 Smart city

As mentioned in the introduction, billions of IoT devices are being linked over time, and completely leveraged IoT in smart cities makes things to secure properly. Some of the IoT threats in smart cities are botnet attacks [13], WSN issues, cloud data confidentiality, and heterogeneity issues. The major concerns for the proliferation of cyber threats in smart city environments are lack of security controls, weak encryption standards, the nonexistent secure device on-boarding services, poor knowledge of social engineering, lack of defense against DoS/DDoS attacks, and fool-proofing the machine learning models.

4   Security Attacks in IoT Layers

IoT security is a major challenge due to its complexity, heterogeneity, and a wide variety of linked resources. The attacker may make an IoT system assault by disrupting or tampering any of the nodes, for example, a physical vulnerability, (or) by using errors in the protocol routing (or) other protocol connected to the network (or) using malicious programs and by encryption techniques such as a cryptographic attack. Based on these vulnerabilities, the attacks are classified into five categories as follows: physical attack, pure software attack, network attack, cryptanalysis attack, and side-channel attack.

4.1 Attacks in IoT Sectors

IoT and its consequences of cyber-attacks in different contexts appear to attract the most coverage in articles and draw attention to research people for its defense. In 2010, Stuxnet malware destroys centrifuges in air-gapped Iranian nuclear facilities. In 2015, a BlackEnergy malware attack on the Ukrainian electrical grid disables 50 substations—destroys hard drives, battery backups, and access to controllers. In 2016, Mirai botnet malware has brought down large-scale networks that stumbled targeting online consumer devices such as IP cameras and home routers and Industoyer attack on the Ukrainian grid—the first autonomous targeted IIoT malware. In 2017, NotPetya (from Russia) and WannaCry (from North Korea) cause billions of dollars of damage for industrial firms. In 2018, a Triton attack on the petrochemical industry in Saudi Arabia disables safety controllers. In recent years, various types of Ransomware attacks target both commercial and industrial sectors and a few are EKANS ransomware attacks on the US natural gas pipeline [14], LockerGoga ransomware strikes 5 firms including Norsk Hydro aluminum and renewable energy company (Norway) causing $70 M+ losses.

Table 1 specifies the possible attacks on different IoT layers [15, 16].

Table 1 Specific attacks on IoT's each layer

S. no.	IoT layers	Possible attacks
1.	Hardware (perception layer)	Node replication, Node capture, Node jamming, Malicious code injection, Side-channel attack, Hardware Trojans
2.	Connectivity management (edge layer)	Spoofing attack, Sniffing attack, Hijacking, MITM
3.	Network communication (network layer)	Sybil attack, Sinkhole attack, Warm hole, Blackhole, Replay attack, Eavesdropping, DoS
4.	Cloud/service (application layer)	Phishing attacks, SQL injection, Malicious scripts, Malware, Buffer overflow, DoS, DDoS

4.2   *Data-Driven Attacks in IoT*

Figure 2 specifies the categorization of various data-driven attacks in IoT and its data loss.

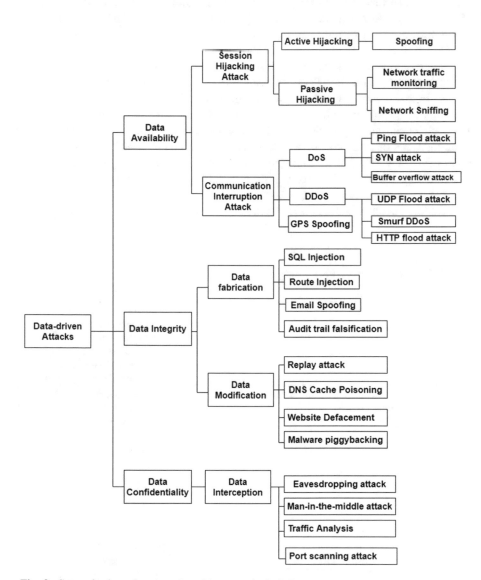

Fig. 2 Categorization of various data-driven attacks in IoT

5 Approaches to Comprehensive IoT Security Solution

Our objective of the proposed architectural framework will provide a generalized comprehensive end-to-end IoT security solution across four different levels with six principles against cyber-attacks, and it encompasses four layers of the IoT reference model.

5.1 Proposed IoT Reference Model and Flow Diagram

The proposed IoT reference model and its building block consist of four layers; the first layer (perception layer) includes physical devices and connectivity technologies along with the intelligent intrusion detection system (IDS) to learn attacks and to defend from attackers, and the second layer (edge layer) consists of edge devices with data storage before being pushed into the cloud through network layer (third layer) and includes intelligent IDS to defend from attackers. Finally, the last layer (cloud layer) at top of the system accumulates data for processing, analyzing, and decision-making. Both the IDS will generate an alarm to the intrusion prevention system (IPS) to take necessary actions against cyber-attacks. Our proposed IoT reference model and the process flow diagram of end-to-end security for IoT devices are illustrated in Figs. 3 and 4, respectively.

By implementing, sensor-level IDS at the hardware-level and edge-level IDS at the edge device will have a double-check security solution to build the system secure. The above-proposed intelligent IDS techniques with ML/DL will detect anomaly attacks in the networks in an adversarial environment too. Developing a comprehensive security solution across various sectors targeting IoT devices requires different levels that incorporate IoT security architecture frameworks across four layers in four different levels with six security principles, and the levels are device security, communication security, cloud security, and secure lifecycle management. This provides a comprehensive end-to-end security solution to mitigate the vulnerabilities and counters any type of attack that makes IoT devices to be more secure and intelligent.

5.2 Security Threat Mitigation Method at Each Layer

Table 2 summarizes the security requirements in IoT layers with recommended security solutions and proposed techniques of IDS [17].

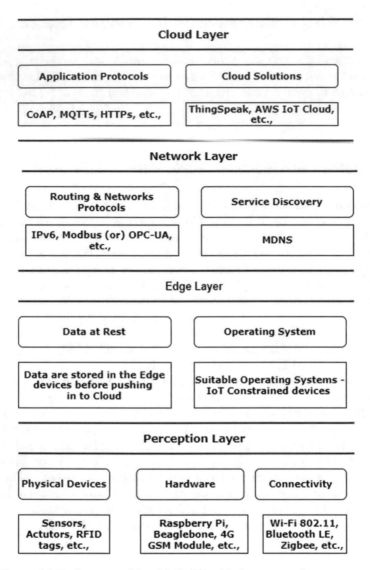

Fig. 3 Proposed IoT reference model and its building blocks—an overview

5.3   *Proposed Architectural Framework*

Figure 5 illustrates the proposed architectural framework—a comprehensive security solution for end-to-end IoT devices against cyber-attacks.

IoT security architecture frameworks across four layers in four different levels with six security principles and the security principles are

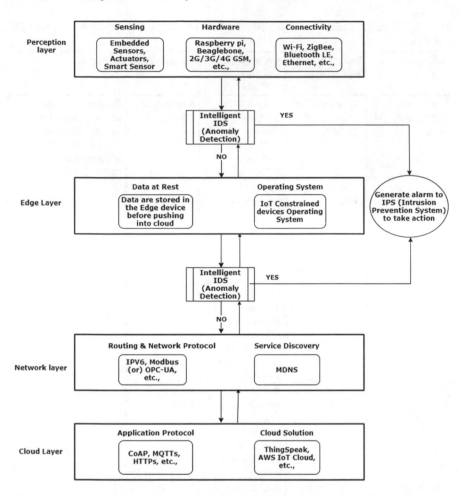

Fig. 4 Process flow diagram of end-to-end security blocks—an overview for IoT devices

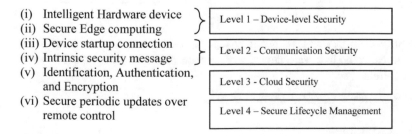

Level 1 Device-level Security. The perception layer belongs to the hardware level of the IoT system which consists of physical things (or) objects with their

Table 2 IoT layer, security requirements, recommended security solutions with the proposed IDS techniques

IoT layers	Security requirements	Recommended security solutions	Proposed IDS technique	
Hardware (perception layer)	Authentication notarization/ signature	Data encryption, IP-Sec. mechanism, Cryptography technology	Sensor-level IDS	Choice of intelligent intrusion detection system (IDS) using ML/DL techniques (discussed in subsection 5.4)
Connectivity management (edge layer)	Data integrity Data confidentiality, Access control	Key management, Secure routing, Symmetric and asymmetric cryptography, AES algorithm (128-bit key length) with diffusion techniques [18], Elliptic curve Diffie–Hellman key agreement protocol [19]		
Network communication (network layer)	Availability/ authentication	End-to-end authentication, Flooding detection (datagram transport layer security) [20]	Edge-level IDS	
Cloud/service (application layer)	Authorization	Biometrics and access control list (ACL), Antivirus and firewalls		

unique identity. The device-level security can be increased by integrating more security features in the hardware device with embedded software by the designers and manufacturers.

Security architecture features at the device level:

- Chip-level security
- Secure booting
- Physical security protection

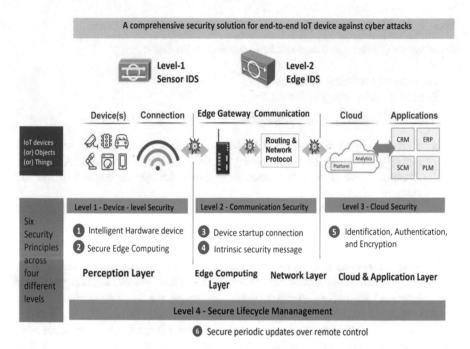

Fig. 5 Architectural framework—a comprehensive security solution for end-to-end IoT devices against cyber-attacks

These physical "objects" (or) "things" have the data transfer, and the device will perform certain specific protection procedures that decide the level of risk. It should be monitored at the initial stage before data enter into edge-level security.

Device-level security principles:

(i) *Intelligent hardware device.* Effective and stable networking needs to be operated by a "smart" system that can manage confidentiality, authentication, encryption, time stamps, caching, proxies, firewalls, connection loss, etc. These intelligent devices have to be stable and ready to work with limited support in a different sector. Along with this hardware security, Level 1 sensor-level IDS will be combined to protect against attackers.

(ii) *Secure edge computing.* The edge device used to process the data locally and filter the data before being pushed into the cloud and applying Level 2 edge-level IDS with ML/DL techniques will secure against IoT edge devices against any kind of anomaly attacks from the intruders.

Level 2 Communication Security. The communication layer relates to the networking of the IoT system that transmits/receives data securely. Lack of secure communication may lead to MITM attacks that are transmitted through Wi-Fi, BLE, LR-WPANs, (or) Ethernet in the perception layer, IPv6, Modbus, (or) open platform communications unified architecture (OPC UA) in the network layer, and

MQTT, CoAP, or Web sockets and HTTPs in the application layer.

Security architecture features at the communication level:

- Data-centric security solutions
- Firewalls and intelligent IDS/IPS

Communication-level security principles:

(iii) *Device start-up connection.* The device has to start a cloud connection initially by avoiding the incoming connections as the first request. A two-way channel connection to the cloud may allow the remote control of the IoT system.

(iv) *Intrinsic security message.* Lightweight message-based protocols are distinguished by many benefits, making it a good choice for IoT applications, such as double encryption choices, queuing, scanning, and even third parties.

Level 3 Cloud Security. The cloud layer acts as a back end for software by storing the enormous amount of data generated from the devices which are analyzed and interpreted to perform actions. When assessing the impact of cloud versus on-site applications, security is the big debate. However, for the IoT, the cloud is viewed as a key facilitator for broad-based adoption. Cloud service providers must play a key role in offering reliable and secure services to protect against data breaches and downtime.

Security architecture features at the cloud:

- Sensitive cloud information should be encrypted.
- Ensure the integrity of other cloud platforms (or) third-party applications.
- Assurance of digital certificates for identification and authentication.

Cloud-level security principle:

(v) *Identification, Authentication, and Encryption.* Digital certificates use the system of asymmetrical, encryption, and authentication which is not only to authenticate a transaction but also to encrypt the channel from the device to the cloud. A digital certificate also has a very complicated cryptographic identity with a user ID/password. To address the inherent challenges to secure IoT edge devices at the device, communication, and cloud level, it is necessary to adhere to these key principles to reduce risk mitigation against IoT edge equipment across various smart city industries.

Level 4—Secure Lifecycle Management. IoT lifecycle management has to be a continuous process which keeps on upgrading the essential software for updates and services to tune the system in to secure place. Security has to be assured start from the device manufacture with design, installation, registration, upgrading the software, auditing, policy procedure, and de-registering the device.

Security architecture features for IoT lifecycle management:

- Secure remote control and
- Active monitoring for patch updates

IoT lifecycle management principle:

(vi) *Secure periodic updates over remote control operation.* The key for the successful IoT deployment is the management of the lifecycle in controlling the device remotely with active monitoring for patch updates securely in allowing remote diagnostics, debugging the software, fine-tuning the machine learning algorithm with a new collection of learning data, adding up additional functionality into the product, and allowing device initiated connection initially followed by bidirectional connection if necessary with right authentication.

5.4   Intelligent Intrusion Detection System (IDS) Using ML/DL in Cyber Security

Intelligent IDS can be developed using the evolving techniques of ML/DL to classify the attack types and detect the malware at the early stage [21–23].

- **Intrusion Detection** System (IDS): Anomaly-based models that use auto-encoders (fully connected/convolutional neural networks (CNNs)/CNN and recurrent neural networks (RNNs)) [24].
- **Malware Detection**: Classification-based models that usefully connected neural networks.
- **Spam and Phishing Detection**: Deep learning methods for natural language processing (unidirectional/bidirectional RNNs), URL classification with fully connected models.
- **Traffic Analysis** (protocol identification): Classification-based fully connected/ RNN models
- **Analysis of Binary Codes**: RNN-based sequence tagging with a subsequent classification of suspicious code segments for identifying known classes of vulnerabilities.
- **Domain Name Generation Algorithm** (DGA) names detection and categorization: RNN/CNN+RNN models as binary classifiers.
- **Long Short-Term Memory** (LSTM): A special kind of **recurrent neural network** (RNN) architecture used for learning sequence prediction problems with order dependence in the deep learning domain.
- **(Semi-supervised) Deep Reinforcement Learning** (DRL): It is capable of interacting with a dynamic environment in solving complex input structure [25, 26].

  - Deep learning: It evaluates complex inputs and selects the best response.

- Reinforcement learning: Agents to learn by trial and error in an interactive environment using feedback from its actions and experiences.

The implementation of IDS using ML/DL against cyber-attacks has to be adopted by considering the following metrics: processing response time—to handle traffic bandwidth, attack mitigation—to identify unknown attacks, and scalability—to adopt as a hardware/software plug-in, resource consumption—to utilize the memory with less constraint.

The choice of IoT-based ML/DL security depends on its learning capabilities to handle anomaly threats in zero-day vulnerabilities with less computation time. Figure 6 depicts the ML/DL taxonomy for IoT security against security threats [27–29]. And, Fig. 7 sketches out the classification of a deep learning-based IDS strategy for IoT security against security threats [30].

6   Conclusion and Future Work

Emerging IoT technology has transformed the world into a customized lifestyle with all in one hand. Moreover, this technology causes threats to users' privacy and security in the cyber world. These security threats enable the defender to search for an effective solution against threat detection and to figure out all possible

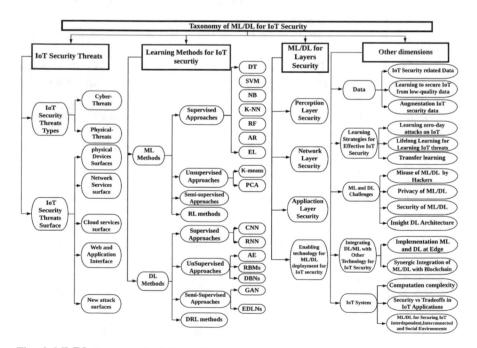

Fig. 6 ML/DL taxonomy for IoT security

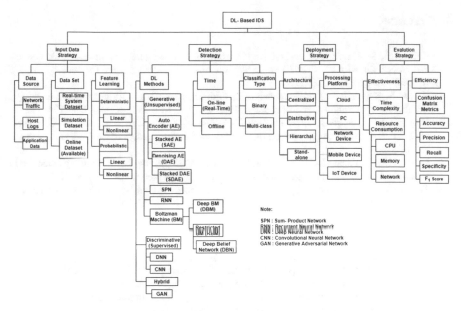

Fig. 7 Classification of deep learning-based IDS

vulnerabilities before being deployed in the IoT domains. Security risks against business and industrial enterprises are on the rise in recent years; they have to be wary of hackers. Cybercriminals will severely harm the business interests and reputations if the defense framework is not professionally planned in IoT systems.

Our proposed comprehensive security architecture framework provides ideology-based solutions for IoT applications against cyber-attacks by considering the network security principles of confidentiality, authentication, integrity, non-repudiation, and access control to the extent of full-fledged end-to-end security. Based on the ideology provided, we are in the development phase of the product for the end customer in different sectors with the right choice of design and security to make it a secure model. It realizes the transformation of the IoT into the next stage of the Secure Internet of Things (SIoT). Still, future research directions can be extended to the identification of new devices with different standards, key management for authentication, trust management hubs, wireless, and software/firmware technologies to resolve upcoming cyber threats in the IoT era.

References

1. Karie NM, Sahri NM, Haskell-Dowland P (2020) IoT threat detection advances, challenges, and future directions. In: 2020 workshop on emerging technologies for security in IoT (ETSecIoT), pp 22–29. IEEE
2. Sisinni E, Saifullah A, Han S, Jennehag U, Gidlund M (2018) Industrial internet of things: challenges, opportunities, and directions. IEEE Trans Industr Inf 14(11):4724–4734
3. Yaacoub J-PA, Salman O, Noura HN, Kaaniche N, Chehab A, Malli M (2020) Cyber-physical systems security: limitations, issues, and future trends. Microprocess Microsyst 77:103201
4. Luo Y, Xiao Y, Cheng L, Peng G, Yao DD (2020) Deep learning- based anomaly detection in cyber-physical systems: progress and opportunities. arXiv preprint arXiv: 2003.13213
5. Tahsien SM, Karimipour H, Spachos P (2020) Machine learning-based solutions for security of Internet of Things (IoT): a survey. J Netw Comput Appl 161:102630
6. Muhammad AN, Aseere AM, Chiroma H, Shah H, Gital AY, Hashem IAT (2020) Deep learning application in smart cities: recent development, taxonomy, challenges, and research prospects. Neural Comput Appl 1–37
7. Ding D, Han Q-L, Xiang Y, Ge X, Zhang X-M (2018) A survey on security control and attack detection for industrial cyber-physical systems. Neurocomputing 275:1674–1683
8. Amrollahi M, Hadayeghparast S, Karimipour H, Derakhshan F, Srivastava G (2020) Enhancing network security via machine learning: opportunities and challenges. In: Handbook of big data privacy, pp 165–189. Springer, Cham
9. Aldowah H, Rehman SU, Umar I (2018) Security in internet of things: issues, challenges, and solutions. In: International conference of reliable information and communication technology, pp 396–405. Springer, Cham
10. Hammi B, Khatoun R, Zeadally S, Fayad A, Khoukhi L (2017) IoT technologies for smart cities. IET Networks 7(1):1–13
11. Rathi R, Sharma N, Manchanda C, Bhushan B, Grover M (2020) Security challenges & controls in cyber-physical system. In: 2020 IEEE 9th international conference on communication systems and network technologies (CSNT), pp 242–247. IEEE
12. Nižetić S, Šolić P, González-de DLI, Patrono L (2020) Internet of Things (IoT): opportunities, issues, and challenges towards a smart and sustainable future. J Cleaner Prod 274:122877
13. Derakhshan F, Ashrafnejad M (2020) The risk of botnets in cyber-physical systems. In: Security of cyber-physical systems, pp 81–106. Springer, Cham
14. Al-Abassi A, Karimipour H, Dehghantanha A, Parizi RM (2020) An ensemble deep learning-based cyber-attack detection in industrial control system. IEEE Access 8:83965–83973
15. Chen K, Zhang S, Li Z, Zhang Y, Deng Q, Ray S, Jin Y (2018) Internet-of-Things security and vulnerabilities: taxonomy, challenges, and practice. J Hardware Syst Secur 2(2):97–110
16. Burhan M, Rehman RA, Khan B, Kim BS (2018) IoT elements, layered architectures, and security issues: a comprehensive survey. Sensors 18(9):2796
17. Cao L, Jiang X, Zhao Y, Wang S, You D, Xu X (2020) A survey of network attacks on cyber-physical systems. IEEE Access 8:44219–44227
18. Kumar Mathi S, Kalyaan P, Kanimozhi S, Bhuvaneshwari S (2017) Integrating non-linear and linear diffusion techniques to prevent fault attacks in advanced encryption standard to enhance security of 4G-LTE networks. Defence Sci J 67(3):276
19. Pallavi S, Narayanan VA (2019) An overview of practical attacks on BLE based IoT devices and their security. In: 2019 5th international conference on advanced computing & communication systems (ICACCS), pp 694–698. IEEE
20. Arvind S, Narayanan VA (2019) An overview of security in CoAP: attack and analysis. In: 2019 5th international conference on advanced computing & communication systems (ICACCS), pp 655–660. IEEE

21. Mohanta BK, Jena D, Satapathy U, Patnaik S (2020) Survey on IoT security: challenges and solution using machine learning, artificial intelligence, and blockchain technology. Internet of Things 11:100227
22. Singh SK, Jeong Y-S, Park JH (2020) A deep learning-based IoT-oriented infrastructure for secure smart city. Sustain Cities Soc 60:102252
23. Kiran KS, Devisetty RNK, Kalyan NP, Mukundini K, Karthi R (2020) Building an intrusion detection system for IoT environment using machine learning techniques. Procedia Comput Sci 171:2372–2379
24. Berman DS, Buczak AL, Chavis JS, Corbett CL (2019) A survey of deep learning methods for cybersecurity. Information 10(4):122
25. Nguyen TT, Reddi VJ (2019) Deep reinforcement learning for cybersecurity. arXiv preprint arXiv:1906.05799
26. Mohammadi M, Al-Fuqaha A, Guizani M, Oh JS (2017)Semisupervised deep reinforcement learning in support of IoT and smart city services. IEEE Internet of Things J 5(2):624–635
27. Khattab A, Youssry N (2020) Machine learning for IoT systems. In: Internet of Things (IoT), pp 105–127. Springer, Cham
28. Gupta R, Tanwar S, Tyagi S, Kumar N (2020) Machine learning models for secure data analytics: a taxonomy and threat model. Comput Commun 153:406–440
29. Aldweesh A, Derhab A, Emam AZ (2020) Deep learning approaches for anomaly-based intrusion detection systems: a survey, taxonomy, and open issues. Knowl-Based Syst 189:105124
30. Kwon D, Kim H, Kim J, Suh SC, Kim I, Kim KJ (2019) A survey of deep learning-based network anomaly detection. Clust Comput 22(1):949–961

Compassion Detection from Text: A Comparative Analysis Using BERT, ULMFiT and DeepMoji

Gourav Awasthi, Rajesh Sabapathy, Chirag Mittal, and Nilanjan Chattopadhyay

1 Introduction

Compassion is fundamental to not just healthy interpersonal relationships and a nurturing tolerant society but is also arguably a key quality of the human condition. According to Arthur Schopenhauer, the renowned German philosopher, "Compassion is the basis of morality" [1]. Across religions, healthcare, education, law and many areas of human endeavor, compassion emerges as a key tenet of higher order human thought, action and overall behavior [2]. To be human is to be compassionate, among other things. As the field of artificial intelligence (AI) progresses in its quest toward artificial general intelligence (AGI), the "hypothetical intelligence of a machine that has the capacity to understand or learn any intellectual task that a human being can" [3], computers have to learn how to detect, measure and exhibit compassion in their dealings with humans.

Teaching computers compassion has several beneficial real-world applications. Compassionate "Chat Bots" or "Voice Agents" in the not-so-distant future could potentially be used to man online fora or suicide helplines, offering "virtual comfort" to sufferers while directing the limited bandwidth that human counselors have toward the most extreme cases. As businesses strive to be increasingly customer-centric, a less futuristic and more immediate application is in call centers where the live call audio could be used to identify when customer care executives may not be displaying the right level of compassion toward an irate customer, triggering intervention by a supervisor or a senior executive, or provide coaching later on. A similar application also lies in the area of letters or emails that com-

G. Awasthi (✉) · R. Sabapathy · C. Mittal · N. Chattopadhyay
Optum Global Solutions (India) Pvt. Ltd., Building 6A, DLF Cyber City,
Gurgaon, Haryana 122010, India
e-mail: gourav_awasthi@optum.com

R. Sabapathy
e-mail: rajesh_sabapathy@optum.com

R. R. Raje et al. (eds.), *Artificial Intelligence and Technologies*,
Lecture Notes in Electrical Engineering 806,
https://doi.org/10.1007/978-981-16-6448-9_47

panies send out to customers to convey news that may not be received well, such as a bank refusing a loan or an insurance company denying a claim. Such letters are typically based on standard templates allowing an opportunity for the language in the templates to be checked for compassion and altered suitably. A reputation for being compassionate in it is dealings and communications could arguably reap a business customer loyalty, and subsequently profits, in the long run.

For computers to understand and exhibit compassion, they clearly must understand human language(s). The founding fathers of the current wave of deep learning —Geoffrey Hinton, Yann LeCun and Yoshua Bengio—have all spoken about the importance of natural language processing (NLP) and made it a focus of their research [4]. The focus of the research community in the preceding years has seen deep learning methods drive significant progress in areas of NLP such as sentiment analysis, text classification, summarization, questions and answering (Q&A) etc. Within sentiment analysis, where the objective is to assign positive, negative and neutral polarities to language, the subfield of emotion detection (ED), which seeks to determine emotions such as happy, sad, fear and surprise from text, speech, video etc. [5], is of particular importance and relevance when it comes to helping computers learn compassion.

ED has largely been aimed at data in the form of image, video and audio [5]. On the contrary, ED from text has not seen a great deal of progress. There are multiple reasons for this: (a) Text data does not provide the visual or tonal cues that are crucial to discerning emotion; (b) Lack of exhaustive knowledge base and dictionaries for text data compared to other types of data when it comes to ED; and (c) Lack of understanding on what methods work well [5]. Moreover, when it comes to detecting compassion from text, which is the focus of this paper, we could not find any prior work in the subfield of ED from text. ED from text largely focuses on the primary or fundamental emotions based on discrete emotion models (DEMs) such as Ekman, Plutchik or OCC [5]. Compassion, on the other hand, is more than just an emotion, comprising of cognitive, affective and behavioral components [2].

The contribution of this paper is twofold: (a) We introduce an annotated compassion detection dataset that can be used for classifying text as compassionate, not compassionate and others; and (b) We provide a comparative analysis of three deep learning-based pretrained models (BERT [6], ULMFiT [7] and DeepMoji [8]), widely used for other NLP tasks, on both above dataset and two other related datasets.

Our choice of the three models mentioned above is based on the following questions: (1) Would a proven model such as DeepMoji, which is a task-specific model aimed at the field of sentiment analysis and one which takes a different approach based on emoji embeddings [8], capture aspects of compassion better than more general pretrained models, which train on just text corpuses?; and (2) For ED tasks, can the performance of models such as DeepMoji and ULMFiT, both of which require significantly lesser infrastructure for model training and prediction, be comparable with a model based on transformer architecture, such as BERT, where the compute requirements are much higher? This would then allow for a performance versus cost tradeoff to be made in industrial applications for firms looking to optimize investment in compute infrastructure without significant compromise in performance.

The rest of this paper is organized as follows: Sect. 2 introduces the specifics of the problem of detecting compassion from text and covers related work across ED and psychology. Section 3 details the data collection and annotation process. Section 4 outlines the experiment setup and methodology. In Sect. 5, results are presented on the Compassion Detection task and briefly discussed. In Sect. 6, we compare the performance of the three approaches (BERT, ULMFiT and DeepMoji) on similar datasets and tasks from previous work in this field, such as on a "politeness" dataset [9] and on an "insults" dataset from Kaggle.[1] Section 7 summarizes conclusions and presents directions for future work.

2 Problem Statement and Related Work

2.1 Problem Statement

The focus of this paper is to detect compassion from text, specifically the kind of text found in online fora in the form of comments or responses on communities, such as those found on the Reddit platform. Such text responses, where users debate politics, or offer support to those in strife, represent a raw form of human interaction, which is useful for detecting the presence or absence of compassion. Methods that work on such text responses can be expected to generalize to other real-word situations.

2.2 Related Work from Clinical Psychology

According to the online Merriam-Webster dictionary [10], compassion is the "sympathetic consciousness of others' distress together with a desire to alleviate it." The clinical psychology literature appears to build on top of this definition. Drawing upon consensus across studies and authors, a recent review [2] proposed a definition of compassion which involves "(1) Recognizing suffering; (2) Understanding the universality of suffering in human experience; (3) Feeling empathy for the person suffering and connecting with the distress (emotional resonance); (4) Tolerating uncomfortable feelings aroused in response to the suffering person (e.g., distress, anger, fear) so remaining open to and accepting of the person suffering; and (5) Motivation to act/acting to alleviate suffering," and that compassion involves cognitive, affective and behavioral components [2]. Additionally,

---

[1] https://www.kaggle.com/c/detecting-insults-in-social-commentary/data.

psychology also differentiates between self-compassion and compassion for others, with studies [2] showing a weak correlation between the two, indicating that they may not be related, requiring further research to establish independence [2]. In this paper, we use the five-part definition of compassion described above while annotating our data and focus only on compassion toward others. Interestingly, the psychology literature also threw up an empirically based model of compassion in a clinical patient-nurse/doctor healthcare setting [11], although this is not of immediate use to the problem at hand, which is detecting compassion from text.

2.3   Related Work from Emotion Detection Literature

The ED from text literature can be viewed at as comprising of three layers: emotion models, datasets and detection approaches [5].

Emotion Models. These are divided into discrete emotion models (DEMs) and dimensional emotion models (DiEMs) [5]. DEMs, such as Ekman [5], Plutchik [5] and OCC [5], consider emotions to be independent and classify each emotion separately with 6, 8 and 22 emotions, respectively. DiEMs, on the other hand, consider emotions to be related and place emotions on a dimensional space with the common dimensions being valence, arousal and dominance [5]. For ED, DEMs have largely been preferred over DiEMs [5].

Datasets. Given the predominance of DEMs in prior ED work, most of the datasets available for research have labels based on the emotions covered in DEMs [5]. The source of these datasets are largely social media platforms such as Twitter [5] and Facebook [5] as well as news websites and blogposts [5]. Some interesting datasets based on transcripts of popular television shows [5] have also been used, as these shows frequently have characters expressing various emotions in day-to-day parlance.

Detection Approaches. Popular approaches for ED from text have been classified into hand-crafted feature engineering based and deep learning based [12]. In the former approach, the key idea is to leverage keywords with emotional value associated with them. Dictionaries and lexicons are organized such that each keyword either has an emotion associated with it as a label or has an emotion probability score attached to it [5]. Statistical features are created around these keywords and then supplied to a downstream classifier to detect a set of emotions. While these methods have high precision, key drawbacks are low recall along with extensive feature engineering [12]. In the deep learning approach, recent work has focused heavily on the use of bidirectional long short-term memory networks (bidirectional LSTMs). Proceedings from SemEval-2019 Task 3: EmoContext Contextual Emotion Detection in Text [12] show a wide variety of architecture choices by participants, with BERT [6] and ULMFiT [7] being among the popular choices as per the analysis of systems study for the task [12]. Additionally, through attention-based bidirectional LSTM without-transformer models such as DeepMoji [8], researchers have employed distant supervision on tweets to learn rich representations of text in the form of emojis to detect emotion to achieve state-of-the-art results in the recent past.

2.4 Related Work Around Empathy Detection

In the clinical psychology literature, empathy is identified as a concept that is close to compassion [2], i.e., empathy means the ability to relate, both cognitively and affectively, to another person's situation [2]. While compassion is seen as comprising of empathy; the key differences are as follows: (a) Compassion arises always in the context of suffering, whereas empathy can be felt in a wider range of situations [2]; and (b) Compassion involves a behavioral aspect as well, which is the desire to ease or eliminate suffering [2].

Despite the above-mentioned differences, key lessons can be learnt by studying the limited literature available on the relatively recent area of empathy detection from text. There is recognition in the literature that empathy, like compassion, is a complicated emotion, comprising of different aspects such as "empathic concern and personal distress" [13] and has "several psychological definitions" [14]. This means that methods and models for annotating data, and subsequently evaluating the annotated data, must be carefully arrived at [13, 14]. One way that the literature has dealt with the problem of accounting for differences in the degree of empathy felt by annotators is to use multi-item scales [13], where annotators provide a rating on a scale as to the extent of empathy they felt, rather than using binary "Yes/No" ratings. Such multi-item scale ratings are averaged to arrive at the final label. The limited literature in this area shows that authors have set up the problem as both a classification task (empathetic vs. not empathetic) [15] and a regression task (predicting the empathy rating on a multi-point scale) [13], with features based on word embeddings and methods based on LSTMs and CNNs [13, 15].

3 Data Collection, Preparation, Annotation and Description

3.1 Data Collection

As described in Sect. 2.3 under datasets, there is no publicly available dataset with labels that can be used for the task of compassion detection from text. We proceeded to curate a text dataset of our own where each piece of text (document) could be labeled with one of three categories: compassionate, not compassionate and others. The steps involved are as follows.

Identifying the Right Source. Given that text for all three categories were needed from the same source, we manually examined novels, religious texts, Tweets, transcripts of TV shows and Reddit communities. Reddit was finalized as the source given that there are several subreddits where we found text for all three categories.

Identifying the Right Sub-Source. Having finalized Reddit, we chose the following subreddits from which to extract original posts and responses: feminism, MensRights, India, cancer, childfree, aww, AskReddit, depression, blog, politics, vegan, todayilearned, news and politics.

Web Crawling. The PRAW library (Python Reddit API Wrapper) was used to crawl 24 threads across the subreddits mentioned above. From each thread, we retrieved original posts and responses from the community which either support or do not support the original post. As we used the above API, the text output from Reddit pages was free from html tags and other content which require cleanup.

3.2 Data Preparation

After excluding documents with less than 10 words, the Web crawling program retrieved a total of 13,663 documents from the subreddits mentioned above—a document could either be an original post or a response. Out of these 13,663 documents, a total of 3,049 documents have been annotated till date and comprise the data for this paper.

3.3 Data Annotation

Annotation of the data was performed by the four authors of this paper. Each document was annotated only once, by one out of the four authors. To annotate the data into the three required categories, following guidelines were agreed upon.

Compassionate. The five-part definition described in Sect. 2.2 was adopted and the scope was fixed as compassion to others, excluding self-compassion. Annotators labeled a document as compassionate if the author of the original post/response displayed: recognition of suffering, empathy toward the sufferer and a desire to help.

Not Compassionate. Any document which had abuse, cruelty, hatred, dislike, apathy, antipathy or selfishness was labeled as not compassionate.

Others. If there was any confusion or if the text was neutral or displayed other emotional/cognitive states, a document was labeled as others.

A total of 1,928 documents annotated by two out of the four authors of this paper were selected to be split into train and dev sets in the ratio 90:10. The remaining

1,121 documents, annotated by the other two authors of this paper, were set aside as the test set. This was done to conserve effort and time without compromising on methods and results, given that each document is sizeable in length and determining the label was not straightforward.

3.4 Data Description

Table 1 shows the distribution of documents across train, dev and test, and across labels.

The following snippets are examples of documents labeled as compassionate and not compassionate.

Compassionate, "I cannot stress enough how brave you are for going through with any of this. I've been assaulted myself and because of my job at the time I was so afraid of asking for help ... I was terrified of it getting out, rumors spreading, and my career taking a massive hit. It is one of my biggest regret. Because you got that done and we're brave as hell for it you have a strong case and I hope with every fiber of my being you get to see justice. Again, hold that head high and be proud of yourself."

Not Compassionate. "... Never in my life have I 'known' someone to be this worthless."

As can be seen from Fig. 1, almost 80% of the documents in train and dev were up to 100 words in length, although this proportion is lower in the test set, which had a higher percentage of longer documents. Figure 2 shows that there is a marginal variation in the proportion of the three target classes across train, dev and test. Figure 3 provides a qualitative sense of the words found in one class that are not found in the other classes.

Given that the dataset was annotated by four different people, and that each document was tagged only once, a set of 98 documents out of the total was annotated by all four annotators to identify the inter-rater agreement. This will help compare the performance of any approach with that of human performance. The Fleiss' Kappa score on inter-rater agreement is 0.699 for the 98 documents described above, which is substantial agreement but not a perfect agreement. Table 2 shows the pair-wise Cohen's Kappa score on inter-rater agreement.

Table 1 Number of documents used for train, dev and test

Label	Train	Dev	Test
Compassionate	516	52	363
Not compassionate	397	50	175
Others	822	91	583

Fig. 1 Distribution of number of words in the documents used for train, dev and test

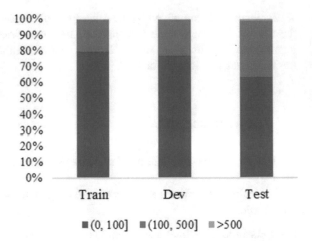

Fig. 2 Distribution of documents labeled across target classes for train, dev and test

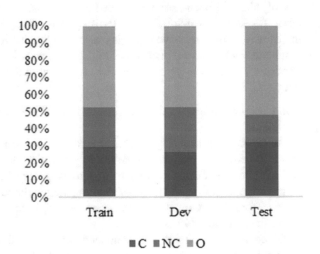

4 Experiment Setup and Methodology

All experiments were performed on a Tesla P100 16 GB RAM GPU server. The server had 64 CPU cores and 512 GB CPU RAM. All the code was written in Python 3.6.7. We conducted training on the train dataset and used the dev dataset (also referred to as validation dataset) to identify the best performing model, in terms of hyperparameters, with comparison being made on the test dataset. For comparing the three approaches, we use precision, recall and $F1$-Score/ micro-averaged $F1$-Score.

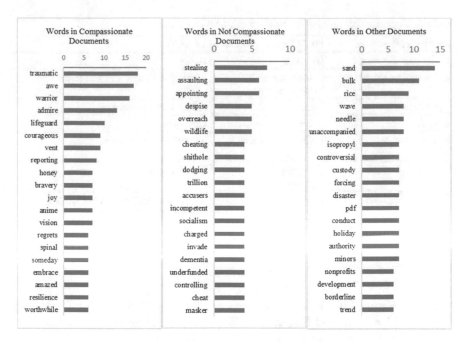

Fig. 3 Figure showcasing top occurring words in compassionate, not compassionate and other documents. Only the uncommon words across these categories were considered

Table 2 Pair-wise Cohen's Kappa score

	Rater 1	Rater 2	Rater 3	Rater 4
Rater 1	1	0.67	0.80	0.81
Rater 2		1	0.63	0.59
Rater 3			1	0.71
Rater 4				1

4.1 ULMFiT Methodology

We used the ULMFiT implementation available through the fastai library to first train the language model and then use the resulting embeddings to train the classifier. As recommended in the ULMFiT documentation, preprocessing is not required; however, out-of-vocabulary words are automatically replaced by the fastai library using special tokens.

Language Model. For the language model, in the first attempt, the final hidden layer was unfrozen and retrained for 10 epochs, with $1e-1$ as the learning rate. Then, the entire network was unfrozen and retrained for 10 epochs, with $1e-3$ as the learning rate.

Table 3 Steps/hyperparameters for BERT and ULMFiT models

Step/hyperparameter	BERT with attention	ULMFiT
Initial learning rate	0.00005	0.003
Learning rate scheduler	Linear scheduler with warmup	Manual with learning rate finder
Optimizer	AdamW	Adam with momentum
Loss function	Cross entropy	Cross entropy
Model selection criterion	Minimum validation set loss	Minimum validation set loss
Library	Hugging face transformers 3.3.1	Fastai 2.0.16
Training iterations	20	20 (10 for last layer, 5 for last 2 layers and 5 for last 3 layers)
Batch size	8	32 (for unfreezing last 1 or last 2 layers), 16 (for unfreezing last 3 layers)
Unfreezing layers	All	Sequential up to last 3 layers

Classifier Model. ULMFit classifier is a deep learning Model having Embedding layer, 3 hidden layers and Output layer. We used gradual unfreezing methodology proposed in ULMFit paper [7] to fine-tune the classifier. We first unfreeze the last hidden layer and fine-tuned it for 10 epochs using a maximum learning rate of 3e-2. We then unfreeze the last 2 hidden layers and fine-tuned it for 5 epochs using a maximum learning rate of 1e-3. Subsequently we unfreeze the last 3 hidden layers and fine-tuned it for 5 epochs using a maximum learning rate of 5e-4. Batch size of 32 was used when we unfroze the last 1 or last 2 layers, and batch size of 16 was used for unfreezing last 3 layers, due to GPU resource constraints. Other hyper-parameters optimized during training are listed in Table 3.

For both the models, the best learning rate was found using the learning rate finder functionality in the fastai library. The cyclical learning rate [16] and fit one cycle policy [17] implementation in fastai was used to fine tune all the layers during training. The best classifier model was identified based on validation set accuracy.

4.2   DeepMoji + Auto ML Methodology

After the text was preprocessed by removing hyperlinks special characters and trailing whitespaces, we first obtained the DeepMoji embeddings using the pyTorch-based torchMoji implementation available from Hugging Face. The pyTorch implementation performs the following text processing internally: conversion to lower case and tokenization to maximum length of 512 tokens with padding through special token. The embeddings were then used to train an ensemble classifier using Auto ML API available in H_2O 3.28.0.3. For our "Compassion Detection dataset," the Auto ML ensemble model had a combination of three individual models across GLM, XGBoost and feed forward neural networks. The Auto ML ensemble model development process was run on an H_2O cluster, which had 50 GB RAM and 10 threads, for 8 h. Target classes were balanced, and fivefold CV was used for model selection.

4.3 BERT Methodology

The text was preprocessed by removing hyperlinks special characters and trailing whitespaces. The max length of tokens was kept to 512, special tokens were added during tokenization and padding was done to maximum token length, and they were masked during training. The "BERT base uncased" model, which we used, is case insensitive. Hence, conversion of string to lower case was not required. The BERT results presented have used the fine-tuning approach, where an additional hidden layer followed by output classification layer is added to the pretrained base uncased BERT model. Table 3 lists the steps and hyperparameters for BERT.

Optimizing Additional Hidden Layer Size. We optimized the size of the additional hidden layer by experimenting as follows. Following hyperparameters were fixed at their default values: optimizer (AdamW), learning rate (1e−5) and learning rate scheduler (constant). Following were varied: hidden layer size (sizes of 25, 50, 75 and 100 were tried) and batch size (4 and 8 were tried; at 16, we ran out of GPU resources). Additionally, the extent of retraining (retrain all layers including additional hidden layer and output classifier layer; retrain only additional hidden layer and output classifier layer) was also varied. Each iteration ran for 10 epochs, after which the validation set loss was noted, based on which batch size of 8 and hidden layer size of 50 were finalized; 10 epochs were sufficient for the validation set loss to plateau. Freezing all layers had the worst performance; hence, all layers were unfrozen during learning for rest of the hyperparameter tuning exercise.

Learning Rate, Optimizer and Learning Schedule. Having finalized the hidden layer size, the batch size and extent of retraining required, we tuned the learning rate (5e−6, 5e−5 and 5e−4 were tried) and learning schedule (constant, linear and cosine learning schedules were tried). During these iterations, we used AdamW as optimizer, based on prior literature and experience, and each iteration was run for 20 epochs after which convergence was usually achieved.

5 Results and Discussion

The results from our experiments on the test set are captured in Table 4.

We find that the combination of embeddings from DeepMoji fed to an ensemble of classifiers works slightly better than BERT and much better than ULMFiT. This possibly indicates that the emoji-based embeddings from DeepMoji could be

Table 4 Precision (P), recall (R) and micro-averaged $F1$ score ($F1\mu$) on test set

Label	BERT			ULMFiT			DeepMoji + Auto ML		
	P	R	$F1\mu$	P	R	$F1\mu$	P	R	$F1\mu$
C	0.75	0.92	0.82	0.62	0.95	0.75	0.82	0.87	0.84
NC	0.38	0.79	0.52	0.33	0.61	0.43	0.40	0.63	0.49
O	0.80	0.77	0.79	0.72	0.82	0.77	0.76	0.80	0.77
All	0.67	0.82	0.739*	0.60	0.83	0.70*	0.70	0.80	0.742*

*—In order to maximize the micro-averaged $F1$ score, the same threshold value was chosen for all 3 classes. The class-level precision, recall and micro-averaged $F1$ score are calculated using the common threshold value and not calculated using optimal class-level threshold values

powerful when it comes to detecting and distinguishing a complex cognitive, affective and behavioral feeling such as compassion. To be sure of this, the effect of the ensemble classifier needs to be separated out in future work. A possible reason for the power of DeepMoji could be that it learns representations for a diverse set of 64 emojis, quite a few of which could have a bearing on compassion, such as the "Folded Hands" emoji or the different kinds of "Angry Face" emojis [8]. We find that the DeepMoji + Auto ML approach distinguishes the compassionate class better, while BERT works better for the other two classes.

In terms of which class is easier to detect, it is clear that all three approaches work well for the compassionate class, whereas all three methods struggle with the not compassionate class. In a way, this has parallels with the analysis of systems from SemEval-2019 Task 3, where systems had best performance for the *Sad* class and the worst performance for the *Happy* class [12].

Examining the precision–recall values, we find that when it comes to detecting compassion, DeepMoji + Auto ML has a more even balance between precision and recall. This finding can be compared with what we find in the literature [12] where it was found that rule-based features and encodings fed to classifiers seem to suffer from a lower recall. BERT and ULMFiT, on the other hand, seem to favor recall over precision.

Finally, we compared the predictions that each approach produced on the sample of 98 documents used for determining the inter-rater agreement. Table 5 describes the results. Given that human annotators disagreed on 28, correct prediction on 70

Table 5 Comparing annotator agreement-disagreement with model performance

No. of annotators who agreed	No. of documents	Approaches predicted the same label as annotators		
		BERT	ULMFiT	DeepMoji + Auto ML
All agreed	70	56	52	63
At least one disagreed	28	16	13	16
Total	98	72	65	79

documents could be construed as a crude upper limit for model performance. We find that out of the 70 documents where the annotators agreed on the label, the output of DeepMoji + Auto ML matched the annotators' label 90% of the time (63 documents out of 70), indicating the potential to detect and discriminate a notion as complicated as compassion.

6 Comparison on Other Datasets

To examine the advantages of DeepMoji + Auto ML vis-à-vis BERT and ULMFiT, we replicated the comparison on two other datasets: a "Politeness" [9] dataset and the "Kaggle Insults" dataset introduced in Sect. 1.

We selected these datasets based on two criteria. The first one was similarity with the task at hand, i.e., detecting compassion from text. The second criterion was that the three methods we are interested in should not have been investigated on the chosen datasets. This criterion ruled out datasets such as those corresponding to emotion detection (SE0714, Olympic), sentiment detection (SE1604, SS-Twitter) and sarcasm detection (SCv1) as these datasets have been extensively investigated.

The two datasets that we chose, as mentioned above, satisfy both criteria. Firstly, the task of detecting politeness in text can be thought of as analogous to detecting compassion, and detecting insults can be thought of as analogous to detecting text that is not compassionate. Secondly, performance benchmarks exist for both these datasets, either in the form of research papers or through Kaggle competitions. Table 6 provides details of these two datasets.

The politeness dataset comprises of sentences derived from the widely used Enron dataset, which contains email conversation exchanges between the employees of the Enron Corporation [18]. The dataset has 1.39 million sentences divided into ten buckets ("P_0" to "P_9") based on the score from a politeness classifier [9]. For the comparison between the three approaches used in this paper, we selected sentences falling in the "P_0" (Impolite) and "P_9" (Polite) buckets and set up a binary classification task. Given that the buckets are based on scores from a pre-existing politeness classifier, the expectation is that the three approaches should show near-perfect performance.

The Kaggle Insults dataset is a social commentary dataset comprising of comments posted in public discussion forums. The task is to predict whether a comment posted during a public discussion is considered insulting to one of the participants

Table 6 Train, dev (validation) and test split details for politeness and Kaggle insults datasets

Politeness dataset				Kaggle insult dataset			
Label	Train	Dev	Test	Label	Train	Dev	Test
P_9	40,000	10,000	26,697	Insult	1,537	151	438
P_0	40,000	10,000	9,521	Non-Insult	2,913	344	799

Table 7 $F1$ score, accuracy and event's precision (P) on test set on politeness and Kaggle insults datasets

Dataset	BERT classifier			ULMFit			DeepMoji + AutoML		
	$F1$	A	P	$F1$	A	P	$F1$	A	P
Politeness	0.99	0.99	0.99	0.97	0.96	0.96	0.97	0.95	0.98
Kaggle insults	0.80	0.86	0.79	0.63	0.72	0.58	0.77	0.83	0.82

Note F1—F1 score; *A*—accuracy; P—precision. Best shortlisted AutoML Models were stacked ensemble models of different XGBoost models. Accuracy and precision were measured at 0.5 threshold

of the larger blog/forum conversation; the task specifically is to not look for insults directed to non-participants (such as celebrities, public figures etc.).

As can be seen in Table 7, in the politeness identification task, while BERT is superior ($F1 = 0.99$), we see that the DeepMoji + AutoML approach only slightly underperforms BERT ($F1 = 0.97$), as is the case with ULMFiT. The performance of all the approaches are close to each other and are near-perfect (nearly equal to 1). This was expected as the labels are from a pre-existing politeness classifier, i.e., what the pre-existing classifier can do, other classifiers should also be able to do, in theory.

In the Kaggle Insults dataset, we observe the same phenomenon. BERT is superior ($F1 = 0.8$) but DeepMoji + AutoML again only slightly underperforms ($F1 = 0.77$); moreover, the latter has a higher precision (0.82 vs. 0.79). ULMFiT is a distant third ($F1 = 0.63$).

Clearly, the usage of an approach combining DeepMoji embeddings with a downstream ensemble classifier holds promise when it comes to tasks which involve "emotions" such as detecting compassion, politeness and insults. Arguably, this could be due to the ability of emoji embeddings to capture signals in text that otherwise are not captured through pretrained models trained only on text corpuses.

7   Conclusions and Future Work

The objective of this paper is twofold. Firstly, we introduce a new original compassion detection dataset which has been curated from Reddit and annotated with three labels (compassionate, not compassionate and others). To our knowledge, no other dataset is available for this task. Secondly, we enrich the emotion detection (ED) from text literature which has a relative shortage of investigations into approaches. Our comparative analysis across three widely used approaches (BERT, ULMFiT and DeepMoji) shows that when it comes to detecting a complex cognitive, affective and behavioral notion such as compassion, an approach combining emoji embeddings from DeepMoji with a downstream ensemble classifier is quite comparable with BERT and performs better than ULMFiT. We confirm this by comparing the three approaches on other related datasets on politeness and insults, where again the classifier based on DeepMoji embeddings is nearly as good as BERT. The key insight from our experiments is that emoji embeddings have possible advantages for tasks which involve an "emotional" aspect and warrant consideration.

As far as future work is concerned, there are three directions that hold promise: (a) Enhance the annotation on our compassion detection dataset with more annotators and having all documents annotated by all annotators; (b) Create newer datasets by leveraging programmatic data-labeling approaches where the labels could go beyond the basic emotions and focus on more complicated human tendencies to help in the quest toward artificial general intelligence; and (c) Combining BERT with emoji embeddings to leverage the best of both worlds for compassion detection.

References

1. Cartwright D (1984) Kant, Schopenhauer, and Nietzsche on the morality of pity. J Hist Ideas 45(1):83–98
2. Strauss C, Lever Taylor B, Gu J, Kuyken W, Baer R, Jones F, Cavanagh K (2016) What is compassion and how can we measure it? A review of definitions and measures. Clin Psychol Rev 47:15–27
3. Artificial general intelligence. https://en.wikipedia.org/wiki/Artificial_general_intelligence. Last Accessed 20 Oct 2020
4. Manning C (2015) Last words: computational linguistics and deep learning, a look at the importance of natural language processing. Comput Linguist 41(4):701–707
5. Acheampong F, Wenyu C, Nunoo-Mensah H (2020) Text-based emotion detection: advances, challenges, and opportunities. Eng Rep 2(7)
6. Devlin J, Chang M, Lee K, Toutanova K (2018) Bert: pre-training of deep bidirectional transformers for language understanding. arXiv preprint arXiv: 1810.04805
7. Howard J, Ruder S (2018) Universal language model fine-tuning for text classification. arXiv preprint arXiv: 1801.06146
8. Felbo B, Mislove A, Søgaard A, Rahwan I, Lehmann S (2017) Using millions of Emoji occurrences to learn any-domain representations for detecting sentiment, emotion and sarcasm. arXiv preprint arXiv: 1708.00524v2
9. Madaan A, Setlur A, Parekh T, Poczos B, Neubig G, Yang Y, Salakhutdinov R, Black AW, Prabhumoye S (2020) Politeness transfer: a tag and generate approach. arXiv preprint arXiv preprint arXiv: 2004.14257v2
10. Merriam-Webster Dictionary. https://www.merriam-webster.com/dictionary/compassion. Last Accessed 20 Oct 2020
11. Sinclair S, Mcclement S, Raffin-Bouchal S, Hack T, Hagen N, Mcconnell S, Chochinov H (2016) Compassion in health care: an empirical model. J Pain Symptom Manage 51:193–203
12. Chatterjee A, Narahari KN, Joshi M, Agrawal P (2019) SemEval-2019 Task 3: EmoContext contextual emotion detection in text. SemEval@NAACL-HLT, pp 39–48
13. Buechel S, Buffone A, Slaff B, Ungar L, Sedoc J (2018) Modeling empathy and distress in reaction to news stories. arXiv preprint arXiv: 1808.10399v1
14. Alam F, Danieli M, Riccardi G (2017) Annotating and modeling empathy in spoken conversations. arXiv preprint arXiv: 1705.04839v3

15. Khanpour H, Caragea C, Biyani P (2017) Identifying empathetic messages in online health communities. In: IJCNLP 2017—proceedings of the 8th international joint conference on natural language processing, 2 (Short Papers)
16. Smith LN (2017) Cyclical learning rates for training neural networks. arXiv preprint arXiv: 1506.01186v6
17. Smith LN (2018) A disciplined approach to neural network hyper-parameters: part 1—learning rate, batch size, momentum, and weight decay. arXiv preprint arXiv: 1803.09820v2
18. Klimt B, Yang Y (2004) Introducing the Enron Corpus. CEAS

A Review: Reversible Information Hiding and Bio-Inspired Optimization

Amishi Mahesh Kapadia and P. Nithyanandam

1 Introduction

Data security and secure communication are a top priority in an age of digital communication, where new technologies are emerging daily. Information sharing using digital means with increased availability of the Internet has made it easy and effortless. However, data protection and security are an ongoing challenge as information sharing takes place over unsecured networks.

Information hiding is also known as data hiding and it can be broadly put across with the varied implications: First, secret information is imperceptible. Secondly, the presence of information is concealed. Third, the recipient and sender details are hidden, and fourth, the transmission channel is hidden [1]. Thus, hiding information is a process in which secret information is hidden in such a way that its existence is not known and thus does not attract the attention of the attackers.

Information hiding is used in applications such as legal, medical and law enforcement military, intelligence agencies, online elections, internet banking, artwork preservation, law enforcement, satellite imagery, and remote sensing.

Information hiding has different branches and is primarily categorized as steganography, copyright marketing, covert channel, subliminal channel, and anonymous communication [1]. Information hiding is applied to text, audio, video, and digital images.

Information hiding techniques for digital images are classified into four categories [1], steganography for covert communication, robust watermarking for copyright protection, fragile watermarking for content authentication and finger-

A. M. Kapadia (✉) · P. Nithyanandam
School of Computing Science and Engineering, VIT, Chennai Campus, Chennai, Tamil Nadu 600127, India
e-mail: amishimahesh.kapadia2013@vit.ac.in

P. Nithyanandam
e-mail: Nithyanandam.P@vit.ac.in

© The Author(s), under exclusive license to Springer Nature Singapore Pte Ltd. 2022   489
R. R. Raje et al. (eds.), *Artificial Intelligence and Technologies*,
Lecture Notes in Electrical Engineering 806,
https://doi.org/10.1007/978-981-16-6448-9_48

printing for transaction tracking. These applications do not require the retrieval of a cover image from a stego image.

There are certain areas where the cover image needs to be retrieved after the secret data has been removed from the stego image. This requirement led to a new category known as reversible hiding of data. Reversible hiding data is divided into two categories: Spatial and Transform domain.

The embedding of the spatial domain technique is achieved by directly changing the pixel values of the cover image and the embedding of the frequency domain is achieved by transforming the coefficients of pixels.

Numerous RDH applications have been developed there and the key categorization based on [2] can be done as follows:

1. Compression-based techniques [3]
2. Difference expansion-based techniques [4, 5]
3. Histogram shifting-based techniques [6–8]
4. Prediction error-based techniques [9–11]

Frequency domain approaches primarily include translating the cover object to the frequency domain using the Discrete Fourier Transform (DFT), the discrete Cosine Transform and the integer wavelet transform (IWT) [12].

In recent years, there has been a further addition to hiding knowledge through Bio-inspired Optimization. These algorithms are mathematical in nature and help design the model in such a way that the resulting final stego image is optimal. There are various bio-inspired algorithms such as Particle Swarm Optimization (PSO), Bee Colony Optimization (BCO), and Cuckoo Algorithm [13].

In this paper, a review of the various techniques of lossless data hiding and how optimization helps further to enhance the process of embedding has been discussed. The sections are divided as follows.

Section 2: Lossless digital image hiding in Spatial Domain, Sect. 3: Lossless digital image hiding in Transform Domain and Sect. 4: Bio-inspired optimization techniques are discussed.

2 Lossless Digital Image Hiding in Spatial Domain

In spatial domain techniques, the pixel of digital images is chosen as the carrier for the transmission of confidential information. These techniques have been developed very early and extensively. Spatial domain models can be categorized into different categories: modulo-based, histogram-based, difference expansion, compression-based, and prediction-based techniques.

2.1 Modulo-Based Addition

This is a patented reversible data hiding scheme by Hosinger et al. [14]. The hidden information is subtracted directly from the pixel value, and both the hidden data and the original data are recovered without any loss. First, the hidden data is convoluted to generate a scrambled image and the image is encoded using the modulo addition technique. Cross-correlation technique is used to extract the information which is recovered by subtracting the secret message from the embedded image in order to obtain the original image. The key embedding is formulated as in (1)

$$\text{Stego image} = (\text{cover image} + \text{payload}) \mod 256 \qquad (1)$$

In this technique, the overflow/underflow issue was addressed using modulo operation. This technique guaranteed reversible operation but added salt and pepper to the embedded image and the embedding capacity is not high.

2.2 RDH by Compression Technique

The concept behind these schemes is to create some space through a lossless compression of the cover image subset and then use the space saved for data embedding. The embedding is carried out by substituting a subset with the compressed form of the cover image and the payload, the capacity is determined by the difference between the subset and its compressed form.

In [15], Fridrich et al., Introduced an embedding methodology in which the hash value of 128 bit is embedded. The optimum bit plane is selected and then the hash value and the compressed cover image are embedded by direct replacement of the pixels in the selected the bits plane.

Goljan et al. proposed an RS scheme in [16]. In this process, the cover image is split into blocks. The function is used to classify blocks as regular, singular, and unusable based on the smoothness of the blocks. The RS blocks are compressed, and the U block is skipped as it can be easily identified. The data embedding flipping function is used to embed a bit in each block R and S, and either it is flipped from R to S and S to R or it remains unchanged. The actual embedded data consists of payload and the compressed RS vector. So, at the receiver end, the payload is extracted and the original files are recovered using compressed RS.

In [3], A generalized least significant bit (G-LSB) compression method was proposed by Celik et al. to increase compression efficiency. Using unaltered portions of cover data as side-information. Rather than changing a bit plane, the lowest levels of raw pixel values are quantified and then used to accommodate payload bits. By a prediction-error entropy coder, G-LSB will change the quantization phase and overwrite the lowest pixel values with the payload bits. This method is better than some previous lossless-compression-based works like [15, 16] due to its scalability along the capacity-distortion curve.

2.3 Difference Expansion Technique

This technique has been proposed by Tian [17]. This method is based on the expansion of the pixel difference value of the neighboring pixels. With this method, one bit is embedded in each pixel. Equation 2 calculates the average a, the difference d is given

$$a = \left\lfloor \frac{x+y}{2} \right\rfloor \tag{2}$$

in Eq. 3 as below.

$$d = x - y \tag{3}$$

The inverse transform is computed as below to retrieve the pixels

$$x = a + \left\lfloor \frac{d+1}{2} \right\rfloor \tag{4}$$

$$y = a - \left\lfloor \frac{d}{2} \right\rfloor \tag{5}$$

The reversible transforms are called the integer transforms of the Haar wavelet, or the S transform. Each bit is left shifted and the bit b is embedded using the LSB as shown below in Eq. 6

$$d' = 2d + b \tag{6}$$

This data embedding operation given in Eq. 5 is known as the reversible difference expansion equation. To avoid overflow and underflow problems so that the pixel range is between ranges of [0, 255], the condition in Eq. 7 is met.

$$d' \leq \min(255 - a), (2a + 1) \tag{7}$$

This technique achieves better embedding efficiency and lower distortion than the above-mentioned lossless compression-based algorithms.

A further variant of this scheme is suggested by Alatter [18]. The Generalized Reversible Integer Transform is used to calculate the average and pixel-wise differences of the elements obtained from the pixels of the image. A vector is used in this process and 4 pixels are picked from different locations. The quad is prepared from adjacent pixels and is divided into three categories.

The first is that which is expandable, the second group includes changeable sets, and the third is that which is not changeable.

A location map is required to track changes and thus perform a reversible retrieval operation. Since this technique has a reversibility option, it can be used recursively to embed more data.

Another variation in the difference expansion is known as the prediction error expansion (PEE) technique. In this method, a pixel predictor is introduced. Consider the pixel and its neighborhood pixel in order to measure its intensity using the formula given in Eq. 8

$$P_e = x - x' \tag{8}$$

where P_e is error expansion x_i the pixel chosen and x' is the predicted value.

Then, one bit of LSB shifting is done as in Eq. 9

$$P'_e = 2P_e + m \tag{9}$$

1 bit per pixel is embedded and therefore has more capacity than the Difference expansion.

Adaptive embedding is another variation in which a smooth image pixel is used to obtain a better image for embedding data from the original cover image. Reversible data embedding is achieved by altering the cover image subset. Pixels located in flat or smoother regions are considered for embedding and noisy or sharp regions are not chosen. Thus, the performance of embedding is enhanced.

2.4 Histogram Shifting

Histogram shifting is another prominent reversible technique for hiding data. The image histogram is generated, and the data is embedded in the histogram such that the pixel value is modified by 1 unit at max. The basic formulation is given in Eq. 10

$$\text{SI}_{i,j} \begin{cases} \{\text{CI}_{i,j} - 1 \text{ if } \text{CI}_{i,j} < a \\ \{\text{CI}_{i,j} - m \text{ if } \text{CI}_{i,j} = a \\ \{\text{CI}_{i,j} - 1 \text{ if } \text{CI}_{i,j} > a \end{cases} \tag{10}$$

where a is the integer and CI_{ij} is a cover image and SI_{ij} is the stego image the integer a is peak of histogram and is used to maximize the capacity. The data is modified based on the below criteria.

If $SI_{i,j} < a - 1$, there is no data hidden and the original pixel value is obtained by $CI_{i,j} + 1$.

$SI_{i,j} \in \{a - 1, 1\}$ The pixel embeds message $m = a - SI_{i,j}$.

$CI_{ij} > a$ No change and original pixel value are CI_{ij}.

The work of Ni et al. [7] has been extensively used in research and several variations have been proposed. Fallahpour et al. [19] applied histogram to blocks of images instead of using the full cover image. Therefore, the histogram is generated for each block derived from the cover image and then the data embedding is performed using the Ni et al. [7] process. This block-based technique increases capacity and reduces distortion.

Further, the advancement in this field is by generating an image difference histogram [20]. This data hiding technique has a high peak histogram point and therefore a natural image correlation is exploited.

There are several extensions to the initial work, and we have only introduced the key work to raise awareness of the concepts for further study.

The key points to be generalized from the above techniques are as follows.

1. The main objective is to obtain a reversible process, and while several efficient techniques have been suggested, such as the location map [7, 21] and the shrinking histogram [22, 23], the concept of overflow/underflow must be carefully considered.
2. Adaptive embedding where properties such as sorting, or pixel selection are the focus for designing the algorithms are very useful for improving the efficiency of RDH algorithms.
3. Histogram shifting (HS), prediction error expansion [11], and difference expansion have been major reversible data hiding techniques in spatial domain, but the performance was improved on implementing them with transform domain.

3 Lossless Digital Image Hiding in Transform Domain

The main drawback of spatial domain techniques is that they are not robust as the embedding is performed in the spatial domain and therefore the distortion or noise generated during the embedding process is high. This need led to the development of transform domain techniques that were both reversible and robust.

3.1 Discrete Cosine Transform

DCT has a good time property frequency but is not used efficiently for the embedding process. The embedding is performed on 8×8 blocks if the image has less complex texture then the embedding capacity can be higher with improved visual quality.

Cox et al. [24] suggested DCT to be applied to the entire image instead of the block-based and the embedding is done using the spread spectrum This algorithm is not based on blocking and performs DCTs on the entire image and adopts a spectrum-dependent information hiding technique. The coefficients visible to the human visual system (HVS) are modified, and thus the information hiding is achieved. The algorithm is resistant to image processing techniques, and thus, RDH is enhanced by using the above technique.

Tao et al.; proposed an adaptive system for hiding information [25]. This scheme carried out data hiding on the alternating current (AC) DCT coefficients. A local classification-based algorithm is implemented, and HVS masking effects are used to determine noise sensitivity. The original image is also needed to extract information at the end of the receiver, thus falling into the category of non-blind retrieval.

Reversible DCT algorithms are primarily based on the following two concepts:

1. Yang et al.,—DCT coefficients based on Histogram modification [26].
2. Chen and Kao—Based on DCT quantized coefficients [27].

3.2 Discrete Fourier Transform

Pereira et al. [28] presented a reversible scheme based on the fast transformation domain of Fourier (FFT). The FFT reduces the computational complexity of the Discrete Fourier Transform (DFT). The marked image consists of two parts, one of which is a template, and the other of which is a spread spectrum signal containing secret information and payload. The template does not contain any information that could be used to detect embedded information in the DFT-transformed image. If there is a template in the image, then we calibrate the image based on the embedded template and then extract the watermark information from the calibrated image. This scheme is robust to various image processing operations such as rotation, scaling, and translation; the security of marked images is not guaranteed as they can be attacked by exchanging their information.

3.3 Discrete Wavelet Transform

The wavelet transform can divide the image into a low-resolution approximation LL image and three low-frequency detail images, i.e., HL (horizontal), LH (vertical), and HH (diagonal).

The first technique for information hiding in the wavelet transform domain was proposed by Corvi et al. [29] generated secret information based on a key that uses randomly generated spectrum spread sequence. Secret information is embedded in the low-frequency part of the wavelet-transformed image, keeping the quality of the reconstructed image intact. The short fall of this algorithm is that the original carrier is needed to detect and extract hidden information.

3.4 Integer Wavelet Transform

Reversible wavelet transform known as the integer wavelet transform which maps integer to integer [30] is used to recover the original image without any distortion. The wavelet transformation can divide the image into a low-resolution approximation LL image and three low-frequency detail images, i.e., HL (horizontal), LH (vertical) and HH (diagonal). Xuan et al. [31] proposed a lossless method to embed data. Data compression is achieved by arithmetic encoding, and in one or more middle-bit planes of the transformed coefficients, the data is hidden.

In addition, the error codes, secret information, and histogram modified book-keeping data are also encoded and retrieved data without loss. The embedding capacity is more than the existing techniques and the resulting image is less distorted.

Xuan et al. [32] a threshold-based histogram embedding algorithm was proposed where information is embedded in the high-frequency sub-bands of IWT to minimize distortion. In order to prevent overflow/underflow problems, small absolute value coefficients are considered for embedding. By splitting the histogram into three parts, the threshold method is applied. The first section is if the data has secret information. The values of coefficients are below and above the threshold value for the center and end section; they are excluded from embedding. Histogram pairs with the maximum threshold are chosen from the embedding which helps to achieve the highest PSNR.

A method by which the image is divided into four non-overlapping blocks was proposed by Agrawal and Kumar [33]. The entropy is calculated, and the block with the highest entropy is dropped for embedding. The IWT is then applied to all selected blocks, and the histogram shift is applied using the peak and zero value pairs (bits are shifted by 1 to the right if the peak point pixel value is less than the zero point value). If the message to be embedded is 1 then the peak point is increased by 1, and no change is made to the zero bit.

Chauhan et al. [34] presented a spread-spectrum-based medical watermarking technique where the Mexican wavelet is used, and the binary watermark is concealed in vertical and horizontal sub-bands using a pseudo-noise (PN) sequence pair. This method has been evaluated for different attacks and is compatible with existing methodologies.

As discussed above, these are the basic concepts implemented in the frequency domain and the key points to be generalized from the above techniques are as follows:

1. Frequency domain techniques are robust for geometric attacks.
2. As the information is hidden in the pixel coefficients, the images are less distorted than the spatial domain techniques.
3. Embedding capacity is also improved compared to spatial domain techniques.
4. These techniques are further enhanced by a hybrid combination of frequency domain and spatial domain methods to provide better solutions.

4 Bio-Inspired Algorithms

A new concept known as bio-inspired algorithms that imitate nature opens a new era in computation for solving optimization problems. There are numerous researches under study to solve complex problems with conventional technique. This research is an emerging field, and the results obtained are impressive. This broadens the scope and viability of bio-inspired algorithms (BIAs) to explore new areas of application and more computing opportunities. Bio-inspired algorithms are classified [35] as Ecology-based (BBO, AWC), Evolution-based (GA, DE) and Swarm-based (PSO, ACO, ABC).

There are many swarm-based algorithms, each with its own advantages and shortfalls. High image quality and robustness are required in medical, military, and other sensitive areas. A bio-inspired algorithm with a proper fitness function design will provide a new insight into the study of the data hiding algorithm. Each of the features of the algorithm and its contribution to the field of hiding information is discussed in the next section.

4.1 Particle Swarm Optimization (PSO)

PSO is a global optimization technique proposed by Kennedy and Eberhart in 1995 [36]. PSO is inspired by the behavior of the flock of birds that are in search of food. In PSO, the term—particles refer to each bird in the entire population. Food is a representation of the solution that has been found. Every particle in the swarm presents a method with four vectors in a high-dimensional space, its current location, the best position found so far, the position found in surrounding area and is best so far and its velocity. In the search space, the location is modified according to the best possible position obtained by itself (pbest) and the best location achieved during the search process by the neighborhood (gbest). A particle's best personal location reflects a particle's cognitive behavior. It is identified as the best particle location to be found. Whenever the particle reaches a location with a greater fitness value than the fitness value of the previous personal best, it will be updated. An individual retains its experience of the past. Social behavior represents the best global position. It is known as the best location to be found in a swarm by all the particles. Once the particle reaches a position with a better fitness value than the fitness value of the previous global position, it will be updated. Each particle learns from swarm participants (population).

Particle update rule

$$p = p + v \tag{11}$$

With

$$v = v + c1 * \text{rand} * (p\text{Best} - p) + c2 * \text{rand} * (g\text{Best} - p) \tag{12}$$

where p: particle's position, v: path direction, $c1$: weight of local information, $c2$: weight of global information, pBest: best position of the particle, gBest: best position of the swarm, rand: random variable.

As per Naheed, Talat et al. [37]. By means of PSO intelligent technique the correlation of image pixel values by multiplying them with optimum weight values, which result in the optimal embedding of information into the calculated interpolation-error values of GA-based image pixels. This approach is fully reversible and the combination of PSO and GA resulted in an improved algorithm with higher embedding capacity.

Another reversible information hiding system for authenticating medical images is proposed by Balasamy et al. [38]. Wavelet transform is applied on the cover image and tent map of the secret image to encrypt the secret information. The operations performed on a binary encoded image are based on the compression algorithm sequences generated from chaotic maps. Particle swarm optimization (PSO) exploits the image pixel correlation of adjacent pixels that provides an optimal balance between embedding capacity and imperceptibility. The proposed method is therefore capable of embedding with low distortion, extracting secret information, and also retrieving the original image. Using the above technique, key aspects of robustness, capacity, and imperceptibility are achieved.

4.2   Cuckoo Search (CS)

Cuckoo Search (CS) is a modern approach to optimization inspired by cuckoo brooding behavior [39]. This algorithm is relatively new, so there is very little insight into the research related to hiding information. An egg symbolizes a solution, and a cuckoo egg is a new solution. The aim is to use a new and potentially better cuckoo as a solution.

1. CS is based on the following three idealized rules;
2. Each cuckoo shall lay one egg at a time and dump the egg in a randomly selected nest; the best nests with high quality eggs will carry on to the next generation;

3. The number of host nests available is fixed and the egg laid by the cuckoo is discovered by the host bird. Discovering operates on some of the worst nests and has discovered solutions dumped from further calculations.

This technique uses a CS to enhance the efficiency of the embedding process. The nest is represented as a sequence of random locations (initially) and optimizes these sequences in the CS optimization algorithm to find the best nest with the maximum effect on the PSNR (Fitness Function). This technique has produced significant changes over successive generations. The outcome of the improved proposed method is the preferred location of the gray scale images in the discrete wavelet transform domain; for the watermark embedding that is used as a corresponding hidden key. It is very clear that the image quality has improved significantly and that this change is done by optimizing CS, enhancing the configuration of the embedding positions while preserving a high degree of robustness.

Cuckoo-inspired optimization of IWT-based reversible watermarking is discussed in [40]. LH, HL, and HH sub-band of integer wavelet transforms are used to hide information that results in a better image quality. Cuckoo optimization algorithm is applied to each sub-band. The nest with the best survival rate, that is, one with the lowest fitness value, is selected and taken to the next iteration. The system further decomposes the LL sub-band and each sub-band is assigned to the cuckoo optimization algorithm. This process will continue until no further improvement has been seen. This algorithm selects a bit plane with a minimum compression ratio and thus increases the imperceptibility of the hidden data.

4.3 Artificial Bee Colony (ABC)

ABC is based on bee behavior and is categorized into two aspects [35]: (1) Mating and (2) Foraging Behavior. The artificial bee colony algorithm simulates the intelligent foraging behavior of a honeybee swarm proposed by Karaboga and Basturk [41]. Bees are predominantly divided into three classes of bees: working bees, scout bees, and bees. Onlooker bees are known as the decision-maker bee waiting in the dance area to choose a food source. Employed bees are the ones who visit the food source, and scout bees perform a random search to find new sources.

A randomly distributed initial population (positions of the food source) is generated. After initialization, the search process cycles are iterated with the onlooker and scout bees employed, respectively. The employed bee modifies the original position when a new food source position is discovered and the old one is deleted from memory. After all employed bees complete the search process, they share the source position information with the onlookers in the dance area. Every onlooker evaluates the nectar information taken from all the bees employed and then chooses a source of food based on the nectar source.

This [42] methodology employs a reversible watermark concept with a hybrid combination of Slantlet transform and Arnold transform. It will use the mean value coefficients of the HL and LH bands for the embedding of the watermark, which provided very good imperceptibility and robustness to the scheme. The embedding strength of the watermarking is controlled with the help of artificial bee colonies (ABCs) in order to obtain an optimal trade-off between invisibility and robustness.

4.4   Firefly Algorithm (FA)

Firefly algorithm proposed by the Yang [43] is a swarm-based heuristic algorithm inspired by the flashing behavior of fireflies. The algorithm is a population-based iterative process with several agents (perceived as fireflies) that simultaneously resolves a given optimization problem. Bio-luminescent glowing capability is used by agents to communicate with each other. The approach for optimization is based on the premise that the solution to the optimization problem is perceived as an agent (firefly) that shines in proportion to its efficiency.

The 3 major rules of this algorithm are:

1. All fireflies are unisex and move toward more attractive and brighter ones, regardless of their sex.
2. The degree of attractiveness of fireflies is proportional to its light.
3. The brightness or light intensity of a firefly is determined by the value of the objective function of the problem.

Few works are available using FA algorithms. In [44], the author has proposed an effective reversible data hiding scheme using firefly algorithm to hide secret data in the cover images. This approach finds the best positions for hiding secret information using FA. Secret data is embedded in the best positions found by histogram modifications. In [45], Ankita kadu et al. propose new strategies for reversible data hiding techniques based on a robust firefly algorithm. The firefly algorithm must find an optimal location for hiding confidential information. The scrambling of the image is applied in the transform frameset to prevent the detection of the hidden secret image.

The decomposition is achieved with a Haar wavelet. It is also used to improve the efficiency of embedding by reducing the distortion of the reversible water-marking database proposed by [46].

In [47], firefly algorithm and PVO-based reversible block-based algorithm are proposed such that embedding is performed with minimum distortion.

As discussed above, there are few bio-inspired techniques used to hide reversible data, and the key points to be generalized from the above techniques are as follows.

Bio-inspired optimization is a new field of research that emerged in the early 2000's and has a great deal of scope for the well-designed fitness function, bio-inspired computing may indeed be useful for the design of information hiding algorithms.

1. As far as the literature is concerned, there are many works in reversible data hiding, but there is a significant amount of scope for optimization since there is little work available on it, and it can help bridge the gap between robustness, capacity, and imperceptibility.
2. Optimization algorithms are flexible in design, provide parameter selection and performance metrics, making them prominent in providing solutions.

A list of techniques is provided in Table 1 summarizing the benefits and drawbacks, and in Table 2, we have the PSNR, embedding capacity and size of image, and summarizing this we conclude that there is a tradeoff between capacity and imperceptibility.

Table 1 Summary of reversible data hiding techniques and its features

Author	Technique	Advantages	Drawbacks
Honsinger et al. [14]	Modulo addition 256		– Salt and pepper noise – Low imperceptibility
Fridrich et al. [15]	Bit plane compression		–
Goljan et al. [16]	RS embedding compression	– High capacity	– Very small and invertible distortion
Celik et al. [3]	G-LSB compression	– Compression efficient – High capacity	– Complex implementation
Tian [17]	Difference expansion	– No compression required – High capacity	
Alattar [18]	Difference expansion	– High capacity	– Low distortion
Aziz et al. [48]	Difference expansion	– High capacity – Improved visual quality	–Appropriate for smooth and textured images
Ni et al. [7]	Histogram-based	– Robust to jpeg compression	– Not completely reversible
Fallahpour [19]	Histogram based	– High capacity	– Increases complexity
Yang et al. [26]	DCT-histogram based	– High capacity	–
Pereira et al. [28]	Template-based DFT	– Robust to rotation	– Cropping and scaling does not give 100% recovery
Xuan et al. [31]	IWT	– IWT results in 100% recovery as all the coefficients are integers so no loss of data	– PSNR is quite low-no insight on robustness
Xuan et al. [32]	IWT-histogram	– Adaptive histogram modification with better performance on overflow/ underflow and capacity then earlier methods	– Minimum threshold does not guarantee to provide improved PSNR

(continued)

Table 1 (continued)

Author	Technique	Advantages	Drawbacks
Agrawal et al. [33]	IWT-entropy and histogram based	– Medical images are used with higher visual quality	– Only coefficients which are smooth are considered for embedding
Golbai et al. [49]	DWT wavelet filter modification	– Robust – High capacity	– Not 100% reversible to attacks
Chauhan [34]	DWT-Mexican wavelet	– Robustness to Gausssian and Poisson noise	– Other attacks like salt and pepper and compression 100% not recovered
Naheed et al. [37]	PSO and GA	– 100% reversibility with high capacity, no visual degradation	–
Balasamy [38]	PSO + wavelet transform	– High robustness, capacity with maximum imperceptibility	– Lossless host image is required at receiver end
Gupta [40]	Cuckoo + wavelet transform	– Mitigate the replay attack on biometric	–
Yang [43]	ABC + slanlet transform	– Improved embedding strength	– Not robust to geometrical attack
Amsaveni [44]	FA + histogram	– Optimal locations to hide the secret data	

Table 2 Summary of reversible data hiding techniques and its performance

Technique	PSNR (dB)	Embedding capacity (bits)	Size of image
Modulo addition 256	–		
Bit plane compression	38.60	4000	256 × 256
RS embedding compression	36.06	8988	256 × 256
G-LSB compression	31.9	34,329	512 × 512
Difference expansion	40.06	101,089	512 × 512
Difference expansion	31.78	569,317	512 × 512
Difference expansion	54.14	131,072	512 × 512
Histogram based	48.2	5460	–
Histogram based	47.29	13,868	–
DCT-histogram based	44 dB (avg)	10,541 17,395	256 × 256 512 × 512
Template-based DFT	–	–	512 × 512
IWT	36.64	85,507	512 × 512
IWT-histogram	60	510,079	512 × 512

(continued)

Table 2 (continued)

Technique	PSNR (dB)	Embedding capacity (bits)	Size of image
IWT-entropy and histogram based	50–55	–	512 × 512
DWT wavelet filter modification	44.37	61,586	512 × 512
DWT-Mexican wavelet	43.9	–	1024 × 1024
PSO and GA	GA-48.8 PSO-48.8	GA-73231 PSO-73206	512 × 512
PSO + wavelet transform	49	38,462	512 × 512
Cuckoo + wavelet transform	45	–	320 × 280 (iris image)
ABC + Slanlet transform	51	16,384	512 × 512
FA + histogram	52.86	–	512 × 512

- Compression-based techniques improve capacity.
- Histogram-based techniques reduce the size of auxiliary data.
- Difference expansion is more robust and does not require any storage of additional data.
- Transform-based techniques are robust compared to the spatial domain and result in improved embedding capacity with low distortion as the data is hidden in pixel coefficients.
- The hybrid combination of spatial and transform domain helps improve results by overcoming the shortfalls of both techniques.
- Optimization techniques are more prevalent as they can be used to solve any complex problem, and their immense source of inspiration derives from the imitation of natural behavior.
- These algorithms are versatile, as they can be modified according to the need to solve the problem.

5 Conclusion

We have presented different means for reversible data hiding in this paper. In addition to spatial and transform domain techniques, we have seen how the bio-inspired optimization is utilized to obtain better performance results. Since a bio-inspired algorithm is inspired by nature, it is flexible and easy to implement.

The prerequisite of a watermark technique is to prove the authenticity of the content and the proof of ownership, and, according to the literature, there are publications on reversible watermarks which are of a robust nature. But in this world of digitization, there is still a need to improve the robustness of attacks, storage capacity, and imperceptibility for security of reversible data hiding.

It has also been noted that there is scope to improve the result with a more efficient reversible data hiding technique. The combination of existing approaches and optimization strategies will help bridge the tradeoff between robustness, efficiency and imperceptibility.

References

1. Lu Z-M, Guo S-Z (2017) Chapter 3—lossless information hiding in images on transform domains, lossless information hiding in images, syngress, pp 143–204. ISBN 9780128120064
2. Feng J-B, Lin I-C, Tsai C-S, Chu Y-P (2006) Reversible watermarking: current status and key issues. Int J Netw Secur 2(3):161–171
3. Celik MU, Sharma G, Tekalp AM, Saber E (2005) Lossless generalized-LSB data embedding. IEEE Trans Image Process 14(2):253–266
4. Dragoi IC, Coltuc D (2014) Local-prediction-based difference expansion reversible watermarking. IEEE Trans Image Process 23(4):1779–1790
5. Al-Qershi OM, Khoo BE (2009) An overview of reversible data hiding schemes based on difference expansion technique. In: Proceedings of the first international conference on software engineering and computer systems, ICSECS09, pp 741–746
6. Chen X, Sun X, Sun H, Xiang L, Yang B (2015) Histogram shifting based reversible data hiding method using directed-prediction scheme. Multimedia Tools Appl 74(15):5747–5765
7. Ni Z, Shi YQ, Ansari N, Su W (2006) Reversible data hiding. IEEE Trans Circuits Syst Video Technol 16(3):354–362
8. Li X, Li B, Yang B, Zeng T (2013) General framework to histogram-shifting based reversible data hiding. IEEE Trans Image Process 22(6):2181–2191
9. Tseng HW, Hsieh CP (2009) Prediction-based reversible data hiding. Inf Sci 179(14):2460–2469
10. Li X, Li J, Li B, Yang B (2013) High-fidelity reversible data hiding scheme based on pixel-value-ordering and prediction-error expansion. Signal Process 93(1):198–205
11. Caciula I, Coanda HG, Coltuc D (2019) Multiple moduli prediction error expansion reversible data hiding. Sig Process Image Commun 1(71):120–127
12. Xuan G, Zhu J, Chen J, Shi YQ, Ni Z, Su W (2002) Distortion less data hiding based on integer wavelet transform. Electron Lett 38(25):1646–1648
13. Huang HC, Chang FC, Chen YH, Chu SC (2015) Survey of bio-inspired computing for information hiding. J Inf Hiding Multimedia Signal Process 6(3):430–443
14. Honsinger CW, Jones PW, Rabbani M, Stoffel JC (2001) U.S. Patent No. 6,278,791. U.S. Patent and Trademark Office, Washington, DC
15. Fridrich J, Goljan M, Du R (2001) Invertible authentication. In: Security and watermarking of multimedia contents III, vol 4314, pp 197–208. International Society for Optics and Photonics
16. Fridrich J, Goljan M, Du R (2001) Distortion-free data embedding. Lect Notes Comput Sci 2137:27–41
17. Tian J (2002) Reversible watermarking by difference expansion. In: Proceedings of workshop on multimedia and security, vol 19
18. Alattar AM (2004) Reversible watermark using the difference expansion of a generalized integer transform. IEEE Trans Image Process 13(8):1147–1156
19. Fallahpour M (2008) Reversible image data hiding based on gradient adjusted prediction. IEICE Electron Express 5(20):870–876
20. Wang J, Mao N, Chen X, Ni J, Wang C, Shi Y (2019) Multiple histograms based reversible data hiding by using FCM clustering. Signal Process 159:193–203
21. Hu Y, Lee HK, Li J (2008) DE-based reversible data hiding with improved overflow location map. IEEE Trans Circuits Syst Video Technol 19(2):250–260

22. Xuan G, Shi YQ, Chai P, Cui X, Ni Z, Tong X (2007) Optimum histogram pair-based image lossless data embedding. In: International workshop on digital watermarking. Springer, Berlin, Heidelberg, pp 264–278
23. Xuan G, Tong X, Teng J, Zhang X, Shi YQ (2012) Optimal histogram-pair and prediction-error based image reversible data hiding. In: International workshop on digital watermarking. Springer, Berlin, Heidelberg, pp 368–383
24. Cox IJ, Kilian J, Leighton FT et al (1997) Secure spread spectrum watermarking for multimedia. IEEE Trans Image Process 6(12):1673e1687
25. Tao B, Dickinson B (1997) Adaptive watermarking in the DCT domain. In: Proceeding of the 1997 IEEE international conference on acoustics, speech, and signal processing, vol 4, no 4, pp 2985–2988
26. Yang B, Schmucker M, Niu X, Busch C, Sun S (2004) Reversible image watermarking by histogram modification for integer DCT coefficients. In: IEEE 6th workshop on multimedia signal processing. IEEE, pp 143–146
27. Chen CC, Kao DS (2007) DCT-based reversible image watermarking approach. In: Third international conference on intelligent information hiding and multimedia signal processing (IIH-MSP 2007), vol 2. IEEE, pp 489–492
28. Pereira S, Ruanaidh JJ, Deguillaume F, Csurka G, Pun T (1999) Template based recovery of Fourier-based watermarks using log-polar and log-log maps. In: Proceedings IEEE international conference on multimedia computing and systems, vol 1. IEEE, pp 870–874
29. Corvi M, Nicchiotti G (1997) Wavelet-based image watermarking for copyright protection. In: SCIA'97: 10th Scandinavian conference on image analysis (Lappeeranta, June 9–11, 1997), pp 157–163
30. Calderbank AR, Daubechies I, Sweldens W, Yeo BL (1998) Wavelet transforms that map integers to integers. Appl Comput Harmon Anal 5(3):332–369
31. Xuan G, Zhu J, Chen J, Shi YQ, Ni Z, Su W (2002) Distortionless data hiding based on integer wavelet transform. Electron Lett 38(25):1646–1648
32. Xuan G, Shi YQ, Chai P, Teng J, Ni Z, Tong X (2009) Optimum histogram pair-based image lossless data embedding. In: Transactions on data hiding and multimedia security IV. Springer, Berlin, Heidelberg, pp 84–102
33. Agrawal S, Kumar M (2016) Reversible data hiding for medical images using integer-to-integer wavelet transform. In: 2016 IEEE students' conference on electrical, electronics and computer science (SCEECS). IEEE, pp 1–5
34. Chauhan DS, Singh AK, Adarsh A, Kumar B, Saini JP (2017) Combining Mexican hat wavelet and spread spectrum for adaptive watermarking and its statistical detection using medical images. Multimedia Tools Appl, 1–15
35. Binitha S, Sathya SS (2012) A survey of bio inspired optimization algorithms. Int J Soft Comput Eng 2(2):137–151
36. Kennedy J, Eberhart R (1942) Particle swarm optimization (PSO). In: Proceedings of the IEEE international conference on neural networks, Perth, Australia, vol 27
37. Naheed T, Usman I, Khan TM, Dar AH, Shafique MF (2014) Intelligent reversible watermarking technique in medical images using GA and PSO. Optik 125(11):2515–2525
38. Balasamy K, Ramakrishnan S (2019) An intelligent reversible watermarking system for authenticating medical images using wavelet and PSO. Clust Comput 22(2):4431–4442
39. Yang XS, Deb S (2010) Engineering optimisation by cuckoo search. Xiv preprint arXiv: 1005.2908
40. Gupta R, Sehgal P (2016) Mitigating iris-based replay attack using cuckoo optimized reversible watermarking. In: Seventh international conference on advances in computing, control, and telecommunication technologies-ACT
41. Karaboga D, Basturk B (2007) A powerful and efficient algorithm for numerical function optimization: artificial bee colony (ABC) algorithm. J Global Optim 39(3):459–471
42. Ansari IA, Pant M, Ahn CW (2017) Artificial bee colony optimized robust-reversible image watermarking. Multimedia Tools Appl 76(17):18001–18025

43. Yang XS (2009) Firefly algorithms for multimodal optimization. In: International symposium on stochastic algorithms. Springer, Berlin, Heidelberg, pp 169–178
44. Amsaveni A, Arunkumar C (2015) An efficient data hiding scheme using firefly algorithm in spatial domain. In: 2015 2nd international conference on electronics and communication systems (ICECS). IEEE, pp 650–655
45. Kadu MA, Kulkarni A, Patil D (2016) Secure data hiding using robust firefly algorithm. Int J Comput Eng Res Trends 3(10):550–553
46. Mamoglu MB, Ulutas M, Ulutas G (2017) A new reversible database watermarking approach with firefly optimization algorithm. Math Prob Eng
47. Abbasi R, Faseeh Qureshi NM, Hassan H, Saba T, Rehman A, Luo B, Bashir AK (2019) Generalized PVO-based dynamic block reversible data hiding for secure transmission using firefly algorithm. Trans Emerg Telecommun Technol e3680
48. Aziz F, Ahmad T, Malik AH, Uddin MI, Ahmad S, Sharaf M (2020) Reversible data hiding techniques with high message embedding capacity in images. Plos One 15(5):e0231602
49. Golabi S, Helfroush MS, Danyali H (2019) Reversible robust data hiding based on wavelet filters modification. Multimedia Tools Appl 78(22):31847–31865

Person Re-identification Using Deep Learning with Mask-RCNN

Aditya Kshatriya, V. M. Nisha, and S. A. Sajidha

1 Introduction

Person re-identification comprises of matching pictures of a particular person captured in a matrix of cameras with non-overlapping fields of view. The task is not the same as the classic id and detection tasks. The identification comprises of ascertaining the identity of a person in an image, whereas the detection comprises of rotoscoping people from their background, in the absence of the identity of their knowledge. Person re-identification answers the question if the given image is of the same person as the query image. The task of detection tells us whether it is a person and the task of identification helps to know who the person is. But the task of re-identification tells us when and where this particular person appeared, respective to a given camera, and using multiple cameras, person re-identification permits the assessment of his/her path for a tiny time period. The task of person re-identification task depends on the pedestrian detection task. In step 1, we form a set of pedestrians by collecting the cropped pedestrian images or extract the pedestrian image signature from every camera seen in the matrix. Next, we compare the similarity or the distance from the query image. At the end, we show the closest images based on the measured similarity.

It is generally considered that the person keeps the same clothes in multiple scenes. Thus, this method works most effectively in a short time period, which makes sure that the visual appearance of the person remains the same. A huge constraint of re-identification is that we cannot apply it to find similarities between people after some days as their appearance may have changed. This is also partly because biometrics like face identification are not exactly possible from far placed

A. Kshatriya · V. M. Nisha (✉) · S. A. Sajidha
School of Computing Sciences and Engineering, VIT Chennai, Chennai, Tamil Nadu, India
e-mail: nishavm@vit.ac.in

S. A. Sajidha
e-mail: sajidha.sa@vit.ac.in

R. R. Raje et al. (eds.), *Artificial Intelligence and Technologies*,
Lecture Notes in Electrical Engineering 806,
https://doi.org/10.1007/978-981-16-6448-9_49

surveillance cameras; hence, the system of person reid is based mostly on the clothing and appearance of the people present in the view.

The cameras are generally spaced over large areas and cover a huge part of the geography of a given place. This provides a lot of coverage and data. Any person leaving a view from one camera must be tracked when he enters another view in another camera, and it is possible that he is surrounded by people who look similar to him visually but are actually different people. Thus, re-identification becomes a suitable method for association of data when the data are given have some sort of continuity to it. This can be used in the long term and also has multiple applications.

2 Literature Survey and Existing Model

In the existing systems, there are four different ways of performing person re-identification. The literature for two of these methods will be presented in this literature survey. In the first section, deep learning techniques are introduced. Next, classical neural networks are introduced and finally, convolutional neural networks (CNN) are introduced, along with some complex applications and variations, including the Siamese neural networks which are often applied in person re-identification. In the second section, the current top tier systems in person re-identification systems will be dealt with. Some representative approaches will be introduced based on feature extraction, deep learning and metric learning.

A neural network is a series of algorithms whose aim is to recognize the invisible relationships that are present in a dataset using a process that imitates the way the human brain operates. In the context of biology, neural networks refer to series of neurons which can be organic or artificial in nature. The advantage of neural networks is that they can adapt to a dynamic input; therefore, the series of algorithms generate an optimum result without having to change or the criteria for the output. The concept of neural networks, which is the core of artificial intelligence systems, is increasing quickly in popularity in the advancing of trading systems.

Human behaviors are determined by central a nervous system which comprises of nerve cells or neurons. There are about 86 billion neurons in the human brain [1]. A neuron passes the information by reacting with other neurons. The membrane potential of the neuron increases slowly, after the dendrites of that neuron receives an input that acts as an excitatory input. If the voltage of the membrane crosses a given threshold, a potential gets initiated and it is further passed along the axon of the neuron to the next synaptic neurons.

The very fundamental idea of neural network is to adjust the basic simple function to fit any given function by tuning the various parameters for the given network. The theorem of universal approximation [2] states that a feed-forward network with one hidden layer including a limited number of neurons can estimate every continuous function based on closely spaced subsets of Rn, given there are some considerations upon the activation function.

The neural network also comprises of three fully connected layers, each of which has 4096 channels for the first two and 1000 channels for the third layer. This results in one channel per class. In addition to this, all the hidden layers in VGG use ReLU which cut training time drastically for the previous AlexNet too. It does not use a Local Response Normalization function either, because it was discovered that it increases the consumption of memory without having any positive impact on accuracy. The Backpropagation method [3] is used a lot and could also be the most universally accepted algorithm used to train neural networks of forward feeding nature. It uses the gradient-descent optimization algorithm with reference to neural networks.

Another method is used by Liu et al. [4] who extracts the Hue Saturation Value histogram, histogram of gradient and the Local Binary Patterns for every patch that is local. Next, a method called local coordinate coding was applied which estimates an input point as a combination of some elements knows as anchored points. In [5], the author uses a bag of words type model where an 11 dimensional color name feature is obtained for every local patch. This feature is then averaged into a global vector by using a bag of word model. Next, a code is generated on the input data and then it is converted into visual words. After this, a TF-IDF model weighs the histogram and each person is allotted a word. In [6], Shen et al. extract dense SIFTs along with Dense Color Histograms from every patch. The authors introduce a correlation model which is able to create correlation patterns among cameras and then tries to match the patches by merging a constraint with the correlation structure which excludes the misalignments of camera positions within images.

Wang et al. [7] describe a model that can consider the context of shape and looks. This method calculates a local histogram of Gradients in the regular film color space. Next, the body of the person is partitioned into regions in two different steps. Step 1 is to distribute images based on their histogram of Gradients look, and Step 2 recognizes parts using a modified context of shape that has already learnt a shape dictionary. Next, labeled images based on regions are created. The shape context is formed using a multi-occurring network that gives out the distributions of probability and their connections over the selected image regions. Further, matching of body parts is completed with the use of L1-norm distance. There are methods which can segment image parts that are more useful like torso, arms, legs, etc., that are semantic in nature and won't be affected by difference of viewpoint. A model that is widely used is which was invented by Farenzena et al. [8].

3   Proposed System Implemetation

3.1   Dataset

There are various datasets available for person re-id. For this work, we choose the Market1501 dataset for basic training and testing. The Market1501 datasets is one of the biggest datasets available with 1501 identities and 6 different cameras.

3.2 Methodology

The system architecture is shown in Fig. 1. The development of the model is done in three phases:

1. Mask-RCNN for removing background
2. Training the model
3. Developing the re-identification model.

Training the Model

We use the neural network called resnet50 to model our feature extractor. Resnet stands for residual neural network. The speciality of resnet50 is that it addresses the problem of vanishing gradient by using a residual function.

Developing the Re-Identification Model

For testing, on our own data, we use a custom code which extracts features using the trained weights and stores them into cuda tensors. Here, we use the extractor method which extracts the features of our own model using the resnet model and the weights that we saved earlier while training.

Then, we create an imagelist for the gallery set. Next, while testing we use the feature extractor on each of the ids to construct feature vectors for each of the ids.

Next, we take the query as input. For the query too, we have to extract the features and save them into numpy arrays. I have taken 4 queries because there are 4 ids. These are stored in the dictionary qdata_dict.

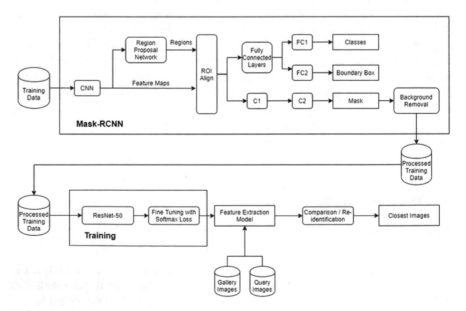

Fig. 1 ResNet mask-based deep learning model for re-identification of person in CCTV images

Further, we compare the feature vectors of query with feature vectors of the images for each id. For this, we define the Euclidean distance and sim_index function to calculate the similarity index. Now, we define the function to get the top 10 closest images. The one who has the lower average distance will be allotted the id of the query image.

ResNet uses a new way of skipping layers in order to have a lower loss and degradation of the gradient-descent function. This was achieved by deciding to use a residual function instead of a learning function that learnt the direct mapping. This meant that some nonlinear layers could be stacked on top of each other, and this should not lower the performance as we could also increase and pile up identity mappings, and due to this, the final architecture would also perform similarly. The residual function is written as $F(x) = H(x) - x$. Here, x stands for the identity function. Thus, it can be written again as $H[x] = F(x) + x$.

It is considered that to optimize this residual function ($F(x)$) is much better and more easy than to optimizing the previous function. Likewise, it is also easier to learn a residual function than to learn the original $H(x)$ function using a stack of convoluted layers.

In order to use this concept in a CNN, the designer of ResNet added something called skip connections. These skip connections added the i/p of the first layer with the o/p of the next two or three layers. Due to this, the new layer has to learn newer features apart from those already present from the previous layers. Due to this, the problem of the vanishing descent also was solved because such a network allows the gradient to go through all the layers without being changed from the top to the bottom. Because of the properties of Resnet, incredibly deep neural networks can be formed. For this project, a ResNet of 50 layers will be used, hence the network is called resnet50.

Layers in ResNet50

Resnet50 is divided into 5 main convolutional layers, each containing some layers of filters between them. All layers have ReLu activation function between them. The skipping occurs for each block to maintain the value for the gradient-descent function. The layers are as follows:

(1) First convolutional layer with a filter of size 7×7, which gives 64 features, with a stride of 2. This gives an output of size 112×112.
(2) Second main convolutional layer has a 3×3 max pooling layer which has a stride of 2 and then three blocks of 3 filter layers, 1×1 64, 3×3 64, and 1×1 256. This main layer outputs a vector of size 56×56.
(3) The third layer again has 4 blocks of 3 filter layers, 1×1 128, 3×3 128 and 1×1 512. The output is a 28×28 vector.
(4) The fourth layer has 6 blocks of 3 filter layers. 1×1 256, 3×3 256 and 1×1 1024. The output is a 14×14 vector.

(5) The last main layer has 3 blocks of 3 layers. 1×1 512, 3×3 512, 1×1 2048. The output is a 7×7 vector.

(6) Finally, we have an average pooling layer which is fully connected and uses softmax loss. The final output of resnet50 is a feature vector of shape [1, 2048] for each image.

4 Results and Discussion

Inorder to identify a person again and again from multiple images that have been captured by different cameras, whose fields of view do not overlap, sample query image is used. The left most single image shown in Fig. 2 and Fig. 3 show sample query images, and right side of Fig. 2 and Fig. 3 show the output images. The Mean Average Precision obtained earlier without mask-RCNN is 68.5%. Then, the model was trained with Mask-RCNN result as input which has segmented images and blacked out backgrounds. The new Mean Average Precision is 70.8%.

The rank 1 through rank 20 score also improved marginally. The comparison is shown in Table 1.

Fig. 2 Sample query image (left) and outputs (right)

Fig. 3 Sample query image 2 (left) and outputs (right)

Table 1 Comparison

Rank accuracy	Without mask-RCNN	With mask-RCNN
Rank-1	85.3	87.4
Rank-5	93.8	94.5
Rank-10	95.7	96.4
Rank-20	97.4	98.3

5 Conclusion

In this paper, the general context of analysis of people in a network of surveillance cameras with non-overlapping views was presented, and then, the person re-identification problem in images was specifically studied. A method to re-identify persons in the gallery set is given. A query was implemented, and a satisfactory result was also obtained. Although the output of Mask-RCNN was obtained, in the initial training and testing phase a low accuracy was obtained. Then, images were upscaled and provided for training, which increased the map slightly. Future improvements for this work may include extra feature extraction models which may also be able to detect other objects along with persons for re-identification.

References

1. https://human-memory.net/brain-neurons-synapses/
2. Hornik K, Stinchcombe M, White H (1989) Multilayer feedforward networks are universal approximators. Neural Netw 2(5):359–366
3. Rumelhart DE, Hinton GE, Williams RJ (1986) Learning representations by back-propagating errors. Nature 323(6088):533
4. Liu X, Song M, Tao D, Zhou X, Chen C, Bu J (2014) Semi-supervised coupled dictionary learning for person re-identification. In: Proceedings of the IEEE conference on computer vision and pattern recognition, pp 3550–3557
5. Zheng L, Shen L, Tian L, Wang S, Wang J, Tian Q (2015) Scalable person re-identification: a benchmark. In: IEEE international conference on computer vision
6. Shen Y, Lin W, Yan J, Xu M, Wu J, Wang J (2015) Person re-identification with correspondence structure learning. In: Proceedings of the IEEE international conference on computer vision, pp 3200–3208
7. Wang X, Doretto G, Sebastian T, Rittscher J, Tu P (2007) Shape and appearance context modeling. In: IEEE 11th international conference on computer vision, 2007 (ICCV 2007). IEEE, pp 1–8
8. Farenzena M, Bazzani L, Perina A, Murino V, Cristani M (2010) Person re-identification by symmetry-driven accumulation of local features. In: Proceedings of the IEEE international conference on computer vision and pattern recognition (CVPR), pp 2360–2367

Efficient Algorithm for CSP Selection Based on Three-Level Architecture

Md. Abdul Quadir, J. Prassanna, J. Christy Jackson, H. Sabireen, and Gagan Gupta

1 Introduction

Cloud computing is a modern and emerging technology in recent times. It is one of the most demanding technologies in today's technology world. Since remarkable advancement in the IT industry [1], cloud technology puts a notable impact on industrial society [2, 3]. Computing assets being hosted on the Internet and supplied to consumers as a service in cloud computing [4], it provides flexible, distributed, shared resources (e.g., processing capacity, space, software) for consumers over the network from remote data centers (e.g., corporation, state, individuals) [5]. Although customers do not influence the fundamental computing assets, they need to ensure that the capacity, accessibility, reliability, security, and performance of these assets are delivered to support such resources [6]. Therefore, cloud computing is the network which provides services that are always available, everywhere and every time on demand [7]. In addition, cloud computing is the platform which provides services on request that are always open, everywhere and every time [8]. Cloud computing, cloud services, and infrastructure are characterized by several features [9]:

Md. Abdul Quadir (✉) · J. Prassanna · J. Christy Jackson · H. Sabireen · G. Gupta
Vellore Institute of Technology, Chennai, India
e-mail: abdulquadir.md@vit.ac.in

J. Prassanna
e-mail: prassanna.j@vit.ac.in

J. Christy Jackson
e-mail: christyjackson.j@vit.ac.in

H. Sabireen
e-mail: sabireen.h2019@vitstudent.ac.in

G. Gupta
e-mail: gagan.gupta2019@vitstudent.ac.in

© The Author(s), under exclusive license to Springer Nature Singapore Pte Ltd. 2022   515
R. R. Raje et al. (eds.), *Artificial Intelligence and Technologies*,
Lecture Notes in Electrical Engineering 806,
https://doi.org/10.1007/978-981-16-6448-9_50

- Cloud networking: Online network storage resources or information.
- Ubiquitous: There are resources and information from anywhere.
- Commodified: The effect is a computer utility system similar to the traditional utility model. As gas and electricity—you charged only for what you use.

Other than a few points of interest in distributed computing, security and protection upset the selection of cloud benefits by different associations and IT industry. Information classification, data security, and trust establishment are viewed as the fundamental security worries for an association moving its information to the cloud stage [10]. Security assumes a focal job in forestalling administration disappointments and developing trust in cloud computing [11]. Specifically, cloud specialist organizations need to make sure about the virtual condition, which empowers them to run administrations for different customers and offer separate administrations for various customers [12]. Vulnerability about information assurance and loss of information control is the significant explanation behind the lessening level of trust in cloud suppliers [13]. Accordingly, it is required to set up a trust in cloud supplier for guaranteeing the information security and getting the assurance about cloud execution [14, 15]. Today, trust building and security are the most dominant factors for cloud.

Computing's success. Control is another significant issue in trust [16, 17]. We trust a framework less when we do not have a lot of authority over our advantages [18]. There are various CSPs available in the market. The cloud consumer always gets very confused to select, which is the best suitable CSP for them. There are multiple CSP selection mechanisms suggested by various researchers [19, 20]. The main concern for choosing the appropriate CSP is to select the CSP, which is secure, good quality service, reliable, and trustworthy [21]. To overcome this issue, we have suggested a new CSP selection architecture. In the proposed architecture, there are four modules through which the best CSP suggests to the cloud consumer. The proposed architecture can be helpful in selecting the appropriate CSP. It takes care of all the aspect of the consumer requirement. The proposed architecture consists of the modules, namely: CSP information administrator, review manager (RM), quality manager (QM), and the selection manager (SM). It is a one above another process through which the best CSP recommends to the consumer. The information in the CSP administrator gathers from two sides. One is from the peer users and the second is from the CSP Company. The inputs are collected in the form of point values from 1 to 5 for the listed attributes. The attributes are security, quality, reliability, trust, and service stability. Each module works in the proper mechanism to select the appropriate CSP.

1.1 Trust Models

There are two types of trust models:

(a) The service provider's frame of reference (SPR) shown in Fig. 1 is basically a type of model in which the trust in the cloud environment is calculated and defined from the service provider's point of view.

(b) Service requester's frame of reference (SRR) shown in Fig. 2 is the trust model in which the trust in the cloud environment is defined from the service requester or cloud user or customer's point of view.

2 Literature Survey

Pirated review and the pirated trust values are the main concerns for that, there is a proposed distributed reputation-based intercloud computing system in [22]. For this, the author used to store the trust values in a distributed manner at various levels of the cloud in the system, through which it winds up workable for each cloud to settle on a free choice to choose cloud reliability. The proposed algorithm helps to detect the pirated reviews or feedbacks from malicious users so that it can be easy to avoid working with the fewer trustworthiness parties, and for this, the main idea is to update the trust and reputation records gathered by the first-hand and second-hand ratings and also update the ratings received by peer users. In another paper [23], the author proposed two new algorithms to overcome the pirated reviews and feedbacks from malicious users those who can affect the trust values proposed the feedback management approach and the feedback filtering algorithm

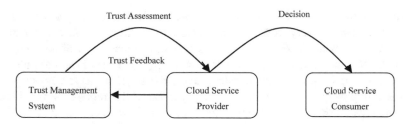

Fig. 1 Service provider's perspective (SPP)

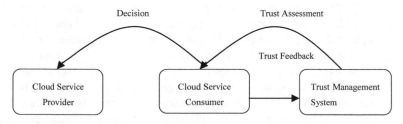

Fig. 2 Service requester's perspective (SRP)

these are given in Eqs. 1 and 2. In this paper, he specifies the trust in the form of a trust association between two fellow entities. The proposed algorithms are based upon the past behaviors and interactions of users. These are personal experiences and feedback filtering approach. The individual encounters signify the direct data of m subsequent to executing with n. It is represented as using Eq. 1:

$$P(m,n) = \langle k, S(m,n) \rangle \qquad (1)$$

Here, $S(m, n)$ denotes the individual contentment of m (customer) regarding n (service provider) based on the quality of service in the kth transaction. This quantity is between 0 and 1, where 0 belongs to not satisfied and 1 belongs to completely satisfied. After this transaction with the service provider, the user provides the rating to the service provider as either -1, which is for unsatisfactory or 1 for satisfactory depending upon the outcome of the transaction. Then these personal are shared with the peer groups as personal satisfaction feedback. The feedback filtering approach is a technique which is used at the point when a client m needs to register the dependability of specialist organization n, at that point client m needs to assess the customized input similitude among m and all other client and after that, it filters the feedback from the other customer that fails to meet the condition of trustworthiness. If N is a set of cloud customers who provide the input to service provider n, to compute the feedback similarity ($S(x, n)$) means between service provider n and its personal experience score is given by:

$$S(x,n) = \sqrt{\frac{|P(m,n) - R(x \in N, n)^2|}{P(m,n) + R(x \in N, n)}} \qquad (2)$$

After completion of the transaction, the cloud customer n will send a new feedback rating to every peer that had given the ratings to the cloud provider m earlier. The cloud classification and middleware services by splitting the cloud environment into various layers are introduced in [24]. These layers are physical, virtual, and application. In the physical layer, the physical server, network, and storage components are included. This layer basically represents the cloud's physical layer. For the virtual layer, these resources are combined. The virtual segment is used to describe the virtual assets serving the material thickness of the previous layer. In a cloud type of IaaS, the customer interacts directly with the virtual layer that hosts the applications of the customer. Virtual computers, virtual networks, and processing are also included. The cloud clients for PaaS and SaaS collaborate with administrations at the application layer. This layer has the customer's applications and is hosted in the virtual tier using resources. The new trust service, the selection mechanism, is introduced [25], which is used to create, incorporate, process, and analyze all of the trust evidence in the framework and to give them an ideal decision of service choice agreeing to others explicit prerequisites for the reliability of the service. The service selection methodology is distributed into five units: The request specification unit is used to gather operational or non-functional specifications for the user's facilities. Users may define the specifications from various attribute factors such as type of service, response time, and performance. The evidence collection unit is used to gather trust evidence. Evidence comes from feedback ratings from the users of cloud services [26].

The trust value assessment layer is used to calculate the trust value of the service and the value of the reputation based on the trust evidence gathered in the previous unit. The service selection unit is the unit in which the cloud customer chooses the service that best fits the user's desires. Then, the assessment review unit is a category where stores are the input scores and the analysis of the real-time change. A new framework is presented in [27], which proposed a three-layer system: customer requester, service manager, and service provider layer. This framework allows both objective and subjective metrics to be measured. The service manager layer is the heart of the system. It involves the reputation management system where a product requester is required to provide trust ratings to a specific service, and a service provider has to pay for the product. This sheet's key component is the general reputation calculator. It is in charge of choosing the service provider. The service requester (user) layer includes the different consumers who use the services in the previous layer. This layer is used to gather the trust feedbacks from the cloud service user. At last, there is a service provider layer which consists of cloud services such as infrastructure as a service (IaaS), platform as a service (PaaS), and software as a service (SaaS).

Some researchers work on the addressed issue. As a result of exploration, we have mentioned some of the related work as the cloud taxonomy, and the operational trust properties have a crucial role in the development of trust [24]. The intercloud computing resource infrastructure is the backbone idea for developing consumer trust in the intercloud Internet computing environment [22]. As an emerging cloud market and high user expectations, the high-level market-oriented cloud architecture is developed [28, 29]. The trust foundation in the hybrid cloud environment using an intercloud broker is also an important work [23]. The main characteristics of cloud computing with respect to the use of it in the scientific scenario are being discussed in [30].

3 Problem Statement

There are a lot of cloud service vendors in the market in which the cloud consumer gets confused, which CSP is best for them. The consumer wants a cloud service which is best suitable according to their requirements. The biggest challenge is to provide a CSP to the consumer, which fulfills all the aspects of the attributes of the cloud requirements [31–33]. The primary concern for the consumer is to select the CSP, which is good in all attributes such as security, quality, reliability, and trustworthy. Though there are some mechanisms available that suggest the CSPs to the consumer, there is a lack of trust and those mechanisms usually focus only on one or two aspects. The primary concern is to gain the trust of consumers and suggest to them the best possible CSPs so that they can select their service provider without any sort of hesitation. In the previously defined systems by other authors, there are some bugs such as those defined systems do not cover every concern aspect of consumer, namely security, trust, quality, stability, and service reliability [34, 35], whereas in the proposed architecture, a system is developed, which can be helpful for the cloud consumer to select the best CSP as per their requirements. In the proposed architecture for CSP selection, we have focused on covering all the possible aspects of cloud consumer's concern so that the best possible CSP can be provided to the consumer

4 Proposed Work

The proposed architecture is totally concerned with the selection of cloud service providers in the manner so that the consumer can get a clear idea about the best possible service provider who can fulfill all the possible attributes and concerns of the cloud consumer. The main idea behind the proposed architecture system shown in Fig. 3 is to filter the available CSPs at various levels and provide the best-filtered cloud service. The filtrations of the services are done based on defined attributes. The proposed architecture is dependent on the selection of appropriate cloud service provider (CSP). It includes the selection criteria for the appropriate CSP. The process of selection of CSP performs in three levels so that the best CSP can be suggested to the consumer. There are modules, namely: review manager, quality manager, selection manager, and the CSP information administrator. Each module works effectively and selects the best CSP among various CSP and sends the

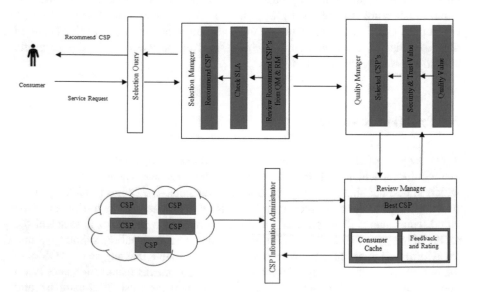

Fig. 3 Proposed three-level CSP selection architecture

suggested CSPs to its upper module. The CSP is suggested based on their security, efficiency, quality, consumer feedbacks, ratings, and the SLA. There is a CSP information administrator who stores all the information about the various CSP. The information about the CSP provided to the modules from the CSP administrator. Each module works effectively, so the best CSPs can be suggested to the cloud consumer.

First of all, the consumer request for the cloud service provider and the request comes as a selection query, and then it goes to the selection manager. The selection manager sends this query to the quality manager (QM), and the quality manager transfers it to the review manager (RM); afterward, RM transfer this query to CSP information administrator. Now the CSP information administrator gathers all the information from the various cloud service providers available in the market. The CSP information gathering algorithms are described below. The information administrator sends information about all the CSPs to the review manager, which are available, and matchs the consumer requirements. After getting all the CSP information from the administrator, the RM uses the consumer cache and the feedback or ratings provided to the received CSPs. The mechanism and the algorithm for the RM working are discussed below. After selecting the best CSPs from the pool of various given CSPs now, the RM sends the best recommendations of CSPs to its upper module called quality manager (QM). The quality manager checks the trust value, security value, and the quality value of all the recommended CSPs efficiently. After calculating all the three discussed values, the next step is to calculate the average of all these values. The method and the algorithm for the calculation of all these values have been discussed in the below sections. The selection manager analyzes all the CSPs that have been recommended by the QM followed by the RM and then checked the service-level agreement (SLA) of the consumer and then provide the best CSP after all the filters to the consumer. The proposed system works on the data and the feedback stored with the information administrator. In the presented architecture, the feedback and ratings of the trial period cloud users are not included. This system works on the real-time feedback and information sharing about cloud service providers with the information administrator.

4.1 CSP Information Administrator

The idea behind the information administrator is to create a pool containing the data and information about several CSPs. More precisely, it can be said that the CSP information administrator works as a central repository for the whole system, which can be helpful for the upper modules to get the correct information about the various attributes of several service providers. It consists of all the information about the various CSP. It consists of all the attributes that define the CSP, like its quality value, feedback, rating, security, and previous user experience for all the CSP. It consists of the information of all the CSP irrespective of their type that is IaaS, PaaS, or SaaS. It is mandatory for every CSP to update its information to the administrator. The main benefit of the information administrator is that it holds the updated value of all the attributes for all the CSPs. The administrator gathers the information in given two ways: One is from the CSP Company and another is from peer users. The data is collected in the form of numeric pointer values. These values are given for the set of specified attributes. These point values are given based on the peer user's experience and CSP Company's point of view about their CSP. Based upon the point values and the below-described mechanism, the administrator gathers information and, on that basis, the best CSP's information sent to the upper modules.

The administrator gathers the information in the following manner:

1. Gather information from CSP Company:

    • The inputs are taken from the CSP Company about the various attributes defined in the form of the points 1 to 5 (1 for the least and 5 for best service).
    • Collect all the information about the CSP.

2. Gather information from consumer:

    • The consumers give points between 1 and 5 for each of the attributes defined by the administrator for every CSP.
    • Here 1 is for the worst experience and 5 for the super experience.

The rating points have to be given on the following attributes: security, quality, reliability, and trust and service stability. After getting the point values from both sides, that is CSP Company, and from the consumers, the next step is to calculate the average of each attribute point value. After calculating the values, the CSP with the top overall pointer value will be sent to the review manager.

Algorithm for gathering the information

1. Gathering from CSP Company:
    1.1. Take input as a rating based from 1 to 5.
    1.2. 1 for the least and 5 for the best service.
    1.3. Take all the information about the company service provider.

2. Gathering from the consumer:
    2.1. Take input as rating pointers from 1 to 5 for each of the specified attributes.
    2.2. 1 for the least and 5 is for the best user experience in each of the specified attribute.

3. After getting the point values from both sides, that is CSP Company, and from the consumers, the next step is to calculate the average of each attribute point value.
    3.1. Calculation of average attribute: The given Eq. 3 calculates the average attribute value.

$$Z_a = \frac{\left(\text{Point value}_{consumer} + \text{Point value}_{company}\right)}{2} \qquad (3)$$

where $a = 1, 2, \ldots, 5$ (The attributes)

4. The next step is to calculate the overall average pointer value by given Eq. 4.

$$\frac{\sum_1^5 z}{5} \qquad (4)$$

5. After calculating the value, the CSP with top overall pointer values will be sent to the review manager for further classification.

4.2 Review Manager

The review manager is responsible for the first filtering of the cloud service providers on the basis of their feedbacks and the ratings given by the peer users in the past. The review manager gathers the CSP information from the CSP information administrator. The review manager will check the feedback and ratings of the CSPs. The review manager checks the ratings given by the various users to the CSP. On the basis of consumer cache, and that is the information about the previously used CSP by the respective consumer, the best CSP is recommended by the review manager. The review manager sends the best CSPs to quality manager as per his examination. After getting the filtered CSP information from the administrator, the review manager checks the existing consumer cache in the following manner:

1. It checks the old record about the previous cloud services used by the consumers and their user experience point values given to them.
2. On the basis of the cache record of the consumer, the review manager selects the appropriate CSP.
3. Another criterion is checking the feedbacks and ratings given by the various consumers to the CSP.
4. The CSP with the highest ratings, and based on the cache record, the review manager selects the appropriate CSP and sends it to the next module that is the quality manager.

Algorithm for review manager

1. Check the consumer details in the CSP information administrator first.
2. If the consumer details matched with the existing consumer details then:
Check the previously used CSPs by the same user.
Check the ratings and the point values given by the user to the CSP.
If the old point values are given as:

if (Security>3 && Quality>3 && Reliability>3 && Trust>3 && service stability>3)

//Point values given by the current user //: Recommended CSP == Previous CSP used by the consumer
else {

if (Security>3 && Quality>3 && Reliability>3 && Trust>3 && service stability>3)

// Point values given by the peer users //: Recommended CSP == New CSP
3. If the consumer details do not match with the existing consumer details then: Check the other peer customer's ratings and feedback of the CSP.
3.2 If (*average point value* $_{specified\ attributes}$>3): Recommended CSP == New CSP with highest average point values.

Send the recommended CSP to the higher-level module that is quality manager (QM).

4.3 Quality Manager

This is a second step that is responsible for filtering the service providers on the basis of quality, security, and trust. The quality manager is the module in which the service providers got separated using the calculated values of quality, security, and trust. After getting the recommended CSP details from the review manager, the quality manager checks the quality value, security value, and the trust value of the recommended CSP. Based upon these three values, that is the quality value, security value, and the trust value; the quality manager selects the best CSP and then recommends the best CSP to its upper module that is the selection manager. The quality manager is responsible for the quality, trust, and security checking of the recommended CSP so that the best CSP can be recommended to the user. For calculating these values, the quality manager uses the point values that have been stored with the CSP information administrator. The following algorithm is given for the working of quality manager:

Algorithm for the quality manager

1. The quality manager collects the point values from the administrator.
2. Checks the quality value of the selected CSP by the following Eq. 5

$$Q = \frac{\left(\text{Quality value}_{consumer} + \text{Quality value}_{company}\right)}{2} \tag{5}$$

3. The next step is to calculate the security value by the given Eq. 6:

$$S = \frac{\left(\text{Security value}_{consumer} + \text{Security value}_{company}\right)}{2} \tag{6}$$

4. Now the trust value is calculated by the given Eq. 7:

$$T = \frac{\left(\text{Trust value}_{consumer} + \text{Trust value}_{company}\right)}{2} \tag{7}$$

5. After the calculation of these three values, that is quality, security, and trust, the next step is to calculate the average of these three values using Eq. 8:

$$\text{Avg} - \frac{Q + S + T}{3} \tag{8}$$

6. In the next step, the quality manager will check

If (Avg > = 3)

{

Recommended CSP = = forward to selection manager

}

Else: Do not forward to selection manager.

7. In the final step, the recommended CSPs list send to the next upper module that is the selection manager for further working.

4.4 Selection Manager

The selection manager is the final step in the proposed system. This step is like a final gateway of the three-step cloud service selection algorithm. This step does the comparative study of all the CSPs in all aspects and recommends the best possible service providers to the consumer. Although the best CSPs already been recommended by the previous managers, still selection manager checks the final attribute values so that there is very little chance of wrong suggestion. The selection manager is responsible for sending the final CSP recommendations to the user. The selection manager reviews all the recommended CSPs by the quality manager and the review manager. After reviewing the CSPs recommended by the previous managers, the selection manager checks the service-level agreement (SLA) on which the consumer wants the CSP. As per the consumer requirements and the SLA, the selection manager provides the best possible CSP to the consumer. The selection manager recommends the best possible CSP to the consumers as per the service-level agreement (SLA) requirement of the consumer.

5   Experiment and Evaluation

5.1   Experimental Setup

The platform used for the performance and evaluation is Windows 64 bit, 8 GB RAM, Intel Core 8th generation, Intel® Core™ i5-8265U @ CPU 1.60 GHz 1.80 GHz, 64-bit operating system, and ×64-based processor. The CloudSim with JavaFX is used as a front end for algorithm testing. The CSP information administrator allows various CSPs to add, change, and delete QoS attributes according to CSP's diverse offerings. The CSP information administrator offers to register and publish their services on two levels. In the first level, they can provide information like service name, provider name, industry, classification, etc.

In the next level, all the CSPs provide the point values for the specified attributes, which are helpful for the cloud consumer to check and select the appropriate CSP. The CSP administrator also permits the cloud consumer to provide the point values for the specified attributes for the CSPs. The feedback and the point values given by the peer users are useful for the calculation of the average point values, which is the base for the proposed system. Here we have taken the data for the five services, namely $S1$, $S2$, $S3$, $S4$, $S5$. Tables 1 and 2 show the attribute values of services from consumer's and company's point of view, respectively. The data and the rating values have taken randomly for the calculation.

5.2   Evaluation

For the calculation, we have taken the specific attribute values for the five services. The average attribute value for each service will be carried out by given Eq. 9:

Table 1 CSP information administrator algorithm (consumer)

Attributes	$S1$	$S2$	$S3$	$S4$	$S5$	Comparison
Security	4.0	3.0	4.5	2.7	4.9	$S5 > S3 > S1 > S2 > S4$
Quality	3.8	3.6	2.8	3.2	2.7	$S1 > S2 > S4 > S3 > S5$
Reliability	4.2	2.8	3.0	1.2	3.4	$S1 > S5 > S3 > S2 > S4$
Trust	4.5	4.0	3.2	1.8	4.8	$S5 > S1 > S2 > S3 > S4$
Service stability	2.9	2.0	4.0	3.5	1.5	$S3 > S4 > S1 > S2 > S5$

Table 2 CSP information administrator algorithm (company)

Attributes	S1	S2	S3	S4	S5	Comparison
Security	4.0	4.8	4.5	4.6	4.9	$S5 > S2 > S4 > S3 > S1$
Quality	5.0	4.0	3.8	3.9	4.5	$S1 > S5 > S2 > S4 > S3$
Reliability	4.2	4.7	4.6	4.5	4.0	$S2 > S3 > S4 > S1 > S5$
Trust	4.9	5.0	4.3	4.2	4.8	$S2 > S1 > S5 > S3 > S4$
Service stability	4.5	4.9	4.8	4.0	3.9	$S2 > S3 > S1 > S4 > S5$

$$Z_a = \frac{\left(\text{Point value}_{consumer} + \text{Point value}_{company} \right)}{2} \tag{9}$$

As per the algorithm in the next stage, after the average attribute calculation, the best CSPs will be sent to review manager (RM). The average attribute values for the cloud services are showing in Table 3. On the basis of these values, the suggested CSPs will be sent to the next level. In this stage, the overall average attribute values begin calculated using $\sum_1^5 Z \div 5$ equation. Table 4 shows the overall average attribute value for the CSP. By analyzing the overall attribute values, the review manager sends the CSP recommendations to the next level, which is the quality manager (QM). For a particular CSP, if the overall average attribute value is more 3, then the CSP is recommended to the QM. Since in the review manager section, the overall attribute average value for the $S4$ service was less than so, the $S4$ is not recommended to the QM by the RM given in Table 5. After getting all the recommended CSPs, the QM calculates the quality value (Q) by using Eq. (5), trust value (T) by Eq. (6), security value (S) by using Eq. (7), and the average value (avg) by using Eq. (8) for all the three specific fields. The CSP with the highest average value will be recommended to the consumer. After completion of all three steps, the final CSP recommendations provide to the user. The intensive filtering of the CSPs makes this algorithm very useful for the CSP selection.

The given Fig. 4 shows the service comparison among all the CSPs on the basis of specific attribute values that have been evaluated in the QM section of the algorithm. These comparisons depict the properties of the CSP so that it can be helpful for the user to select the appropriate CSP. Figure 5 represents the CSP service ranking as per the average-specific attribute on which basis the CSP with the highest avg value is recommended to use.

Table 3 CSP information administrator algorithm (average attribute values)

Average attributes	S1	S2	S3	S4	S5	Comparison
Z_1	4.0	3.9	4.5	4.65	4.9	$S5 > S4 > S3 > S1 > S2$
Z_2	4.4	3.8	3.3	3.55	3.6	$S1 > S2 > S5 > S4 > S3$
Z_3	4.2	3.75	3.8	2.85	3.7	$S1 > S3 > S2 > S5 > S4$
Z_4	4.7	4.5	3.75	3.0	4.8	$S5 > S1 > S2 > S3 > S4$
Z_5	3.7	3.45	4.4	3.75	2.7	$S3 > S4 > S1 > S2 > S5$

Table 4 Review manager (RM)

Services	Overall average
S1	4.2
S2	3.88
S3	3.95
S4	2.98
S5	3.94

Table 5 Quality manager (QM)

Specific value check	S1	S2	S3	S4
Q	4.4	3.8	3.3	3.5
T	4.7	4.5	3.7	3.0
S	4.0	3.9	4.5	4.6
Avg	4.3	4.06	3.8	3.7

Fig. 4 Service comparison on the basis of the specific attribute value

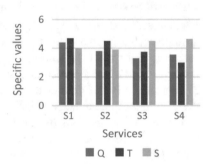

Fig. 5 Service ranking

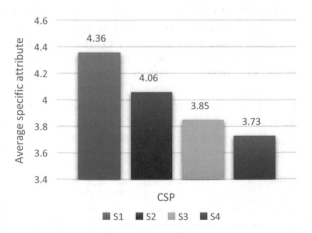

6   Conclusion and Future Work

The presented work consists of a comparative study among the various papers on the trust management system and selection of appropriate CSP. The presented work is based upon the filtering approach in which the best filtered CSPs can be recommended to the consumer. This work is presented for the ease of appropriate service provider selection by the consumer. After analyzing the various related works, the new architecture for the CSP selection has been introduced in the paper which can be helpful for the cloud consumer to choose the appropriate CSP as per the requirement and this architecture takes care of all the attributes that are needed to be kept in mind while selecting the best CSP for the consumer. The proposed architecture algorithm is a three-step process in which CSPs being filtered based on their performance and other specified attributes. In the presented paper, we tried to focus on all the aspects of the cloud service attribute. The testing of proposed work is done on small dataset; so, the immediate applicability of this system in the industry is not recommended without extensive evaluation. However, in the future, we are planning to extend this work in which there will be feedback and other data of the trial period cloud consumer will also be there so that the more appropriate CSP can be recommended to the cloud user.

References

1. Buyya R, Yeo CS, Venugopal S, Broberg J, Brandic I (2009) Cloud computing and emerging IT platforms: vision, hype, and reality for delivering computing as the 5th utility. Future Gener Comput Syst 25(6):599–616
2. Zhou J, Abdullah NA, Shi Z (2011) A hybrid P2P approach to service discovery in the cloud. Int J Inf Technol Comput Sci 3(1):1–9, 29
3. Zeng W, Zhao Y, Zeng J (2009) Cloud service and service selection algorithm research. In: Proceedings of the first ACM/SIGEVO summit on genetic and evolutionary computation. ACM, pp 1045–1048
4. Saaty TL (1980) The analytic hierarchy process for decision in a complex world. RWS Publications, Pittsburgh
5. Salomoni D (2016) Scientific clouds. In: Grid and cloud computing: concepts and practical applications, vol 192, p 31
6. Karim R, Ding C, Miri A (2013) An end-to-end QoS mapping approach for cloud service selection. In: 2013 IEEE ninth world congress on services (SERVICES). IEEE, pp 341–348
7. Manuel P (2015) A trust model of cloud computing based on quality of service. Ann Oper Res 233(1):281–292
8. Wu X, Zhang R, Zeng B, Zhou S (2013) A trust evaluation model for cloud computing. Procedia Comput Sci 17:1170–1177
9. Firdhous M, Hassan S, Ghazali O (2013) A comprehensive survey on quality of service implementations in cloud computing. Int J Sci Eng Res
10. Md AQ, Varadarajan V, Mandal K (2019) Efficient algorithm for identification and cache based discovery of cloud services. Mob Netw Appl 1–17

11. Tian LQ, Lin C, Ni Y (2010) Evaluation of user behavior trust in cloud computing. In: 2010 international conference on computer application and system modeling (ICCASM), vol 7. IEEE, pp V7–567
12. Firdhous M, Ghazali O, Hassan S (2012) Trust management in cloud computing: a critical review. arXiv preprint arXiv: 1211.3979
13. Firdhous M, Ghazali O, Hassan S (2011) Applying bees algorithm for trust management in cloud computing. International conference on bio-inspired models of network, information, and computing systems. Springer, Berlin, Heidelberg, pp 224–229
14. Eymann T, König S, Matros R (2008) A framework for trust and reputation in grid environments. J Grid Comput 6(3).225–237
15. Thampi SM, Bhargava B, Atrey PK (eds) (2013) Managing trust in cyberspace. CRC Press
16. CSMIC. SMI framework. http://betawww.cloudcommons.com/servicemeasurementindex
17. Goscinski A, Brock M (2010) Toward ease of discovery, selection and use of clusters within a cloud. In: 2010 IEEE 3rd international conference on cloud computing (CLOUD). IEEE, pp 289–296
18. Saaty TL (1996) Decisions with the analytic network process (ANP). University of Pittsburgh (USA), ISAHP, p 96
19. Govindaraj P, Jaisankar N (2017) A review on various trust models in cloud environment. J Eng Sci Technol Rev 10(2)
20. Georgiopoulou Z, Lambrinoudakis C (2016) Literature review of trust models for cloud computing. In: 2016 15th international symposium on parallel and distributed computing (ISPDC). IEEE, pp 208–213
21. Habib SM, Ries S, Muhlhauser M (2010) Cloud computing landscape and research challenges regarding trust and reputation. In: 2010 7th international conference on ubiquitous intelligence & computing and 7th international conference on autonomic & trusted computing. IEEE, pp 410–415
22. Abawajy J (2009) Determining service trustworthiness in intercloud computing environments. In: 2009 10th international symposium on pervasive systems, algorithms, and networks. IEEE, pp 784–788
23. Abawajy J (2011) Establishing trust in hybrid cloud computing environments. In: 2011 IEEE 10th international conference on trust, security and privacy in computing and communications. IEEE, pp 118–125
24. Abbadi IM, Martin A (2011) Trust in the cloud information security technical report 16(3–4):108–114
25. Fan WJ, Yang SL, Perros H, Pei J (2015) A multi-dimensional trust-aware cloud service selection mechanism based on evidential reasoning approach. Int J Autom Comput 12(2): 208–219
26. Buyya R, Yeo CS, Venugopal S (2008) Market-oriented cloud computing: vision, hype, and reality for delivering it services as computing utilities. In: 10th IEEE international conference on high performance computing and communications, 2008. HPCC'08. IEEE, pp 5–13
27. Filali FZ, Yagoubi B (2015) A general trust management framework for provider selection in cloud environment. East European conference on advances in databases and information systems. Springer, Cham, pp 446–457
28. Habib SM, Hauke S, Ries S, Mühlhäuser M (2012) Trust as a facilitator in cloud computing: a survey. J Cloud Comput Adv Syst Appl 1(1):19
29. Ranjan R, Zhao L, Wu X, Liu A, Quiroz A, Parashar M (2010) Peer-to-peer cloud provisioning: service discovery and loadbalancing. Cloud computing. Springer, London, pp 195–217
30. Md AQ, Varadarajan V, Mandal K (2019) Efficient algorithm for identification and cache based discovery of cloud services. Mobile Netw Appl 24(4):1181–1197
31. Sun L, Dong H, Hussain FK, Hussain OK, Chang E (2014) Cloud service selection: state-of-the-art and future research directions. J Netw Comput Appl 45(134–150):32
32. Churchman CW, Ackoff RL, Arnoff EL (1957) Introduction to operations research

33. Godse M, Mulik S (2009) An approach for selecting software-as-a-service (SaaS) product. In: IEEE international conference on cloud computing, 2009. CLOUD'09. IEEE, pp 155–158
34. Zeng C, Guo X, Ou W, Han D (2009) Cloud computing service composition and search based on semantic. IEEE international conference on cloud computing. Springer, Berlin, Heidelberg, pp 290–300
35. Md AQ, Vijayakumar V (2019) Dynamic ranking of cloud services for web-based cloud communities: efficient algorithm for rating-based discovery and multi-level ranking of cloud services. Int J Web Based Commun 15(3):248–270

A Smart Device to Identify the Pandemic of Chronic Obstructive Pulmonary Disease

J. Bethanney Janney, T. Sudhakar, G. UmaShankar, L. Caroline Chriselda, and H. Chandana

1 Introduction

World Health Organization's analysis reports that 300 million people worldwide have suffered from COPD and asthma, and that by 2025 this will rise to 400 million. In India, untreated lung disorders have grown by 11% of the population including children, pregnant mothers, elders and the aged struggle often. Smoking and air contamination are the key cause of the crisis. In the Intensive Care Unit, patients suffer from breathing issues because of no ventilation [1].

There are six vital signs. Respiratory rates are elevated here. In general, the rate of respiration is the amount of breaths per minute. Such levels of breath can differ from environmental factors such as temperature, humidity, strain and composition of chemicals. We need to test a patient's respiratory rate in repose. A person's natural respiratory rate varies between 15 and 20. Breathing intensity lower than twelve is considered bradypnea and higher than thirty is considered tachypnea. When a patient requires assistance from a physician/medical provider to control the respiratory rate, the specialist is not regularly available for health check; hence, the patient has failed their checkup [2, 3].

Embedded and networking systems are advancing; a modern, increasing Internet of Things (IoT) system is helpful without any human help to track the sufferer's breathing rate. The respiratory rate of the patient is captured using a medical sensor that is based on the amount of air inhaled and exhaled. The result is then viewed on an internet site. The proposed work therefore developed a remote real-time cough detection method with integrated healthcare systems which can measure and showcase observed physiological conditions, including cough sequence, heart rate, respiratory rate, blood oxygen saturation (SpO_2), blood pressure and temperature [4]. The proposed device should achieve primary goals such as detecting, pro-

J. Bethanney Janney (✉) · T. Sudhakar · G. UmaShankar · L. C. Chriselda · H. Chandana
Department of Biomedical Engineering, Sathyabama Institute of Science and Technology, Chennai 600119, India

cessing, monitoring, transmitting and should be capable of providing an automated alarm depending on different control parameters analytical threshold values. Auxiliary specifications such as reliability, safety, low energy consumption and price should also be met [5].

2   Related Works

Naranjo Hernández et al. (2018) developed the first approach to the conception through capacitive sensing of a portable system to track the respiratory rate without touch. The structure and functional creation of an e-Health based on the Internet of Things paradigm are which provides an expandable result for remote surveillance through this sensor and other tool [6].

Zhou et al. (2019) proposed a handheld smartphone spirometer, which uses a respiratory signal acquisition head with a Lilly type sensor, to transmit information to smart phone or other mobile terminals for the transfer of a bluetooth signal to display the data and the effect. The 12 case study for individuals with the spirometer showed that the COPD and asthma at a basic level could effectively used by clinicians in the treatment point test [7].

Badnević et al. (2018) proposed a specialized diagnostic system that can distinguish between asthma, COPD or regular lung movement patients on the basis of lung function measurements and patient symptoms details. The tests indicate that asthma and COPD can be accurately diagnosed by the expert diagnostic system [8].

Windmon et al. (2018) reported a chronic obstructive pulmonary disease (COPD) demonstrating the possibility of exploiting cough samples recorded with a mobile microphone and decoding the related sound signals using machine learning algorithms to diagnose COPD-indicated cough waveform [9].

Sidi Ahmed et al. (2015) identified a new device that disassembles an information acquisition experiment, and software to exploit the data received for cough detection, visual projection and classification. Sensor's importance is assessed in three functions: shared knowledge acquired through the apps, ability to differentiate cough from certain types of incidents, and ability to identify cough incidents [10].

3   Materials and Methods

The proposed device uses a smart app to recognize early signs of chronic obstructive pulmonary disorder (COPD). It uses a pressure sensor, respiratory sensor and oxygen saturation sensor to distinguish and diagnose regular cough or diseased cough. This cough detection system is a real-time system. The proposed framework is shown in Fig. 1.

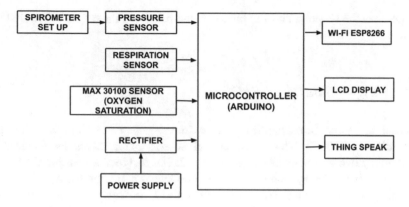

Fig. 1 Proposed block diagram

3.1 Arduino

Arduino is an individual panel microcontroller comprises of a basic open source hardware board built for an 8-bit Atmel AVR microcontroller, or a 32-bit Atmel ARM. The software consists of a common programming language compiler and a boot loader running on handy hardware and software package. It contains an Arduino Integrated Development Environment (IDE) circuit board which is programmable and ready-made software for writing and transferring the programming code to the physical device [11].

3.2 MPX2050 Pressure Sensor

The devices of the MPX2050 series are silicon piezo resistive pressure sensors which provide a highly specific and linear voltage output, directly proportional to the input pressure. The sensor is a standard, monolithic silicone diaphragm with a strain gauge and an optimized on-chip thin-film resistor network [12].

3.3 Respiratory Sensor or Sound Detection Sensor

The sound sensor module offers a basic sound detection system and is commonly used for sound intensity detection. The microphone provides an amplifier, high-level detector and buffer with the audio. When the sensor detects a sound, the signal voltage is processed and sent to a microcontroller and processed. The Arduino sound sensor module detects whether the sound reaches the threshold value. The Sound is recorded and fed to an op-amp LM393. If the sound level is greater than the set point an LED blinks on the panel and the output is lowered [13, 14].

3.4   Oxygen Saturation SpO$_2$

IIcart Rate Button hears built-in pulse oximetry and a heart rate monitor from Maxim's MAX30100. This is an optical sensor that receives its readings by transmitting two wavelengths of light by two LEDs and then measuring the pulsing blood absorbance using a photo detector. The signal is interpreted by a low-noise analog signal processing device and transmitted through the Micro BUS I2C interface to the intended MCU [15].

3.5   Liquid Crystal Display (LCD)

Liquid crystal displays (LCDs) consists of components that integrate the properties of liquids and crystals. It incorporates two glass panels between them in liquid crystal material. The liquid crystal substance would be aligned in certain line if appropriate voltage was used for electrodes. The polarizers will rotate the light rays via the LCD, resulting in the desired characterizations being activated/revealed [16].

3.6   Wi-Fi Module

ESP8266EX is also connected with individual devices via its GPIOs. It provides complete and self-loading Wi-Fi networking features from additional device while this program is powered by ESP8266EX, it boots directly from an external disk. Alternatively, wireless Internet connectivity may be applied to any device based on a microcontroller as a Wi-Fi adapter [16, 17].

3.7   Power Source

This work describes the use of filters, rectifiers and then voltage controls to operate power supply circuits. A steady dc voltage is achieved, beginning with an ac voltage, by correcting the ac voltage, removing the dc voltage and then controlling it in order to attain the preferred fixed dc voltage. The control is typically achieved from an IC voltage controller which holds a dc voltage and gives the same dc voltage a little lower even when the voltage differs input or the output load linked to the dc voltage changes [18].

3.8 Spirometer Setup

Spirometry refers to a collection of basic measurements of an individual's respiratory ability. Spirometer is a calculation of the amount of oxygen that the lungs inspire and expire. This system is attached directly to the pressure sensor and the average value is measured to the P1-P2 microcontroller by spirometer ratio FEV1/FVC (in percentage) [18, 19].

3.9 Simulation Tool

The Arduino IDE uses software to interchange the executable code in the Arduino board by means of a loader program in the system firmware for hexadecimal encoding. The Internet of Things which is now Internet-connected applies to the thousands of physical objects in the world which process and transfer information. Sensors can be linked and attached to all those different objects so that they can demonstrate artificial intelligence on computers, excluding the requirement of a human being, in real time. The Internet of Things makes us aware and sensitizes the world surrounding us [4, 7]. Thing Speak is a software and open source IoT to store and retrieve data from the web or the local area of a network using the HTTP protocol. Thing Speak enables sensor monitoring, role monitoring and social network applications with status alerts to be created [20].

4 Results and Discussion

The system was tested on 50 subjects between the 19–45 age group containing 32 males and 18 females. The sensors were positioned in the patient body region, respectively. The critical parameters such as cough sequence, oxygen level, SpO_2, BP, heart rate and respiration rate are also calculated using various sensor types and output is analyzed by cloud computing Think Speak. The performance is captured in normal condition and during cough. During the cough period the peak variance in the waveform and values is observed. The output of the sensors is interfaced with the controller and their parameters are collected in a single point and displayed on a LCD monitor. The patient data is stored in the cloud computing platform and the output of the abnormal waveform and values is sent to the doctor via mobile app. When an excessive cough is caused, the LED will either flash or turn on and the buzzer will give alarm.

The simulation result was analyzed and the findings were evaluated. The design of hardware is shown in Fig. 2. The normal waveform without cough generated in the Think speak is displayed in Figs. 3 and 4 is the waveform that is been detected when there is excessive cough. From the obtained waveform difference in peak was observed during cough sequence than normal condition. Figures 5 and 6 show the heart rate and SpO₂ level, respectively. The output of respiration sensor and pressure sensor is displayed in the LCD monitor.

The regular and cough sequence of cold was reported and the findings were discussed in Table 1. The table shows the outputs of our smart system that are in normal condition and diseased state. It is found that the decibel range of cough episodes increases and the heart rate rises for COPD patients as opposed to the

Fig. 2 Design of hardware

Fig. 3 Normal waveform of decibel values without cough

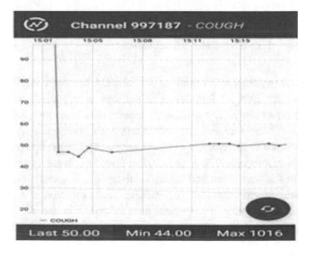

Fig. 4 Waveform of decibel values with cough

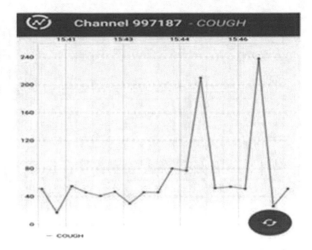

Fig. 5 Waveform of heart rate

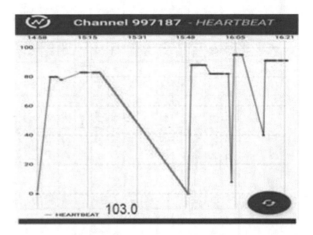

Fig. 6 Waveform obtained from oxygen saturation sensor

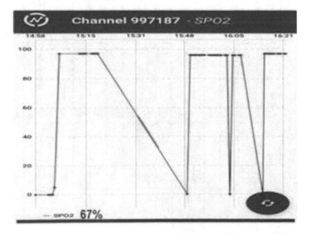

Table 1 Output of the system for cough detection

S. No.	Cough episodes (Db)	Spirometry ratio (%)	Heart rate (beats per minute)	SpO$_2$	Result	S. No	Cough episodes (Db)	Spirometry ratio (%)	Heart rate (beats per minute)	SpO$_2$	Result
01	50	95	71	97	NORMAL	26	220	48	146	50	COPD
02	51	82	78	98	NORMAL	27	241	39	141	51	COPD
03	65	87	77	91	NORMAL	28	233	43	152	65	COPD
04	51	96	75	98	NORMAL	29	218	57	161	51	COPD
05	80	86	79	97	NORMAL	30	237	52	156	80	COPD
06	56	92	74	91	NORMAL	31	249	64	149	56	COPD
07	58	88	76	92	NORMAL	32	245	53	141	58	COPD
08	54	87	73	98	NORMAL	33	237	67	137	54	COPD
09	78	91	76	98	NORMAL	34	222	42	138	78	COPD
10	47	89	72	94	NORMAL	35	217	51	108	71	COPD
11	55	86	74	91	NORMAL	36	255	55	135	59	COPD
12	32	91	77	93	NORMAL	37	220	32	103	67	COPD
13	68	84	79	97	NORMAL	38	238	68	143	72	COPD
14	54	89	71	99	NORMAL	39	218	54	112	66	COPD
15	47	87	80	92	NORMAL	40	227	47	147	58	COPD
16	34	91	73	90	NORMAL	41	234	34	156	62	COPD
17	45	93	72	94	NORMAL	42	218	45	149	64	COPD
18	51	98	71	96	NORMAL	43	221	51	157	57	COPD
19	43	87	77	91	NORMAL	44	231	43	152	47	COPD
20	58	88	78	99	NORMAL	45	247	58	138	67	COPD
21	47	96	74	98	NORMAL	46	212	54	134	49	COPD
22	55	86	76	93	NORMAL	47	247	49	144	51	COPD

(continued)

Table 1 (continued)

S. No.	Cough episodes (Db)	Spirometry ratio (%)	Heart rate (beats per minute)	SpO$_2$	Result	S. No	Cough episodes (Db)	Spirometry ratio (%)	Heart rate (beats per minute)	SpO$_2$	Result
23	32	92	73	98	NORMAL	48	243	42	151	57	COPD
24	68	88	76	100	NORMAL	49	231	47	157	63	COPD
25	54	87	74	93	NORMAL	50	247	53	145	62	COPD

typical event. Yet for COPD diagnosis the level of oxygen and the amount of SpO$_2$ decreases.

Hence, an effective, compact cough system is the monitoring device for early detection and treatment of pulmonary problems. Therefore, the experiment's output provides the cough waveform, the flow amounts, the heart rate and the saturation of the oxygen. It even tracks the body's most important signs. The 12 bit resolution sensors are attached to Arduino using ADC. Cloud computing provides for sensor performance. The performance helps differentiate the normal and irregular waveform of the patient. This approach has been widely used in clinical practice and has been advanced in the quantitative analysis of cough.

The proposed system adds more value and importance than the previous methods. Although the [7] existing system contains sensors and controls for tracking individuals only in an offline mode which won't help doctors foresee on a daily basis. Mohammadi et al. (2013) strive for the recognition of the impact on exhaustion, everyday life activities and quality of life for home-based nursing pulmonary rehabilitation of COPD patients. Randomize patients into case and control groups using the block randomization process. Before and after intervention in both classes, exhaustion, ADL and QOL were measured. Findings revealed that the mean fatigue scores decreased significantly [21]. Windmon et al. (2018) reported a chronic obstructive pulmonary disease (COPD) demonstrating the possibility of exploiting cough samples recorded with a mobile microphone and decoding the related audio signals using machine learning algorithms to diagnose COPD-indicated cough patterns. The experimental results gave lesser accuracy.

The new framework provides a robust and efficient personal healthcare network. Remote health management program may be introduced to track the patient's health condition via the internet and prevent dangerous situations. This platform integrates an Arduino board; two sensor modules may be used at remote places and calculates the heart beat. It is also important for future analysis and review of the status of the patient's safety. This model can be developed further by adding blood pressure control devices to make it more effective in hospitals as a very powerful and devoted patient monitoring system and respiration rate along with Wi-Fi transmitter.

5 Conclusion

This proposed work recommends a portable device for identifying cough at an early stage to treat pulmonary illnesses. The device senses the body's critical signals such as oxygen saturation, heart rhythm, respiration rate and cough sequence. The result is recorded and viewed in a live manner on the thing speak platform which produces the output as waveforms which is further easier to distinguish from the normal waveform. A Wi-Fi transmitter has been added so that the output is stored in the cloud for telemetry purpose and monitored anywhere through the application.

The proposed method designed a low-cost spirometer and installed with a differential pressure transducer at the core of the system, with basic experimental configurations, to test its viability and reliability.

References

1. Grosbois JM, Heluain-Robiquet J, Machuron F, Terce G, Chenivesse C, Wallaert B, Le Rouzic O (2019) Influence of socioeconomic deprivation on short-and long-term outcomes of home-based pulmonary rehabilitation in patients with chronic obstructive pulmonary disease. Int J Chron Obstruct Pulmon Dis 14:2441–2449
2. Magdy DM, Metwally A (2020) Effect of average volume-assured pressure support treatment on health-related quality of life in COPD patients with chronic hypercapnic respiratory failure: a randomized trial. Respir Res 21:64–72
3. Radogna AV, Siciliano PA, Sabina S, Sabato E, Capone S (2020) A low-cost breath analyzer module in domiciliary non-invasive mechanical ventilation for remote COPD patient monitoring. Sensors (Basel) 20(3):653–662
4. Fasidi FO, Adebayo OT (2019) Development of a mobile monitoring system for asthmatic patients. J Health Med Informat 10:324–332
5. Song WJ, Millqvist E, Morice AH (2019) New ERS cough guidelines: a clinical framework for refining the patient management strategy. Asia Pac Allergy 9(4):1–12
6. Naranjo-Hernández D, Talaminos-Barroso A, Reina-Tosina J, Roa LM, Barbarov-Rostan G, Cejudo-Ramos P, Márquez-Martín E, Ortega-Ruiz F (2018) Smart vest for respiratory rate monitoring of COPD patients based on non-contact capacitive sensing. Sensors 18(7):2144–2152
7. Zhou P, Yang L, Huang Y-X (2019) A smart phone based handheld wireless spirometer with functions and precision comparable to laboratory spirometers. Sensors (Basel) 43:1235–1264
8. Badnjevic A, Gurbeta L, Custovic E (2018) An expert diagnostic system to automatically identify asthma and chronic obstructive pulmonary disease in clinical settings. Sci Rep 8:123–134
9. Windmon A, Minakshi M, Chellappan S, Athilingam P, Johansson M, Jenkins BA (2018) On detecting chronic obstructive pulmonary disease (COPD) cough using audio signals recorded from smart-phones. HEALTHINF, 329–338
10. Sidi Ahmed M, Paulo Da Cunha P, Thierry R, Thomas D, Ricardo C, Thierry D, Carlos V (2015) Sensor-based system for automatic cough detection and classification. In: 8th ICT innovations conference ELEMENT 2015—enhanced living environments, Macedonia, 24–26 September, pp 270–279
11. Soliński M, Łepek M, Kołtowski Ł (2020) Automatic cough detection based on airflow signals for portable spirometry system. Inf Med Unlocked 18:313–320
12. Duarte AG, Tung L, Zhang W, Hsu ES, Kuo YF, Sharma G (2019) Spirometry measurement of peak inspiratory flow identifies suboptimal use of dry powder inhalers in ambulatory patients with COPD. J COPD Found 6(3):246–255
13. Jun Z, Li C, Zhu W, Zhou H, Han X (2020) Wearable respiratory strain monitoring system based on textile-based capacitive strain sensor. In: International conference on advanced algorithms and control engineering (ICAACE), Zhangjiajie, China, 24–26 April, pp 1–11
14. Rudraraju G, Palreddy SD, Mamidgi B, Sripada NR, Sai YP, Vodnala NK, Haranath SP (2020) Cough sound analysis and objective correlation with spirometry and clinical diagnosis. Inf Med Unlocked 19:1–19
15. Jagannath M, Madan Mohan C, Kumar A, Aswathy MA, Nathiya N (2019) Design and testing of a spirometer for pulmonary functional analysis. Int J Innov Technol Exp Eng 8 (4):1–9

16. Imran A, Posokhova I, Qureshi HN, Masood U, Riaz S, Ali K, John CN, Hussain I, Nabeel M (2020) AI4COVID-19: AI enabled preliminary diagnosis for COVID-19 from cough samples via an app. Inf Med Unlocked 100378:1–11

17. Pramono RXA, Imtiaz SA, Rodriguez-Villegas E (2016) A cough-based algorithm for automatic diagnosis of pertussis. PLoS ONE 11(9):1–9

18. Hoesterey D, Das N, Janssens W (2019) Spirometric indices of early airflow impairment in individuals at risk of developing COPD: spirometry beyond FEV1/FVC. Respir Med 156:58–68

19. Fukuhara A, Saito J, Birring SS, Sato S, Uematsu M, Suzuki Y, Rikimaru M, Watanabe N, Saito M, Kawamata T, Umeda T (2020) Clinical characteristics of cough frequency patterns in patients with and without asthma. J Allergy Clin Immunol Pract 8(2):654–666

20. Drugman T, Urbain J, Bauwens N (2013) Objective study of sensor relevance for automatic cough detection. IEEE J Biomed Health Informat 17(3):699–707

21. Mohammadi F, Jowkar Z, Reza Khankeh H, Fallah TS (2013) Effect of home-based nursing pulmonary rehabilitation on patients with chronic obstructive pulmonary disease: a randomised clinical trial. Br J Community Nurs 398:400–403

A Novel Approach for Initializing Centroid at K-Means Clustering in Paradigm of Computational Geometry

Tuhin Kr. Biswas and Kinsuk Giri

1 Introduction

In modern days of science, the advanced technology has covered all the aspects of our life in every sphere. The main advantage of using computational geometry is that it can be described through its various geometrical models applying on particular computing algorithms. In this paper, certain types of those geometrical models such as convex hull, voronoi diagram and empty circles are efficiently used to choose the initial cluster centers for k-means clustering. On the other hand in the field of computer sciences, the data analysis and data clustering are an important aspect for organization of data into some certain defined characteristics. This work focuses on the improvement of traditional k-means clustering algorithm through the selection of initial clusters centroids from the circumference points of voronoi or empty circles. A short review of certain related works is given below.

Several attempts have been done with various research group to improve the performances of k-means clustering. Recently, Franti and Sieranoja [1] have shown that to improve the k-means clustering the local fine-tuning capability is much more effective. A new algorithm on the basis of trial and error method, viz., k-means++ was proposed by David et al. [2]. On the basis of the weighted probability distribution, they have computed the distance between data points and nearest selected center/s, and subsequently, the selection of each center was done using the probability proportional to the square of computed distance between selected center and data points. Niknam et al. [3] elaborated an algorithm on modified imperialist competitive algorithm (MICA) where they have adopted a hybrid approach using k-means and MICA, called k-MICA. Here, the traditional k-means were used to initialize the MICA.

T. Kr. Biswas · K. Giri (✉)
Department of Computer Science and Engineering, National Institute of Technical Teachers' Training and Research Kolkata, Block-FC, Sector-III, Salt Lake City, Kolkata 700106, India
e-mail: kinsuk@nitttrkol.ac.in

© The Author(s), under exclusive license to Springer Nature Singapore Pte Ltd. 2022   545
R. R. Raje et al. (eds.), *Artificial Intelligence and Technologies*,
Lecture Notes in Electrical Engineering 806,
https://doi.org/10.1007/978-981-16-6448-9_52

Celebi et al. [4] have proposed a modified k-means clustering approach with respect to the initialization, where the algorithm works on traditional two partitioning method, viz., Var-part and PCA-part, that was previously proposed by Su and Dy [5]. In context of computational geometry, Reddy et al. [6, 7] have attempted to modify k-means algorithm using a new initialization technique with the help of voronoi diagram and subsequent circumference points of largest and smallest voronoi circles.

In our work, to initialize the traditional k-means clustering the circumference points of empty or voronoi circles are used here instead of random initialization as in traditional k-means. Novelty of this technique is that to compute the circumference points of empty circles, initially the convex hull and voronoi diagram are constructed along with voronoi circles. From here, the circumference points of those empty circles are selected as initial centroid of k-means. The details of the proposed algorithm are described in Sect. 3. Moreover to validate our proposed method, we have compared our results with traditional k-means with a synthetic data.

This paper has been organized in the following way. In Sect. 2, we discuss the preliminaries. The proposed method has been described Sect. 3. In Sect. 4, we report and discuss the experimental results. We have drawn our conclusions in Sect. 5 along with the relevant references at last section.

2 Preliminaries

In this section, we will discuss about some basic concepts and terms those are used in this paper.

2.1 Convex Hull and Voronoi Diagram

For a given data set, the convex hull (hereafter CH) is the boundary among a set of data points such that all the points remain inside the boundary, with some points lying on the boundary line, considering as hull points. Basically, CH is a convex polygon that contains maximum data points in its interior region and the rest points are the vertices of the polygon itself. The popular algorithm, viz., Graham Scan [8] efficiently compute the convex hull for a set of data points. On the other hand, the Russian mathematician Georgy Voronoi invented the voronoi diagram [9] (hereafter VD). This diagram makes partition of a given data set in such a manner that every point is separated and closer to its generating point. The intersection of two partitioning edge is considered as the voronoi vertices (hereafter VVs).

2.2 Voronoi Circle/Empty Circle

The empty circle (hereafter EC) or the voronoi circle (hereafter VC) is a circle that can be drawn for a given set of points in a plane such that no other point will remain in the interior of the circle. These circles are constructed using the VVs considering as possible candidate centers of those circle. Preparata and Shomos [10] presented the EC problem in O(nlogn) time with the help of voronoi diagram.

2.3 K-Means Clustering

The k-means clustering (hereafter KM) is a centroid-based clustering [11] method that falls under the unsupervised learning approaches. KM clustering is widely used to organize the data into various groups or clusters according to some certain defined characteristics, due its simplest form of operation and understandability. In this method, the cluster is formed based on the choice of initial centroid that is drawn in a random manner. Starting from the initial centroids KM algorithm assigns each data point into various clusters having the lowest distance form centroid and repeatedly assign the data points until its convergence. In other words, depending upon the initial centers the KM algorithm repeatedly finds the suitable centroid for clustering in an iterative way.

2.4 Silhouette Score

The Silhouette Score [12] (hereafter SH) is a distance metric that is used in this paper to measure the performance of clustering in qualitative nature. The SH score is basically a ratio between inter- and intra-cluster distance, that means how the final clustered data points are related with each other. The higher value of SH proofs better quality of clustering.

2.5 Euclidean Distance

Euclidean distance (hereafter ED) captures the same by aggregating the squared difference in each variable. ED is the most used metric to measure the distance between data points. In this work, we have used the ED to measure the distance between the data points within one single cluster as well as between the various formed clusters to validate that the better quality of clustering. The lower value of ED represents better quality of clustering.

2.6 Completeness

Completeness (hereafter CM) for any clustering is a measure of perfection that indicates the data points from the same label of class should be clustered or grouped together into the same cluster. In a simple word, completeness defines how much the elements in one cluster are strongly connected with each other rather than the elements in another clusters. Hence, the completeness indicates a measure toward the clustering quality. In this work with ED and SH as defined previously, we also used the CM as a performance metrics for clustering. The higher value of CM means better quality of clustering.

3 The Proposed Method

Let, for any given data set P, initially the convex hull CH (P) and voronoi diagram VD (P) are constructed. Then, all the VVs from the VD (P) and hull points from the CH (P) are stored into two different arrays called vor_ver and hull_pts, respectively. We have selected those VVs which are interior to the CH. Now, for all the VVs we compute the convex hull for VVs and the given data set. If in any step the new computed hull is same as exact the previous one then it is considered that voronoi vertex as interior to the hull.

After getting the VVs inside the interior hull, we consider those VVs as the possible candidate center of ECs. Next, we compute distances from VVs to each data point and find the minimum distances from each of the said VVs. Those minimum distances are radius of the corresponding ECs. Hence, we construct all the ECs having the center inside the CH.

Finally, we sort all these ECs with respect to the descending order of their corresponding radii. Next, we keep first five ECs from the array and catch their circumference points. It is observed that there may exist few common circumference points for any two consecutive ECs. Hence, the common circumference points are removed from the sorted list of circumference points and we stored them into an array called circum_final. So, in a word, the array circum_final contains all the different circumference points of first five ECs. As per the traditional KM, we need to supply the cluster number k as an input parameter. The traditional KM initially selects the k number of centers in a random manner, but, here our proposed method takes first k number of points from this circum_final. After selecting k initial centers using ECs, we proceed via traditional KM clustering algorithm to get our output. The pseudo code for our proposed algorithm is given in Fig. 1.

Following functions and variables have been used in the algorithm.

CH (P): function to construct convex hull for the set of n data points, VD (P): function to construct VD for the set of n data points, vor_ver: variable to store the all VVs, inside_hull: function to compute VVs interior to the convex hull, inside_vor: array to store the all VVs interior to the convex hull, circle_in: function to

Pseudo Code *for proposed algorithm*	

Input: A data set P containing n data points, the number of clusters k.

Output: The data set P grouped into k number of clusters, the final k number of centroids for each clusters and the label of clustering for each instances of P.

Step-1: Call *CH (P)*

Step-2: Call *VD (P)*

Step-3: Store all the voronoi vertices in *vor_ver*.

Step-4: Call *inside_hull* & assign *inside_vor*.

Step-5: Call *circle_in*.

Step-6: Sort all the circle in non-increasing order of their corresponding radii and assign *sorted_circle*.

Step-7: Compute circumference points for each *sorted_circle, remove the repeated points (if any)* and assign *circum_final*.

Step-8: Take first k number of points from *circum_final* and supply to the *k-means* clustering as initialization.

Step-9: Call *KM (n, k)* with the k initial cluster centers chosen from *(circum_final)*.

Fig. 1 Pseudo code for our proposed algorithm

compute the empty circle for each inside_vor, sorted_circle: array to assign the empty circle in non-increasing order of their radius, sorted_circum: array to store the all circumference points of sorted_circle, circum_final: array to store all the circumference points of sorted_circle removing the duplicate points.

KM(n, k): function to compute KM with set P of n data points and k number of initial centers.

4 Results and Discussions

We implemented our method using PYTHON. In order to construct CH and VD for any data set, the libraries are available. Nevertheless, no package is available to construct ECs. Therefore, first we developed a code to extract all the ECs inside the CH for any given data set. Our experiment was carried out with a synthetic data set having three attributes and five classes generated with a Gaussian distribution. Figure 2a is the scatter plot of generated synthetic data set. Next, we derived CH, VD and ECs inside CH, shown as in Fig. 2b. As we mentioned in Sect. 3, here we show first five ECs. Once we get all the distinct circumference points for first five ECs, we select first five points together as initial centroid of our proposed modified KM and run it. For comparison, we also run traditional KM with random initial

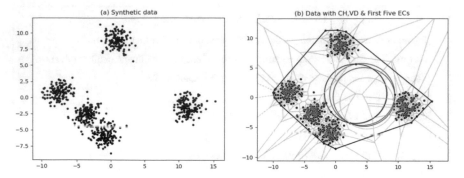

Fig. 2 **a** Generated synthetic data points with three attributes and five groups, **b** CH with data points, VD and first five ECs

Table 1 Multi-parameter analysis performed in three consecutive runs at our experiment

Run indices		Methods	IT	SH	CM	ED
Runs	1	KM	9	0.60	0.59	9640
		Our proposed modified KM	**6**	**0.70**	**0.96**	**887**
	2	KM	5	0.60	0.89	5821
		Our proposed modified KM	**6**	**0.70**	**0.96**	**887**
	3	KM	7	0.61	0.77	9332
		Our proposed modified KM	**6**	**0.7**	**0.96**	**887**

We computed IT, SH, CM and ED (see text for details)
Bold faced are the results of our method

centers. Each algorithm was run for three times. We also computed few standard indices and distance metrics. These are (a) Iteration numbers (hereafter IT), (b) Silhouette scores (SH), (c) Completeness (CM) and (d) Euclidean distances (ED). In Table 1, the complete results of the KM and our proposed modified KM are given.

Out of the five successive runs, we here show the result of our first run. Figure 3a, b shows the clustering results of KM and our proposed modified KM, respectively, for our first run. In both of Fig. 3a, b, the violet "filled circles" indicates the initial centers while the red "+" symbols indicates the final centroid of the respective clusters formed. We have compared our proposed method with traditional KM in the view of both qualitative and quantitative nature of clustering. In this regard, we have used some well-known performance metrics to identify the quantitative and qualitative measure of clustering, and hence, their values represent the novelty and validity of our proposed method over KM clustering.

From Table 1, it is clearly visible that our proposed method exceeds KM on (a) getting low ED, (b) getting high SH, CM and also (c) improving iteration numbers, as at 1st and 3rd runs, the number of iterations (IT) used by our proposed modified KM is decreased with respect to KM. Here, the less number of iterations

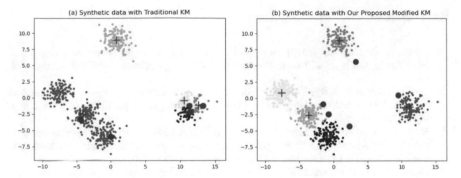

Fig. 3 Clustering for **a** KM and **b** our modified KM. Violet filled circle represents the initial centers, while the red "+" symbols represent final centroid of the respective formed clusters

taken by our method indicates the better quantitative nature of clustering as well as low ED and high SH, CM for our proposed method indicates the better qualitative nature of clustering than the traditional KM. Hence, it is seen that the proposed algorithm performs better than traditional KM.

5   Conclusions

The clustering approaches are the most important area of research over the decades. The KM clustering is one of the simplest forms of clustering which is widely used in various fields of data science research to find and extract the important characteristics of grouped data. Our proposed method is a modified approach on the traditional KM algorithm. In this work, instead of random initialization as in KM, we propose a new initialization method based on some certain concept of computational geometry, mainly the concept of CH, VD and ECs. It is well known that in traditional KM random initialization may tend the algorithm up to a long time to converge. Therefore, to overcome this limitation of randomness of traditional KM, our proposed method selects the initial cluster centers from the circumference points of ECs and these circumference points remain same even if the algorithm is run several times. In case of real-world data set with large number of attributes and clusters, our modified k-means may also play a vital role for betterment of data clustering. In future, we will do this analysis.

References

1. Fränti P, Sieranoja S (2019) How much can k-means be improved by using better initialization and repeats? Pattern Recogn 93:95–112. https://doi.org/10.1016/j.patcog.2019. 04.014
2. David A, Sergei V (2007) K-means++: the advantages of careful seeding. In: Proceedings of the annual ACM-SIAM symposium on discrete algorithms, vol 8, pp 1027–1035. https://doi. org/10.1.1.360.7427
3. Niknam T, Fard E, Pourjafarian N, Rousta A (2011) An efficient hybrid algorithm based on modified imperialist competitive algorithm and K-means for data clustering. Eng Appl Artif Intell 24:306–317. https://doi.org/10.1016/j.engappai.2010.10.001
4. Celebi M, Hassan E, Hassan K (2013) Deterministic initialization of the K-means algorithm using hierarchical clustering. Int J Pattern Recogn Artif Intell 26:10. https://doi.org/10.1142/ S0218001412500188
5. Su T, Dy J (2007) In search of deterministic methods for initializing K-means and Gaussian mixture clustering. Intelligent data analysis, vol 11, pp 319–338. https://doi.org/10.5555/ 1367948.1367950
6. Reddy D, Jana P (2012) Initialization for K-means clustering using Voronoi diagram. Procedia Technol 4:395–400. https://doi.org/10.1016/j.protcy.2012.05.061
7. Reddy D, Jana P (2014) A new clustering algorithm based on Voronoi diagram. Int J Data Mining 6:49–64. https://doi.org/10.1504/IJDMMM.2014.059977
8. Graham R (1972) An efficient algorithm for determining the convex hull of a finite planar set. Inf Process Lett 1:132–133. https://doi.org/10.1016/00200190%2872%2990045-2
9. Georges V (1908) Nouvelles applications des paramètres continus à la théorie des forms quadratiques. Premier mémoire. Sur quelques propriétés des formes quadratiques positives parfaites. Journal für die Reine und Angewandte Mathematik 133:97–178. https://doi.org/10. 1515/crll.1908.133.97
10. Preparata F, Shamos M (1985) Computational geometry. Springer. ISBN 978-1-4612-1098-6
11. Steinhaus H (1957) Sur la division des corps matériels en parties. Bull Acad Polon Sci (in French) 4:801–804
12. Rousseeuw J (1987) Silhouettes: a graphical aid to the interpretation and validation of cluster analysis. Comput Appl Math 20:53–65. https://doi.org/10.1016/0377-0427(87)90125-7

Intelligent Forecasting Strategy for COVID-19 Pandemic Trend in India: A Statistical Approach

Siddharth Nair, Ganesan Ckm, R. Varsha, Sankhasubhra Ghosal, M. Vergin, and L. Jani Anbarasi

1 Introduction

As per the World Health Organization (WHO), we have come across a developing illness namely novel coronavirus (COVID-19) which is answerable for tainting a large number of individuals and killing thousands worldwide. There have been 1.29 M confirmed corona cases as of September 30, 2020 [1]. Although 817 K people have recovered, there have 97,497 confirmed deaths. India is ranked 2nd with 7,706,946 confirmed cases in the world according to the WHO. Various studies have been conducted to study the effects of COVID-19 on people. Predictions have been made regarding the future, and the reasons have been discovered in the process. The first case confirmed in India was in Kerala on January 30, 2020. Since then, India has witnessed a flood of affirmed coronavirus cases because individuals have not been taking the lockdowns seriously. It is important to understand what the future looks like for the country assuming there is no change in the current trend in the spread of COVID-19 [2]. Thus, this research work aims at accurately predicting the rise of confirmed cases and deaths in the Indian subcontinent by performing ML and time-series forecasting algorithms on the official dataset acquired from the World Health Organization [3].

It is possible to apply various supervised machine learning algorithms to do better predictions for applications such as predictive maintenance, healthcare monitoring, and smart agriculture [4]. According to [5], the authors made use of the ARIMA model to predict a trend of COVID-19 in Italy, Spain, and France as they were the most affected countries of Europe. The results of the paper concluded that ARIMA model was the most suitable and efficient model in predicting the spread of the COVID-19 for the future. In continuation to the above work, in [6], the authors had paid special attention to predict how long will COVID-19 last and what can be the approximate date of the COVID-19 ending. They predicted that there will be a

S. Nair (✉) · G. Ckm · R. Varsha · S. Ghosal · M. Vergin · L. J. Anbarasi
Vellore Institute of Technology Chennai Campus, Chennai, Tamil Nadu 600127, India

© The Author(s), under exclusive license to Springer Nature Singapore Pte Ltd. 2022
R. R. Raje et al. (eds.), *Artificial Intelligence and Technologies*,
Lecture Notes in Electrical Engineering 806,
https://doi.org/10.1007/978-981-16-6448-9_53

second wave of the COVID-19 around the end of 2020. In this paper, they made use of the SARIMA model which again proved to give the best results. In [7], with the help of simple heuristic, the researchers had forecasted the number of cases for the year 2020. They approximated it using an exponential curve, and the accuracy was quite satisfactory. In continuation to the above, in [8], the researchers had predicted the number of reported cases which would be visible in the near future. They identified the isolation of unreported cases and the impact of asymptomatic infectious cases. In [9], the analyst proposed a hybrid AI model for the prediction of COVID-19. They used natural language processing along with LSTM to build an artificial intelligence model which was used to predict more efficient models. In [10], the authors made use of SIR model that tracks the transmission and the recovery time. They used two approaches to find efficient results.

2 Proposed Methodology

The prediction of the spread of COVID-19 is a process with many possible statistical and artificial intelligence techniques. Intelligent algorithms with respect to machine learning, ensemble learning, and time-series forecasting are preferred in this proposed method to achieve better accurate COVID-19 prediction.

2.1 Machine Learning

The proposed machine learning model for pandemic trend prediction includes acquiring data source and prediction based on various quantitative prediction models such as linear regression and support vector regression.

2.1.1 Linear Regression

In this quantitative prediction model, one variable is taken as an explanatory variable, which makes the other a dependent variable. It is vital to note whether a relationship between the variables exist or not. This can be determined using a scatterplot. If there is no relation between the two variables, then the scatterplot will remain a constant and fitting a model of linear regression will be of no use. The strength of the association of the two variables can be achieved when the correlation coefficient is between -1 and 1. The equation can be represented as follows:

$$Z = j + hV \tag{1}$$

In Eq. (1), V stands for the explanatory variable, 'Z' stands for the dependent variable, and 'j', 'h' are the constants.

2.1.2 Support Vector Regression

Support vector machines in addition to linear classification can also perform non-linear classification. This is done using the kernel trick method. If the data is unlabeled, supervised learning is not possible and hence, an unsupervised learning method is followed, which clusters the data to groups, and classifies the test set to the groups formed. The **support vector clustering** uses the statistics of support vector machines which involves forming a plane and this plane (hyperplane) acts as boundary. If our hyperplane has an equation of $y = wx + b$, then the decision boundaries would be

$$Wx + b = +a \text{ and } Wx + b = -a, \text{ thus } -a < y - wx < a \tag{2}$$

2.2 Ensemble Learning Models

Quantitative forecasting of COVID-19 data is also accomplished by applying an ensemble supervised machine learning approach namely random forest regression which is described below.

2.2.1 Random Forest Regression

Random forest is a method in which classification and regression techniques are implemented using many decision trees. This procedure is called bootstrap aggregation, ordinarily known as bagging. In this model, each decision tree is trained with varying samples of the dataset. Row and feature sampling are done in a random manner so that it forms different data samples for each model. During computation of the end result, the individual trees are all combined instead of relying on one particular tree. In regular trees, the splitting of each node is based on the best split of each and every variable, whereas in random forest, the splitting of each node is decided by a randomly allocated subset of predictors. This method turns out to offer better results in certain cases than algorithms such as support vector machines and is sturdy against overfitting. Interpreting the random forest however is not as straightforward as that of a regular classification tree where a particular variable's decision influencing power depends on its position in the tree. So, variable importance in random forests has to be calculated differently. One such variable importance calculation method is to keep a count of the frequency of selection of variables by each individual tree.

2.3   Time-Series Models

Time-series models are appropriate forecasting strategies to predict and implement measures to respond to the impact of the COVID-19 outbreak. Autoregressive integrated moving average (ARIMA) and Holt-Winters exponential smoothing models are developed to forecast the pandemic predictions.

In comparison with statistical models including simple exponential smoothing and linear regression, ARIMA offers a higher flexibility. Some simple exponential smoothing models are instances or special cases of ARIMA models. On the other hand, certain advanced exponential smoothing algorithms such as Holt-Winters forecasting have ARIMA analogues which makes it another ideal candidate for time-series forecasting. These models are designed to capture the seasonality of data which traditional machine learning and ensemble learning models fail to do. In general, models such as ARIMA and Holt-Winters forecasting offer the middle ground of not being complicated enough to overfit the dataset while thereby being flexible so as to capture patterns and other relationships that could be found within the dataset.

2.3.1   ARIMA and SARIMAX

The autoregressive integrated moving average model is a time-series forecasting algorithm which is used to get a better analysis of time-series data or to forecast data points. In certain situations, ARIMA models are used to demonstrate non-stationarity of data.

The 'autoregressive' portion of the model involves using lagged or prior values in a regression equation, as depicted by equation the equation below, to predict the result in subsequent stages.

$$y_k = a + \sum_{j=1}^{n} \partial_j y_{k-j} + \epsilon_k \tag{3}$$

The 'moving average' portion of the model, as depicted by equation the equation below, refers to generating a sequence of averages of varying subsets of the entire dataset. This is used to smooth out data thereby generating regularly updated averages which are taken over a period of specified time.

$$y_k = a + \epsilon_k + \partial_1 \epsilon_{k-1} + \cdots + \partial_f \epsilon_{k-f} \tag{4}$$

The 'Integrated' portion of the model refers to the stationarity of data. In case of non-stationarity, the data is replaced with the differences between the current and previous values. This process can be repeated multiple times as stationarity is important for the data to fit the model well. ARIMA models consist of three major parameters which influence its performance. The factors 'p', 'd', and 'q' correspond to the order of the AR model, degree of differencing, and order of MA model, respectively. The parameter 'p' refers to the autocorrelation factor (ACF) of the dataset whereas the parameter 'q' refers to the partial autocorrelation factor (PACF) of the model which are calculated separately once the data is made stationary.

The seasonal ARIMA model is the same as the ARIMA model with the only difference being the seasonality factor. This factor includes four parameters of which the first three mirror the values of 'p', 'd', and 'q' of the ARIMA model. The fourth parameter 's' is known as the periodicity of the cycle of seasonality of the data.

2.3.2 Holt-Winter Exponential Smoothing

Holt-Winters is a time-series behavior model. For forecasting a model is always required, and Holt-Winters paves a way to model a classic value-mean, a slope—difference over period of time, and a pattern—seasonality which are the three main features of time series model. Exponential smoothing is used in Holt-Winter which encodes lots of values from the past and uses the same to predict future and present trends. Since it has three aspects of time series model, it is also called triple exponential smoothing. Using these three parameters, the models predict future and the present trend. Several parameters are required for these models: three-one for each of the exponential smoothing (a, β, γ), the season length, and the number of periods in the season. In Eqs. (5) and (6), $\{x_t\}$ represents the sequence of raw data which starts at time $t = 0$ and $\{s_t\}$ represents the results obtained by the exponential smoothing algorithm which may be regarded as a great estimate of what the succeeding x value will. If the sequence starts at $t = 0$, exponential smoothing can be simply explained as follows

$$s_0 = x_0 \tag{5}$$

$$s_t = a * x_t + (1 - a)s_{t-1} \tag{6}$$

where 'a' is the smoothing factor.

The Holt-Winters process involves four-stage performances. The first step includes the decomposition of the time-series data into seasonal, trends, and reminder components. The subsequent phases include the generation of iterations of the components, forecasting the forthcoming time periods and the identification of seasonal patterns.

Table 1 Accuracy and error comparison of algorithms

India	Type	Prediction models	R-squared	MSE	RMSE	MAPE	MAE
Cumulative cases	Time series	ARIMA	**0.986**	13,292,849,806.2317	115,294.621	3.81	72,709.964
		Holt-Winters	0.984	47,404,427,534.3858	217,725.578	.93	95,262.572
	ML	SVR	0.85	75,367,441,372.5	274,531.314	16.92	978,451.59
		Linear regression	0.50	136,782,368,912.6	369,841.004	50.47	185,567.25
	EL	Random forest	0.82	84,298,224,318.96	297,080.169	18.45	100,345.87
Cumulative deaths	Time series	ARIMA	**0.990**	991,657.149	995.819	0.859	705.648
		Holt-Winters	0.984	1,626,527.840	1275.354	2.67	701.05
	ML	SVR	0.81	94,280,552.76	9709.81	53.48	8923.16
		Linear regression	0.60	2,032,623,005.92	45,084.62	102.43	42,567.82
	EL	Random Forest	0.79	102,258,746.47	10,112.30	69.28	15,196.35

The bold values signify the highest R-squared values achieved for cumulative cases and cumulative deaths respectively. It highlights that the ARIMA model outperforms the others in terms of accuracy

3 Results

The proposed model generated accurate results by achieving an R-squared value of 99% for ARIMA, 98% for Holt-Winters exponential smoothing, 82% for random forest regression, 50% for linear regression, and 85% for support vector regression. The results show that the time series models such as ARIMA and Holt-Winters outperform the others (Table 1).

Figures 1 and 2 are the results obtained when the dataset is trained and tested with time-series models. It is clear from the graphs that the values time-series models predict are similar to the actual values. The implementation and complexity of this solution varies as the spread of COVID-19 differ across states. This solution is extremely practical as it learns from previously known data, thereby providing more accurate forecasts with increasing amounts of data.

Fig. 1 Cumulative cases versus date

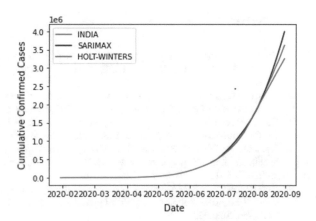

Fig. 2 Cumulative deaths versus date

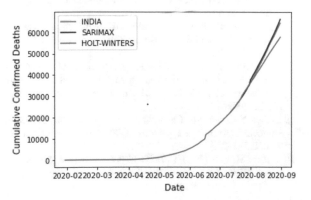

4   Conclusion

ARIMA and Holt-Winters exponential smoothing in time-series models, linear regression, and SVR in machine learning and random forest regression in ensemble learning are used in this proposed work for forecasting the COVID-19 cases. It is clear from graphical and statistical results that the errors in ML and EL algorithms are too high and their accuracy score (R-squared) is significantly lower than that of time-series forecasting models such as ARIMA and Holt-Winters. SVR outperforms the other ML and EL algorithms with an average R-squared score of 0.83 but does not meet the required standards to predict the rate of COVID-19 in India. Among the time-series algorithms, ARIMA offers the highest accuracy (0.986 and 0.99 for COVID-19 confirmed case in India) which is followed by Holt-Winters (0.984 and 0.984 for COVID-19 confirmed deaths in India).

References

1. Hu S et al (2020) Weakly supervised deep learning for COVID-19 infection detection and classification from CT images. IEEE Access 8:118869–118883. https://doi.org/10.1109/ACCESS.2020.3005510
2. Gupta S, Raghuwanshi GS, Chanda A (2020) Effect of weather on COVID-19 spread in the us: a prediction model for India in 2020. Sci Total Environ 138860
3. Casella F (2021) Can the COVID-19 epidemic be controlled on the basis of daily test reports? IEEE Control Syst Lett 5(3):1079–1084. https://doi.org/10.1109/LCSYS.2020.3009912
4. Gondalia A, Dixit D, Parashar S, Raghava V, Sengupta A, Sarobin VR (2018) IoT-based healthcare monitoring system for war soldiers using machine learning. Procedia Comput Sci 133:1005–1013
5. Kufel T (2020) ARIMA-based forecasting of the dynamics of confirmed Covid-19 cases for selected European countries. Equilibr Quart J Econ Policy 15(2):181–204. https://doi.org/10.24136/eq.2020.009
6. Malkia Z, Atlamb E-S, Ewisc A, Dagnewe G, Alzighaibia AR, Ghadaa EL, Elhosseinif A, Hassanieng AE, Gada I. ARIMA models for predicting the end of COVID-19 pandemic and the risk of a second rebound
7. Shi F et al. Review of artificial intelligence techniques in imaging data acquisition, segmentation and diagnosis for COVID-19. IEEE Rev Biomed Eng. https://doi.org/10.1109/RBME.2020.2987975
8. Staszkiewicz P, Chomiak-Orsa I, Staszkiewicz I (2020) Dynamics of the COVID-19 contagion and mortality: country factors, social media, and market response evidence from a global panel analysis. IEEE Access 8:106009–106022. https://doi.org/10.1109/ACCESS.2020.2999614
9. Chen Y-C, Lu P-E, Chang C-S, Liu T-H. A time-dependent SIR model for COVID-19 with undetectable infected persons. IEEE Trans Netw Sci Eng. https://doi.org/10.1109/TNSE.2020.3024723
10. Horry MJ et al (2020) COVID-19 detection through transfer learning using multimodal imaging data. IEEE Access 8:149808–149824. https://doi.org/10.1109/ACCESS.2020.3016780

Design of Infusion Device for Disabled Patients

G. Umashankar, V. Akshya, Sindu Divakaran, J. Bethanney Janney,
T. Sudhakar, J. Premkumar, and S. Krishnakumar

1 Introduction

Doctors are generally using medicines for treating the patients. But the patients with
disabilities cannot take drugs as other patients without disabilities taking in a
controlled manner. Hence, the proposed system introduces a drug injecting
mechanisms in an appropriate dosage level in controlled manner. The current study
avoids the patients risk by giving the drugs in a scheduled time with exact dosage. It
can be categorized into manual and automatic types [1]. Manual system consists of
two liquid reservoirs and multiple way stop clock to maintain the flow level, and
precision should be maintained when it is automatic. The pressure level of the
infusion device is not depending on the gravitational force. To achieve the high
precision rate, the proposed system require syringe infusion system, especially in
pediatric cases or in intensive therapies where small amounts of high concentration
medicines are used for long periods of time [2]. Microcontroller controls all processes in the proposed system. Using the SIM card allows users of GSM networks
to move their phone number quickly from one GSM phone to another by simply
moving the SIM card. A syringe mechanism was used in this work to act as the
reservoir of fluids and to establish the required pressure for infusion [3]. This type
of infusion generates a high-precision, continuous flow. DC motor is used in the
patient's body to administer the correct amount of insulin. That way it can be
stopped over or under dosage. Microcontroller controls all operations. We can be
configured to offer correct prescription volumes at specific tariffs. Sensors such as
cuff sensor track the patient's blood pressure level and heart rate, using cuff and
temperature sensor to control the patient's body temperature [4]. Doctor is told
about the status of the medication administered to the patient, and whether there is

G. Umashankar (✉) · V. Akshya · S. Divakaran · J. B. Janney · T. Sudhakar ·
J. Premkumar · S. Krishnakumar
Department of Biomedical Engineering, Sathyabama Institute of Science and Technology,
Chennai 600119, India

R. R. Raje et al. (eds.), *Artificial Intelligence and Technologies*,
Lecture Notes in Electrical Engineering 806,
https://doi.org/10.1007/978-981-16-6448-9_54

any injection issue. Even outside the hospital [5], it is easy to use and therefore the infusion pump becomes a helpful home tool. The interface would enable patients to save money by avoiding the home nursing costs.

2 Related Works

Abo Zahhad et al. (2014) deal with telemedicine that is monitoring the physiological parameters of homecare and elderly patients by various sensors using mobile communication, and the methods of transmitting data and signals have been studied for our knowledge. The use of mobile transmission of data has been proven to be an effective tool and helpful to the society [4].

Magnani et al. (2015), describes the realization of a sensor capable of measuring the volume of a drop in free fall. The sensor realized is made of a simple low-cost red laser and a photodiode and optics to focus the beam on the light sensor. But the Web server is used for transmission in which the availability of Internet is necessary [6].

Calcagnini et al. (2008) investigated about changes in the risk of EMI and to extend the risk assessment of EMI to newer telecommunications products where medical devices are malfunctioning due to electromagnetic interference from various electronic devices. It reduces the rate of failure (from 58 to 30%) [7].

Ranjitha Pragnya et al. (2013) explain about the wireless sensor network-based monitoring for elderly people. It measures the motion of the person, pressure, and temperature. As wireless technology is used, it is not reliable for proper transmission of readouts, and the readouts are not displayed to the control. It makes difficult for the patient's attender reducing their personal and daily activities inhibiting their range of motion, as the wireless transmission is a short range of communication technology [8].

Vidyasagar et al. (2006) discussed about the multidrug infusion device developed to provide drug delivery at a precisely controlled rate. It provides an accurate, consistent drug delivery but the status of the patient is not monitored, and intimation is not done so that error would happen [3].

3 Materials and Methods

3.1 GSM

The proposed system appears in Fig. 1. GSM is a cellular network based on TDMA [1] technology developed in Europe and used in most parts of the world. GSM phones use a SIM card to mark the account that the user has. Using the SIM card allows users of GSM networks [9] to transfer their phone number easily from one

GSM phone to another by simply switching the SIM card. The message [10] includes the insulin quantity and exact injection time. This message gets via UART to PIC microcontroller. GSM is the most acknowledged standard in broadcast communication, and it is actualized all inclusive. GSM is a circuit exchange framework that partitions each 200 kHz channel into eight 25 kHz scheduled vacancies [11].

3.2 RTC

PIC microcontroller also operates real-time clock (RTC) power. A real-time clock (RTC) is a computer clock that keeps tracking the current time (most often in the form of an integrated circuit). RTCs are usually interfaced with I2C device controllers. The most prevalent available RTC is DS1307. Using the I2C protocol, also called as the two wires interface, the device can be interface which enables data to be read and written in serial form (SCL and SCA). DS1307 comes with a built-in power sensing circuit which senses power failures automatically to switch on back up supplies. Timekeeping operation continues while the part keeps operating with the backup supply. DS1307 RTC uses a 32.768 kHz crystal oscillator, and it does not require any external resistors or capacitors to operate [12].

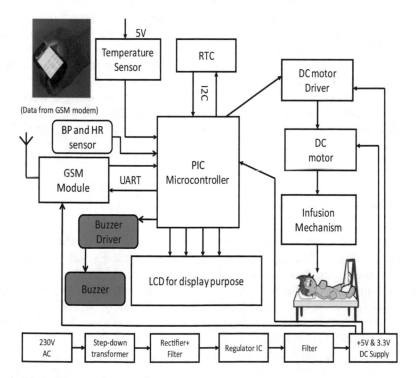

Fig. 1 Block diagram of proposed system

3.3   Buzzer

Buzzer driver controls buzzer function. Buzzer requires either higher current or voltage, or both, which can be supplied by the output of the previous stage. The buzzer produces a same noisy sound irrespective of the voltage variation applied to it. It consists of piezo-crystals between two conductors. When a potential is applied across these crystals, they push on one conductor and pull on the other. This, push and pull action results in a sound wave. Most buzzers produce sound in the range of 2–4 kHz [13]. Therefore, it requires a driver circuit or power emitter follower or power CC stage to let the buzzer ring. This alarm monitors the precision of the infusion being set—parameters fall beyond the planned operating period in the instant off. Alarms indicate misuse or wear out of some parts, which help to guarantee the patient's health and safety.

3.4   DC Motor

This process is controlled through the DC motor. Operation is based on basic electromagnetism, in any electric motor. DC motor is powered by PIC microcontroller-operated DC motor driver. It supports control of direction as well as distance. High performance (especially at low speeds), high power density, simplicity of control, and a large installed base help determine applications for DC motors. DC motors develop full-load torque at low speed. This, combined with low inertias, results in excellent performance from DC motors [14]. The DC motor driver drives both the forward and the reverse direction of the DC motor which controls the injection part. A 230 V AC power supply is provided for transformer step-down. It reduces up to 9v which is given to the rectifier which converts AC to DC. Then the filter which filters the unwanted noise is given to. Regulator IC uses two types of IC.

3.5   Blood Pressure Sensor

A blood pressure sensor is a blood pressure measurement device [15] composed of an inflatable cuff to limit blood flow and mercury or mechanical pressure measurement manometer. An occluding cuff is placed on the left arm and is connected to an air pump and a pressure sensor. Cuff is inflated until a pressure greater than the typical systolic value is reached, then the cuff is slowly deflated. As the cuff deflates, when systolic pressure value approaches, pulsations start to appear. These pulsations represent the pressure changes due to heart ventricle contraction and can be used to calculate the heartbeat rate [16]. This is often used in tandem with a means to assess at what blood pressure flow is just beginning, and at what pressure the unimpeded blood pressure sensor is being used to track the patient's blood pressure [8]. If there are any changes in the blood pressure of the patient, PIC microcontroller will automatically send a message to the doctor via UART PROTOCOL. This way this protocol is used to communicate dually. By that time the PIC microcontroller must interrupt the infusion process. The doctor will then adjust the amount of insulin to be administered, depending on the patient's pain level then the cycle will begin.

3.6 Temperature Sensor

The LM35 has a merit over linear temperature sensors [17] adjusted in ° Kelvin, because the consumer is no need to do any alteration with the constant voltage from its output in order to obtain convenient centigrade scale. The low-output impe dance, linear output, and precise inherent calibration of the LM35 device make interfacing to readout or control circuitry especially easy [18].

3.7 UART

A universal asynchronous receiver/transmitter (UART) is transceiver which controls the transmission and receiving the serial data. It converts the received serial data from the digital systems to a single serial bit of information for outbound transmission along parallel circuits [19].

3.8 LCD

The message sent to doctor will be shown on the LCD. This monitor displays information about the infusion rate, the total amount of drug to be given, the flow rate, the total and pending duration, and some warning information. There are two registers to this LCD: command and data. A buzzer is an audio signaling device that will announce the arrival of the alert [20].

4 Results and Discussion

The performance of the proposed method is illustrated as simulation and is evaluated and is shown in Fig. 2. When the system or controller is triggered by receiving a text message through GSM from the doctor, the message is displayed in the LCD and is shown in Fig. 3.

Fig. 2 Working of hardware assembly

Fig. 3 First stage of simulation and LCD display of hardware

Fig. 4 Display of message details from the input control

RTC records the drip rate and time. An alarm is set to alert the patient or the attender the time of infusion 15 min prior to the drip. The physiological conditions of the patient are measured and indicated to the input control, which is shown in Fig. 4.

The end result of drip depends on these measurements; when the sensor senses abnormal values of pressure, heart rate, or temperature, the controller cancels the drip as it may be dexterous to the patients to administer condition and the status of the drug during these types of condition and the status of the patient is intimated to the input control and it is shown in Figs. 5 and 6 and to a neighbor or a relative which will be a very helpful means to homecare or elderly patients who are alone at home without any supervision.

Fig. 5 Display of patient status the input controller

Fig. 6 Commands and intimated results of the input controller

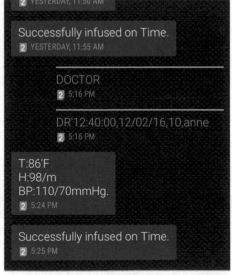

The above figure shows the intimation of all the parameters through text messages to the doctors

Fig. 7 Successful infusion of
drug is displayed

Whereas when the readings are normal as per the preset rates then the flow of system takes place. All the measurements and readings will be displayed in the LCD simultaneously. The successful infusion of drug is displayed in Fig. 7. This system offers better hardware design for the medical professionals. The type of motor can be varied as per its reliability. This proposed approach can be implemented to deliver the appropriate doses of different variety of drugs for the patients in an adequate manner to all the departments in the hospitals. This system provides a great scope in miniaturization. It can be implemented at micro-electromechanical systems (MEMS) level. System with microneedles can be used, and a wearable wrist or arm size device could be designed. The type of motor can be varied as per its reliability. There are wide range of use for infusion pumps in different areas where various drugs are used such as the departments of chemotherapy, pain management, total parental nutrition, and anesthesia/sedation.

5 Conclusion

From the above discussions, the proposed system is designed and verified for persons with disabilities. All the mechanisms have been incorporated to the infusion device properly, and communication system is designed to interact with the doctors about the dosages. Finally, whenever doctor sends the message of appropriate amount of insulin or the necessary drug, the biological parameters are measured and intimated to the input control, injection will be automatically injected into patient's body in the appropriate time. It can also able to measure heart activities, body temperature, and blood pressure variations in controlled manner. It is a strong fact that the final product will improve the health of countless patients by decreasing the compliances of complex self-drug infusion.

References

1. Assunção R, Barbosa P, Ruge R, Guimarães PS, Alves J, Silva I, Marques MA (2014) Developing the control system of a syringe infusion pump. In: 11th international conference on remote engineering and virtual instrumentation
2. Cho ST, Wise KD (1993) A high-performance micro flow meter with built-in self test, vol 36, pp 47–55
3. Vidyasagar KEC, Phani T, Sudesh S (2011) Design and development of wireless intravenous multidrug delivery system, vol 2(3), pp 269–271
4. Abo Zahhad M, Ahmed SM, Elnahas O (2014) A wireless emergency telemedicine system for patients monitoring and diagnosis
5. Muller A, Scwartzbach MI (2006) An introduction to XML and WEB technologies. Addison-Wesley
6. Magnani A, Pesatori A, Melchion D, Norgiatitled M (2015) Control system of the syringe infusion pump in biomedical application. IEEE Trans 64(9):96–99
7. Calcagnini G, Censi F, Triventi M, Mattei E, Bartolini P (2008) Electromagnetic immunity of infusion pumps to GSM phones, vol 43, no 3, pp 225–228
8. Ranjitha Pragnya K, Krishna Chaitanya J (2013) Wireless sensor network based healthcare monitoring system for homely elders, vol 6, no 5, pp 2078–2083
9. Muller A, Scwartzbach MI (2006) Medico hospital equipment and maintenance management. Addison-Wesley
10. Morville P, Rosenfeld L (2002) Information architecture for the world wide web: designing large-scale web sites, 2nd edn. O'Reilly
11. ur Rahman Z (2017) GSM technology: architecture, security and future challenges, Feb 2017 (Bacha Khan University, Charsadda)
12. Bindu B (2020) MikroC pro for PIC tutorials, interfacting the DS1307 real time clock with PIC microcontroller, pp 1–2, 25th Feb 2020
13. Bhatt A (2010) Piezo buzzer, 28 July 2010
14. Kimbrell J. Drives, motors & motion, automation direct, Dc motors
15. Webster JG (2009) Medical instrumentation application and design, 4th edn. Wiley
16. Lopez S. Blood pressure monitor fundamentals and design. Document Number: AN4328; Application Note Rev. 2
17. Bronzino JD (2006) The biomedical engineering handbook—medical devices and systems, 3rd edn. CRS Press
18. Texas Instrumentation. LM35 precision centigrade temperature sensors, SNIS159H—August 1999–Revised December 2017
19. George L. Hi-Tech C, microcontroller, PIC, Proteus, RS-232, USART, 08 Aug
20. Chen H-W, Lee J-H, Lin B-Y, Chen S, Wu S-T (2017) Liquid crystal display and organic light-emitting diode display: present status and future perspectives, 01 Dec 2017

Identifying Mood in Music Using Deep Learning

Shikha Rani, Manoj Kaushik, and Vrinda Yadav

1 Introduction

Music plays an indispensable role in our day-to-day life with its applications in entertainment, therapeutic, enthusiastic motivation, neuron chillness, etc. Generally, we prefer listening to music in accordance with our mood. This paper focuses on to classify the songs based on mood or underlying emotions, so that the music search on the basis of preference can be simplified. Hence, the paper proposes an automatic deep learning-based method for identifying mood in music. Techniques in unsupervised learning are based on pattern matching in the area of songs classification, while supervised learning techniques do not perform well in the context of known songs. Machine learning techniques can forecast the mood of a song with the help of low dimensionality [1], whereas the deep learning methods using convolutional neural network can achieve the same objective through dimension reduction, and for this reason, less capturing power can be processed with the high-dimensional signal.

In literature, feature extraction in machine learning is performed as the preprocessing task [2–4]. Certain features like mel-frequency cepstral coefficients (MFCC) work on timbral information, and signal processing acoustic feature works on rhythmic and pitch information. MFCC loses temporal structure at the time of conversion of an audio signal to a vector signal. Convolutional neural network (CNN) is employed to extract features automatically. CNN uses spectrogram to extract the temporal structure. Spectrogram represents both the frequency and temporal structure. It is an effective representation of an audio signal than the MFCC because it contains adequate information about frequency, time, and amplitude. Spectrogram can be expressed as in Eq. (1).

S. Rani (✉) · M. Kaushik · V. Yadav
Computer Science and Engineering, Centre for Advanced Studies, Lucknow 226031, India
e-mail: vrinda@cas.res.in

$$\text{STFT}\{x[n] [\imath, w]\} = X [\imath, w] = \sum_{n=-\infty}^{n=\infty} x[n]\omega(n-m)\text{e}^{-jwn} \qquad (1)$$

where $x[n]$ represents input signal, (n) represents windows size, and $X[\imath, w]$ represents DTFT of the windowed data.

The remaining paper is categorized as following manner. Section 2 discusses about the related work. Section 3 discusses about the methodology used in the paper including the CNN architecture. The consecutive section is about the achieved results and the conclusions there after.

2   Related Work

Many research papers are dedicated to signal processing acoustic feature. Acoustic feature can be classified as [5] rhythmic features [3] like beat, loudness, DWT and timbral texture feature [2] like MFCC that are used to discriminate music [6], pitch feature [7] that deals with detection of the fear emotion.

In a study by Asuman et al. [3], classification is achieved using multilayer perceptron network for the songs taxonomy by genres. You et al. presented a method to classify music in an effective way based on vocalists [6]. In music information retrieval [8–10], authors employed music information retrieval to identify songs based on emotions. Patra et al. [11] proposed an algorithm on mood taxonomy for songs based on lyrics and audio signals.

Another work discusses the objective of extracting and predicting songs on the basis of mood using machine learning approaches like k-mean, SVM, and Naive Bayes. These approaches work on three features, namely MCCF, peak difference, and frame energy. Machine learning approach requisite structural data to perceive the partition between the songs and low-dimensional feature are used while CNN approach has the potential to categorize the music data using numerous layers in the network.

The main contribution of this paper is the development of a deep learning method to classify songs based on their moods. Proposed method demonstrates better results in terms of accuracy on created dataset and also eliminates the problem of high dimensionality. The proposed method is computationally fast in identifying mood in music automatically.

3   Methodology

In the preprocessing phase, an audio signal is first converted into a spectrogram using short-time Fourier transform (STFT) [6]. CNN can easily recognize the pattern of spectrogram with the help of multiple layers. Next, we will discuss about the dataset, preprocessing of the data, and the model architecture.

3.1 Dataset

The dataset used in the paper consists of 240 Bollywood songs. Each song is of 3–6 min time duration. Each music file is divided into two segments having 30 s of time duration, wherein the first music clip of 30 s is from the starting point of the original song and the next music clip of 30 s after 2 min of the original song. A total number of music clips in the dataset are 480. Music clips categorized as relaxed and loud each category consist of 240 audio clips. The dataset was illustrated in Table 1.

Dataset is divided into two parts: training and testing; the training dataset is 75% and the testing dataset is 25% of the entire dataset. We used one hot encoding [12] to label the data before training and testing. CNN performs well with one hot encoding.

3.2 Data Preprocessing

The raw audio clips are patternized before feeding to the convolution neural network. In the preliminary step, extraction of the audio segment is performed. Every audio file consists of 2048 Hz frequency of the fast furious transform, and sample rate is 44100 Hz. In the second step, audio clip has to be trimmed with the help of 60 dB threshold. Then, normalization approach has been applied onto trimmed data.

Normalization helps to remove the effect of different range value in the numeric column of database. In next step, zero data padding and STFT transformation have been performed. STFT processes frequency information over time interval, and it divides time signal into segment of equal duration. Then, Fourier transformation is applied on every segment of the signal.

Spectrogram is an output of STFT transformation. Visualization of STFT is usually realized using the spectrogram, which is an intensity show case of STFT magnitude over time. In one of the previous work [8], spectrogram plays a major role in the song classification problem. Spectrogram works as an image representation with time in x-axis and frequency in y-axis. Spectrogram represents energy of frequency component, which varies over the time period. Data preprocessing flowchart is represented in Fig. 1.

Table 1 Dataset specification

Property	Value
No. of songs considered	240
No. of audio clip in the dataset	480
Bit_rate	32 kbps
Sample_rate	4.4 kHz

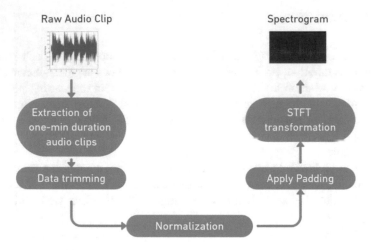

Fig. 1 Preprocessing of dataset

3.3 Model Architecture

CNN is the popular neural network used in the area of image classification. The basic idea behind CNN is to understand the image in a proper way. Another benefit is time reduction in learning as well as training with better accuracy of the model. Earlier CNN architecture has been used for vocalist classification in audio signals. The proposed model consists of five different layers, i.e., convolution layer, max pooling layer, dropout layer, flatten layer, and dense layer. CNN architecture with different layers is shown in Fig. 2.

Two convolution layers are used in the proposed model, and every convolution layer has 16 filter layers with the input shape (257, 79, 1), whereas the output shape of first and second convolution layer has (none, 257, 77, 16) and (none, 125, 36, 16), respectively ('none' express batch size). Two maximum pooling layers are

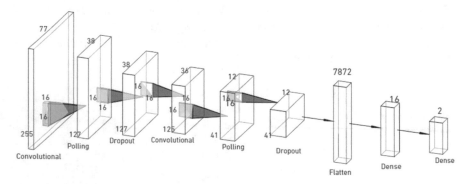

Fig. 2 Architecture of the proposed model

used in the proposed model such that every maximum pooling layer exists in 2 * 2 size, and also there are two dropout layers [13]. Dropout rate of first layer is 40% and 45% of the second layer. This reduces the number of parameters and the complexity by preventing the model from overfitting. One flatten layer is used to serialize all 7872 neurons into a single layer.

The proposed model uses two dense layers: first dense layer of all 7872 neurons that are fully connected to 16 neurons of the second dense layer that uses sigmoid activation function. The second dense layer is connected to 2 neurons using softmax activation function for classification of the songs. Table 2 represents parameters of the CNN architecture.

In this paper, we used binary cross-entropy loss function; for binary classification, it gives probability output. Table 3 shows the various hyperparameters used in the CNN, and the loss function is given in Eq. (2).

$$H(q) = -\frac{1}{N}\sum_{i=1}^{N} y \log(\hat{y}) + (1-y)\log(1-\hat{y}) \qquad (2)$$

where y represents true value and $\hat{y}$ represents predicted value.

4 Results

We trained our model for 45 epochs with 350 batch sizes per epoch for training generator and 150 batch sizes per epoch for validation generator. The training and testing sets are 75% and 25%, respectively, of entire dataset. The achieved accuracy for this experiment is 90% for mood classification. The training time of our model is 9 h 32 min and 31 s. Table 4 describes the training and testing splitting of the datasets.

Table 2 CNN architecture of proposed model

Layers	Species	Output shape	Parameters
1	Convolution 2D layer-1	(None, 255, 77, 16)	160
2	Max pooling 2D layer-1	(None, 127, 38, 16)	0
3	Dropout layer-1	(None, 127, 38, 16)	0
4	Convolution 2D layer-2	(None, 125, 38, 16)	2320
5	Max pooling 2D layer-2	(None, 41, 12, 16)	0
6	Dropout layer-2	(None, 41, 12, 16)	0
7	Flatten layer-1	(None, 7872)	0
8	Dense layer-1	(None, 16)	125,968
9	Dense layer-2	(None, 2)	34

Table 3 Hyperparameters for proposed CNN

Hyperparameter	Value
Loss	'Binary cross-entropy'
Batch size	350
Epoch	45
Padding	'Valid'
Optimization algorithm	Adam
Multiprocessing	'False'

Table 4 Dataset classes description

Classes	Training set	Testing set
Loud	180	60
Relaxed	180	60
Total	**360**	**120**

Fig. 3 Accuracy graph of trained model

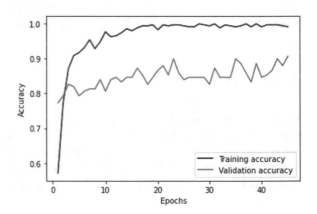

Experiments were conducted on device having Intel(R) Core(TM) i5-5300U CPU @2.30 GHz, 2.30 GHz with 4 logical processor(s), 512 GB RAM. Google-Colab has been used for development environment and the programming language used is Python with the Keras deep learning [14] and TensorFlow libraries [15].

Figures 3 and 4 are shown the accuracy curve and loss curve, respectively, for the proposed model while training. In Fig. 5, confusion matrix is represented for the predicted classes indicating how many outcomes are correct and how many are wrong as identified by the model.

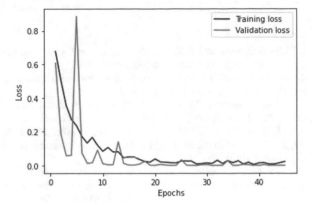

Fig. 4 Loss graph of trained model

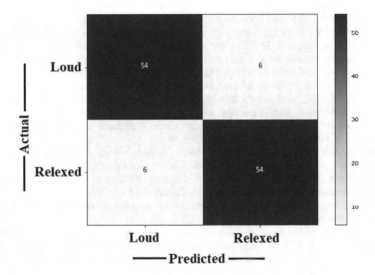

Fig. 5 Confusion matrix

5 Conclusions

The paper focuses on identifying mood in a song using a deep learning model CNN and achieved an accuracy of 90%. The songs are categorized as either loud or relaxed. The machine learning approaches based on acoustic feature and MFCC feature suffer from high-dimensional problem which can be handled with the deep learning approaches. There is a possibility to further categorize music in future for various purposes with enhanced accuracy of the proposed work using appropriate methodology.

References

1. Whitman B, Flake G, Lawrence S (2001) Artist detection in music with minnowmatch. In: IEEE signal processing society workshop. IEEE, pp 559–568
2. Komal K (2016) Mood based hindi songs classification system invariance of MFCC feature extraction technique. Int J Eng Dev Res 4:642–647
3. Anshuman G, Mohd S, Sarfaraz M, Aadam S (2014) Genre classification of songs using neural network. In: 2014 international conference on computer and communication technology (ICCCT), Allahabad, pp 285–289. https://doi.org/10.1109/ICCCT.2014.7001506
4. Tzanetakis G, Cook P (2002) Musical genre classification of audio signals. IEEE Trans Speech Audio Process 10(5):293–298
5. Vyas G, Dutta MK (2014) Automatic mood detection of Indian music using MFCCS and k-means algorithm. In: 2014 seventh international conference on contemporary computing (IC3), Noida, pp 117–122. https://doi.org/10.1109/IC3.2014.6897159
6. You SD, Liu C, Chen W (2018) Comparative study of singing voice detection based on deep neural networks and ensemble learning. Hum Cent Comput Inf Sci 8:34. https://doi.org/10.1186/s13673-018-0158-1
7. Chebbi S, Ben Jebara S (2018) On the use of pitch-based features for fear emotion detection from speech. In: 2018 4th international conference on advanced technologies for signal and image processing (ATSIP), Sousse, pp 1–6. https://doi.org/10.1109/ATSIP.2018.8364512
8. Sarkar R, Choudhury S, Dutta S, Roy A, Saha S (2019) Recognition of emotion in music based on deep convolutional neural network. Multimedia Tools Appl 79:765–783
9. Lokhande PS, Tiple BS (2017) A framework for emotion identification in music: deep learning approach. In: 2017 international conference on intelligent computing and control systems (ICICCS), Madurai, pp 262–266. https://doi.org/10.1109/ICCONS.2017.8250723
10. Dang T-T, Shirai K (2009) Machine learning approaches for mood classification of songs toward music search engine. In: 2009 international conference on knowledge and systems engineering. https://doi.org/10.1109/kse.2009.10
11. Patra BG, Das D, Bandyopadhyay S (2016) Multimodal mood classification framework for hindi songs. Computación y Sistemas 20:515–526. https://doi.org/10.13053/CyS-20-3-2461
12. Sklearn.preprocessing.OneHotEncoder. https://scikit-learn.org/stable/modules/generated/sklearn.preprocessing.OneHotEncoder.html. Accessed 29 Oct 2020
13. Srivastava N et al (2014) Dropout: a simple way to prevent neural networks from overtting. J Mach Learn Res
14. Chollet F et al (2015) Keras. GitHub. Retrieved from https://github.com/fchollet/keras
15. Ertam F, Aydın G (2017) Data classification with deep learning using Tensorflow. In: 2017 international conference on computer science and engineering (UBMK), Antalya, pp 755–758. https://doi.org/10.1109/UBMK.2017.8093521

Consanguinity in Risk Assessment of Retinoblastoma Using Machine Learning

S. Ashwini and R. I. Minu

1 Introduction

Ocular tumors are mass of cells that develop in the eye or the surrounding areas of the eye orbit. These tumors can be benign or malignant in nature. Tumors arising on ocular surface include many forms of stromal, epithelial, and other tumors [1]. Retinoblastoma is an embryonic intraocular tumor arising in the retina of the eye. It is one of the common intraocular malignancies of childhood leading to vision loss and can become fatal with occurrence of 1 in 15,000 live births worldwide [2]. The diagnosis of retinoblastoma is mostly made before 5 years of age in the children [3]. The 40% of retinoblastoma patients have hereditary form, while remaining 60% have non hereditary retinoblastoma [4].

Hereditary retinoblastoma mostly manifests as bilateral and multifocal unilateral forms, whereas non heritable form manifests as unilateral retinoblastoma in children [2]. There are 90% chance for appearance of retinoblastoma phenotype with individuals carrying the disease causing genotype. The familial retinoblastoma is inherited as autosomal dominant disorder where it is possible for retinoblastoma affected child to have unaffected parents in a family due to reduced penetrance [5].

Retinoblastoma occurs in individuals as a result of mutations in retinoblastoma 1 (RB1) gene leading to synthesis of faulty retinoblastoma (pRB) protein in the cells. The genetic complexity of this disease makes it difficult for individuals to understand the disease risk in the family members and significance of genetic testing in such scenarios. Genetic counseling helps people to comprehend the genetic basis of the clinical diagnosis. The aim of any genetic counseling process is to help people decide what to do once the diagnosis is being made, i.e., to determine whether the family requires genetic testing or not. This has to be carried out along with psychosocial and supportive counseling in accordance with the principles of clinical

S. Ashwini (✉) · R. I. Minu
SRM Institute of Science and Technology, Chennai, India
e-mail: as9792@srmist.edu.in

health care ethics which includes autonomy, beneficence, and non-maleficence and justice [6]. It is important to proffer genetic counseling to the persons having a family history of retinoblastoma [2]. Based on analysis of the patient's family history and personal history, genetic direct and indirect studies should be carried out in order to predict the appropriate predisposition risk and disease risk which might perhaps, influence the prognostic outcome of the patient treatment.

The environmental and social factors that tend to influence the retinoblastoma risk in families are parental age and nutrition, consanguinity (more prevalent in developing countries), HPV infections, assisted reproductive technology (IVF) and more. The mother's intake of multivitamins and father's diet with more calcium might have some associations with the offspring's risk of new germline mutations [7]. The maternal diet with low-fried foods and cured meats might reduce the risk of unilateral retinoblastoma in her child [8]. The bilateral retinoblastoma is attributed to maternal sexually transmitted disease during pregnancy which is suspected 7 to be a marker for HPV infection [9]. The parental age especially fathers older age were associated to increase risk of bilateral retinoblastoma in India [9].

Consanguineous marriages, predominantly found and culturally favored in Indian communities, tend to play a significant role in inheritance of many genetic disorders particularly autosomal recessive disorders in which the mutant recessive genes get inherited within the family through asymptomatic individuals [10]. In contrast, the association between consanguinity and its effect on inheritance of mutant dominant allele is yet to be explored precisely with respect to many autosomal dominant disorders. The association of consanguineous marriages with eye diseases has been studied earlier by Nirmalan [11]. It is reported that there is a correlation of consanguinity with bilateral retinoblastoma [12]. Thus, the need to study more about the effect of consanguinity on retinoblastoma is emphasized.

This study is to establish the effect of consanguinity in risk assessment of retinoblastoma by analyzing the family history and pedigree chart of the recorded families of retinoblastoma patients by comprising the family information for at least three generations in the pedigree to minimize the patient's risks for having affected offspring and to offer evidence-based risk assessment by using machine learning techniques for consanguineous couples seeking preconceptional genetic counseling for retinoblastoma.

2 Background

Retinoblastoma, identified in 1500 s was the first tumor in which the genetic basis of the childhood malignancies was revealed [13, 14]. According to American literature, the first case of retinoblastoma was reported in 1818 at the New York Hospital. In 1854, the term fungus haematodes coined by Hey of Leeds, England, was used to refer to retinoblastoma which was then identified and named as glioma of the retina by the popular German pathologist, Virchow [15].

The clinical diagnosis of retinoblastoma is established by the appearance of white reflex seen in children. The clinical manifestation of retinoblastoma varies in individuals based on the genetic mutation causing the disease similar to any other cancer. The differential diagnosis of retinoblastoma includes many conditions and associated syndromes. The common clinical presenting features of retinoblastoma [16] include Leukocoria (56%), Strabismus (20%) [14], Red painful eye (7%), Poor vision (5%).

The other manifestations found in association with retinoblastoma include hyphema orbital cellulitis, heterochromia iridis and unilateral mydriasis [16].

The unilateral form of retinoblastoma usually appears at a later age during diagnosis compared to bilateral form [17]. But this is not true in all cases of retinoblastoma in which it is possible for a child to have diffused infiltrating tumor at a younger age. This is generally due to delay in diagnosis [16]. Also, unilateral and bilateral forms of retinoblastoma have their differences due to impact of social demographic factors. The histopathological findings of tumor are required for staging the tumor which helps to establish useful treatment modalities for the patient for better prognosis [18].

3 Epidemiology of Retinoblastoma

The basic epidemiological information on pediatric malignancies explains the understanding of childhood malignancies and also the efforts to develop new cancer control strategies. Retinoblastoma contributes 4% of all pediatric cancers [14]. Globally, India holds 20% of retinoblastoma cases [13]. The retinoblastoma incidence is approximately 1 in 15,000 live births around the world [2]. In the patients afflicted with unilateral form without a family history have 85% of non heritable retinoblastoma and 15% have retinoblastoma that can be inherited [4]. The retinoblastoma education program has shown to cause decline in the proportion of advanced stages of retinoblastoma significantly. This shows that improvement in early detection of retinoblastoma and cancer care services for patients is necessary for early intervention and better clinical outcomes in the children who are affected with this malignancy in the developing countries.

4 Genetics of Retinoblastoma

The development of retinoblastoma will occur only when the two hits happen in the developing retinal cells. In case of heritable retinoblastoma, the first mutation is inherited due to which the individual is predisposed to retinoblastoma, and the second mutation takes place as an independent event in the developing retina. The hereditary retinoblastoma usually takes bilateral form appearing at an early age. Non heritable retinoblastoma is acquired due to both mutations occurring in the

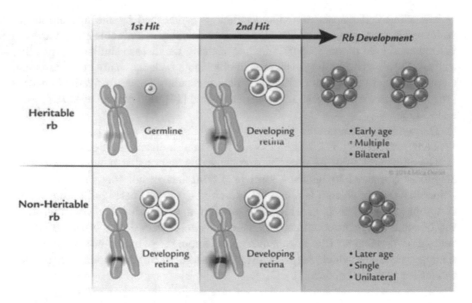

Fig. 1 Two hits in retinoblastoma development (*Source* Grossniklaus [14])

developing retina (See Fig. 1). The non heritable retinoblastoma tumors appear as unilateral form at a later age in children [4].

The two hit mechanism for retinoblastoma development was proposed by Knudson in 1971 based on the graph of the percentage of undiagnosed cases plotted against the age at diagnosis. The linear curve was corresponding to bilateral patients with inherited single copy of the mutated gene, and second mutation was acquired in the developing retinal cells. The exponential curve was corresponding to patients with unilateral tumors and 18 theoretically proposed that both the mutations occur in the developing retina [19]. A similar analysis of the RB patients in Indian population also was in concordance with the two hit hypothesis of Knudson.

Dryja and colleagues cloned and sequenced the RB1 gene in the laboratory which is the first tumor suppressor gene to be cloned in 1984 [14]. The RB1 gene has 27 exons dispersed along the gene of size 180 kb approximately. The RB1 germline mutation spectrum includes 40% of nonsense mutations, 20% of frameshift mutations including insertions and deletions of certain number of bases, 20% of altered splicing mutations and 10% of rare translocations (chromosomal rearrangements), deletion or duplication of exons or the complete gene [5]. The remaining 10% comprises Missense mutations and promoter region hypermethylation ([5]. These mutations are also seen in the tumor tissues. The recent investigations reveal that the MYCN gene amplification may also be a cause for retinoblastoma when there are no RB1 gene mutations identified.

The line graph is done based on the details acquired from the Rustam H. Usmanov and Tero Kivela. The Asia–Pacific countries in which predicted annual incidence of retinoblastoma in 2013 graph shows that India has the maximum

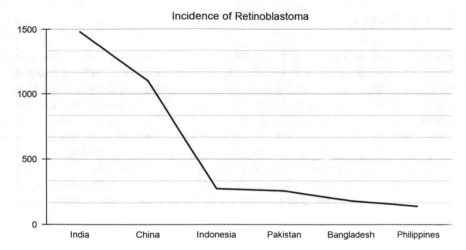

Fig. 2 Countries in which predicted annual incidence of Retinoblastoma in 2013 [20]

number of retinoblastoma affected children in the Asia–Pacific with about 1486 cases. It was reported that every year more than 5000 new cases are diagnosed around the World, of which over 1500 are diagnosed in India (Fig. 2).

5 Methodology

See Fig. 3.

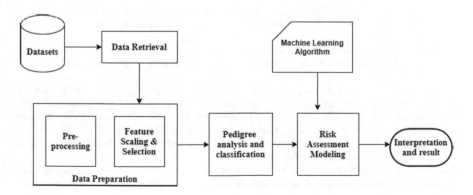

Fig. 3 Overview of the work plan

6 Conclusion

Retinoblastoma is the common childhood ocular malignancy worldwide. The incidence of retinoblastoma is 1 in 20,000 children worldwide. India is found to have 20% of the retinoblastoma patients in the world. Consanguinity is found to be involved in risk assessment and genetic counseling in many genetic eye diseases. Several studies have explored the association of eye diseases with consanguinity [11]. The aim of this research is to study the effect of consanguinity on retinoblastoma patients and their families. To achieve the research aim, a methodology was designed. When completed, it would fill the gaps in the research aim. One of the main contributions of our study is to express the consideration of consanguinity as a factor that influences transmission of retinoblastoma among generations in future.

References

1. Ali MJ, Reddy VAP, Honavar SG, Naik M (2016) Orbital retinoblastoma: where do we go from here? J Cancer Res Therapeutics 7(1) (January–March 2011)
2. Aerts I, Lumbroso-Le Rouic L, Gauthier-Villars M, Brisse H, Doz F, Desjardins L (2006) Retinoblastoma. Orphanet J Rare Dis
3. Sharifzadeh M, Ghassemi F, Amoli FA, Rahmanikhah E, Tabatabaie SZ (2014) Retinoblastoma in adults: a case report and literature review. J Ophthalmic Vis Res 9 (3):388–391
4. Corson TW, Gallie BL (2007) One hit, two hits, three hits, more? Genomic changes in the development of retinoblastoma. Genes Chromosom Cancer 46:617–634
5. Houdayer C, Gauthier-Villars M, Castéra L, Desjardins L, Doz F, Stoppa-Lyonnet D (2012) Retinoblastoma—genetic counseling and molecular diagnosis, retinoblastoma: an update on clinical, genetic counseling, epidemiology and molecular tumor biology. In: Kumaramanickavel G (ed) InTech, ISBN: 978-953-51-0435-3. Available from: http://www.intechopen.com/books/retinoblastoma-an-update-on-clinical-genetic-counseling-epidemiology-andmolecular-tumor-biology/retinoblastoma-genetic-counseling-protocols-and-molecular-diagnostic-protocols
6. Hodgson J, Spriggs M (2005) A practical account of autonomy: why genetic counseling is especially well suited to the facilitation of informed autonomous decision making. J Genet Counsel 14(2)
7. Bunin GR, Li Y, Ganguly A, Meadowsa AT, Tseng M (2013) Parental nutrient intake and risk of retinoblastoma resulting from new germline RB1 mutation. Cancer Causes Control 24 (2):343–355
8. Lombardi C, Ganguly A, Bunin GR, Azary S, Alfonso V, Ritz B, Heck JE (2015) Maternal diet during pregnancy and unilateral retinoblastoma. Cancer Causes Control 26(3):387–397
9. Heck JE, Lombardi CA, Meyers TJ, Cockburn M, Wilhelm M, Ritz B (2012) Perinatal characteristics and retinoblastoma. Cancer Causes Control 23(9):1567–1575
10. Hamamy H (2012) Consanguineous marriages-Preconception consultation in primary health care settings. J Commun Genet 3:185–192
11. Nirmalan PK (2006) Consanguinity and eye diseases with a potential genetic etiology. Data from a prevalence study in Andhra Pradesh, India. Ophthalmic Epidemiol 13:7–13

12. Subha L, Reddy ASR (2015) A clinical study of retinoblastoma. J Pharm Bioallied Sci 7 (Suppl 1):S2–S3
13. Thériault BL, Dimaras H, Gallie BL, Corson TW (2013) The genomic landscape of retinoblastoma: a review. Clin Exp Ophthalmol. https://doi.org/10.1111/ceo.12132
14. Grossniklaus HE (2014) Retinoblastoma. Fifty Years of Progress. The LXXI Edward Jackson Memorial Lecture. Am J Ophthalmol 158(5):875–891
15. Wardrop J (2013) Observations on the fungus haematodes or soft cancer. George Ramsay and Co, Edinburgh, p 1809
16. Pandey AN (2014) Retinoblastoma: an overview. Saudi J Ophthalmol 28:310–315
17. Othman IS (2012) Retinoblastoma major review with updates on Middle East management protocols. Saudi J Ophthalmol 26:163–175
18. Mendoza et al (2015) Histopathologic grading of anaplasia in retinoblastoma. Am J Ophthalmol 159(4):764–776
19. Knudson Jr AG (1971) Mutation and cancer: statistical study of retinoblastoma. Proc Nat Acad Sci USA 68(4):820–823
20. Usmanov RH, Kivelä T (2014) Authors' reply for "predicted trends in the incidence of retinoblastoma in the Asia-pacific region". Asia-Pacific J Ophthalmol 3(5):326–327

Detecting Human Emotions Through Physiological Signals Using Machine Learning

R. Balamurali, Priyansh Brannen Lall, Krati Taneja,
and Gautam Krishna

1 Introduction

Finding the relationship between human emotions and actions has been an active research topic for philosophers, psychologists, and neurobiologists for centuries. Emotions contribute significantly to the processes of action generation as well as action execution and control. Emotions are an integral part of the process of reasoning and decision making. Emotions are displayed by visual, vocal, and other physiological means [1]. Different types of emotions affect an individual's decision-making and rational thinking process in various ways. For example, an angry person is more certain to adopt destructive methods for the successive events; a sad person's perspective is more likely to be biased by negativity. Similarly, while making decisions, individuals are often influenced by their affective mental states. Autistic children struggle greatly to express stressful emotions. Consequently, they might experience a breakdown after a long time of continuous ignorance. This can result in aggression toward others and even self-injury. So to make it known they sometimes end up having aggression outwardly and even to the point of self-injury. Teachers and other caregivers can have a hard time anticipating and preventing meltdowns. Detecting their emotions will help the guardians to understand and help them. It is also

R. Balamurali
Department of CSE, IcfaiTech (Faculty of Science and Technology),
The ICFAI Foundation for Higher Education, Hyderabad 501203, India
e-mail: balamurali@ifheindia.org

P. B. Lall (✉)
Bhilai Institute of Technology, Durg 491001, India
e-mail: priyansh.blessed@gmail.com

K. Taneja
ABES College of Engineering, Ghaziabad 201009, India

G. Krishna
Vidhya Academy of Science and Technology, Kerala 680501, India

© The Author(s), under exclusive license to Springer Nature Singapore Pte Ltd. 2022   587
R. R. Raje et al. (eds.), *Artificial Intelligence and Technologies*,
Lecture Notes in Electrical Engineering 806,
https://doi.org/10.1007/978-981-16-6448-9_57

important to monitor emotions as they are perceptions of bodily changes and can help in identifying matters of concern at an early stage before they become more serious. This task toward emotion detection is dedicated to such consumers who can benefit from early detection of their affected states. Detection of emotions can be done in various ways for example using text, speech, facial expressions, body movements, gestures, and physiological signals. One can fake emotions using text, speech, or facial expressions but not by their pulse rate since emotions are beyond voluntary control and are associated with patterns of bodily changes.

There are six basic emotions—happy, sad, fear, surprise, anger, and disgust that are widely accepted and used till date. These emotions were later expanded and included amusement, contempt, contentment, embarrassment, excitement, guilt, and pride in achievement, relief, satisfaction, sensory pleasure, and shame. This research work is about collecting physiological signals—body temperature, heart rate, skin resistance, and pulse wave corresponding to four types of emotions— happy, normal (relaxed), angry, and sad from 22 individuals of age group 20–22 years and to classify it using various machine learning and deep learning techniques. People of this age group tend to relax at a high pulse rate than the individuals of the age group 60–70 and show distinguishable changes in the pattern of the pulse during different emotions [2].

2 Literature Review

Here, we provide a brief overview of the previous work done in the area of emotion detection using various techniques. Emotion detection is a task of recognizing a consumer's mental state. Emotion plays a major role in decision-making and rational thinking process of an individual. The communication between computers and human beings would have been more effective if computers were able to perceive and respond to non-verbal communications. Emotions and facial expressions are the most natural ways to communicate our feelings. Automatic emotion detection using facial expressions recognition is now an active area of research in various fields such as computer science, health care, and psychology.

Cohen et al. [1] worked on involuntary facial expression recognition from live video input using temporal cues. A new architecture of hidden Markov models (HMMs) has been proposed for automatically segmenting and recognizing human facial expression from videos. The originality of this architecture is that both segmentation and recognition are done automatically using a multilevel hidden Markov models (HMM) architecture while increasing the differentiating power between the different classes. Quazi [2] developed an emotion recognition system based on physiological signals. The physiological signals considered are skin temperature, heart rate, and skin conductance. Data is collected from 16 individuals (male and female) in the age group of 18–72. K-means clustering has been used to cluster the data into four groups (emotions). The system gives an accuracy of 86.25%. Deep learning-based facial emotion recognition (FER) approaches using deep networks enabling "end-to-end" learning—a review of publicly available evaluation metrics—which are a standard for a quantitative comparison of FER researches are discussed by Chul Ko [3]. Speech and text also convey emotions and are used by researchers to detect emotions. Busso et al. [4] suggested two approaches to fuse the two modalities, i.e., emotion detection based on facial expression and emotion detection based on the speech to increase the accuracy of the system. The approaches use decision-level and feature-level integration for the fusion. Emotions do not arise through free will and are often accompanied by physiological signals. Li and Chen [5] use four physiological signals, i.e., electrocardiogram, skin temperature, skin conductance, and respiration to extract different feature for emotion detection. Canonical correlation analysis has been used as pattern classifier and achieved an accuracy of 85.3%. Gouizi et al. [6] use electromyogram signal (EMG), blood volume pulse (BVP), respiratory volume (RV), skin temperature (SKT), skin conductance (SKC), and heart rate (HR) and achieved the accuracy of 85% for different emotional states. Wioleta [7] presented a brief review of the progress of research in emotion recognition using physiological signals. Future challenges and enhancements are also described. Ragot et al. [8] compared the accuracy using laboratory and wearable sensors, which displayed similar accuracy. This shows recognition of emotion is dependable even outside a laboratory. A brief review of emotion detection methodology and advancements along with future challenges for better emotion detection has been described by Kołakowska et al. [9]. Kim and André [10] use musical induction method that directly leads the individual to the emotional states without any laboratory setting. Four-channel biosensors have been used to measure respiration changes, skin conductivity, electrocardiogram, and electromyogram. The best features for analysis were extracted and classified. The recognition accuracy increased to 95 and 70% for subject-dependent and subject-independent classification, respectively, when emotion-specific multilevel dichotomous classification (EMDC) scheme is used. Soleymani et al. [11] use non-verbal cues to detect emotions. Jenke et al. [12] estimate emotions from physiological signals (EEG) using ridge regression and compared the results using different evaluation measures. The evaluation measure —bandwidth accuracy—is identified as more appropriate. Khalili and Moradi [13] proposed a multimodal fusion between brain and peripheral signals for emotion detection. For further references, emotion detection using physiological signals is also discussed in [14–23]. Priya Muthusamy [25] proposed specific patterns of biological responses that can be used to predict emotions using biomedical measurements. He used photoplethysmography (PPG), electrodermal activity (EDA), and skin temperature (SKT) for feature extraction and used it in common wearables such as wearable sensing glove and Android smartphone. Shu et al. [26] discussed about various kinds of biosensors used to detect the physiological signals. She used classification techniques such as support vector machine to extract 22 features with 76% of correctly classified test cases. Our paper, on the other hand, describes the model which extracts four features from biosensors and registers an accuracy of 82.55% using the random forest tree classification.

3 System Design

Various physiological signals for the basic four types of emotions, i.e., happy, normal, sad, and angry are collected using the four types of biological sensors namely the GSR sensor, skin temperature sensor, pulse wave sensor, and pulse rate sensor (see Fig. 1). Emotions play a central role in our decision making, rational thinking, and behavior. People's actions are most likely to affect when they are happy, sad, or angry. A happy person is more likely to select happy events going around him; similarly, a person with sad and angry emotions is more likely to select events that will make him more sad or aggressive. Being in the same emotion more frequently can cause serious effects to our mental as well as physical health. Therefore, the prediction of emotion is important for not only human–human interaction but also human–machine interaction. Now this prediction of emotion can be done by various methods like emotion detection using speech, emotion detection using facial expressions, emotion detection using text, emotion detection using body movements, and emotion detection using physiological signals. We used emotion detection using physiological signals because one can fake emotions in text, speech, or facial expressions by voluntary action but the reading of their body temperature, pulse, or heart rate will correctly identify the actual emotion. Working of different sensors is as follows.

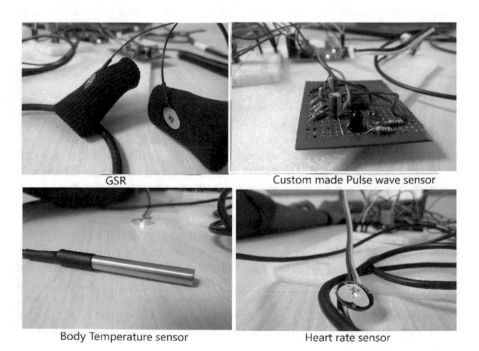

Fig. 1 Various sensors used in this work

3.1 Galvanic Skin Response (GSR) Sensor

The emotional arousal of an individual can be measured through the GSR sensor which measures the sweat gland activity. To measure skin resistance, the electrical properties of skin are taken into account. Specifically, how the skin resistance varies with sweat gland activity, i.e., the greater the sweat gland activity, the more perspiration, and therefore, less skin resistance. Most commonly, GSR measures conductance and not resistance. Conductance is the inverse of resistance and is measured in Siemens (Conductance = 1/Resistance).

3.2 Skin Temperature Sensor

Skin temperature sensors use the skin as an indicator to measure the continuous body temperature. They have a lead length of 2 m and are designed to operate over the range 0–50 °C. Skin temperature falls when stress or other sensations occur because blood flow decreases due to factors like blood vessel constriction. This is most noticeable at extremities such as fingertips and nose.

3.3 Heart Rate Sensor

The front of the pulse rate sensor is the side with the heart logo. This is the side that makes with the skin with a small round hole LED. This LED shines light into the fingertip or other capillary tissues, and this sensor reads the light that bounces back. The back of the pulse rate sensor is the part where the rest of the parts of the sensors are mounted.

3.4 Pulse Wave Sensor

We have made a custom-designed pulse wave sensor (see Fig. 2). The output of the pulse wave sensor is filtered, and the corresponding reading is also taken as an attribute.

All the data collected from these signals are fed into a signal conditioning circuit, i.e., the microcontroller. These signals are amplified and processed by the microcontroller. This whole circuit is connected to the computer via a USB cable. After uploading the program to the Arduino, the readings from the sensors were collected in an Excel sheet using PLX-DAQ. PLX-DAQ is a Parallax microcontroller data acquisition add-on tool for Microsoft Excel. Any number of sensors can be connected to the microcontroller, and the serial port of a computer can now send data

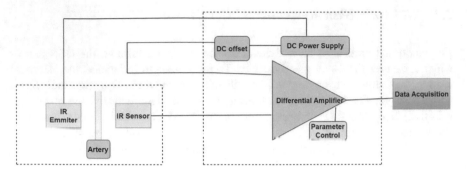

Fig. 2 Schematic of pulse wave sensor

directly into Excel. We collected the data in Microsoft Excel for four attributes, namely heart rate, body temperature, skin resistance, and pulse. The rate of blood flow in our body is different for each emotion, which makes it the deciding factor for the data collection.

3.5 Data Collection

Several researchers in this field have used six categories of emotions—happiness, sadness, fear, disgust, surprise, and anger for the categorization of emotion displayed by individuals. To make it well distinguishable, we chose four out of these six emotions—happy, sad, anger, and normal as the basis of detection (see Fig. 3), after which we classified the data in Weka 3.8 using different algorithms to attain maximum accuracy.

It does sound easy to collect just four emotions of 22 different individuals but it is the trickiest and the most critical part of the work. We limited data collection to 22 individuals consisting of an equal number of males and females, i.e., 11. We collected the data from subjects having a similar age group 20–22 years. Collecting data from the subjects, we were already acquainted with proved to be a lot beneficial. Unique individuals have diverse perspectives and further varying standpoints, so keeping a flexible approach as opposed to a generalized way of deciding methods is advantageous. But we also collected some videos and songs ahead of time according to each emotion for the display, such as funny children and animals' videos for happy; emotional short stories for sad; violent scenes, versus battle gaming and arm wrestling for anger and relaxing music for normal.

We had a separate room for the collection of data to avoid any kind of change or disturbance in room temperature affecting our information from physiological sensors. Before collecting the data, we made them aware of the hardware and their contribution to this research work. All the four emotions were calculated from the same individual; therefore, it is necessary to decide the order of the emotions to be

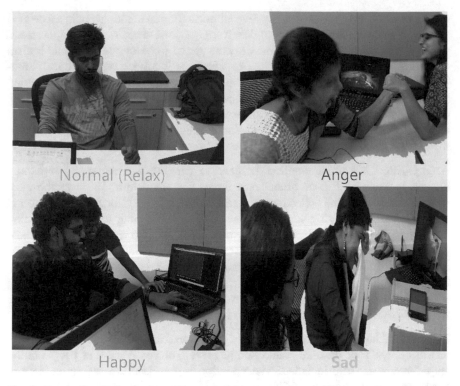

Fig. 3 Data collected for the four different emotions

calculated. We chose to collect data for normal emotion at first since it would be easy to inculcate subsequent emotion after normal. Moreover, we wanted the individual to relax in the beginning and be familiar with the process. The next one after normal is the sad emotion followed by angry and then happy emotions.

Normal (Relaxed). Relaxation is a reduction in physical or mental activity. That is when an individual maintains a state of peace with himself, showing no emotions and unaffected by emotions, calmly thinking peacefully. A light and peaceful conversation helped to lower down the anxiety level and make it a hospitable environment. Questions like, "How was your day?", "Where are you from?" were asked to start a friendly conversation and lighten their mood. After which, depending on the individual, soothing music or video was played to obtain their data. For 2–3 min, test readings were taken to check the proper functioning and positioning of the sensors. It also helped to develop the individual's familiarity with the setup before taking the actual readings. After playing each video for emotion induction, the subject was monitored if the individual really felt the emotion and accurate information is obtained. All four sensors were connected to a person's right hand and remained intact for 5 min. In 5 min, approximately 250 values were recorded for each emotion.

Sad. Sadness is the best-known negative emotion. It is an emotional pain associated with or characterized by, feelings of disadvantage, loss, despair, grief, helplessness, disappointment, and sorrow. Sadness is considered to be one of the basic human emotions, and it is a natural response to situations involving psychological, emotional, or physical pain. The sad emotion is induced in individuals by displaying various videos. Emotional videos of a father–son relationship, touching scenes from different movies like Pursuit of Happiness, glimpses of Syrian children's suffering, and some sad songs were also played. Some individuals even played the songs on their cell phones which they thought would bring back sad memories.

Angry. Anger is a strong emotion designed to send the clear message "something has got to change." It is an urgent plea for justice and action. This was the most troublesome part of the collection of data because it is easy for individuals to get relaxed, sad, or happy by watching some videos or listening to music but being actually angry for 5 min is a demanding task. For girls, arm wrestling and a debate on a topic close to their heart were initiated. For boys, gaming, fight scene videos, and arm wrestling worked. Playing computer games is fun, at first boys were enjoying playing with one hand but later after 2–3 failures, they got angry. Losing with one hand made them furious, and we collected that data for 5 min. Wrestling with girls even pushed them to try even harder and increased the pulse rate.

Happy. Making the subjects/individuals happy was the easiest part. Some funny videos on YouTube of some pranks and gags were played. Videos of people doing stupid stuff were also included. The variation in the heart rate of the subject in happy emotion is eminently visible in comparison to his relaxed emotion. There is a gap of at least 5 min between initiations of each successive emotion. For each emotion, the heart rate, body temperature, GSR, and pulse wave signals are collected for 5 min that roughly provided 250 datasets. The collected data was labeled accordingly as happy/normal/sad/anger. Likewise, data is collected from 22 individuals. Finally, we collected datasets of 5299 rows corresponding to happy emotion, 5464 rows corresponding to normal emotion, 5012 rows corresponding to sad emotion, and 4911 rows corresponding to angry emotion. The overall dataset is 20,686 (see Table 1). A sample view of the dataset is shown (see Fig. 4).

4   Results and Discussion

Simple K-means clustering has been applied in Weka for combined as well as individual dataset to compare and interpret the results. For combined dataset, body temperature is high for anger, mid to high for happy, mid to low for normal, and

Table 1 Data collection—overview

Happy	Normal	Sad	Angry
5299	5464	5012	4911

Fig. 4 Sample dataset view

No.	1: Heart rate Numeric	2: Temperature Numeric	3: GSR Numeric	4: Pulse Numeric	5: Emotion Nominal
1	60.0	34.75	481.0	0.03	happy
2	60.0	34.75	364.0	0.06	happy
3	60.0	34.75	202.0	0.02	happy
4	60.0	34.75	192.0	0.05	happy
5	47.0	34.75	516.0	0.06	happy
6	49.0	34.75	481.0	-0.03	happy
7	61.0	34.75	466.0	0.03	happy
8	98.0	34.75	271.0	0.05	happy
9	146.0	34.75	206.0	-0.05	happy
10	133.0	34.75	344.0	-0.02	happy
11	120.0	34.75	466.0	-0.02	happy
12	122.0	34.75	499.0	-0.06	happy
13	99.0	34.75	459.0	0.0	happy
14	99.0	34.69	406.0	-0.01	happy
15	99.0	34.75	435.0	-0.04	happy
16	185.0	34.75	508.0	0.01	happy
17	175.0	34.75	398.0	0.03	happy
18	176.0	34.75	419.0	-0.02	happy
19	189.0	34.69	485.0	0.0	happy
20	179.0	34.69	410.0	-0.02	happy
21	165.0	34.75	427.0	-0.06	happy
22	149.0	34.75	511.0	0.01	happy
23	131.0	34.75	458.0	0.0	happy
24	114.0	34.75	348.0	-0.04	happy
25	114.0	34.75	684.0	0.0	happy
26	90.0	34.75	422.0	0.04	happy
27	98.0	34.75	451.0	-0.06	happy
28	85.0	34.75	500.0	-0.07	happy
29	78.0	34.69	320.0	0.03	happy

high for sad. Skin resistance is found to be low for anger, mid to high for happy, low for normal, and high for sad. Heart rate is low for anger, mid to high for happy, low to high for normal, and low for sad (see Fig. 5).

For a random individual dataset, the body temperature is very distinguishable, being low for happy and anger, high for normal, and mid for sad emotion. The skin resistance is from low to mid for sad, normal and anger, and higher for happy. The heart rate is low for sad, from mid to high for happy, from low to mid for normal, and from low to high for anger (see Fig. 6).

The collected dataset has been analyzed with various machine learning algorithms using Weka for data classification based on the four emotions, i.e., normal (relaxed), happy, sad, and anger. The datasets were analyzed with 12-fold cross-validation using four classification algorithms, i.e., J48, LMT, decision table, and random forest tree. The overall classification accuracy for the combined dataset for J48 is 78.74% (see Fig. 7), and the accuracy produced by LMT is 79.32% (see Fig. 8). The accuracy achieved through decision table is 71.30% (see Fig. 9), and the accuracy achieved through random forest tree is 82.55% (see Fig. 10).

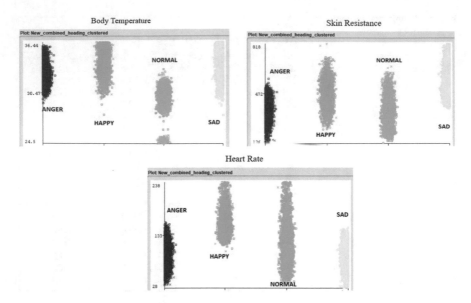

Fig. 5 Simple K-means clustering for combined data

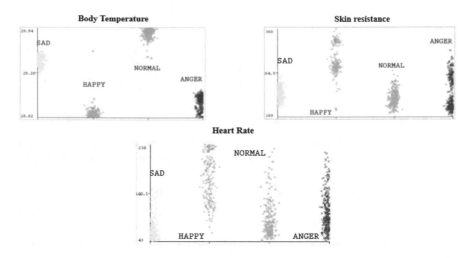

Fig. 6 Simple K-means clustering for individual dataset

For individual dataset, we got an accuracy of 99.9% using random forest tree algorithm. The random forest tree algorithm gives a better result and proved to be the most appropriate algorithm for emotion detection using physiological signals (see Figs. 11 and 12).

Table 2 shows the overall accuracy and accuracy of individual emotions with respect to various algorithms. We have also made a fully connected neural network

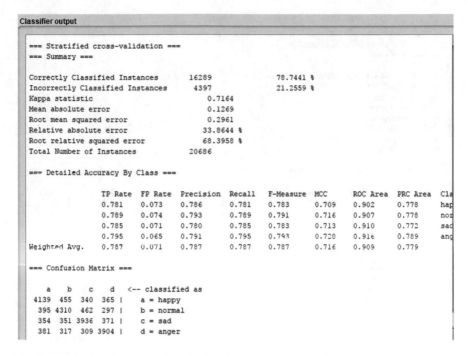

Classifier output

```
=== Stratified cross-validation ===
=== Summary ===

Correctly Classified Instances      16289            78.7441 %
Incorrectly Classified Instances     4397            21.2559 %
Kappa statistic                        0.7164
Mean absolute error                    0.1269
Root mean squared error                0.2961
Relative absolute error               33.8644 %
Root relative squared error           68.3958 %
Total Number of Instances          20686

=== Detailed Accuracy By Class ===

               TP Rate  FP Rate  Precision  Recall  F-Measure  MCC    ROC Area  PRC Area  Cla
               0.781    0.073    0.786      0.781   0.783      0.709  0.902     0.778     hap
               0.789    0.074    0.793      0.789   0.791      0.716  0.907     0.778     nor
               0.785    0.071    0.780      0.785   0.783      0.713  0.910     0.772     sad
               0.795    0.065    0.791      0.795   0.793      0.720  0.916     0.789     ang
Weighted Avg.  0.787    0.071    0.787      0.787   0.787      0.716  0.909     0.779

=== Confusion Matrix ===

   a    b    c    d   <-- classified as
 4139  455  340  365 |   a = happy
  395 4310  462  297 |   b = normal
  354  351 3936  371 |   c = sad
  381  317  309 3904 |   d = anger
```

Fig. 7 J48 classification output from Weka

Classifier output

```
=== Stratified cross-validation ===
=== Summary ===

Correctly Classified Instances      16417            79.3629 %
Incorrectly Classified Instances     4269            20.6371 %
Kappa statistic                        0.7247
Mean absolute error                    0.134
Root mean squared error                0.2815
Relative absolute error               35.7448 %
Root relative squared error           65.0234 %
Total Number of Instances          20686

=== Detailed Accuracy By Class ===

               TP Rate  FP Rate  Precision  Recall  F-Measure  MCC    ROC Area  PRC Area  Cla
               0.788    0.073    0.788      0.788   0.788      0.715  0.934     0.834     hap
               0.796    0.070    0.802      0.796   0.799      0.727  0.938     0.855     nor
               0.796    0.072    0.780      0.796   0.788      0.719  0.938     0.831     sad
               0.795    0.060    0.805      0.795   0.800      0.738  0.944     0.854     ang
Weighted Avg.  0.794    0.069    0.794      0.794   0.794      0.725  0.938     0.844

=== Confusion Matrix ===

   a    b    c    d   <-- classified as
 4177  446  335  341 |   a = happy
  383 4347  471  263 |   b = normal
  359  320 3990  343 |   c = sad
  385  304  319 3903 |   d = anger
```

Fig. 8 LMT classification output from Weka

```
Classifier output

=== Stratified cross-validation ===
=== Summary ===

Correctly Classified Instances      14750                71.3043 %
Incorrectly Classified Instances     5936                28.6957 %
Kappa statistic                         0.617
Mean absolute error                     0.2164
Root mean squared error                 0.3141
Relative absolute error                57.7488 %
Root relative squared error            72.5617 %
Total Number of Instances           20686

=== Detailed Accuracy By Class ===

                 TP Rate  FP Rate  Precision  Recall  F-Measure  MCC     ROC Area  PRC Area  Cl
                 0.753    0.127    0.672      0.753   0.710      0.604   0.913     0.804     ha
                 0.701    0.098    0.719      0.701   0.710      0.608   0.906     0.808     no
                 0.684    0.077    0.740      0.684   0.711      0.624   0.914     0.789     sa
                 0.713    0.082    0.731      0.713   0.722      0.637   0.921     0.806     an
Weighted Avg.    0.713    0.096    0.715      0.713   0.713      0.618   0.913     0.802

=== Confusion Matrix ===

    a    b    c    d   <-- classified as
 3988  602  301  408 |   a = happy
  672 3832  575  385 |   b = normal
  650  438 3428  496 |   c = sad
  627  454  328 3502 |   d = anger
```

Fig. 9 Decision table classification output from Weka

with 10 hidden layers using TensorFlow. Adam optimizer has been used for optimization, and softmax function is used as the last layer (classifier). For the combined dataset, we achieved an accuracy of 33.8%, and for individual dataset, we got an accuracy of 98.75% (The average accuracy of three individual datasets).

5 Conclusion and Future Scope

In this work, we have analyzed the data for four emotion categories, i.e., happy, normal, sad, and anger. This research work was aimed to design a prototype for consumer electronics using non-pervasive wearable sensors for emotion detection. After applying four machine learning algorithms in Weka and by a thorough comparative study, the random forest tree algorithm in Weka provided the highest accuracy of 82.55% for the combined dataset and for the individual dataset, the highest accuracy attained is 99.9%.

Physiological signals were found to be the best approach for emotion detection after an extensive literature review, as they cannot be voluntarily manipulated. The data collection process proved to be unique for each individual for the inducement of a particular emotion. The heart rate sensor is found to give at least 50 fluctuating readings before it could stabilize. The data is not very consistent for each individual

Classifier output

```
=== Stratified cross-validation ===
=== Summary ===

Correctly Classified Instances        17077              82.5534 %
Incorrectly Classified Instances      3609               17.4466 %
Kappa statistic                       0.7672
Mean absolute error                   0.1293
Root mean squared error               0.2493
Relative absolute error               34.4936 %
Root relative squared error           57.5922 %
Total Number of Instances             20686

=== Detailed Accuracy By Class ===

                TP Rate  FP Rate  Precision  Recall  F-Measure  MCC    ROC Area  PRC Area  Cla
                0.822    0.063    0.817      0.822   0.819      0.757  0.960     0.902     hap
                0.834    0.061    0.831      0.834   0.832      0.772  0.965     0.918     nor
                0.824    0.057    0.822      0.824   0.823      0.766  0.965     0.905     sac
                0.822    0.051    0.833      0.822   0.828      0.774  0.966     0.911     anc
Weighted Avg.   0.826    0.058    0.826      0.826   0.826      0.767  0.964     0.909

=== Confusion Matrix ===

    a    b    c    d    <-- classified as
  4354  372  271  302 |   a = happy
   322 4555  364  223 |   b = normal
   300  297 4130  285 |   c = sad
   353  260  260 4038 |   d = anger
```

Fig. 10 Random forest classification output from Weka

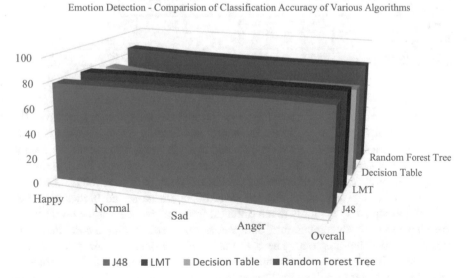

Fig. 11 Comparison of various data classification techniques (3D area)

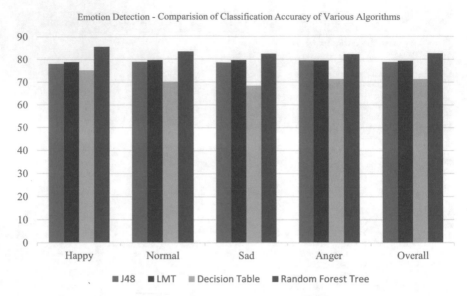

Fig. 12 Comparison of various data classification techniques (Columns)

Table 2 Algorithms and classification accuracy

Algorithm	Classification accuracy (%)				
	Happy	Normal	Sad	Anger	Overall
J48	78.10	78.87	78.53	79.49	78.74
LMT	78.82	79.55	79.60	79.47	79.32
Decision table	75.25	70.13	68.39	71.30	71.30
Random forest tree	85.55	83.33	82.40	82.22	82.55

for all four emotions. We observed high variance at the start of every reading. The heart rate is distinguishably higher for happy and normal emotions than it is for anger and sadness. The skin resistance for sad is the highest, mid-ranged for happy, and lower for anger and normal. The body temperature is lower for normal and ambiguous for the other three emotions. Better inducement of emotions and a larger dataset may have given results that are more distinguishable and higher accuracy.

Each individual's reading is much different from one another, some sweat more, some have lower body temperature, and people were experiencing a combination of emotions rather than just one. The data collected is limited to four emotions and cannot identify any other emotion that the model is not trained with. The amount of data collected is very small and is limited to the age group of 20–22 yrs. This data is insufficient to predict the data of people outside this age group. An increase in the amount of data collected will lead to better accuracy.

For future improvement and development of the system, the use of other physiological sensors such as blood pressure sensor, respiratory rate sensor, EEG sensor, and other sensors is required. The addition of these sensors together with the existing pulse rate sensor, body temperature sensor, and skin conductance sensors will provide more information about a consumer's physiological parameters, which in turn can improve the accuracy of the emotion recognition system. Also, with the added sensors, the system will be able to recognize other emotions such as stress, fear, excitement, panic, and so on.

The size of the system can be miniaturized with industrial collaboration, preferably as a wearable gadget, and introduced in the market for consumer electronics. The system can be incorporated into a smartwatch and other handheld devices such as a computer mouse and can then be used to identify emotions in real time. This could help to identify the emotional state of a consumer and can prevent him/her from performing a task that might be negatively affected by his/her emotional state. Tasks such as driving and performing an important surgery need the consumer to be emotionally stable in order to complete the task without any mishaps. The system can also be used to keep track of elderly consumers and monitor their health. As mentioned in the Introduction section, the emotions of autistic consumers can be identified using this wearable system.

References

1. Cohen I, Sebe N, Garg A, Chen LS, Huang TS (2003) Facial expression recognition from video sequences: Temporal and static modelling. Comput Vis Image Underst 91(1–2): 160–187
2. Quazi MT (2012) Human emotion recognition using smart sensors. M.S. Thesis, Department of Electronics and Communication Engineering, Massey University, PM, NZ
3. Chul Ko B (2018) A brief review of facial emotion recognition based on visual information. Sensors 8(2)
4. Busso C, Lee S, Narayanan S (2009) Analysis of emotionally salient aspects of fundamental frequency for emotion detection. IEEE Trans Audio Speech Lang Process 17(4):582–596
5. Li L, Chen JH (2006) Emotion recognition using physiological signals from multiple subjects. Presented at International Conference on Intelligent Information Hiding and Multimedia, Pasadena, CA, USA
6. Gouizi K, Bereksi Reguig F, Maaoui C (2011) Emotion recognition from physiological signals. J Med Eng Technol 35(6–7):300–307
7. Wioleta S (2013) Using physiological signals for emotion recognition. Presented at 6th International Conference on Human System Interactions (HSI), Sopot, Poland
8. Ragot M, Martin N, Em S, Pallamin N, Diverrez J-M (2017) Emotion recognition using physiological signals: laboratory vs. wearable sensors. Presented at AHFE 2017 International Conference on Advances in Human Factors and Wearable Technologies, Los Angeles, California, USA, 2017, pp 15–22
9. Kołakowska A, Landowska A, Szwoch M, Szwoch W, Wróbel M (2013) Emotion recognition and its application in software engineering. In: 6th international conference on human system interactions, HIS, Sopot, Poland, pp 532–539
10. Kim J, André E (2008) Emotion recognition based on physiological changes in listening music. IEEE Trans Pattern Anal Mach Intell 30(12):1–17

11. Soleymani M, Koelstra S, Patras I, Pun T (2011) Continuous emotion detection in response to music videos. In: IEEE conference on face and gesture, pp 803–80

12. Jenke R, Peer A, Buss M (2013) A comparison of evaluation measures for emotion recognition in dimensional space. In: Humaine association conference on affective computing and intelligent interaction, Geneva, Switzerland

13. Khalili Z, Moradi MH (2009) Emotion recognition system using brain and peripheral signals: Using correlation dimension to improve the results of EEG. In: International joint conference on neural networks, Atlanta, GA, USA, pp1571–1575

14. Murugappan M, Rizon M, Nagarajan R (2008) Lifting scheme for human emotion recognition using EEG. In: International symposium on information technology, Kuala Lumpur, Malaysia

15. Leon E, Clarke G, Sepulveda F, Callaghan V (2006) Real-time physiological emotion detection mechanisms: effects of exercise and affect intensity. In: IEEE engineering in medicine and biology 27th annual conference, Shanghai, China

16. Jang EH, Park BJ, Kim SH, Eum Y, Sohn JH (2011) Identification of the optimal emotion recognition algorithm using physiological signals. In: 2nd international conference on engineering and industries (ICEI), Jeju, South Korea

17. Murali S, Rincon F, Atienza D (2015) A wearable device for physical and emotional health monitoring. In: Computing in Cardiology Conference (CinC), Nice, France

18. Wijeratne U, Perera U (2013) Intelligent emotion recognition system using electroencephalography and active shape models.In: IEEE-EMBS conference on biomedical engineering and sciences, Langkawi, Malaysia

19. Cheng K-S, Chen Y-S, Wang T (2012) Physiological parameters assessment for emotion recognition. In: IEEE-EMBS conference on biomedical engineering and sciences, Langkawi, Malaysia

20. Zheng BS, Murugappan M, Yaacob S (2012) Human emotional stress assessment through Heart Rate Detection in a customized protocol experiment. In: IEEE symposium on industrial electronics and applications, Bandung, Indonesia

21. Safta I, Grigore O, Caruntu C (2011) Emotion detection using psycho-physiological signal processing. In: 7th international symposium on advanced topics in electrical engineering (ATEE), Bucharest, Romania

22. Maaoui C, Pruski A (2018) Unsupervised stress detection from remote physiological signal. In: IEEE international conference on industrial technology (ICIT), Lyon, France

23. Cheng Z, Shu L, Xie J, Philip Chen CL (2017) A novel ECG-based real-time detection method of negative emotions in wearable applications, Shenzhen, China

24. Eibe F, Hall MA, Witten IH (2016) The WEKA Workbench. Online appendix for data mining: practical machine learning tools and techniques, Morgan Kaufmann, Fourth Edition

25. Priya Muthusamy R (2018) Emotion recognition from physiological signals using Bio-sensors. https://diuf.unifr.ch/main/diva/sites/diuf.unifr.ch.main.diva/files/T4.pdf. last accessed 2018/07/23

26. Shu L, Xie J, Yang M, Li Z, Li Z, Liao D, Xu X, Yang X (2018) A review of emotion recognition using physiological signals. Sensors 18(7):2074s

Retinal Vessel Segmentation and Disc Detection from Color Fundus Images Using Inception Module and Residual Connection

Mithun Kumar Kar, Malaya Kumar Nath, and Madhusudhan Mishra

1 Introduction

The recent technological advances in image processing and computer vision field developed opportunities to the biomedical engineers and scientists to meet the requirements of clinical practice. Retinal fundus images provide significant diagnostic and quantifiable information about eye-related diseases. Mainly the disc region of the fundus image contains important diagnostic information, whereas the blood vessels analysis provides information about eye pressure-related disease. The measurement of optic disc and retinal nerve fiber layer (RNFL) is the substantial source for glaucoma diagnosis, which is a chronic and irreversible eye disease [1–3].

The vessel segmentation algorithms can be divided into two categories (traditional instruction-based algorithms and automatic machine learning algorithms). In instruction-based methods, different image processing algorithms like filters, edge detections, morphological, tracking approaches and region-based segmentation, etc., are used. The traditional machine learning-based approaches generally used supervised learning algorithms. It needs manual efforts for labeling the context used for training the datasets. But, recently developed deep learning algorithms give satisfactory results and eliminate the manual segmentation of vessels. In last couple of years, a lot of work has been done on vessel segmentation using deep learning techniques. Many authors use different deep neural network architectures for vessel segmentation, among which some networks become popular due to their robustness and accuracy. For segmenting biomedical images, UNet [4] is used by the researchers. It is an encoder–decoder architecture, which reveals admirable performance.

M. K. Kar (✉) · M. K. Nath
Department of Electronics and Communication Engineering,
National Institute of Technology Puducherry, Karaikal 609609, India

M. Mishra
Department of Electronics and Communication Engineering, North Eastern Regional Institute
of Science and Technology (NERIST), Arunachal Pradesh, Nirjuli 791109, India

© The Author(s), under exclusive license to Springer Nature Singapore Pte Ltd. 2022   603
R. R. Raje et al. (eds.), *Artificial Intelligence and Technologies*,
Lecture Notes in Electrical Engineering 806,
https://doi.org/10.1007/978-981-16-6448-9_58

Uysal1 et al. [5] implemented fully connected convolutional neural network (CNN) for vessel segmentation from grayscale fundus images. Authors have used gray-level normalization, contrast limited adaptive histogram equalization (CLAHE), and gamma correction in the preprocessing step to make the dataset appropriate for training. They have used CNN layer followed by batch normalization and ReLU. Softmax loss function is used for segmenting the data into two classes, vessel and non-vessel pixels. Gu et al. [6] proposed a context encoder network (CE-NET) to confine more complex high-level spatial information for 2D vessel segmentation. They have used a context extractor module between encoder and decoder module. It captures additional high-level features and encompasses essential spatial information.

Yan et al. [7] train the UNet-based model with a joint-loss which utilizes pixelwise loss and a segment-level loss. They introduced a feature fusion module with a multiscale convolution block to capture more semantic information. The fusion module preserves the spatial information by combining spatial path with a large kernel. Huazhu and their co-authors [8] proposed a deep vessel architecture which consists of multilevel CNN with side output layers to learn a rich hierarchical representation. Additionally, they used conditional random fields (CRF) to model the long-range interactions between pixels which increase the accuracy of the segmented results.

Hu et al. [9] segmented the blood vessels from color fundus images using CNN and fully connected CRFs. It uses multiscale CNN architecture with an improved cross-entropy loss function to train DRIVE and STARE databases. Shin et al. [10] proposed a vessel graph network (VGN) by linking CNN architecture and graph neural network (GNN). This network jointly models both local appearances and global vessel structures by utilizing semi-regular graph nodes. Authors divided the method into three parts such as (i) generating pixel wise features and vessel probabilities, (ii) extract features to reflect vascular connectivity using GNN, and (iii) inference module to produce the final segmentation map.

The above-mentioned method did not emphasize the blood vessels in the disc region for glaucoma analysis. In this paper, blood vessels segmentation and disc regions detection are performed by fully connected residual model with inception modules in UNet backbone. The rest part of the paper is organized as follows. Methodology for blood vessel segmentation and disc detection is discussed in Sect. 2. Training, testing, and hyperparameter selection for the proposed model are presented in Sect. 3. Experimental results and analysis are conferred in Sect. 4. Finally, Sect. 5 summarizes the work and presents a road map for future directives.

2 Methodology

In the proposed method, two parallel networks are designed for vessel segmentation and disc region detection simultaneously. For vessel segmentation, a modified ResNet [11]-based network with inception module is proposed. For disc

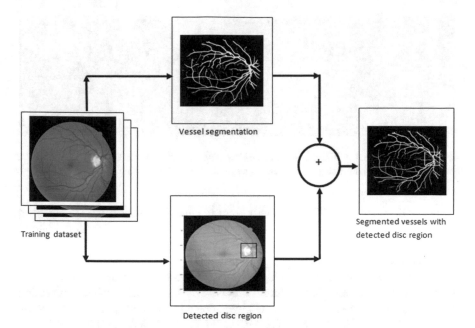

Fig. 1 Proposed block diagram for blood vessel segmentation and disc region detection

segmentation, simple CNN network with skip connection is used. The block diagram is shown in Fig. 1.

Preprocessing is performed prior to the image segmentation and disc detection. Here, gray-level transformation is performed. This is achieved by adding or multiplying a constant value to the gray value in the color channels of a given image. RGB color fundus image is split into three color channels (R, G, B). The blood vessels are more predominant in green channel compared to red and blue channels. This has been shown in Fig. 2. Due to this the color image is converted to grayscale using the transformation given in Eq. (1) [12].

$$Y(i,j) = 0.333R(i,j) + 0.5G(i,j) + 0.166B(i,j) \tag{1}$$

where R, G, and B represent the red channel, green channel, and blue channel of the color fundus image. Data augmentation techniques like horizontal and vertical flipping are incorporated in preprocessing stage to make the data size appropriate for training.

2.1 Network Architecture for Vessel Segmentation

In the proposed method, inception modules with UNet backbone are adopted with a joint loss to accomplish end-to-end segmentation. The model consists of encoder

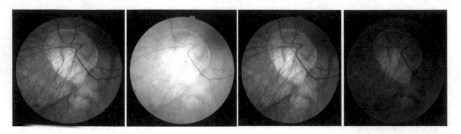

Fig. 2 Representation of different color channels for visualization of blood vessel features. From left to right: color fundus image, red channel, green channel, and blue channel images, respectively

stage and decoder stage. 32 layers are present in encoding section, and 25 layers are present in decoder section. The first two layers of encoder path consist of convolution layers followed by elementwise nonlinear activation function (ReLU) layer along with batch normalization and a max pooling layer. A bottleneck or identity module is created with inception block having residual connection. It consists of inception module with two layers of convolution followed by ReLU and batch normalization. The output of the residual module is concatenated with its input to preserve the activation from previous layer. After each residual block again a set of convolution layer followed by ReLU, batch normalization and dropout are applied to preserve the fine details corresponding to vessel segments or pixels. This residual block with inception module is repeated for five times. Dropout is applied after each inception module to reduce complexity of CNN network and to utilize the dominant features.

Figure 3 represents the network architecture for vessel segmentation. It has one encoder path and one decoder path. In the decoder path, the activations are up-sampled with deconvolution and concatenated with the layers of encoder path of same size. This provides the positional re-occurrence of the segmented feature. Likewise, there are five units having series of convolutional layer followed by ReLU and batch normalization layer in the expansion path. Each unit is up-sampled with an up-sampling layer and concatenated with the encoder section. It makes use of feature maps from both the lower level with the relevant contracting path.

Batch normalization is used to reduce internal co-variate shift and provides faster convergence. It is performed through subtracting mean from the mini batch output and normalized by standard deviation of the mini batch. The batch normalized activation is given in Eq. (2).

$$A_i = \frac{a_i - a_{\text{mean}}}{\sqrt{\sigma_b^2 + c}} \qquad (2)$$

where a_{mean} represents mini batch mean, σ_b^2 is the mini batch variance, and c is a numeric constant used for numerical stability.

During training, the scaling and shifting parameters are updated with every epochs, which makes the distribution shifts convenient and hence results in faster

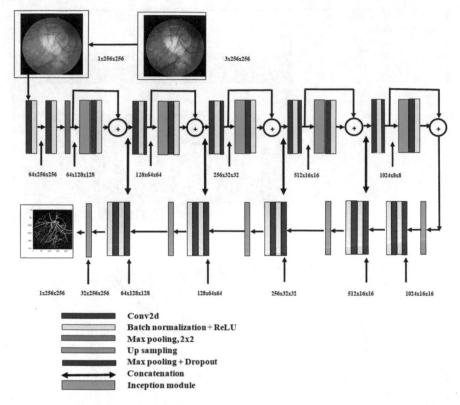

Fig. 3 Network architecture for vessel segmentation

convergence. ReLU function is a nonlinear activation function denoted as $f(x) = \max(0, x)$. It increases the training performance by taking the positive values only. This makes gradient computation faster. Max pooling operation is performed for down-sampling the data. It helps to avoid overfitting. Mathematically, max pooling is represented in Eq. (3). Dropout rate of 0.5 is employed with max pooling to reduce the overfitting [13].

$$P(i,j) = \sum_{u=0}^{1} \sum_{v=0}^{1} C(2i - u, 2j - v) \tag{3}$$

Figure 4 shows the internal structure of inception module, which is used to extract multiscale features of vessel segments from different scales. Kernel sizes 1×1, 1×3, 3×1, and 3×3 are implemented to extract multiscale features with dimension reduction. This module expands representational bottleneck by preserving the loss of information due to deeper network. The vessel and non-vessel pixels of the fundus image are not evenly distributed. This imbalanced distribution can be overcome by using weighted binary cross-entropy (WBCE) loss function along with dice loss. The WBCE loss is defined in Eq. (4).

Fig. 4 Internal structure of
inception module

Next layer

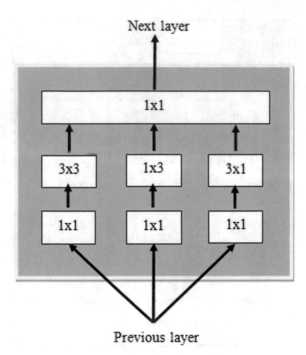

Previous layer

$$L_{\text{WBCE}}(y, \hat{y}) = -(y\beta \log(\hat{y}) + (1 - y) \log(1 - (\hat{y}))) \tag{4}$$

Here, $\hat{y}$ represents the predicted value by the prediction model.

The dice loss function gives the measure of similarity between true value and predicted value. It is represented in Eq. (5).

$$L_{\text{Dice}}(y, \hat{y}) = 1 - \frac{2y\hat{y} + 1}{y + \hat{y} + 1} \tag{5}$$

The estimated joint loss function is represented in Eq. (6)

$$\text{Loss} = L_{\text{WBCE}} + L_{\text{Dice}}(y, \hat{y}) \tag{6}$$

2.2 Network Architecture for Vessel Segmentation

The disc detection task is taken as object localization task with supervised learning. The labels are taken as center of the disc in csv file. The training data is expanded using different augmentation techniques such as: horizontal flipping, vertical flipping, translation, and resizing. In the same way, labels are updated. The prediction problem is taken as predicting the center x and y coordinate in the bounding box of the fundus image. A model is developed, which consists of several convolutional layers and residual layers. The proposed disc region detection architecture consists of eight section of convolutional layer followed by ReLU layer and batch normalization layer with four skip connections. After each section, a max pooling layer is used to down-sample the data. Finally a fully connected layer is used with adaptive average pooling [14] to classify the data into two classes. The input of each section layer is added to the output of that layer using skip connection. The number of input channels, in case of skip connections, will be the sum of output channels from the previous layer and the skip layer. This helps to preserve the information of previous layer. The final convolution layer output is flattened and applied to fully connected layer. As coordinate values are predicted, no activation is required for the final layer. The output of fully connected layer passing through adaptive average pooling denotes the number of classes, which is shown in Fig. 5.

The loss function used for this detection task is taken as Huber loss [15]. It is described in Eq. (7).

$$L(y, \hat{y}) = \frac{1}{n} \sum_i Z_i \qquad (7)$$

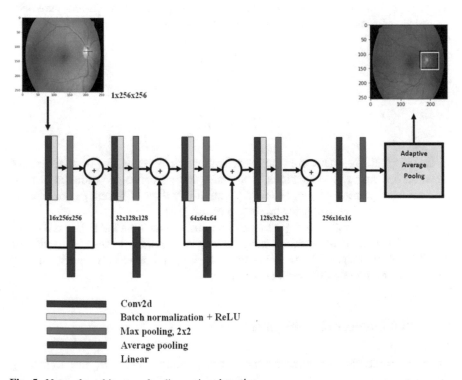

Fig. 5　Network architecture for disc region detection

where

$$Z_i = \begin{cases} 0.5(y - \hat{y})^2, & \text{if } |y - \hat{y}| < 1 \\ |y - \hat{y}| - 0.5, & \text{otherwise} \end{cases} \tag{8}$$

3   Training, Testing, and Hyperparameter Selection

Both the models for vessel segmentation and disc region detection are trained using standard backpropagation algorithm. Adam optimizer [16] is implemented for optimization of learning parameters. It is a gradient-based optimization algorithm considering stochastic objective functions, based on adaptive estimates of lower-order moments. It has less memory requirements and computationally efficient for large data. Its hyperparameters have perceptive interpretations and usually necessitate slight tuning. The Adam optimizer is updated by the given in Eq. (9) [17].

$$y_{t+1} = y_t - \frac{\alpha}{\sqrt{V_t} + \xi} M_t \tag{9}$$

where y_{t+1} is the updated parameter vector which depends on the current parameter vector y_t. M_t and V_t are the bias-corrected first and second moment estimates described by

$$M_t = \frac{m_t}{1 - \beta_1^t} \tag{10}$$

$$V_t = \frac{v_t}{1 - \beta_2^t} \tag{11}$$

where β_1 and β_2 stand for the exponential decay rates for first-order and second-order moment estimations. The first- and second-order moments of the gradient are represented by m_t and v_t, respectively. Values of β_1 and β_2 are taken as 0.9 and 0.999, respectively, for training purpose [18].

The proposed parallel model is trained on publicly available DRIVE and STARE databases. The network is trained 80% on grayscale fundus images and tested on rest 20% of the images. The model is trained with maximum of 50 epochs, having initial learning rate 0.0001 and batch size of 30.

4   Experimental Results and Analysis

All experiments in this research work are accomplished with the following: Windows 10 Pro operating system, Intel (R) Core (TM) CPU @ 1.8 GHz, 16 GB RAM, and GTX 1660 GPU card. Then, the system was tested using Digital Retinal

Images for Vessel Extraction (DRIVE) database and Structured Analysis of the Retina (STARE) database. Performance of the system has been evaluated using accuracy, specificity, sensitivity, and precision. The details about databases, experimental results, and analysis are conferred in the subsequent subsections.

4.1 Databases

The evaluation is conducted on two publicly available DRIVE database [19] and STARE database [20] for retinal vessel segmentation and disc region detection from the fundus images. DRIVE database comprises of 40 images as vasculature ground truth, of which 20 images are employed as training set and rest as the testing set. All the images are captured with a 45° field of view (FOV), having resolution of 565×584 pixels. With the test database, two manually segmented ground truths are provided, where one set is used as gold standard and the other can be used to evaluate computer generated segmentation for comparison.

STARE database contains 20 images with two sets of manually labeled vessel segmented ground truths which are captured with 35° field of view and having 700×605 pixel resolution. Among the two sets, the first one is labeled as ground truth whereas the other is taken as gold standard.

4.2 Performance Measures

The difference between actual and predicted value is calculated to evaluate the performance of a network. The segmented vessel and non-vessel pixels are represented in binary. The correctly segmented vessel pixels are treated as true positives (TP_{vessel}), whereas wrongly segmented vessel pixels are called false negatives (FN_{vessel}). True negative ($TN_{\text{non-vessel}}$) and false positive ($FP_{\text{non-vessel}}$) represent the correctly segmented non-vessel pixels and incorrectly segmented non-vessel pixels, respectively. TP, FP, FN, and TN are used to calculate the different performance measures such as accuracy (Acc), sensitivity (Sen), specificity (Spe), and precision (Pre).

Accuracy (Acc) can be defined as the ratio of correctly segmented blood vessel pixels to the total number of pixels in the image. It is the gold standard metric for all types' segmentation. Specificity (Spe) measures the correct segmented non-vessel pixels. Sensitivity (Sen) is a measure to compute the segmented vessel pixels correctly by the model. Precision (Pre) measures the fraction of perfectly segmented blood vessel pixels to the total number of segmented blood vessel pixels. These measures are represented in Eqs. (12–15).

$$Acc = \frac{TP_{\text{vessel}} + TN_{\text{non-vessel}}}{TP_{\text{vessel}} + TN_{\text{non-vessel}} + FP_{\text{non-vessel}} + FN_{\text{vessel}}} \times 100 \qquad (12)$$

$$Sen = \frac{TP_{\text{vessel}}}{TP_{\text{vessel}} + FN_{\text{vessel}}} \times 100 \qquad (13)$$

$$Spe = \frac{TN_{\text{non-vessel}}}{TN_{\text{non vessel}} + FP_{\text{non-vessel}}} \times 100 \qquad (14)$$

$$Pre = \frac{TP_{\text{vessel}}}{TP_{\text{vessel}} + FP_{\text{non-vessel}}} \times 100 \qquad (15)$$

4.3 Results

For DRIVE and STARE databases, using data augmentation techniques, a total of 900 images is taken for training and testing. The detected disc region is shown in Fig. 6. Here the bounding box size is chosen empirically to be 50×50.

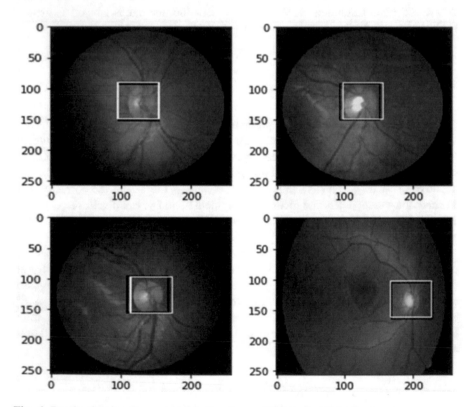

Fig. 6 Results showing disc region detection from arbitrary selected fundus images

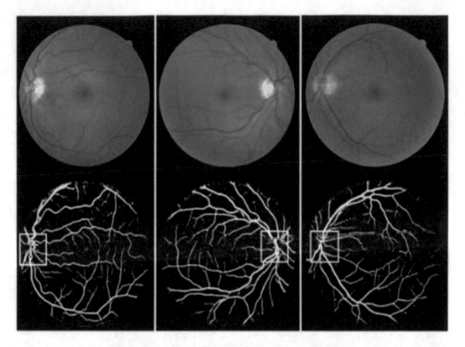

Fig. 7 Segmented vessels in disc region from DRIVE database and STARE database

The segmented blood vessels with detected disc region are shown in Fig. 7 (DRIVE database and STARE database). The final results of combining segmented blood vessels in the disc region with ground truth are shown in Figs. 8 and 9. Five arbitrary images are taken from DRIVE database, and their corresponding segmented blood vessels along with ground truths are shown in Fig. 8. Figure 9 corresponds to the segmented blood vessels along with their ground truths and original images for STARE database, respectively. The overall performance measures for both databases are shown in Table 1. These matrices are computed by considering the total segmented blood vessels in the image. The accuracy of the segmented blood vessels in detected optic disc region is noticed to be 0.99 and 0.98 for DRIVE and STARE databases, respectively. It may help in the analysis of blood vessels in the disc region which is an important area for diagnosis of reproducible glaucomatous visual field defects.

5 Conclusion

In this paper, a novel method for retinal vessel segmentation with disc region detection is proposed using inception module in UNet backbone structure. It uses a joint loss function to detect more details about the vessels in optic disc region.

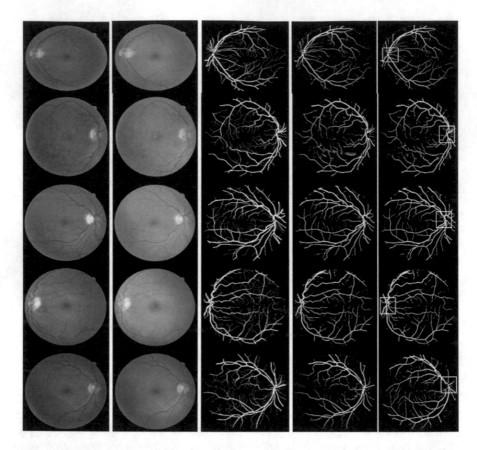

Fig. 8 Disc region of segmented vessels from DRIVE database: first, second, and third column represent original color fundus image along with its corresponding grayscale image and the ground truth for blood vessels, whereas fourth and fifth columns represent the segmented blood vessels and detected disc region, respectively

The disc region is detected using residual layers with skip connections. The proposed method achieved an accuracy of 84.90% and 79.80% for DRIVE and STARE databases, respectively. This segmentation accuracy approaches to maximum value in the detected disc region. The blood vessel in the disc region gives an indication of progressive damage of optic nerves with respect to normal anatomical variation and their subjective nature. This may be used as a measure for assessment of retinal nerve fiber layer thickness and optic nerve head structures for clinical evaluation. The segmented vessel in the optic disc region may help the researchers in analysis of thinning of the retinal nerve fiber layer and alteration in the structural appearance of the optic nerve head.

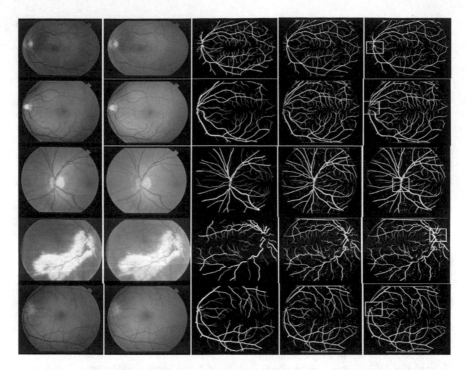

Fig. 9 Disc region of segmented vessels from STARE database: first, second, and third columns represent original color fundus image along with its corresponding grayscale image and the ground truth for blood vessels, whereas fourth and fifth columns represent the segmented blood vessels and detected disc region, respectively

Table 1 Performance metrics for segmented blood vessels from retinal images

Databases	Acc	Pre	Sen	Spe
DRIVE	0.8490	0.7650	0.8530	0.8050
STARE	0.7980	0.6500	0.7630	0.6965

References

1. Nath MK, Dandapat S (2012) Differential entropy in wavelet sub band for assessment of glaucoma. Int J Imaging Syst Technol 22:161–165
2. Nath MK, Dandapat S (2013) Multiscale ICA for fundus image analysis. Int J Imaging Syst Technol 23:327–337
3. Kar MK, Ravichandran G, Elangovan P, Nath MK (2019) Analysis of diagnostic features from fundus image using multiscale wavelet decomposition. ICIC Expr Lett Part B Appl 10:175–184
4. Ronneberger O, Fischer P, Brox T (2017) U-Net: convolutional networks for biomedical image segmentation. In: International conference on medical image computing and computer-assisted intervention

5. Uysal E, Guraksin GE (2020) Computer-aided retinal vessel segmentation in retinal images: convolutional neural networks. Multimedia Tools Appl 1929–1958
6. Gu Z, Cheng J, Fu H, Zhou K, Hao H, Zhao Y, Zhang T, Gao S, Liu J (2019) CE-Net: context encoder network for 2D medical image segmentation. IEEE Trans Med Imaging
7. Yan Z, Yang X, Cheng KT (2018) Joint segment-level and pixel wise losses for deep learning based retinal vessel segmentation. IEEE Trans Biomed Eng 65:1912–11923
8. Fu H, Xu Y, Lin S, Wong DWK, Liu J (2013) Deep vessel: retinal vessel segmentation via deep learning and conditional random field. In: Proceedings of IEEE international conference intelligent transportation systems (ITSC)
9. Hu K, Zhang Z, Niu X, Zhang Y, Cao C, Xiao F, Gao X (2018) Retinal vessel segmentation of color fundus images using multiscale convolutional neural network with an improved cross entropy loss function. J Neurocomput
10. Shin SY, Lee S, Yun ID, Lee KM (2019) Deep vessel segmentation by learning graphical connectivity. Med Image Anal
11. He K, Zhang X, Ren S, Sun J (2016) Deep residual learning for image recognition. In: Proceedings of the IEEE conference on computer vision and pattern recognition, pp 770–778
12. Kumar T, Verma K (2010) A theory based on conversion of rgb image to gray image. Int J Comput Appl 0975–8887
13. Srivastava N, Hinton G, Krizhevsky A, Sutskever I, Salakhutdinov R (2014) Dropout: a simple way to prevent neural net networks from over fitting. J Mach Learn Res 15(1):1929–1958
14. Liu Y, Zhang YM, Zhang XY, Liu CL (2016) Adaptive spatial pooling for image classification. Pattern Recogn 1–10
15. Huber PJ (2016) Robust estimation of a location parameter. Annal Math Stat
16. Kingma DP, Ba JL (2015) Adam: a method for stochastic optimization. In: International conference on learning representations
17. Ruder S (2016) An overview of gradient descent optimization algorithms, CORR. arXiv preprint arXiv:1609.04747
18. Elangovan P, Nath MK (2020) Glaucoma assessment from color fundus images using convolutional neural network. Int J Imaging Syst Technol 1–17
19. Niemeijer M, Ginneken B, Loog M (2004) Comparative study of retinal vessel segmentation methods on a new publicly available database. Proc SPIE Int Soc Opt Eng 5370:648–657
20. Hoover A, Kouznetsova V, Goldbaum M (2000) Locating blood vessels in retinal images by piece-wise threshold probing of a matched filter response. In: IEEE Trans Med Imaging 65:203–210

Human Emotion Detection Through Hybrid Approach

Krishna Mohan Kudiri and Hitham Seddiq Alhassan Alhussian

1 Introduction

Emotion detection for real-world conditions is the process of recognising human mental states through speech and facial expressions. Basic human emotions for real-world conditions are finite, namely anger, happiness, sadness, disgust, boredom and surprise [1, 2]. Currently, emotion detection is possible through facial expression, speech and physiological changes, as imparted from electrocardiogram (ECG), electroencephalogram (EEG), skin conductance, etc. [3, 4]. Correspondingly, the impediment in real-world conditions is that human emotion detection is restricted only through speech and facial expressions, because human communication is essentially, a mixture of speech and facial expressions. In dealing with modalities, the issue at hand for real -world condition is that facial expression always dominates speech data in terms of emotion detection accuracy [1, 5]. According to [6, 7], in the real-world condition, human emotions on the face is not always universal. The human face for a single emotion is different for both deliberate facial expressions, and facial expressions during speech which can be illustrated in Fig. 1.

In Fig. 1, facial expression of deliberate face of a person is different from facial expression during speech with high intensity with angry as emotion. In fact, deliberate facial expressions are universally standard and easy to understand for different intensity of emotion.

On the other hand, facial expressions during the speech process (non-deliberate) with different intensity of emotions can change from person to person, depending on their speech behaviour and can affect the emotion detection accuracy and robustness [8, 9]. As asserted by [10, 11], the issue of emotion detection through facial expressions in real-world conditions (such as different intensity of emotions (low, medium and high)) is how to detect emotions when human facial expression is different for the deliberate and non-deliberate ones.

K. M. Kudiri (✉) · H. S. A. Alhussian
UniversitiTeknologi PETRONAS, Seri Iskandar, Malaysia

© The Author(s), under exclusive license to Springer Nature Singapore Pte Ltd. 2022   617
R. R. Raje et al. (eds.), *Artificial Intelligence and Technologies*,
Lecture Notes in Electrical Engineering 806,
https://doi.org/10.1007/978-981-16-6448-9_59

Fig. 1 Facial expressions of
the emotion—anger;
a deliberate face **b** face during
speech at high intensity [6]

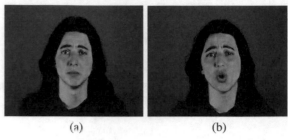

(a)   (b)

Fig. 2 Image resolution issue
for facial expressions [6]

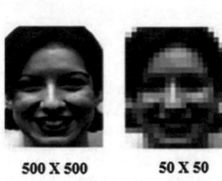

500 X 500   **50 X 50**

Occasionally, the type of noise occurs during emotion detection through facial expressions in real-world conditions is image resolution which can be illustrated in Fig. 2. Here, 500 × 500 is clearer than 50 × 50. Emotion detection through facial expression with poor resolution image can hence affect the accuracy and robustness. Various emotion detection techniques through facial expressions, however, can preserve emotion detection accuracy and robustness in real-world conditions.

According to [2], data loss during emotion detection through speech signal may be an error condition by which data can be destroyed by noise. As asserted by [12], noise can be increased due to numerous reasons such as random additive noise issue and data masking issue, which is possible for different intensity of emotions (low, medium and high) for speech signal. Emotion detection through speech with noise (random additive noise issue and data masking issue) can hence affect the accuracy and robustness. Various emotion detection techniques through speech signal, however, can preserve emotion detection accuracy and robustness in real-world conditions.

According to [12], information fusion for emotion detection through speech and facial expressions is not only correlated but also not synchronous (asynchronus) to each other. As asserted by [13], synchronisation issue, data correlation issue and relationship between emotions issue can be complex due to facial expressions in speech process with different intensity of emotions (such as low, medium and high) and can hence affect the accuracy and robustness. Various emotion detection techniques through speech and facial expressions using, however, can preserve emotion detection accuracy and robustness in real-world conditions.

2 Background Study

Emotion detection is the ability to discern emotions in real-world conditions without compromising accuracy and robustness [14]. In real-world conditions, human communications are essentially a mixture of speech and facial expressions where facial expressions are relatively dominating over speech while conveying emotion [14]. It is hence easy to detect emotion from deliberate facial expressions [1, 7]. Facial expressions during a speech process, on the other hand, may instigate on fusion for detecting emotions. Robust and accurate emotion detection is needed to create a model for detect emotions while people interacting with each other in real-world conditions.

Various efforts have been made to negate the drawbacks in emotion detection under the real-world conditions. For instance, the geometric-based feature extraction technique has been used to deal with facial expressions during speech process with different intensity of emotions. Yet, one limitation in terms of real-world conditions is having to maintain physical objects immediate to the human face for diagnosing facial expressions [15]. Consequently, appearance-based feature extraction technique for emotion detection through facial expression has missed the mark when involving correlation between the facial regions [16].

A hybrid feature extraction technique (geometric-based and appearance-based) is unable to detect emotions by automatically focussing on only a few regions of the face during a speech process for poor resolution input images [17]. Feature extraction technique via speech signal does have its disadvantages. Random additive noise and data masking issue due to different intensity of emotions can lead to poor emotion detection accuracy and robustness as found in existing prosodic feature extraction techniques [18].

Existing fusion techniques for both speech and facial expressions also have limitations. Feature-level fusion is unable to subjugate the synchronisation problem between both modalities (speech and facial expressions) [19]. Decision-level fusion between both speech and facial expression often neglect the correlation data between them [4]. Hybrid fusion (feature level fusion and decision level fusion), however, is unsuccessful in discerning the relationship among the emotions.

3 Methodology

Based on the review in Chap. 2, Fig. 3 shows the proposed hybrid approach to provide solution to counter the primary setbacks of multimodal emotion detection revolving around accuracy and robustness factors.

Figure 3 shows the flow of the proposed hybrid approach, where a video file is defined as an input. The input is a video file with a frame rate of 30 frames in one second. Firstly, the proposed approach will generate two feature vectors, one from speech and the other from facial expression, from the input video file for every

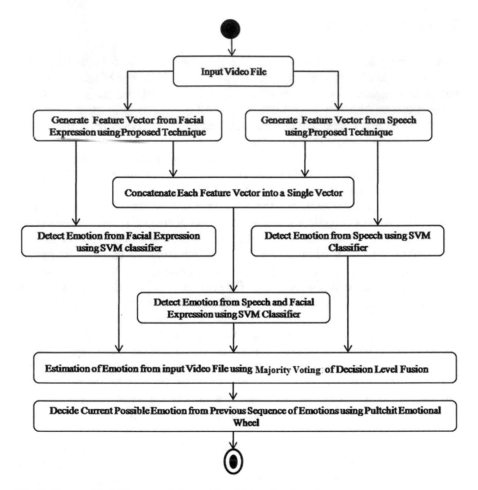

Fig. 3 Proposed hybrid approach for multimodal emotion detection

frame of 33.3 ms each (1 s = 1000 ms, 1000 ms/30 frames = 33.3 ms for each frame per second). The purpose of generating two feature vectors for every frame of the video file is to understand human facial expressions during speech process through micro expression, which can immensely change within seconds as clearly depicted in Fig. 4.

In Fig. 5, the expression of anger emotion from the face evinced from the 3rd–7th seconds and is collectively recognised as a set of micro expressions of angry emotion through facial expressions. Secondly, the two feature vectors will transpire another feature vector by concatenation, which is a part of the feature-level fusion technique as mentioned in Chap. 2 and is illustrated in Fig. 4. Here, A, B, C, D, E and F represent values in the feature vector. Consequently, these three feature vectors will act as inputs to a set of three classifiers, separately. In this case, SVM is

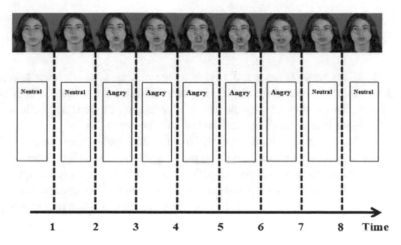

Fig. 4 Changing of human facial expression for every second in a single video file [20]

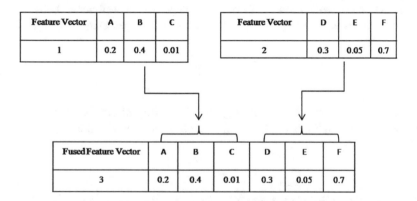

Fig. 5 Concatenation of feature vectors to produce another feature vector

used, which employs radial basis kernel to detect complex emotions (which was discussed in Chap. 2). Next, the output of all these classifiers will in turn become inputs to a final decision unit, which applies a majority voting technique as the decision-level fusion to estimate the emotion for a sequence of frames for each second from the input video file (as aforementioned in Chap. 2). Finally, the last step is to estimate the current emotion through previous emotions with the aid of Plutchik's emotion circle for each second of the video file input.

The different components of the proposed hybrid approach are further described in the following subsections of this chapter. To simplify the understanding of the workings of each part of the proposed approach, various examples are given at the end of each sub-section.

3.1   Relative Sub-Image-Based Feature Extraction Through Facial Expression

In this algorithm, step (1) reads a 2D input image of a face, which is cropped by a face detector from an input image, if any. Step (2) resizes the input image into a fixed length. Here, resize input image into $N \times N$ (pixels), hence it will be helpful to divide the input image into fixed number of sub-images.

N is number of pixels. For instance, in this case, the input image is resized to 500 $\times$ 500 pixels. Next, step (3) divides the input image into total of 5×5 sub-images. Thus, the size of each sub-image is 100×100 pixels. Step (4) computes the average pixel intensities for each sub-image. Finally, step (4) computes the relative difference between each sub-image with all other sub-images using its average pixel intensity values. Concept of Algorithm (3.1) is further explained in Example 3.1 (page 86).

Step 1. Read an image and resize. Step 2. Divide an input image into sub-images (image segmentation). Step 3. Calculate the average pixel intensities for each sub-image or segmented image. Step 4. Calculate relative difference between each sub-image to its adjacent sub-images or segmented images using Eq. (1).

$$\text{Relative difference} = \left[\frac{(x_i - x_j)}{(x_i + x_j)} \right], \ x_i \neq x_j \tag{1}$$

Here, x_i and x_j represent the average pixel intensity of two adjacent segmented images of given input. Also, x_i and x_j cannot be with equal values with opposite signs. If so, output will be zero.

3.2   Relative Bin Frequecy Coefficients Through Speech

In this algorithm, step (1) reads the input speech signal of the acoustic type. Step (2) generates the frequency spectrum of the input speech signal using Fast Fourier Transform (FFT) and also helps to remove inaudible frequencies from the input signal (which is also called as pre-processing of the signal). Here, this signal (in frequency spectrum) starts from 0 to 3400 Hz (but processing will start from 50 to 3450 Hz to eliminate the unaudible frequencies (0–49 Hz) from the input signal.

In this case, all frequency samples are positive (which is done by abs() function in MATLAB). Next, step (3) divides the frequency spectrum into frequency bins of 100 Hz each because speech signal often varies for every 10 ms. Calculate mean frequency for each bin. Finally, a feature vector is developed by computing the relative differences of each frequency bin with the other frequency bins by means of Eq. (1).

Step 1. Read speech signal. Step 2. Compute the frequency spectrum from the above using FFT. Step 3. Compute frequency bins of 100 Hz each and calculate mean frequency of each bin. Step 4. Compute relative difference for each frequency bin with other frequency bins using Eq. (1). Here, x_i and x_j represent the mean frequency of two adjacent bins. Hence, these bin frequencies cannot be equal with opposite signs.

3.3 Hybrid Approach

The hybrid approach (feature level fusion and decision level fusion) is a fusion technique for speech and facial expressions [21]. The third part of the proposed hybrid approach focuses on fusion between both the modalities, namely speech and facial expressions. Here, the input is a video file. This process involves several sub-operations, such as feature level fusion, decision level fusion and the relationship between emotions technique. This hybrid technique for fusion is unique such that it detects emotions not only for high intensity of emotions (for both deliberate and non-deliberate conditions), but also for low and medium intensity of emotions as well, with the help of the relationship between emotions technique for a given video file. Figure 6 illustrates the block diagram of the hybrid approach (for vectors fusion), which is sub-part of the proposed hybrid approach.

4 Results and Discussion

The accuracy of emotion detection of the proposed hybrid approach was evaluated by measuring the average overall accuracy for all basic emotions, as offered by its various parts, namely, feature extraction part from speech, feature extraction part

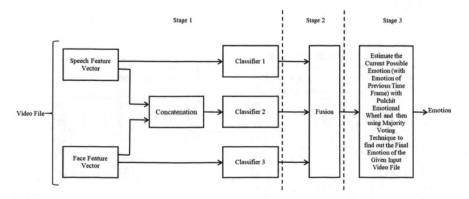

Fig. 6 Block diagram for the proposed hybrid approach for vectors fusion

from facial expressions and fusion between them (speech and facial expressions), the high average overall accuracy reflects high emotion detection accuracy. In fact, the deliberate facial expressions and facial expressions during speech process in real-world conditions are the main causes for low emotion detection accuracy for automatic emotion detection approaches. Thus, this section reviews the strength of feature extraction and fusion of said hybrid approach in the form of emotion detection accuracy for the deliberate facial expressions and facial expressions during speech process in real-world conditions.

In fact, human emotions detection in real world is possible only with two modalities, namely speech and facial expressions. Nonetheless, facial expressions always dominate speech while detecting emotions. In fact, speech is needed while dealing with facial expressions during speech process in order to detect emotions [2]. Henceforth, this research has chosen them (facial expressions, speech and fusion between them) and their performances have been compared with each other.

4.1 Facial Expressions

See Table 1.

4.2 Speech

See Table 2.

4.3 Hybrid Approach

See Table 3.

Table 1 Emotion detection through face

Database	Appearance-based technique		Hybrid technique		
	[22]	[23]	[24]	[5]	RSB
ENTERFACE'05	40	49	51	57	82
SAL	58	55	78	75	84
DaFEx	47	40	55	60	79
Cohn-Kanade	62	60	77	75	81
Yale	65	61	70	68	78
JAFEE	67	60	73	75	85

Table 2 Emotion detection through speech

Database	[25]	[26]	[27]	[28]	RBFC
Emo-Db	62	65	72	75	79
CSC Corpus	65	68	77	75	82
DaFEx	55	56	59	70	80
ENTERFACE'05	52	54	57	71	78
SAL	58	58	71	69	75

Table 3 Emotion detection through face and speech

Database	Hybrid technique			
	[29]	[30]	[31]	Proposed Fusion
ENTERFACE'05	79	75	74	85
CK + CSC Corpus	75	75	76	87
SAL	81	77	81	87
DaFEx	73	72	70	82
IITK + Emo-Db	75	74	70	82

As per the literature, human emotion detection (for real-world conditions) is a multimodal through speech and facial expressions, and thus, fusion between them (speech and facial expressions) is also an essential aspect for bimodal emotion detection techniques. The hybrid technique for fusion, which is a mixture of both feature level and decision-level fusion techniques, offers high average overall accuracy for real-world applications, and thus, this research compares the proposed hybrid approach (with the proposed hybrid technique for fusion) with conventional techniques in terms of their potential to maximise their average overall accuracy. The average overall accuracy was calculated, and the results are shown in Tables 1, 2 and 3. Here, DaFEx, CK + CSC Corpus, SAL, IITK + Emo-Db and ENTERFACE'05 databases (of high emotional intensity) are used.

The average overall accuracy provided by various emotion detection techniques of hybrid method for fusion through both facial expressions and speech is presented. Several tests among all the available techniques are carried out in order to find the most suitable one for handling both deliberate facial expressions and non-deliberate facial expressions with high emotion intensity during speech. Furthermore, Tables 1, 2 and 3 demonstrate that with every technique, the performance of ENTERFACE'05 and DaFEx databases was lower than other databases due to non-deliberate facial expressions.

Besides, from Table 3, the conventional emotion detection techniques (with hybrid type fusion) by achieving a maximum average overall accuracy ranging from 70 to 81%, performed inferior to the proposed hybrid technique (which uses the proposed hybrid type fusion). This makes these techniques unsuitable for real-world applications. It can be seen from Table 3 that the proposed hybrid technique (with proposed hybrid type fusion) achieved a superior average overall accuracy (ranging from 82 to 87%) in all circumstances, which makes its usage for fusion through both facial expressions and speech a suitable component of the proposed approach in real-world applications.

5 Conclusion and Future Work

According to the results obtained by proposed hybrid approach, we strongly believe that hybrid approach can be implemented practically in real-world applications due to the high emotion detection accuracy and robustness. However, there are still some limitations in proposed hybrid approach, we summarise them below for further future works:

1. Although proposed hybrid approach can achieve high emotion detection accuracy and robustness, but still we believe that it can further improved by incorporating more diversity features for training set.
2. Another possibility of new future work direction is trying to reduce the complexity overhead of features optimization which will enhance the performance of proposed hybrid approach in terms of execution times.
3. Since the proposed hybrid approach produces high emotion detection accuracy as well as stable robustness, it can be used in practical real-time platforms using Raspberry Pi or Arduino Uno circuits.

References

1. Tanveer A, Taskeed J, Ui-Pil C (2013) Facial expression recognition using local transitional pattern on gabor filtered facial images. IETE Tech Rev 30:47–52
2. Aastha (2013) Speech emotion recognition using combined features of HMM & SVM Algorithm. Int J Adv Res Comput Sci Softw Eng 3:387–393
3. Gunes H, Pantic M (2010) Automatic, dimensional and continuous emotion recognition. Int J Synth Emot 1:68–99
4. Sethu V, Ambikairajah E, Epps J (2009) Pitch contour parameterisation based on linear stylisation for emotion recognition. In: Annual conference of the international speech communication association, pp 2011–2014
5. Bermani AK, Ghalwash Z, Youssif AA (2012) Automatic facial expression recognition based on hybrid approach. Int J Adv Comput Sci Appl 102–108
6. Srivastava A, Mane S, Shah A, Shrivastava N, Thakare B (2017) A survey of face detection algorithms. In: International conference on inventive systems and control, pp 1–4
7. Ashraf AB, Lucey S, Cohn JF, Chen T, Ambadar Z, Prkachin KM, Solomon PE (2009) The painful face—pain expression recognition using active appearance models. Image Vis Comput 27:1788–1796
8. Metallinou B, Lee S, Narayanan S (2010) Decision level combination of multiple modalities for recognition and analysis of emotional expression. In: International conference on acoustics, speech, and signal processing, pp 1–8

9. Oh D, Osherson DN, Todorov A (2015) Robustness of emotional expression recognition under low visibility, Thesis
10. Eero V (2014) Emotion recognition from speech using prosodic features. Acta Universitatis oluensis Publications, Finland
11. Bai Y, Guo L, Jin L, Huang Q (2009) A novel feature extraction method using pyramid histogram of orientation gradients for smile recognition. In: International conference on image processing, pp 38–39
12. Wu CH, Lin JC, Wei WL (2013) Two-level hierarchical alignment for semi-coupled HMM-based audiovisual emotion recognition with temporal course. IEEE Trans Multimedia 15(8):1880–1895
13. Horii T, Nagai Y, Asada M (2015) Emotion recognition and generation through multimodal restricted boltzmann machines. In: International conference on intelligent robots and systems, Hamburg, pp 1–2
14. Sidorova J (2009) Speech emotion recognition with TGI+.2 classifier. In: Conference of European chapter of the association for computational linguistics, pp 1–12
15. Caifeng, Shaogang GB, Peter WM (2009) Facial expression recognition based on local binary patterns: a comprehensive study. Image Vis Comput 27:803–816
16. Kumbhar M, Jadhav A, Patil M (2012) Facial expression recognition based on image feature. Int J Comput Commun Eng 117–119
17. Rabie A, Lang C, Hanheide M, Castrillón-Santana M, Sagerer G (2008) Automatic initialization for facial analysis in interactive robotics. Comput Vis Syst 5008:517–526
18. Abdelwahab M, Busso C (2015) Supervised domain adaptation for emotion recognition from speech. In: 40th IEEE international conference on acoustics, speech and signal processing, Brisbane, Australia
19. Castellano G, Kessous L, Caridakis G (2008) Emotion recognition through multiple modalities: face, body gesture, speech. Affect Emot Hum Comput Interact 4868:92–103
20. Bylsma L (2006) Cohn-Kanade (CK and CK+) database. [Blog] Available at: http://psychology.pitt.edu/emotion-depression-and-development. Accessed 7 May 2014
21. Anish Menon R, Anurag D, Priyanka Reddy B, Sukanya B (2016) Various approaches for human emotion recognition: a study. Int J Soft Comput Artif Intell 4:43–47
22. Mlakar F, Slovenia M (2015) Automated facial expression recognition based on histograms of oriented gradient feature vector differences. Signal Image Video Process 9:245, 253
23. Deshpande G, Apte SD (2014) Facial emotion recognition using gabor features. Int J Electr Commun Comput Eng 5:1–5
24. Sezgin ZC, Gunsel B, Kurt GK (2012) Perceptual audio features for emotion detection. J Audio Speech Music Process 1–21
25. Bishop ZC (2013) Pattern recognition and machine learning. Springer Publication
26. Hassan A, Damper R (2010) Emotion recognition from speech using extended feature selection and a simple classifier. In: Conference of the international speech communication association, pp 2043–2046
27. Aswin KM, Keerthi V, Keerthana S, Sreekutty IK (2016) HERS: human emotion recognition system. In: International Conference on Information Science, pp 176–179
28. Vogt T, André E, Bee N (2008) EmoVoice—a framework for online recognition of emotions from voice. Percept Multimodal Dialogue Syst 5076:188–199
29. Shah SK, Khanna V (2015) Facial expression recognition for color images using gabor, log gabor filters and PCA. Int J Comput Appl 42–46
30. Valstar MF, Pantic M (2010) Induced disgust, happiness and surprise: an addition to the MMI facial expression database. In: International language resources and evaluation conference, pp 52–65
31. Mohammadh M (2013) Automated facial expression recognition using local transitional pattern, Thesis

Computation of Biconditional Cordial Labeling of Super Subdivision of Graphs

M. Kalaimathi, B. J. Balamurugan,
and Jonnalagadda Venkateswara Rao

1 Introduction

In graph theory, labeling of a graph is an extensive and potential area of research. It explains the labeling techniques of a graph based on mathematical functions and conditions. The latest updates and progress in the area of graph labeling are referred from Gallian [2]. The labeled graphs had been proposed by Rosa [8], and the applications of graph labeling are mentioned in [5]. In this paper, the textbooks Harary [3] and West [10] have been cited for the concepts and terminologies of graphs. Cahit [1] introduced the notion of cordial labeling in the year 1990. In 2015, Murali et al. [6] introduced the concept of biconditional cordial labeling of graphs and investigated different cases of cycle graphs which admit the biconditional cordial labeling. Further, in the year 2016, Murali et al. [7] extended this labeling to ladder and $K_n + nK_1$ graphs. In addition to that, we extended the existence of this labeling to some cycle-related graphs in [4]. In the year 2001, Sethuraman and Selvaraju [9] introduced a graph operation called super subdivision of graph, which generates variety of graph pictures. Super subdivision of a graph can be used as a powerful operation to get larger graphs from a given graph. In this paper, we prove that the super subdivision of ladder and grid graphs admits the biconditional cordial labeling.

M. Kalaimathi · B. J. Balamurugan (✉)
School of Advanced Sciences, VIT University, Chennai Campus,
Vandalur–Kelambakkam Road, Chennai, Tamil Nadu 600127, India
e-mail: balamurugan.bj@vit.ac.in

M. Kalaimathi
e-mail: kalaimathi.m2018@vitstudent.ac.in

J. V. Rao
Department of Mathematics, School of Science and Technology,
United States International University, Nairobi, Kenya
e-mail: jvrao@usiu.ac.ke

© The Author(s), under exclusive license to Springer Nature Singapore Pte Ltd. 2022   629
R. R. Raje et al. (eds.), *Artificial Intelligence and Technologies*,
Lecture Notes in Electrical Engineering 806,
https://doi.org/10.1007/978-981-16-6448-9_60

2 Preliminaries

We adopt the following notations in this paper.

(i) $v_f(0)$ and $v_f(1)$ denote, respectively, the number of vertices having labels 0 and the number of vertices having labels 1 of the graph G under the function f.

(ii) $e_{f^*}(0)$ and $e_{f^*}(1)$ denote, respectively, the number of edges having labels 0 and the number of edges having labels 1 of the graph G under the function f^*.

2.1 Definition

A ladder graph $L_n = P_n \times K_2$ is a planar, undirected, connected graph with $2n$ vertices and $n + 2(n - 1) = 3n - 2$ edges.

2.2 Definition

A graph obtained through the Cartesian product of two paths P_m and P_n, denoted as $P_m \times P_n$ is known as a grid graph. This graph has mn vertices and $2mn - (m + n)$ edges.

2.3 Definition

Let G be a graph with n vertices and t edges. A graph H is said to be a super subdivision of G if H is obtained from G by replacing every edge e_i of G by a complete bipartite graph K_{2,m_i} for some m_i, $1 \leq i \leq t$ in such a way that the ends of e_i are merged with the two vertices of the 2-vertices part of $K_{2,m}$ after removing the edge e_i from G.

2.4 Remark

In this paper, we call the "super subdivision" of graphs as "SD" graphs for simplicity.

3 Biconditional Cordial Labeling of Graphs

3.1 Definition

"Let $G = (V, E)$ be a graph. A function $f : V \rightarrow \{0, 1\}$ of the graph G is called a biconditional cordial labeling of G if an induced edge function $f^* : E \rightarrow \{0, 1\}$ defined by

$$f^*(uv) = \begin{cases} 1, & \text{if } f(u) = f(v) \\ 0, & \text{if } f(u) \neq f(v) \end{cases}$$

satisfies the following two conditions

$$|v_f(0) - v_f(1)| < 1$$

$$|e_{(f^*)}(0) - e_{(f^*)}(1)| \leq 1.$$

3.2 Definition

A graph which admits the biconditional cordial labeling is called a biconditional cordial graph.

3.3 Example

A biconditional cordial labeling of graph G is shown in Fig. 1.

Fig. 1 Biconditional cordial labeling of a graph G

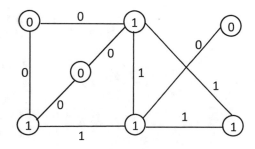

4   New Results

In this section, we show the existence of the biconditional cordial labeling to the SD of ladder graph L_n and grid graph $P_m \times P_n$.

4.1   Theorem

The $SD(L_n)$ for $n \geq 2$ admits a biconditional cordial labeling.

Proof Let

$$V = \{u_i : 1 \leq i \leq n\} \cup \{v_i : 1 \leq i \leq n\} \cup \{u_{i(i+1)}^k : 1 \leq i \leq n-1, 1 \leq k \leq t\}$$

$$\cup \{w_i^k : 1 \leq i \leq n, 1 \leq k \leq t\} \cup \{v_{i(i+1)}^k : 1 \leq i \leq n-1, 1 \leq k \leq t\} \text{ and}$$

$$E = \{u_i u_{i(i+1)}^k : 1 \leq i \leq n-1, 1 \leq k \leq t\} \cup \{u_{i+1} u_{i(i+1)}^k : 1 \leq i \leq n-1, 1 \leq k \leq t\}$$

$$\cup \{u_i w_i^k : 1 \leq i \leq n, 1 \leq k \leq t\} \cup \{v_i w_i^k : 1 \leq i \leq n, 1 \leq k \leq t\}$$

$$\cup \{v_i v_{i(i+1)}^k : 1 \leq i \leq n-1, 1 \leq k \leq t\} \cup \{v_{i+1} v_{i(i+1)}^k : 1 \leq i \leq n-1, 1 \leq k \leq t\} \text{ be}$$

the vertex set and edge set of $SD(L_n)$. The $SD(L_n)$ has $3nt + 2(n - t)$ vertices and $6nt - 4t$ edges.

Define a function $f : V \to \{0, 1\}$ such that

$$f(u_i) = \begin{cases} 0, & \text{if } i \text{ is odd} \\ 1, & \text{if } i \text{ is even} \end{cases}$$

$$f(v_i) = \begin{cases} 0, & \text{if } i \text{ is even} \\ 1, & \text{if } i \text{ is odd} \end{cases}$$

$$f\left(u_{i(i+1)}^k\right) = \begin{cases} 0, & \text{if } i \text{ is odd, } k \text{ is odd} \\ 1, & \text{if } i \text{ is odd, } k \text{ is even} \end{cases}$$

$$f\left(u_{i(i+1)}^k\right) = \begin{cases} 0, & \text{if } i \text{ is even, } k \text{ is even} \\ 1, & \text{if } i \text{ is even, } k \text{ is odd} \end{cases}$$

$$f\left(w_i^k\right) = \begin{cases} 0, & \text{if } i \text{ is odd, } k \text{ is odd} \\ 1, & \text{if } i \text{ is odd, } k \text{ is even} \end{cases}$$

$$f\left(w_i^k\right) = \begin{cases} 0, & \text{if } i \text{ is even, } k \text{ is even} \\ 1, & \text{if } i \text{ is even, } k \text{ is odd} \end{cases}$$

$$f\left(v_{i(i+1)}^{k}\right) = \begin{cases} 0, & \text{if } i \text{ is odd, } k \text{ is even} \\ 1, & \text{if } i \text{ is odd, } k \text{ is odd} \end{cases}$$

$$f\left(v_{i(i+1)}^{k}\right) = \begin{cases} 0, & \text{if } i \text{ is even, } k \text{ is odd} \\ 1, & \text{if } i \text{ is even, } k \text{ is even} \end{cases}$$

Case (i) When n is odd, t is odd, we have

$$v_f(0) = \frac{3nt + 2(n-t) + 1}{2}, v_f(1) = \frac{3nt + 2(n-t) - 1}{2} \text{ and}$$

$$|v_f(0) - v_f(1)| = \left| \frac{3nt + 2(n-t) + 1}{2} - \frac{3nt + 2(n-t) - 1}{2} \right| = 1.$$

Therefore, $|v_f(0) - v_f(1)| \leq 1$ is satisfied.

Case (ii) When n is odd, t is even; n is even, t is odd; and n is even, t is even,

we have $v_f(0) = \frac{3nt + 2(n-t)}{2}$, $v_f(1) = \frac{3nt + 2(n-t)}{2}$ and

$$|v_f(0) - v_f(1)| = \left| \frac{3nt + 2(n-t)}{2} - \frac{3nt + 2(n-t)}{2} \right| = 0.$$

Therefore, $|v_f(0) - v_f(1)| \leq 1$ is satisfied.
Define an induced edge function $f^* : E \to \{0, 1\}$ such that

$$f^*\left(u_i u_{i(i+1)}^{k}\right) = \begin{cases} 0, & \text{if } k \text{ is even, } 1 \leq i \leq n \\ 1, & \text{if } k \text{ is odd, } 1 \leq i \leq n \end{cases}$$

$$f^*\left(u_{i+1} u_{i(i+1)}^{k}\right) = \begin{cases} 0, & \text{if } k \text{ is odd, } 1 \leq i \leq n \\ 1, & \text{if } k \text{ is even, } 1 \leq i \leq n \end{cases}$$

$$f^*\left(u_i w_i^{k}\right) = \begin{cases} 0, & \text{if } k \text{ is even, } 1 \leq i \leq n \\ 1, & \text{if } k \text{ is odd, } 1 \leq i \leq n \end{cases}$$

$$f^*\left(v_i w_i^{k}\right) = \begin{cases} 0, & \text{if } k \text{ is odd, } 1 \leq i \leq n \\ 1, & \text{if } k \text{ is even, } 1 \leq i \leq n \end{cases}$$

$$f^*\left(v_i v_{i(i+1)}^{k}\right) = \begin{cases} 0, & \text{if } k \text{ is even, } 1 \leq i \leq n \\ 1, & \text{if } k \text{ is odd, } 1 \leq i \leq n \end{cases}$$

$$f^*\left(v_{i+1}v_{i(i+1)}^k\right) = \begin{cases} 0, & \text{if } k \text{ is odd}, 1 \leq i \leq n \\ 1, & \text{if } k \text{ is even}, 1 \leq i \leq n \end{cases}$$

When $n \geq 2$, $e_{f^*}(0) = \frac{6nt-4t}{2}$ and $e_{f^*}(1) = \frac{6nt-4t}{2}$ and

$$\left|e_{f^*}(0) - e_{f^*}(1)\right| = \left|\frac{6nt-4t}{2} - \frac{6nt-4t}{2}\right| = 0.$$

Therefore, $\left|e_{f^*}(0) - e_{f^*}(1)\right| \leq 1$ is satisfied. Hence, the $SD(L_n)$ for $n \geq 2$ admits a biconditional cordial labeling.

4.2 Example

A biconditional cordial labeling of $SD(L_5)$ with $t = 3$ is shown in Fig. 2.

4.3 Theorem

The $SD(P_m \times P_n)$ for $m, n \geq 2$ admits a biconditional cordial labeling.

Proof Let $V = \{u_{ij} : 1 \leq i \leq m, 1 \leq j \leq n\}$ $\cup \{u_{ij(ij+1)}^k : 1 \leq i \leq m,$ $1 \leq j \leq n-1, 1 \leq k \leq t\}$ $\cup \{u_{ij(i+1j)}^k : 1 \leq i \leq m-1, 1 \leq j \leq n, 1 \leq k \leq t\}$ and $E = \{u_{ij}u_{ij(ij+1)}^k : 1 \leq i \leq m, 1 \leq j \leq n-1, 1 \leq k \leq t\}$ $\cup \{u_{ij+1}u_{ij(ij+1)}^k : 1 \leq i \leq m,$ $1 \leq j \leq n-1, 1 \leq k \leq t\}$ $\cup \{u_{ij}u_{ij(i+1j)}^k : 1 \leq i \leq m-1, 1 \leq j \leq n, 1 \leq k \leq t\}$

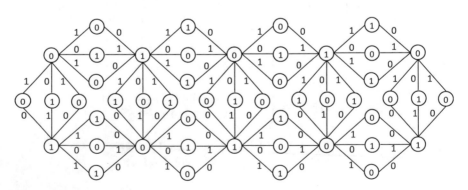

Fig. 2 Biconditional cordial labeling of $SD(L_5)$

$\cup \left\{ u_{i+1j} u^k_{ij(i+1j)} : 1 \le i \le m-1,\ 1 \le j \le n, 1 \le k \le t \right\}$ be the vertex set and edge set of $SD(P_m \times P_n)$. The $SD(P_m \times P_n)$ has $mn + t(2mn - m - n)$ vertices and $4tmn - 2t(m+n)$ edges.

Define a function $f : V \to \{0, 1\}$ such that

$$f(u_{ij}) = \begin{cases} 0, & \text{if } i \text{ is odd}, j \text{ is odd} \\ 1, & \text{if } i \text{ is odd}, j \text{ is even} \end{cases}$$

$$f(u_{ij}) = \begin{cases} 0, & \text{if } i \text{ is even}, j \text{ is even} \\ 1, & \text{if } i \text{ is even}, j \text{ is odd} \end{cases}$$

$$f\left(u^k_{ij(ij+1)}\right) = \begin{cases} 0, & \text{if } i \text{ is odd}, j \text{ is odd}, k \text{ is odd} \\ 1, & \text{if } i \text{ is odd}, j \text{ is odd}, k \text{ is even} \end{cases}$$

$$f\left(u^k_{ij(ij+1)}\right) = \begin{cases} 0, & \text{if } i \text{ is odd}, j \text{ is even}, k \text{ is even} \\ 1, & \text{if } i \text{ is odd}, j \text{ is even}, k \text{ is odd} \end{cases}$$

$$f\left(u^k_{ij(ij+1)}\right) = \begin{cases} 0, & \text{if } i \text{ is even}, j \text{ is odd}, k \text{ is even} \\ 1, & \text{if } i \text{ is even}, j \text{ is odd}, k \text{ is odd} \end{cases}$$

$$f\left(u^k_{ij(ij+1)}\right) = \begin{cases} 0, & \text{if } i \text{ is even}, j \text{ is even}, k \text{ is odd} \\ 1, & \text{if } i \text{ is even}, j \text{ is even}, k \text{ is even} \end{cases}$$

$$f\left(u^k_{ij(i+1j)}\right) = \begin{cases} 0, & \text{if } i \text{ is odd}, j \text{ is odd}, k \text{ is odd} \\ 1, & \text{if } i \text{ is odd}, j \text{ is odd}, k \text{ is even} \end{cases}$$

$$f\left(u^k_{ij(i+1j)}\right) = \begin{cases} 0, & \text{if } i \text{ is odd}, j \text{ is even}, k \text{ is even} \\ 1, & \text{if } i \text{ is odd}, j \text{ is even}, k \text{ is odd} \end{cases}$$

$$f\left(u^k_{ij(i+1j)}\right) = \begin{cases} 0, & \text{if } i \text{ is even}, j \text{ is odd}, k \text{ is even} \\ 1, & \text{if } i \text{ is even}, j \text{ is odd}, k \text{ is odd} \end{cases}$$

$$f\left(u^k_{ij(i+1j)}\right) = \begin{cases} 0, & \text{if } i \text{ is even}, j \text{ is even}, k \text{ is odd} \\ 1, & \text{if } i \text{ is even}, j \text{ is even}, k \text{ is even} \end{cases}$$

Case (i) When n is even, m is even, t is odd or even, we have

$$v_f(0) = \frac{mn + t(2mn - m - n)}{2}, v_f(1) = \frac{mn + t(2mn - m - n)}{2} \text{ and}$$

$$\left| v_f(0) - v_f(1) \right| = \left| \frac{mn + t(2mn - m - n)}{2} - \frac{mn + t(2mn - m - n)}{2} \right| = 0.$$

Therefore, $\left|v_f(0) - v_f(1)\right| \leq 1$ is satisfied.

Case (ii) When n is odd, m is odd, t is odd or even, we have

$$v_f(0) = \tfrac{mn+t(2mn-m-n)+1}{2}, \; v_f(1) = \tfrac{mn+t(2mn-m-n)-1}{2} \text{ and}$$
$$\left|v_f(0) - v_f(1)\right| = \left|\frac{mn+t(2mn-m-n)+1}{2} - \frac{mn+t(2mn-m-n)-1}{2}\right| = 1.$$

Therefore, $\left|v_f(0) - v_f(1)\right| \leq 1$ is satisfied.

Case (iii) When n is odd, m is even, t is even and n is even, m is odd, t is even, we have $v_f(0) = \tfrac{mn+t(2mn-m-n)}{2}, \; v_f(1) = \tfrac{mn+t(2mn-m-n)}{2}$ and

$$\left|v_f(0) - v_f(1)\right| = \left|\frac{mn+t(2mn-m-n)}{2} - \frac{mn+t(2mn-m-n)}{2}\right| = 0.$$

Therefore, $\left|v_f(0) - v_f(1)\right| \leq 1$ is satisfied.

Case (iv) When n is odd, m is even, t is odd and n is even, m is odd, t is odd, we have $v_f(0) = \tfrac{mn+t(2mn-m-n)+1}{2}, \; v_f(1) = \tfrac{mn+t(2mn-m-n)-1}{2}$ and

$$\left|v_f(0) - v_f(1)\right| = \left|\frac{mn+t(2mn-m-n)+1}{2} - \frac{mn+t(2mn-m-n)-1}{2}\right| = 1.$$

Therefore, $\left|v_f(0) - v_f(1)\right| \leq 1$ is satisfied.
Define an induced edge function $f^* : E \to \{0,1\}$ such that

$$f^*\left(u_{ij}u^k_{ij(ij+1)}\right) = \begin{cases} 0, & \text{if } t \text{ is even}, 1 \leq i \leq m, 1 \leq j \leq n \\ 1, & \text{if } t \text{ is odd}, 1 \leq i \leq m, 1 \leq j \leq n \end{cases}$$

$$f^*\left(u_{ij+1}u^k_{ij(ij+1)}\right) = \begin{cases} 0, & \text{if } t \text{ is odd}, 1 \leq i \leq m, 1 \leq j \leq n \\ 1, & \text{if } t \text{ is even}, 1 \leq i \leq m, 1 \leq j \leq n \end{cases}$$

$$f^*\left(u_{ij}u^k_{ij(i+1j)}\right) = \begin{cases} 0, & \text{if } t \text{ is even}, 1 \leq i \leq m, 1 \leq j \leq n \\ 1, & \text{if } t \text{ is odd}, 1 \leq i \leq m, 1 \leq j \leq n \end{cases}$$

$$f^*\left(u_{i+1j}u^k_{ij(i+1j)}\right) = \begin{cases} 0, & \text{if } t \text{ is odd}, 1 \leq i \leq m, 1 \leq j \leq n \\ 1, & \text{if } t \text{ is even}, 1 \leq i \leq m, 1 \leq j \leq n \end{cases}$$

When $m, n \geq 2$, $e_{f^*}(0) = \tfrac{4tmn-2t(m+n)}{2}$ and $e_{f^*}(1) = \tfrac{4tmn-2t(m+n)}{2}$ and

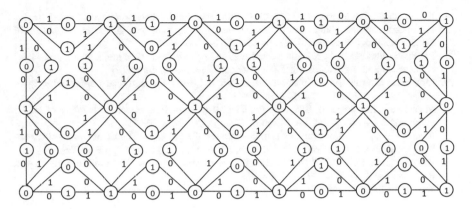

Fig. 3 Biconditional cordial labeling of $SD(P_3 \times P_6)$

$$\left|e_{f^*}(0) - e_{f^*}(1)\right| = \left|\frac{4tmn - 2t(m+n)}{2} - \frac{4tmn - 2t(m+n)}{2}\right| = 0.$$

Therefore, $\left|e_{f^*}(0) - e_{f^*}(1)\right| \le 1$ is satisfied. Hence, the $SD(P_m \times P_n)$ for $m, n \ge 2$ admits a biconditional cordial labeling.

4.4 Example

A biconditional cordial labeling of $SD(P_3 \times P_6)$ with $t = 2$ is shown in Fig. 3.

5 Conclusion

In this paper, we proved that the super subdivision of ladder graph L_n and grid graph $P_m \times P_n$ admit a biconditional cordial labeling. Investigating the existence of the biconditional cordial labeling of super subdivision of other graphs is potential area of research and future scope too.

References

1. Cahit I (1990) On cordial and 3-equitable labeling of graph. Utilitas Math 370:189–198
2. Gallian JA (2019) A dynamic survey of graph labeling. Electr J Comb 17:DS6
3. Harary F (1972) Graph theory. Addison Wesley

4. Kalaimathi M, Balamurugan BJ (2020) Biconditional cordial labeling of cycle related graphs. In: Fourth national conference on recent trends in mathematics & its applications AIP Conference Proceedings, vol 2282, pp 020017-1–020017-10
5. Prasana L, Saravanthi N, Nagalla Sudhakar K (2014) Applications of graph labeling in Major Areas of computing science. Int J Res Comput Commun Technol 3(8)
6. Murali BJ, Thirusangu K, Madura Meenakshi R (2015) Biconditional cordial labeling of cycles. Int J Appl Eng Res 10(80):188–191
7. Murali BJ, Thirusangu K, Balamurugan BJ (2016) Biconditional cordial labeling of ladder graph and $K_2 + nK_1$ graph. Int J Control Theory Appl 9(51):37–45
8. Rosa A (1967) On certain valuations of the vertices of a graph. In: Theory of Graphs (Internat. Sympos. Rome. 1966), Gordan and Breach, Newyork, Dunod, Paris, pp 349–359
9. Sethuraman G, Selvaraju P (2001) Gracefulness of arbitrary of super subdivision of graphs. Indian J Pure Appl Math 32(7):1059–1064
10. West DB (2009) Introduction to graph theory. PHI Learning Private Limited, 2nd edn

COVID-19 Pandemic Review: Future Directions on Detection of Coronavirus Using Imaging Modalities and Computational Intelligence

Ch. Jayalakshmi, R. Kumar, Dhanalakshmi Samiappan, and G. N. Swamy

1 Introduction

Coronavirus, an irresistible illness is brought about by serious intense respiratory condition (SARS-CoV-2) [1] and labeled COVID-19 because of visual appearance (under an electron magnifying lens) to sunlight-based corona (like a crown). The battle in opposition of COVID-19 has propelled scientists all over the world to investigate, comprehend, and devise new indicative and therapy strategies to complete this risk to our age. In this article, we look at how the PC vision network is battling with this peril by proposing new sorts of techniques, repairing viability, and acceleration of the current undertakings [2].

An easiest way to transmit COVID-19 is through the air and physical contact, such as hand contact with an infected person. The virus inserts itself into the lung cells through the respiratory system and replicates there, destroying these cells. COVID-19 comprises RNA and is very difficult to diagnose and treat due to its mutation characteristics. The most well-known indications of COVID-19 incorporate fever, hack, windedness, dazedness, migraine, and muscle throbs. The virus is so perilous and can provoke the death of people with weakened immune systems. Infectious disease specialists and physicians around the world are working to discover a treatment for the disease. COVID-19 is currently the leading cause of death for thousands of countries worldwide, including the USA, Spain, Italy, China, the United Kingdom, Iran, and others [3].

R. Kumar (✉) · D. Samiappan
SRM Institute of Science and Technology, SRM Nagar, Kattankulathur, Kancheepuram, Chennai, TN 603203, India
e-mail: kumarr@srmist.edu.in

Ch. Jayalakshmi (✉) · G. N. Swamy
Department of Electronics and Instrumentation Engineering, VR Siddhartha Engineering College, Kanuru, Vijayawada, AP 520007, India
e-mail: jayalashmichandra@vrsiddhartha.ac.in

© The Author(s), under exclusive license to Springer Nature Singapore Pte Ltd. 2022   639
R. R. Raje et al. (eds.), *Artificial Intelligence and Technologies*,
Lecture Notes in Electrical Engineering 806,
https://doi.org/10.1007/978-981-16-6448-9_61

2   Diagnostic Approaches for Corona Detection

The 2019 novel COVID named COVID-19 by World Health Organization is getting bunches of consideration as of late on the grounds that it is another sort of COVID that is exceptionally infectious and has not been seen among human previously.

The virus started in China and rapidly spread around countries [1, 2]. China adopted many technologies to suppress the spread of the virus, from applications that localize people who have the virus to unmanned vehicles that deliver medical supplies, sanitizer robots, food making robots, and delivery robots. Scientists are trying to tackle the pandemic by collecting databases and proposing algorithms that learn from them.

As per WHO, patients suffering from COVID-19 were categorized into symptomatic, pre-symptomatic, and asymptomatic. Once the person is suspected, then he/she needs to undergo the testing process for the confirmation whether the person was affected by the coronavirus or not. There are several diagnostic approaches that are available to test a person for the confirmation whether the person was affected or not as shown in Fig. 1.

These procedures utilize the fanatical changes in the tainted individual's organ by imaging like CT, or viral nucleic destructive like RT-PCR utilizing at least a quality, or resulting generation sequencing entire genome, immunological atoms created by the tainted individual, or by the infection in the patient's body-antigen–immunizer response-based tests like ELISA, and using of these end approaches is shown in Fig. 2.

Among the mentioned detection procedures, RT-PCR test is considered as reliable test for detection of coronavirus but still insufficient number of RT-PCR kits and also the delay in generating the results made the researchers and also government to look for other types of detection methods. Out of which, imaging approaches was considered as alternate answer to recognize if the individual was

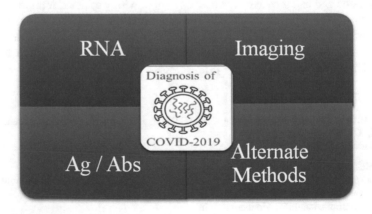

Fig. 1 Approaches for diagnosis of COVID-19

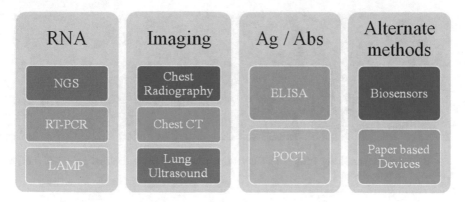

Fig. 2 COVID-19 detection methods using various approaches

contaminated by infection. As of late, with the fast advancement of man-made reasoning [4], AI and profound learning-based frameworks can supplant people by giving a precise determination for the detection of various diseases. The ideal determination can spare radiologists' time and can be practical than standard tests for COVID-19. Deep learning innovation has been broadly utilized in clinical image processing because of its ground-breaking highlight portrayal. A few strategies dependent on deep learning have distributed to identify COVID-19 pneumonia from CT images and X-ray imaging.

3 Different Imaging Modalities

3.1 Chest Radiography

Chest radiography [5] is the most promptly accessible methodology for the identification of lung irregularities in numerous focuses. Contrasted with chest figured tomography (CT), chest radiographs are less expensive and are related with a lower radiation portion. Nonetheless, the absence of particularity, just as affectability for discovery of COVID-19, especially in patients with no or just minor manifestations, must be considered. The standard utilization of chest radiographs, be that as it may, is not viewed as shown in stable, incubated patients with COVID-19. The chest X-ray (CXR) commonly shows equal attacks yet might be typical in beginning illness. Sample chest radiograph images at different days of infection were shown below in Fig. 3.

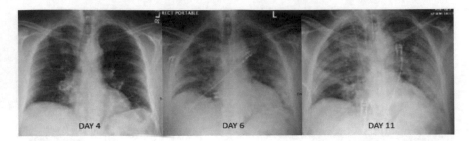

Fig. 3 Sample chest X-ray images of COVID patient at different days of infection [Image Courtesy: Google Images]

3.2 Chest Ultrasound

Chest ultrasound [5] is basically acted in certain focuses to emergency patients, to screen treatment impacts, and for diagnosing intricacies of COVID-19 pneumonias, for example, pleural effusions. Likewise, with all other imaging methodology in COVID-19 patients, ultrasound assessments ought to be kept to a base to maintain a strategic distance from the danger of disease of the clinical faculty. Chest ultrasound images are widely used to diagnosis the lesions and various chest diseases. With reference to COVID-19, the impressions of chest ultrasound is similar to chest CT. Sample chest ultrasound image is shown below in Fig. 4.

3.3 Chest CT

Contrasted with X-rays, CT screening is extensively preferred because of its legitimacy and three-dimensional perspective on the lung. In ongoing examinations [6], the normal indications of contamination can be seen from CT cuts, e.g., ground-glass haziness (GGO) in the beginning phase and pneumonic solidification in the final stage. Sample chest CT images with disease progression were shown in

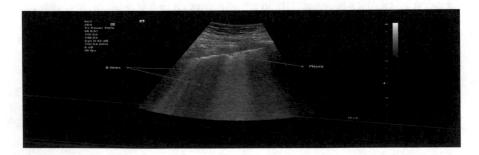

Fig. 4 Sample chest ultrasound image [Image Courtesy: Google Images]

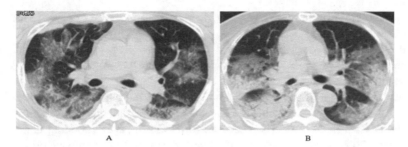

Fig. 5 Sample chest CT images of critically ill COVID-19 patient indicating the progression of disease [Image Courtesy: Google Images]

Fig. 5. The subjective assessment of contamination and longitudinal changes in CT cuts could along these lines give valuable and significant data in battling in opposition of COVID 19. Nonetheless, the manual framework of lung contaminations is monotonous and tedious process.

The computerized tomography is delicate and absolute [5]. CT imaging overall shows attack, area of increased attenuation in the lung on computed tomography, and sub fragmental mix. Here, comparably astounding in patients with no symptoms/patients with no medical proof of lower respiratory bundle thought or any infection in the lungs are considered as non-covid patients. As a general rule, weird CT investigates have been employed to separate corona virus in conjecture with negative sub-atomic finding cases; a tremendous part of these patients had positive sub-nuclear tests on keep testing.

Developing algorithms for CT image analysis is the current hot research aims to contribute to the automatic detection of the coronavirus and also for its progression of disease. It is also suggested that algorithms that were applied to identify lung cancer and lung collapse using X-ray images will also be useful for identifying abnormal cases from COVID-19 patients.

4 Progression Stages of COVID-19 on Chest CT

A few examinations ordered COVID-19 CT discoveries into a few phases dependent on time since the beginning of symptoms [7–11]. Totally progression of COVID-19 was portrayed into four principal stages: initial, progression, advanced and final stage [9].

- Initial phase (0–2 days): The majority infected individual have negative chest CT (56%). The rest of the patients have dominatingly 44% of ground-glass opacities and 17% of consolidation. Imaging disclosures when present were every now and again one sided [9]. Similar disclosures were portrayed in another examination, which displayed that in starting disorder 16.1% of cases present with a solitary physical issue found in around 70% of cases in the right lower lobe of the lungs.

- Progression stage (3–5 days): As the infirmity propels, more ground-glass opacities and consolidations are identified in both the lung lobe. Around 9% of tested patients have negative chest CT [9].
- Advanced stage (6–12 days): A large portion of the patients in this stage have positive chest CT disclosures. More than 60% of patients have consolidation and GGO around 90%. The imaging revelations are corresponding in 88% and around 70% of cases were identified with peripheral predominant [9].
- Final stage (>14 days): Patients in this stage were observed nearer GGO and consolidation values similar to the advanced stage. The imaging revelations are individual in 88% and peripheral predominant in 72% of cases [12].

5   Related Work

This segment examines the sorts of works that are generally identified with division in chest CT, machine learning, and deep learning processes for the classification of corona virus infected patients.

5.1   Segmentation in Chest CT

CT imaging is a famous procedure for the detection of lung illness [13]. By and by, fragmenting various human parts and injuries from chest CT cuts can give critical data to specialists to analyze and measure lung illnesses. As of late, numerous works have been given and acquired promising exhibitions. These figurings consistently use a classifier with removed features for knob division in chest CT. For instance, Zhou et al. [14] did a novel deep learning calculation to solve the large-scene-small-object issue, which disintegrates the 3D division issue into three 2D ones, and accordingly diminishes the model unpredictability by a significant degree and, simultaneously, essentially improves the division accuracy. Shen et al. [15] introduced a robotized lung fragmenting framework reliant on bidirectional chain code to improve the presentation. Regardless, the near visual appearances of knobs and foundation make it hard for eliminating the knob regions. To defeat this hazardous circumstance, a couple of profound learning calculations have been proposed to pick up capability with pivotal visual depictions. Jin et al. [16] utilized GAN-consolidated data to raise the planning of a discriminative model for obsessive lung division. Jiang et al. [17] planned two deep networks to portion lung tumors from CT cuts by including different remaining floods of shifting goals. To boost the learning capacity and accomplish a better performance Fan et al. [18] utilizes lung infection segmentation deep network (Inf-Net) to subsequently perceive corrupted territories from chest CT cuts and equivalent deficient decoder is employed to add up to the huge level features and produce a worldwide guide.

Coronavirus examination can be addressed as a picture fragmentation issue to eliminate the fundamental highlights of the illness. Guan et al. [19] initiated that CT check assessment inserted individual aspiratory parenchyma ground glass and consolidative pneumonic opacities, from time to time with a balanced morphology and a fringe lung conveyance. This fragmentation issue can be understood by building up an algorithm that can extricate the littler comparable areas that can show contamination with the COVID-19 infection.

Table 1 summarizes the performance of segmentation results of various architectures using CT scan images.

5.2 Classification of COVID 19

As of late, deep learning systems have been proposed to distinguish patients contaminated with COVID-19 by means of radiological imaging. For instance, a practical and effective deep learning-based chest radiograph classification (DL-CRC) structure to recognize the COVID-19 cases with good precision from other odd and normal cases. An unprecedented dataset is set up from four straightforwardly open sources including the posteroanterior (PA) chest perspective on X-ray information for COVID-19, pneumonia, and standard cases [20]. Sethi et al. [21] proposed four unmistakable profound CNN models were inspected on pictures of chest X-rays for finding of COVID-19. These models have been pre-arranged on the ImageNet information base as such reducing the necessity for huge preparing sets as they have pre-prepared loads.

Convolutional neural networks have a high computational unpredictability just as a huge memory necessity to actualize with the goal that Gozes et al. [22] introduced a framework that uses 2D and 3D deep learning models, changed and adjusted current deep network representative and consolidated them with medical comprehension. A feebly regulated deep learning system for recognizing and ordering COVID-19 disease from CT images is introduced by Hu et al. [23] that can limit the necessities of manual labeling of CT images yet at the same time have the option to get exact contamination identification and recognize COVID-19 from non-COVID-19 cases.

Oh et al. [24] presents a patch-based convolution neural organization technique with a generally modest number of teachable boundaries for COVID-19 detection. This technique is roused by our measurable examination of the potential imaging biomarkers of the CXR radiographs. While Rajaraman et al. [25] shows the usage of iteratively pruned deep learning model social affairs for identifying aspiratory indication of COVID-19 with chest X-rays. A custom convolutional neural organization and an assurance of ImageNet pre-prepared models are arranged and surveyed at quiet level on freely accessible CXR assortments to learn methodology explicit element portrayals.

Table 1 Summary of the performance of segmentation results of various architectures using CT scan images

Reference	Dataset used	Methods proposed	2/3D	Automated (Y/N)	Performance
Zhou et al. [14]	Harbin Riyadh	Machine-agnostic segmentation	2.5D	Y	Proposed method exhibits a sensitivity of 77.6%, dice similarity coefficient (DSC) of 78.3%, and the worst case performance of dice is 55.7%. Proposed method was compared with other architectures such as 2D U-NET, H-DU Net, MPU net,3D U-Net and 3D V-net
Shen et al. [15]	LIDC	Bidirectional coding method combined with SVM classifier	2D	Y	Proposed method presents an average of overlap as 97.3%, average of over segmentation as 0.3%, average of under segmentation as 2.4%, and re-inclusion rate as 92.6%
Jiang et al. [17]	TCIA MSKCC LIDC	Incremental MRRN Dense MRRN (multiple resolution residually connected network)	2D	–	Incremental MRRN exhibits superiority over dense MRRN in terms of DSC, HD95, and sensitivity except for precision. **DSC:** TCIA: 70%, MSKCC: 75%, LIDC: 68%. **HD95:** TCIA: 7.94 mm. MSKCC: 5.85 mm, LIDC: 2.60 mm. **Sensitivity**: TCIA: 80%, MSKCC: 82%, LIDC: 85%. **Precision**: TCIA: 73%, MSKCC: 72%, LIDC: 70%. Proposed methods were compared with other architectures such as U-Net, Segnet, RF + fCRF and FRRN

(continued)

Table 1 (continued)

Reference	Dataset used	Methods proposed	2/3D	Automated (Y/N)	Performance
Fan et al. [18]	COVID-19 segmentation dataset	Inf-Net Semi-Inf-Net	2D	Y	Semi-Inf-Net exhibits the precedence over Inf-Net in terms of dice (73.9%), sensitivity (72.5%), specificity (96%), structure measure (80%), enhanced alignment measure (89.4%), and mean absolute error (6.4%). Proposed methods are compared with U Net, attention—U-Net, Gated U-Net, Dense U-Net, U-Net++

The primary advantage of these AI-based platforms is to quicken the cycle of finding and treatment of the COVID-19 disease. Barabas et al. [26] presents device prototype which can help in two different ways—through automatic assessment of bodily temperature at different checkpoints and by authorizing appropriate cleanliness norms identified with face covers.

Wang et al. [27] model spotlights on transfer learning, model integration, and arrange chest X-ray pictures as demonstrated by three marks: ordinary, COVID-19, and viral pneumonia. According to the exactness and hardship regard, pick the models ResNet-101 and ResNet-152 with extraordinary effect for mix and intensely improve their weight extent during the arrangement cycle. Subsequent to preparing, the model can achieve 96.1% of the sorts of chest X-ray pictures exactness on the test set. This development has higher affectability than radiologists in the screening and finding of lung nodules. El-kenawy et al. [28] proposes a system for COVID-19 characterization with three fell stages.

In the main stage, the progressive element portrayal is consequently removed from the preparation CT pictures by the CNN model of AlexNet. A short time later, the proposed highlight choice calculation, utilizing SFS and Guided WOA strategies, is implemented to choose highlights in the subsequent stage. Wang et al. [29] presents a deep learning-based model for COVID-19 classification on chest CT is valuable to counter the episode of SARS-CoV-2. 3D CT volumes were obtained by using weakly supervised deep learning structure for COVID-19 plan and injury restriction. Yang et al. [30] presents the LSTM algorithm from the start to anticipate the tainted populace in China. Notwithstanding, it does not clarify the elements of diffusion process, and the drawn out forecast mistake is excessively enormous. Han et al. [31] presents another endeavor of pitifully managed screening of COVID-19 using deep 3D multiple instance learning (AD3D-MIL). Proposed method for the screening of COVID-19 with delicate marks has yet high interpretability. AD3D-MIL joins a deep model generator to make deep 3D events thusly.

Sun et al. [32] presents a versatile element choice guided deep timberland for COVID-19 versus CAP classification by utilizing the chest CT images. In particular, the AFS-DF utilizes the deep backwoods to get familiar with the significant level representation dependent on the area explicit features. In the interim, a versatile feature selection operation is utilized to decrease the repetition of features dependent on the trained forest. Vrbancic et al. [33] utilize another adjusted characterization technique that can identify COVID-19 patient's dependent on a chest X-ray, and afterward receive a nearby interpretable model—agnostic clarifications way to provide the insights. The characterization technique utilizes a dark wolf optimizer algorithm to upgrade hyperparameter esteems inside transfer learning tuning of a CNN. Rahimzadeh and Attar [34] deep convolution networks with introduced preparing strategies for ordering X-ray images into three categories, which are COVID-19, customary, and pneumonia, by using two available datasets. The maker presents some preparation systems that help the organization learn better when we have an unbalanced dataset. Samiappan and Chakrapani [35] proposed an ANN classifier for an ultrasound images. It is a feed forward multilayer backpropagation Network (BPN) using Levenberg–Marquardt training to overcome the problem of local minima in the back propagation network. Latha et al. [36] have summarized various segmentation and classification methods which are suitable for ultrasound images. Bridge et al. [37] proposed initiation work on an openly accessible dataset and remotely favor on a dataset containing 1909 strong chest X-rays and 84 COVID-19 X-beams. The technique presented brings an improved locale under the beneficiary working trademark (DeLong's p-value <0.05) contrasted and the sigmoid initiation.

6   Some Performance Metrics Used

Performance of any machine learning and deep learning networks was analyzed through a confusion matrix. Using the confusion matrix, the following metrics such as sensitivity, specificity, F1 score, and accuracy are computed as shown below [24, 28, 32, 34, 36]:

$$\text{Accuracy } (A) = \frac{(\text{TN} + \text{TP})}{(\text{TN} + \text{TP} + \text{FN} + \text{FP})} \tag{1}$$

$$\text{Sensitivity } (\text{Sn}) = \frac{(\text{TP})}{(\text{TP} + \text{FN})} \tag{2}$$

$$\text{Specificity } (\text{Sp}) = \frac{(\text{TN})}{(\text{TN} + \text{FP})} \tag{3}$$

$$\text{Precision } (P) = \frac{(\text{TP})}{(\text{TP} + \text{FP})} \tag{4}$$

$$F1 \text{ score}(F1) = \frac{2(P \times \text{Sn})}{(P + \text{Sn})} \tag{5}$$

$$\text{Area Under the ROC Curve (AUC)} = \frac{(\text{Sn} + \text{Sp})}{2} \tag{6}$$

where

TP True positive value
TN True negative value
FP False positive value
FN False negative value.

Most of the cases, the above-mentioned parameters are compared among various networks to check the performance of the proposed works. Table 2 summarizes the performance metrics of different classification architecture proposed by various researchers.

7 Future Directions

By observing the above literature, it is evident that most of the work was carried out using both X-ray and as well as CT images for the purpose of two class (i.e., COVID or non-COVID) or three class (i.e., normal, pneumonia, COVID). So, still there is a necessity to explore in the direction of developing classifiers indicating the disease progression stages. Also, there is a possibility of developing a classifier such that it detects the infected person by feeding the input image of the person under testing in the form of either chest X-ray or chest CT. And also, there is other possibility of investigation the long-term effects of COVID-19 patients using the imaging modalities.

8 Conclusion

Coronavirus is a rising pandemic infection that, in a brief timeframe, can seriously imperil the soundness of numerous individuals all through the world. It legitimately influences the lung cells, and if not precisely analyzed early, can cause irreversible harm, including passing. The nitty-gritty synopses of primer agent work, including accessible assets to encourage further research and development is talked about.

Table 2 The performance metrics of different classification architectures

Reference	Image modality	Image dataset used	Methods	Accuracy (%)	AUC (%)	Sensitivity (%)	Specificity (%)	Precision (%)	F1 score (%)
Sethi et al. [21]	X-ray	Non-COVID-5928 COVID-320	Inception V3	98.1	–	81.1	99.2	85.9	83.4
			ResNet50	82.5		24.2	99.7	95.6	38.6
			MobileNet	98.6		87.8	99.3	87.8	87.8
			Xception	97.4		62.2	99.5	88.9	73.2
Gozes et al. [22]	CT	Normal-150 COVID-120	DL with 2D slice and 3D volume analysis	–	99.6	98.2	92.2	–	–
Oh et al. [24]	X-ray	Normal-191 Bacterial-54 TB-57 Viral-20 COVID-180	U-Net	85.9		84	95.3	82.3	82.5
			FC-Dense Net 67	81.8		76.6	91.5	73.1	74.3
			FC-Dense Net 103	88.9		85.9	96.4	83.4	84.4
Rajaraman et al. [25]	X-ray	Pediatric RSNA Twitter COVID-19 Montreal COVID-19	VGG16-U	97.02	99.77	97.02	–	97.09	97.02
			VGG-16-P	97.22	99.38	97.22		97.25	97.22
			VGG19-U	95.63	99.36	95.63		95.88	95.63
			VGG-19-P	97.62	99.72	97.62		97.67	97.62
			IncV3-U	97.42	99.69	97.42		97.46	97.42
			IncV3 -P	98.41	99.62	98.41		98.41	98.41
EL-Kenawy et al. [28]	CT	Normal-794 COVID-334	Alex Net	79.00	–	81.00	77.3	75.00	77.88
			VGG 16	58.21		95.08	28.44	51.75	67.02
			VGG 19	77.17		62.00	89.92	83.78	71.26
			Google Net	73.06		50.00	92.44	84.75	62.89
			ResNet50	77.17		62.50	88.62	81.08	70.59
Han et al. [31]	CT	COVID Pneumonia Non-COVID	C3D	89.7	97.1	85.0	–	88.2	
			DeCovNet	90.6	97.5	84.1		93.7	
			AD3DMIL	94.3	98.8	90.5		95.9	

(continued)

Table 2 (continued)

Reference	Image modality	Image dataset used	Methods	Accuracy (%)	AUC (%)	Sensitivity (%)	Specificity (%)	Precision (%)	F1 score (%)
Sun et al. [32]	CT	COVID-1495 Community Acquired Pneumonia-1027	Logistic Regression	89.81	95.52	91.64	87.10	91.22	91.42
			SVM	89.97	95.62	91.51	87.70	91.38	91.43
			Random Forest	89.41	95.40	90.51	87.85	91.21	90.85
				89.96	95.73	92.66	86.04	90.65	91.64
				91.79	96.35	93.05	89.95	93.10	93.07
Vrbancic et al. [33]	X-ray	COVID-182 Other-660	VGG-19	76.73	50.00	.00	–	76.73	86.83
			Transfer Learning(TL)	91.56	85.64	96.76		92.71	94.62
			Grey wolf Optimizer for TL tuning	94.76	93.47	95.83		97.42	96.52
Rahim zadeh and Attar [34]	X-ray	COVID-180 Pneumonia-6054 Normal-8851	Xception	91.31	–	73.35	99.55	–	–
			ResNet50 V2	89.79		74.02	99.33		
			Xception + ResNet50 V2	91.40		80.53	99.56		

This paper means to diagram the as of late created frameworks dependent on deep learning strategies utilizing distinctive clinical imaging modalities like computer tomography (CT) and X-ray. This survey explicitly talks about the frameworks created for COVID-19 analysis utilizing deep learning techniques and gives insights on notable datasets used to prepare these networks and the techniques utilized for segmentation of lung images. Deep learning-based methodologies are helpful for the programmed discovery of corona infected patients utilizing X-rays and CT images. CT images are likewise useful in discovering the illness seriousness which should be mulled over so as to lessen the casualty rate over the globe. The significant point is to build the quantity of datasets for corona infected patients and utilizing advance deep learning algorithms to accomplish better execution for the detection and prediction of corona virus.

References

1. Paules CI, Marston HD, Fauci AS (2020) Coronavirus infections—more than just the common cold. JAMA 323(8):707–708
2. Chen Y, Liu Q, Guo D (2020) Emerging coronaviruses: genome structure, replication, and pathogenesis. J Med Virol 92(4):418–423
3. Hui DS, Azhar IE, Madani TA, Ntoumi F, Kock R, Dar O, Ippolito G, Mchugh TD, Memish ZA, Drosten C, Zumla A, Petersen E (2020) The continuing 2019-nCoV epidemic threat of novel coronaviruses to global health—The latest 2019 novel coronavirus outbreak in Wuhan, China. Int J Infect Dis 91:264–266
4. Yan Q, Zhang L, Liu Y, Zhu Y, Sun J, Shi Q, Zhang Y (2020) Deep hdr imaging via a non-local network. IEEE Trans Image Process 29:4308–4322
5. Jajodiaa GA, Ebnerb L, Heidingerc B, Chaturvedia A, Proschc H (2020) Imaging in corona virus disease 2019 (COVID-19)—A Scoping review. Eur J Radiol Open 7:100237
6. Ye Z, Zhang Y, Wang Y, Huang Z, Song B (2020) Chest CT manifestations of new coronavirus disease 2019 (COVID-19): a pictorial review. Eur Radiol 2019(37):1–9
7. Pan L, Mu M, Yang P, Sun Y, Wang R, Yan J, Li P, Hu B, Wang J, Hu C, Jin Y, Niu X, Ping R, Du Y, Li T, Xu G, Hu Q, Tu L (2020) Clinical characteristics of COVID-19 patients with digestive symptoms in Hubei, China: a descriptive, cross-sectional, multicenter study. Am J Gastroenterol
8. Shi H, Han X, Jiang N, Cao Y, Alwalid O, Gu J, Fan Y, Zheng C (2020) Radiological findings from 81 patients with COVID-19 pneumonia in Wuhan, China: a descriptive study. Lancet Infect Dis 20(4):425–434
9. Bernheim A, Mei X, Huang M, Yang Y, Fayad ZA, Zhang N, Diao K, Lin B, Zhu X, Li K, Li S, Shan H, Jacobi A, Chung M (2020) Chest CT findings in coronavirus Disease-19 (COVID-19): relationship to duration of infection. Radiology 200463
10. Wang K, Kang S, Tian R, Zhang X, Zhang X, Wang Y (2020) Imaging manifestations and diagnostic value of chest CT of coronavirus disease 2019 (COVID-19) in the Xiaogan area. Clin Radiol
11. Zhou S, Wang Y, Zhu T, Xia L (2020) CT features of coronavirus disease 2019 (COVID-19) Pneumonia in 62 patients in Wuhan, China. AJR Am J Roentgenol 1–8
12. Pan F, Ye T, Sun P, Gui S, Liang B, Li L, Zheng D, Wang J, Hesketh RL, Yang L, Zheng C (2020) Time course of lung changes on chest CT during recovery from 2019 novel coronavirus (COVID-19) Pneumonia. Radiology 200370

13. Sluimer I, Schilham A, Prokop M, Van Ginneken B (2006) Computer analysis of computed tomography scans of the lung: a survey. IEEE Trans Med Imaging 25(4):385–405

14. Zhou L, Li Z, Zhou J, Li H, Chen Y, Huang Y, Xie D, Zhao L, Fan M, Hashmi S, Abdelkareem F, Eiada R, Xiao X, Li L, Qiu Z, Gao X (2020) A rapid, accurate and machine-agnostic segmentation and quantification method for CT-Based COVID-19 diagnosis. IEEE Trans Med Imaging 39(8)

15. Shen S, Bui AA, Cong J, Hsu W (2015) An automated lung segmentation approach using bidirectional chain codes to improve nodule detection accuracy. Comput Biol Med 57:139–149

16. Jin D, Xu Z, Tang Y, Harrison AP, Mollura DJ (2018) CT-realistic lung nodule simulation from 3D conditional generative adversarial networks for robust lung segmentation. In: MICCAI. Springer, pp 732–740

17. Jiang J, Hu Y-C et al (2018) Multiple resolution residually connected feature streams for automatic lung tumor segmentation from CT images. IEEE Trans Med Imaging 38(1):134–144

18. Fan D-P, Zhou T, Ji G-P, Zhou Y, Chen G, Fu H, Shen J, Shao L (2020) Inf-Net: automatic COVID-19 lung infection segmentation from CT images. IEEE Trans Med Imaging 39(8)

19. Guan CS, Lv ZB, Yan S, Du YN, Chen H, Wei LG, Xie RM, Chen BD (2020) 'Imaging features of coronavirus disease 2019 (COVID-19): evaluation on thin-section CT.' Acad Radiol 27(5):609–613

20. Sakib S, Tazrin T, Fouda MM, Fadlullah ZM, Guizani M (2020) DL-CRC: deep learning-based chest radiograph classification for COVID-19 detection: a novel approach. IEEE Access, vol 8

21. Sethi R, Mehrotra M, Sethi D (2020) Deep learning based diagnosis recommendation for COVID-19 using Chest X-Rays Images. In: Proceedings of the second international conference on inventive research in computing applications (ICIRCA-2020) IEEE Xplore Part Number: CFP20N67-ART; ISBN: 978-1-7281-5374-2, 978-1-7281-5374-2/20/$31.00 ©2020 IEEE

22. Gozes O et al (2020) Rapid ai development cycle for the coronavirus (COVID-19) pandemic: Initial results for automated detection & patient monitoring using deep learning ct image analysis. arXiv preprint arXiv:2003.05037

23. Hu S, Gao Y, Niu Z, Jiang Y, Li L, Xiao X, Wang M, Fang EF, Menpes-Smith W, Xia J, Ye H, Yang G (2020) Weakly supervised deep learning for COVID-19 infection detection and classification from CT images. IEEE Access vol 8. https://doi.org/10.1109/ACCESS.2020.3005510

24. Oh Y, Park S, Ye JC (2020) Deep learning COVID-19 features on CXR using limited training data sets. In: IEEE Transactions on Medical Imaging, pp 0278–0062 (c) 2020 IEEE

25. Rajaraman S, Siegelman J, Alderson PO, Folio LS, Folio LR, Antani SK (2020) Iteratively Pruned Deep learning ensembles for COVID-19 detection in chest X-rays. IEEE Access

26. Barabas J, Zalman R, Kochlan M (2020) Automated evaluation of COVID-19 risk factors coupled with real-time, indoor, personal localization data for potential disease identification, prevention and smart quarantining. 978-1-7281-6376-5/20/$31.00 ©2020 IEEE

27. Wang N, Liu H, Xu C (2020) Deep learning for the detection of COVID-19 using transfer learning and model integration, © IEEE 2020

28. El-Kenawy E-SM, Ibrahim A, Mirjalili S, Eid MM, Hussein SE (2016) Novel feature selection and voting classifier algorithms for COVID-19 classification in CT images. IEEE Access, vol 4. https://doi.org/10.1109/ACCESS.2020.3028012

29. Wang X, Deng X, Fu Q, Zhou Q, Feng J, Ma H, Liu W, Zheng C (2020) A weakly-supervised framework for COVID-19 classification and lesion localization from chest CT. IEEE Trans Med Imaging 39(8)

30. Yang Y, Yu W, Chen D (2020) Prediction of COVID-19 spread via LSTM and the deterministic SEIR model. In: Proceedings of the 39th Chinese control conference July 27–29, 2020

31. Han Z, Wei B, Hong Y, Li T, Cong J, Zhu X, Wei H, Zhang W (2020) Accurate screening of COVID-19 using attention-based deep 3D multiple instance learning. IEEE Trans Med Imaging 39(8)
32. Sun L, Mo Z, Yan F, Xia L, Shan F, Ding Z, Song B, Gao W, Shao W, Shi F, Yuan H, Jiang H, Wu D, Wei Y, Gao Y, Sui H, Zhang D, Shen D (2020) Adaptive feature selection guided deep forest for COVID-19 classification with Chest CT. IEEE J Biomed Health Inform, © IEEE
33. Vrbancic G, Pecnik S, Podgorelec V (2020) Identification of COVID-19 X-ray images using CNN with optimized tuning of transfer learning. 978-1-7281-6799-21201$31.00 ©2020 IEEE
34. Rahimzadeh M, Attar A (2020) A modified deep convolutional neural network for detecting COVID-19 and pneumonia from chest X-ray images based on the concatenation of Xception and ResNet50V2 © 2020 The Authors. Published by Elsevier Ltd.
35. Samiappan D, Chakrapani V (2016) Classification of carotid artery abnormalities in ultrasound images using an artificial neural classifier. Int Arab J Inf Technol 13(6A)
36. Latha S, Samiappan D, Kumar R (2020) Carotid artery ultrasound imageanalysis: a review of the literature. J Eng Med, 1–27, ©IMechE https://doi.org/10.1177/0954411919900720
37. Bridge J, Meng Y, Zhao Y, Du Y, Zhao M, Sun R, Zheng Y (2020) Introducing the GEV activation function for highly unbalanced data to develop COVID-19 diagnostic models. IEEE J Biomed Health Inform 24(10):2776–2786. https://doi.org/10.1109/JBHI.2020.3012383

Survey on Fusion of Audiovisual Information for Multimedia Event Recognition

S. L. Jayalakshmi⊙, S. L. Jothilakshmi⊙, V. G. Ranjith,
and Siddharth Jain

1 Introduction

Recognition of video activities remains a critical problem in both vision and machine learning. They have numerous prospective applications such as autonomous surveillance [1, 2], video tagging, and multimedia information retrieval [3]. Most of the literature on machine learning is focusing on the visual modality and the fusion of other modalities, such as audio. In reality, event awareness and recognition are exhaustive, as they are based on simultaneous sensations occurring generally at the same time through the human sensory organs. Using a fusion of audio and video data improves the efficiency of multiple applications, such as automatic video event detection, video retrieval, or emotional recognition [4]. In comparison, the current analysis has found that audio information provides complementary information to capture information contained in the videos. For the two separate modalities to be fused, we must integrate the techniques used in the modalities. In this paper, we focus on a brief survey of features and classifiers used for fusion of audiovisual event recognition.

The main focus of audio event recognition is to identify different types of audio events present in the real-time environment. It plays a vital role in many applica-

S. L. Jayalakshmi (✉) · S. Jain
Vellore Institute of Technology, Chennai, Tamil Nadu, India
e-mail: jayalakshmi.sl@vit.ac.in

S. Jain
e-mail: siddharth.jain2019@vitstudent.ac.in

S. L. Jothilakshmi (✉)
Amrita College of Engineering and Technology, Nagercoil, Tamil Nadu, India
e-mail: sl_jothilakshmi@amrita.edu.in

V. G. Ranjith
Noorul Islam Centre for Higher Education, Nagercoil, Tamil Nadu, India
e-mail: ranjith@niuniv.com

© The Author(s), under exclusive license to Springer Nature Singapore Pte Ltd. 2022
R. R. Raje et al. (eds.), *Artificial Intelligence and Technologies*,
Lecture Notes in Electrical Engineering 806,
https://doi.org/10.1007/978-981-16-6448-9_62

tions such as audio surveillance, audio scene classification, ambient-assisted living, and smart home monitoring. Similarly, the goal of video event recognition is to identify spatialtemporal visual patterns of video events. Some of the applications of video event recognition involve video surveillance, medical diagnosis, sports, consumer video, human–computer interaction, and health monitoring system [5].

In the case of joint audiovisual event recognition, the audio features like Mel-frequency cepstral coefficients (MFCCs), log Mel energies, and video features like color and motion features are computed and from the moving blocks of each frame of the video clip. Subsequently, the fusion of audio and visual elements is achieved by merely linking all features from both the modalities. There is a range of application areas using multimodal data convergence and fusion. Some of the applications include the following: (i) biomedical systems for emergency care, (ii) health monitoring system, (iii) smart outdoor environment monitoring [6], (iv) multimodal video retrieval, and (v) emotion recognition [7].

Some of the unique challenges of joint (fusion) audiovisual event recognition tasks include the following: (i) how the modalities complement each other, (ii) modalities both communicate but may not harmonize with each other, and (iii) absence of one of the modality at the time of the testing, but it is available in the training phase.

This paper provides a detailed analysis of the fusion of audiovisual event recognition roles in multimedia applications. We discuss the following three relevant issues: (i) the impact of audiovisual features on recognition, (ii) the availability of various audiovisual datasets, and (iii) different modeling techniques for multimedia event recognition tasks. The remaining part of this paper is structured as follows: Sect. 2 presents the description of the audiovisual system. Section 3 deals with various applications that use the fusion of audiovisual information.

2 Audiovisual System Description

In audiovisual fusion, the audio data can be either taken from only one microphone or an array of microphones (indoor environment). In the case of a microphone array, the scenario has composed of a fixed number of omnidirectional audio sensors and stereo cameras. The multimodal system identifies the sound sources and the movement of the objects in the indoor environment using the audiovisual information available from the sensors. Generally, there are two phases involved in the audiovisual fusion system. The first phase discusses the essential features derived from each modality based on the type of application used. The collection of features plays a vital role in the audiovisual multimedia event recognition. The second phase conveys the different fusion methods for combining the knowledge acquired from audiovisual modalities. The fusion of features has given to the classifier for recognition and is shown in Fig. 1.

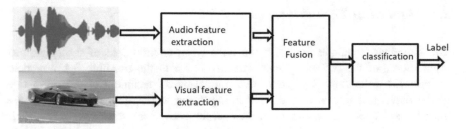

Fig. 1 Block diagram of audiovisual event recognition (early fusion) system

2.1 Features

Representation of audiovisual modalities is an essential step before they are combined. Some of the well-known audio features used widely, in the audio and speech applications, include the following: zero crossing rate, MFCCs, MFCC derivatives, linear prediction cepstral coefficients, spectral components, and time delay of arrival [8, 9].

On the other hand, it is difficult to find suitable visual features from video inputs. Generally, application-specific visual features are extracted from the eye and mouth regions. Some of the visual features available in the literature include model-based, motion-based, image-based, and geometry-based features [5, 6, 8]. In most instances, the modality information is merged only after the feature extraction.

2.2 Fusion Approaches

Fusion may take place at various stages and perform before the modeling process by incorporating the aspects of both modalities. Multimodal convergence is mostly achieved through three distinct types of fusion approaches as follows: (i) early fusion (conducted at the feature level), (ii) middle fusion (balanced between early fusion and late fusion), and (iii) late fusion (performed at the decision level). Principal component analysis, linear discriminant analysis, and Adaboost are the most important fusion techniques used in the literature for audiovisual event recognition [10]. Modality conditioning is the effect of a modality on the behavior of the mechanism of another one. There are two types of modality conditioning based on the: (i) attention model (eyes movement) and (ii) shape and patterns (object shape).

2.3 Modeling Techniques

The modeling technique identifies the underlying environment of the given test input. Some of the modeling techniques available in the machine learning field include the following [9]: generative model-based techniques, discriminative model-based techniques, and deep learning-based techniques. In the generative model-based approach, a model is created for each class separately using their examples. However, some of the commonly used generative models are hidden Markov models (HMMs), Gaussian mixture models (GMMs), and their variances. Support vector machines (SVMs) and artificial neural network are some of the discriminative model-based classifiers, which look for building the linearly separable hyperplane [8].

Deep learning-based techniques such as deep neural network (DNN) and convolutional neural network (CNN) handle a large amount of data for complex recognition applications [11]. The hybrid model-based technique was implemented using combination of generative, discriminative, and deep learning models. On the other hand, the estimation-based techniques were proposed for fusing multiple modalities by using Kalman and particle filter approaches.

2.4 Datasets

There are many audiovisual datasets available for implementing the fusion of audiovisual applications. Some of the benchmark datasets available in the literature are PETS (object tracking) [12], TRECVID (video retrieval) [13], eNTERFACE [11], VIRAT video dataset [6], and acted facial expressions in the wild (AFEW) [8]. The performance of various approaches is evaluated with the help of benchmark datasets using the following metrics: accuracy [11, 14] and equal error rate [15].

3 Applications of Audiovisual Fusion Information

This section summarizes about some of audiovisual fusion applications such as emotion recognition, object/human tracking, and automated surveillance. These applications used deep learning methods for achieving successful multimedia event recognition. Other applications such as video retrieval, biomedical applications, transportation system, and multimedia analysis, also have utilized the audiovisual fusion information for better performance.

3.1 Emotion Recognition

The recognition of emotion plays an important role in user interface design and has been widely researched for audiovisual fusion applications. Ortega et al. [7] assessed a hybrid classifier with mixed audio, video, and text features for audiovisual emotion recognition. The late-fusion approach incorporates vocal signal characteristics, facial features, and the vocal signal's textual transcript. DNN architecture can acquire a robust mapping of the spontaneous and natural behavior of a subject and has emotional state from multiple sources of information in sentimental research.

Hossain and Muhammad [11] proposed an emotion recognition system using a deep learning method. The proposed system extracted Mel-spectrogram from the speech signal and is represented as an image. Similarly, from video signals, frames were extracted for each video segment. The extracted audio features and video features were combined by using an individual CNN. The fusion of two CNNs outputs was combined using two consecutive extreme learning machines (ELMs).

A method for extracting visual features using the combination of local binary pattern on three orthogonal planes (LBP-TOP), an ensemble's of CNN, and bi-directional long short-term memory (BLSTM) was proposed by Cai et al. [8]. Then, audio features are extracted using OpenSmile toolkit. Using feature-level fusion (linear SVM) and model-level fusion (Bayesian network [BN]), the visual features and audio features are combined to find the emotion of a person.

3.2 Object/Human Tracking

Object/human tracking is the method of associating the position of the object/human from one frame to another in a series. Liu et al. [6] used the optical flow generator and a color-based active-learning histogram matcher to detect and monitor the movement of various precious targets for suspicious behavior detection through the fusion of multi-intelligence data. It is a user-based voice-to-text color feature descriptor approach combined with automatic extraction of hue features from image pixels. The cloud-based audio–video (CAV) fusion approach detects and monitors multiple moving targets from airborne videos combined with audio support in multiple moving object environments. Castrillon-Santana and Fenu [15] demonstrated a deep intermediate fusion strategy of audiovisual biometric data for audiovisual user re-identification and verification.

3.3   *Automated Surveillance*

Automated surveillance is the process of monitoring the environment to find any abnormalities in the video sequence. Jesus et al. [1] reviewed various approaches to perform automatic human violence detection. Exploring two-stage CNN model with benchmark datasets given better recognition accuracy for human activity detection. Cristani et al. [16] proposed a system to fuse audio and visual information for automated surveillance using only one camera and one microphone. The audiovisual concurrence matrix was used as a feature representation for performing online training without the need for training examples. A smart surveillance system called CASSANDRA was proposed by Zajdel et al. [17] for finding human aggressive actions. In another work, Brousmiche et al. [5] proposed the fusion of several architectures for the audiovisual recognition task using concatenation, addition, or multimodal compact bilinear pooling methods. To combine the functionality of audiovisual information, several feature-wise linear modulation (FiLM) layers were fused.

Table 1 summarizes the widely used audiovisual fusion applications with their state-of-the-art approaches. It is clear that in some of the approaches, the use of facial expressions is not connected with the visual articulators. Similarly, in some implementations, the audio signal is blended with the visual signal. To conclude, the main limitation of audiovisual fusion is the phenomenon of vagueness due to the presence of several moving objects and sound occurrences at the same moment [9]. Generally, feature fusion-based approaches involve more time to recognize real-time audiovisual events. The combination of cepstral features and visual deep features provides better discrimination to classify complex events [5]. In classification, instead of the conventional generative model (HMM and GMM) approach, the more overlapping classes were modeled by the deep learning-based approaches [11, 15]. However, deep models require more training data than the discriminative models (SVM) [8]. In the future, a hybrid model-based approach using a lesser amount of training examples will be developed for improving the accuracy of the audiovisual system.

4   Conclusion

In this paper, we have presented a detailed review of recent developments in the fusion of audiovisual information for multimedia event recognition tasks. First, we presented a description of an audiovisual fusion system with features, modeling techniques, and datasets. Then, we presented a brief review of various types of applications that use the fusion of audiovisual information. A few important future directions include the development of robust audiovisual fusion representation, a good generalization modeling framework, and a universal audiovisual dataset for complex real-time multimedia event recognition tasks.

Table 1 Comparison of different state-of-the-art-methods involved in fusion of audiovisual applications

Audiovisual application	Ref	Audio features	Video features	Modeling techniques	Dataset
Emotion recognition	[7]	Segment-level acoustic features	Normalized face orientation in various degrees	DNN architecture	AVEC SEWA database and RECOLA dataset
	[8]	Audio features extracted using OpenSmile tools	LBP-TOP, an ensemble's of CNNs, and BLSTM	Feature-level fusion (linear SVM) and model-level fusion (BN)	AFEW dataset
	[11]	Mel-spectrogram	Visual frames	CNN and ELMs	eNTERFACE dataset
Object/ human tracking	[6]	Voice-to-text color feature	Hue histogram	Active-learning histogram	VIRAT video dataset
	[15]	Voice spectrogram	Multimodal face representations	Matcher (AHM) DNN architecture	AveRobot dataset
Automated surveillance	[16]	Power spectral density Cochleograms	Visual features extracted using per-pixel mixture of Gaussians model and histogram	k-nearest neighbors (KNN) and hierarchical clustering	Real video dataset
	[17]	Cochleograms	Optical flow features	Dynamic BN	Real-time data from train station
	[5]	Mel-band extraction and CNN	Visual deep features using DenseNet	FiLM	Kinetics dataset

References

1. Jesus T, Duarte J, Ferreira D, Dur˜aes D, Marcondes F, Santos F, Gomes M, Novais P, Gon‚calves F, Fonseca J et al (2020) Review of trends in automatic human activity recognition using synthetic audio-visual data. In: International conference on intelligent data engineering and automated learning. Springer, pp 549–560
2. Qian X (2020) Multi-target localization and tracking using audio-visual signals. PhD thesis, Queen Mary University of London
3. Fayek HM, Kumar A (2020) Large scale audiovisual learning of sounds with weakly labeled data. arXiv preprint arXiv:2006.01595
4. Parthasarathy S, Sundaram S (2020) Training strategies to handle missing modalities for audio-visual expression recognition. arXiv preprint arXiv:2010.00734

5. Brousmiche M, Rouat J, Dupont S (2019) Audio-visual fusion and conditioning with neural networks for event recognition. In: 2019 IEEE 29th international workshop on machine learning for signal processing (MLSP). IEEE, pp 1–6
6. Liu K, Liu B, Blasch E, Shen D, Wang Z, Ling H, Chen G (2015) A cloud infrastructure for target detection and tracking using audio and video fusion. In: Proceedings of the IEEE conference on computer vision and pattern recognition workshops, pp 74–81
7. Ortega JD, Senoussaoui M, Granger E, Pedersoli M, Cardinal P, Koerich AL (2019) Multimodal fusion with deep neural networks for audio-video emotion recognition. arXiv preprint arXiv:1907.03196
8. Cai J, Meng Z, Khan AS, Li Z, O'Reilly J, Han S, Liu P, Chen M, Tong Y (2019) Feature-level and model-level audiovisual fusion for emotion recognition in the wild. In: 2019 IEEE conference on multimedia information processing and retrieval (MIPR). IEEE, pp 443–448
9. Chandrakala S, Jayalakshmi S (2019) Environmental audio scene and sound event recognition for autonomous surveillance: A survey and comparative studies. ACM Comput Surv (CSUR) 52(3):1–34
10. Atrey PK, Hossain MA, El Saddik A, Kankanhalli MS (2010) Multimodal fusion for multimedia analysis: a survey. Multimedia Syst 16(6):345–379
11. Hossain MS, Muhammad G (2019) Emotion recognition using deep learning approach from audio–visual emotional big data. Inf Fusion 49:69–78
12. Chan A, Vasconcelos N (2009) People counting data for pets 2009 dataset, vol 18. University of California, San Diego March, p 2014
13. Katsaggelos AK, Bahaadini S, Molina R (2015) Audiovisual fusion: challenges and new approaches. Proc IEEE 103(9):1635–1653
14. Ortega JD, Cardinal P, Koerich AL (2019) Emotion recognition using fusion of audio and video features. In: 2019 IEEE international conference on systems, man and cybernetics (SMC). IEEE, pp 3847–3852
15. Castrillon-Santana M, Fenu G (2020) Deep multi-biometric fusion for audio-visual user re-identification and verification. In: Pattern recognition applications and methods: 8th international conference, ICPRAM 2019, Prague, Czech Republic, February 19–21, 2019, revised selected papers, vol 11996. Springer Nature, p 136
16. Cristani M, Bicego M, Murino V (2007) Audio-visual event recognition in surveillance video sequences. IEEE Trans Multimedia 9(2):257–267
17. Zajdel W, Krijnders JD, Andringa T, Gavrila DM (2007) Cassandra: audio-video sensor fusion for aggression detection. In: 2007 IEEE conference on advanced video and signal based surveillance. IEEE, pp 200–205

A Hybrid Ensemble Prediction Method for Analyzing Air Quality Data

Apeksha Aggarwal and Ajay Agarwal

1 Introduction

Air pollution is one of the biggest environmental challenges that world is facing in this era. Major pollutants contributing to air pollution include particulate matter, oxides of carbon, ozone, lead, and oxides of nitrogen. For our study, we have selected nitrogen dioxide (NO_2) for analysis. NO_2 when inhaled causes severe health problems such as respiratory problems, eye infections, and so on [1]. In the present work, we have analyzed NO_2 because it is among one of the primary air pollutants in Delhi [2]. Delhi is the capital city of India, and it lies in the list of one of the highest polluted locations of the world. Government of India is taking major steps to cure air pollution because of this critical environment condition. Hence, analysis of trends of air pollution for the city is quite necessary.

In the past, varied time series analysis models have been proposed, to model global trends in the time series. But, in this study, we have focused on capturing localized trends using partitioning. Specifically, k-means is chosen because it segregates out similar trends effectively and efficiently. Thus, forecasting is performed on the considering the similarity in trends. In addition to forecasting the time series data using already existing methods, exponential smoothing is further performed to capture the overall historical effect of the time series.

In this work, time series data for a few locations of Delhi have been analyzed. A detailed description about these locations is provided further in Sect. 3. Main aim is to study the general trend of NO_2 concentrations for different locations.

A. Aggarwal (✉)
Department of Computer Science and Engineering, SEAS, Bennett University,
Greater Noida, India
e-mail: apeksha.aggarwal@bennett.edu.in

A. Agarwal
Department of Information Technology, KIET Group of Institutions, Ghaziabad, India
e-mail: ajay.agarwal@kiet.edu

© The Author(s), under exclusive license to Springer Nature Singapore Pte Ltd. 2022
R. R. Raje et al. (eds.), *Artificial Intelligence and Technologies*,
Lecture Notes in Electrical Engineering 806,
https://doi.org/10.1007/978-981-16-6448-9_63

Henceforth, we have used a hybrid partitioning approach along with exponential weighted moving average (EWMA) [3] time series technique. Details are given in further sections. Section 2 provides related work. Section 3 presents a detailed discussion on experiments and methods. Section 4 illustrates results, and Sect. 5 provides the conclusion.

2 Related Work

Varied air pollution forecasting approaches have classified air pollution into different categories. One of these categories for classifying them includes the type of the data collection method used, which can be captured from a variety of sources such as air quality monitoring station, sensors, mobile, handheld devices, and so on [2]. Similarly, a variety of methods and techniques is available to study air pollution. In this work, we specifically focus upon the use of data mining [4] concepts and techniques to study air pollution. Many researchers in the past [1] utilized linear regression to forecast concentrations of various pollutants such as NO_2. Feng et al. [1] is a notable work in the study of air pollution, in which time series of particulate matter concentration is decomposed into few subseries by wavelet transform, and multilayer perceptron model of ANN [5] is applied to each of these subseries to sum up the results of prediction strategy.

Time series analysis is specifically done in various popular studies for analyzing air pollution in various countries [6–9]. Time series analysis [10] has also been utilized in various other applications such as, in weather monitoring [11], medical applications [12], economics [12], and so on.

Various models [13] utilized adaptive network-based fuzzy inference system to model nonlinear functions and forecast the time series over air quality. Results presented by Zheng et al. [14] are quite significant in predicting air quality using meteorological features. In addition to prediction, relationship of spatial and temporal features with air quality is explained in this work [14]. Other important works in air quality include inferring pollution from vehicles [15], monitoring indoor air quality [16] and so on [15, 17, 18]. Further studying air quality, a variety of deep learning techniques has recently came into existence [7, 8]. But most of the deep learning models require extremely huge amounts of data to train deep complex networks. Such huge datasets for air quality of developing country India are not readily available for use. Secondly, building deep learning models requires huge infrastructures and computing capabilities. Our proposed approach shows promising results utilizing the small dataset over much smaller computing architecture.

3 Method and Experiments

This section provides a detailed discussion on the data and tools. Experiments done for time series smoothing have been discussed in this section in detail.

3.1 Data and Tools

Data analyzed for the present work are time series data taken Web site of Government of India, consisting of concentrations of NO_2 (in $\mu g/m^3$) for every 15 min for the year 2016 (as per the availability) [9]. Four locations of Delhi are analyzed in this work. Dataset comprises of four locations of Delhi, viz. Dwarka, Mandir Marg, Punjabi Bagh, and Anand Vihar. Each of these locations are denoted by notations L2, L3, L4, and L1, respectively, in further sections. Temporal partitions are specified monthwise, annotated as T1, T2…Tn.

Locations analyzed are depicted in Fig. 1. R Studios Version 1.0. has been used for experimental work. Packages of R such as qcc and stats have been used specifically for time series analysis and k-means clustering algorithm have been used for partition-based clustering [9]. Preprocessing of data is done by removing tuples with missing values and noise. Also, data are in the form of time series, consisting of 96 values per day. Note that 95% of data are selected as training set, and we have forecasted next four values per day.

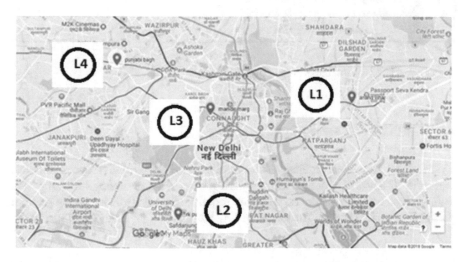

Fig. 1 Four locations L1, L2, L3, and L4 analyzed are annotated on Google Maps

3.2   *Method*

Partitioning-based clustering algorithm is used in our work, to divide data into bins
[9, 19]. We have used, k-means clustering algorithm, to cluster out similar sets of
time series representing NO_2 concentration per day as in [9]. For each month, data
are divided into set of clusters representing similar days. Algorithm for k-means is
given as per algo 1. Value of k is selected using elbow curve, which came out to be
different for every partition. The method is already presented in [9]. Seasonal and
Trend decomposition using Loess (STL) decomposition of the series is done to
identify the periodicity. Specifically, STL is chosen because using this method we
can identify the general trend in the time series. Further, the STL method has
already been a popular choice in the past. EWMA is a weighted smoothing method,
formula for EWMA in recursion is given as per (1), where coefficient α represents
the degree of weighting decrease, whose value lies between 0 and 1. S_t is the value
of the exponential moving average (EMA) at any time period t.

$$S = \begin{cases} Y_1, & \text{if } t = 1 \\ \alpha Y_t + (1 - \alpha).S_{t-1}, & \text{if } t > 1 \end{cases} \tag{1}$$

Input: Set of n points
Output: k clusters

1 Initialize means (e.g. by picking k samples at random);
2 **repeat**
3   Assign each point to nearest mean;
4   Move mean to center of its cluster;
5 **until** *cluster centers do not change*;

Algorithm 1: K-means clustering algorithm.

4   Results and Discussions

EWMA [20] is applied on each of the partitioned clusters. Individual monthwise
variances of EWMA with and without partitioning-based clustering are compared.
Table 1 depicts all the variances, of all the months, when EWMA is applied without
partitioning. EWMA chart for locations L1, L2, L3, and L4 is depicted in Fig. 2.

Table 1 Variances for varied partitions

Month\location	L1	L2	L3	L4
T1	7.36	5.39	4.43	1.60
T2	8.61	5.36	4.66	1.79
T3	8.72	5.90	4.17	1.56
T4	11.63	6.79	3.81	1.83
T5	9.77	7.51	3.51	2.21
T6	6.95	6.46	2.47	1.76
T7	7.74	4.74	2.24	1.16
T8	3.93	4.09	2.5	1.90
T9	6.71	4.50	1.84	2.35
T10	8.65	5.76	2.64	1.95

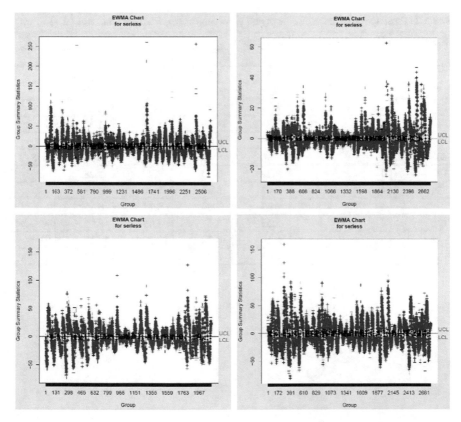

Fig. 2 Graph depicting moving average trend fitting for EWMA method over all the four locations

4.1 EWMA

Table 2 shows individual variances after clustering is applied over L1. From tables, it is quite clear that our proposed method gave better results, i.e., more variances and better smoothing over time series than the standard method. Similar results have been found for other locations of L2, L3, and L4. Furthermore, Fig. 3 shows the plotted variances for L1 and L2.

4.2 Forecasting

To check the stationarity of the time series, Dickey–Fuller (DF) test is applied in this work. For most of the months, DF test rejects the null hypothesis with a p value

Table 2 Clusterwise variance after applying clustering over EWMA for location L1

Month	No. of clusters	Cluster no.				
		1	2	3	4	5
T1	4	7.53	7.59	7.53	7.28	–
T2	4	8.61	8.61	10.65	7.83	–
T3	4	8.72	8.70	8.72	8.82	–
T4	4	11.29	11.59	11.57	11.63	–
T5	4	9.89	9.77	8.81	7.96	–
T6	5	5.96	6.46	6.61	6.95	6.95
T7	4	7.19	7.23	7.28	7.23	–
T8	4	3.93	3.97	3.58	3.93	–
T9	4	6.63	7.04	6.75	6.71	–
T10	4	8.55	8.67	8.65	8.65	–
T11	5	10.19	10.22	10.19	10.19	9.99
T12	4	6.63	6.26	7.26	6.69	–

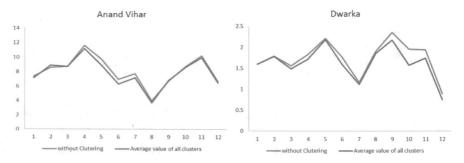

Fig. 3 Graph depicting graph depicting moving average variance trend for with and without EWMA method over two locations of L1 and L2

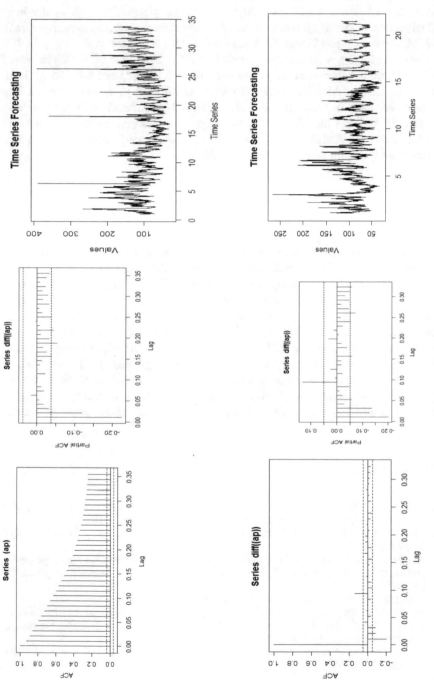

Fig. 4 Figures illustrate sample ACF, PACF, and forecasted values for next 5 days of L1 for the partitions of T1 and T2. Forecasted trends are depicted in red color

of 0.01 depicting the stationarity of the series. Autocorrelation functions (ACF) and partial autocorrelation functions (PACF) have been plotted for each of the individual months to identify p and q values. Autoregressive integrated moving average (ARIMA) model is used to forecast next five values in the interval.

For example, Fig. 4 shows a sample plots of ACF, PACF for L1 for the month of T1 and T3, along with forecasted values over ARIMA (0, 0, 1) and ARIMA (0, 1, 1). Similarly, different ARIMA models are fitted for different p, d, and q values. The average root mean square error (RMSE) for L1 is calculated as 10.29 which is calculated in terms of mean absolute percentage error (MAPE) as 8.44%, which was quite low. For L3, RMSE was 10.30 and MAPE was found to be 9.92%. Similar results were found for other locations as well.

5 Conclusion

Air pollution is one of the major environmental concerns these days. There is an extreme need to study air pollution trends in environment. This work studies concentrations of NO_2 for every 15 min. Data analyzed is time series data, consisting of nitrogen dioxide concentrations for a period of 1 year. In developing countries like India, much less work is done in this regard, so we have analyzed data of four locations of Delhi where air quality stations are already installed. Time series data of Delhi are analyzed and trends have been identified using STL and EWMA method. Partition-based EWMA method is suggested in this work, which makes use of k-means clustering to forecast future concentrations of NO_2. Results are evaluated on smoothing fit with variance. Furthermore, in future, we plan to extend our study for several cities of India as well as for various other pollutants.

References

1. Donnelly A, Misstear B, Broderick B (2015) Real time air quality forecasting using integrated parametric and non-parametric regression techniques. Atmos Environ 103:53–65
2. Zheng Y, Capra L, Wolfson O, Yang H (2014) Urban computing. ACM Trans Intell Syst Technol 5(3):1–55
3. Hunter J (1990) Exponentially weighted moving average control schemes: properties and enhancements: discussion. Technometrics, 32(1):21. Available: https://doi.org/10.2307/1269839
4. Kumar V (2016) Introduction to data mining. Pearson India
5. Feng X, Li Q, Zhu Y, Hou J, Jin L, Wang J (2015) Artificial neural networks forecasting of PM2.5 pollution using air mass trajectory based geographic model and wavelet transformation. Atmos Environ 107:118–128
6. Kramer H, Schmid L (1997) Ewma charts for multivariate time series. Seq Anal 16(2):131–154

7. Keerthana R (2020) Forecasting of the air pollution based on meteorological data and air pollutants using deep learning: a novel review. Int J Adv Trends Comput Sci Eng 9(1): 801–807
8. Freeman B, Taylor G, Gharabaghi B, Thé J (2018) Forecasting air quality time series using deep learning. J Air Waste Manag Assoc 68(8):866–886
9. Aggarwal A, Toshniwal D (2019) Detection of anomalous nitrogen dioxide (NO_2) concentration in urban air of India using proximity and clustering methods. J Air Waste Manag Assoc 69(7):805–822
10. Breitung J, Hamilton J (1995) Time series analysis. Contemp Sociol 24(2):271
11. Eymen A, Köylü Ü (2018) Seasonal trend analysis and ARIMA modeling of relative humidity and wind speed time series around Yamula Dam. Meteorol Atmos Phys
12. Incidence of infective endocarditis in England, 2000–13: a secular trend, interrupted time-series analysis. BDJ 218(5):291–291 (2015)
13. Prasad K, Gorai A, Goyal P (2016) Development of ANFIS models for air quality forecasting and input optimization for reducing the computational cost and time. Atmos Environ 128:246–262
14. Zheng C, Jin A, Sun Q (2014) Indoor air quality monitoring system for smart buildings. In: Proceedings of the 2014 ACM International joint conference on pervasive and ubiquitous computing
15. Zheng Y, Liu F, Hsieh HP (2013) U-Air: when urban air quality inference meets big data. In: Proceedings of 19th SIGKDD conference on knowledge discovery and data mining
16. Friedman C, Sandow S (2011) Utility-based learning from data. CRC Press, Boca Raton, FL
17. Zheng Y, Chen X, Jin Q, Chen Y, Qu X, Liu X, Chang E, Ma W-Y, Rui Y, Sun W (2014) A cloud-based knowledge discovery system for monitoring fine-grained air quality
18. Tian Y et al (2018) Association between ambient air pollution and daily hospital admissions for ischemic stroke: A nationwide time-series analysis. PLOS Med 15(10):e1002668
19. Kamber M, Pei J (2006) Data mining, 2nd edn. Morgan Kaufmann
20. Li H, Xu Q, He Y, Fan X, Li S (2020) Modeling and predicting reservoir landslide displacement with deep belief network and EWMA control charts: a case study in Three Gorges Reservoir. Landslides 17:693–707

Printed in the United States
by Baker & Taylor Publisher Services